Lecture Notes in Mathematics 1474

S. Jackowski B. Oliver K. Pawałowski (Eds.)

Algebraic Topology Poznań 1989

Proceedings of a Conference held in Poznań, Poland, June 22-27, 1989

Springer-Verlag
Berlin Heidelberg New York
London Paris Tokyo
Hong Kong Barcelona
Budapest

Editors

Stefan Jackowski
Instytut Matematyki
Uniwersytet Warszawski
ul. Banacha 2
PL-00-913 Warszawa 59, Poland

Bob Oliver
Matematisk Institut
Ny Munkegade
8000 Aarhus C, Denmark

Krzystof Pawałowski
Instytut Matematyki UAM
ul. Matejki 48/49
60-769 Poznań, Poland

Mathematics Subject Classification (1980): 57-06, 55-06, 19-06

ISBN 3-540-54098-9 Springer-Verlag Berlin Heidelberg New York
ISBN 0-387-54098-9 Springer-Verlag New York Berlin Heidelberg

Printing and binding: Druckhaus Beltz, Hemsbach/Bergstr.
46/3140-543210 - Printed on acid-free paper

Preface

In June, 1989, the International Conference on Algebraic Topology was held in Poznań, Poland. The conference was part of the scientific activity in connection with the 70-th anniversary of the Adam Mickiewicz University in Poznań. It was supported by the Adam Mickiewicz University, Warsaw University, and Polish government grant RP.I.10.

There were many of our colleagues and students from both Poznań and Warszawa who helped to contribute to the success of the conference. We would especially like to mention Agnieszka Bojanowska, Adam Neugebauer and Bogdan Szydło, who helped with the organizational work, and the two conference secretaries Danuta Marciniak and Katarzyna Kacperska-Panek.

The conference consisted of 10 plenary talks, as well as 49 talks in special sessions in various fields. These proceedings contain papers presented at the conference, as well as some other papers (mostly) submitted by conference participants. We tried—and with some success—to encourage the submission of survey papers.

All papers in the volume have been refereed. We would like to thank the referees for their work, and Andrzej Weber for proofreading of several manuscripts which had to be retyped during the editorial process.

Stefan Jackowski
Bob Oliver
Krzysztof Pawałowski

Warszawa/Århus/Poznań,
November 1990

Table of Contents

SOME APPLICATIONS OF SHIFTED SUBGROUPS IN TRANSFORMATION GROUPS

by C. Allday and V. Puppe

If a torus G of rank r acts on a compact space X, and if all isotropy subgroups have rank at most s, then there is a subtorus $K \subseteq G$ of rank $r - s$ such that the action of K on X is almost–free. When G is an elementary abelian p–group (i.e., $G \cong (\mathbb{Z}/(p))^r$, where p is a prime number), then there is no immediate analogue of the very useful fact above, since a finite number of proper subgroups can cover G. In order to overcome this difficulty, and others, shifted subgroups (to be defined in detail in Section 2 below), which have been used in modular representation theory for some time (see, e.g., [Benson, 1984]), have been introduced into the cohomological study of finite transformation groups. The use of shifted subgroups is quite natural; and indeed they seem to have appeared in transformation groups through the work of at least four different authors: A. Adem introduced them explicitly in his thesis ([Adem, 1986], and see [Adem, 1988]); they also appear explicitly in the work of A. Assadi ([Assadi, 1988], [Assadi, 1989a], [Assadi, 1989b] and [Assadi]); and shifted subgroups of rank one appeared implicitly in our paper [Allday, Puppe, 1985].

In this paper we intend to give a survey of some of these applications of shifted subgroups. We shall concentrate on the work of A. Adem and ourselves and closely related results. Since it would require a substantial amount of background material, we have not included Assadi's work concerning applications of the theory of varieties of G–modules in transformation groups: for this see Assadi's papers cited above. We have included one of Adem's theorems (Theorem (4.14) below), the proof of which makes substantial use of varieties of G–modules, but, for the same reason, we have not included Adem's proof. Otherwise, for the most part, we have included proofs, although we have referred some proofs, especially the proofs of some technical details, to our forthcoming book ([Allday, Puppe]).

In the first two sections we summarize some background material from algebra, including, in Section 2, the definition and some of the main properties of shifted subgroups. In the third section we give some of the basic topological notations and definitions which we shall use. We have chosen to work with paracompact finitistic spaces. There is only a small amount of technical difficulty in extending the results given here from finite–dimensional G–CW–complexes to paracompact finitistic G–spaces; and yet many more applications are included amongst the latter, for example, continuous actions on topological manifolds.

The last three sections give some of the applications of shifted subgroups. In Section 4 we treat equivariant Tate cohomology (as defined by R. Swan), in Section 5 we give an application in the manner of P. A. Smith's original method, and in Section 6 we give an application to equivariant cohomology (as defined by A. Borel).

1. k[G]–modules

Here we collect a few useful facts about k[G]–modules. Throughout this section G will be a finite group and k will be a field.

(1.1) **Theorem.** (1) A k[G]–module is projective if and only if it is injective.
(2) Any product of projective k[G]–modules is projective.

Proof. (2) follows at once from (1).
(1) follows from [Brown, 1982], Chap. VI, Corollaries (2.2) and (2.3). (1) also follows since k[G] is a symmetric algebra, and hence Frobenius: see, for example, [Fuller, 1989].

(1.2) **Corollary**. If M is a projective k[G]–module, then the dual module $M^* := \mathrm{Hom}_{k[G]}(M, k[G])$ is also projective.

(1.3) **Definition**. We shall say that a k[G]–module M is Tate acyclic if $\hat{H}^*(G; M) = 0$. (This is a slight simplification of the notion of a cohomologically trivial module: see [Brown, 1982], Chap. VI, Sec. 8.)

(1.4) **Theorem.** Suppose that G is a finite p–group, where p is any prime number, and that k is a field of characteristic p. Then the following conditions on a k[G]–module M are equivalent.

(1) M is free.
(2) M is projective.
(3) M is Tate acyclic.

(4) $\hat{H}^i(G; M) = 0$ for at least one $i \in \mathbb{Z}$.

Proof. This is contained explicitly in [Brown, 1982], Chap. VI, Theorem (8.5).

(1.5) **Corollary**. Let G and k be as in Theorem (1.4). Then
(1) any direct limit of free k[G]–modules is free; and
(2) if k′ is an extension field of k, if M is a k[G]–module, and if $M \otimes_k k'$ is a free k′[G]–module, then M is a free k[G]–module.

Proof. Since G has a complete resolution of finite type, P_*, say, for any $i \in \mathbb{Z}$, and k[G]–module M, $\mathrm{Hom}_{k[G]}(P_i, M) \cong P_i^* \otimes_{k[G]} M$. Hence, if $\{M_j \mid j \in J\}$ is a directed system of k[G]–modules, then $\hat{H}^*(G; \varinjlim_j M_j) \cong \varinjlim_j \hat{H}^*(G; M_j)$.

Similarly for (2), $\hat{H}^*(G; M \otimes_k k') \cong \hat{H}^*(G; M) \otimes_k k'$.

2. Shifted subgroups

In this section we recall the defnition of shifted subgroups and state some of their basic properties. Throughout this section G will be an elementary abelian p–group (also known as a p–torus), where p is a prime number; i.e. $G \cong (\mathbb{Z}/(p))^r$ for some $r \geq 0$: and k will be a field of characteristic p. The number r is called the rank of G, denoted rk G.

Suppose that G is generated by $g_1, \dots, g_r$. For $1 \leq i \leq r$, let $\tau_i = 1 - g_i \in k[G]$. Let $\nu_i = (1 - g_i)^{p-1}$ for $1 \leq i \leq r$. Since $\tau_i^p = 0$, it follows that the homomorphism from the polynomial ring $k[X_1, \dots, X_r] \longrightarrow k[G]$ given by $X_i \longrightarrow \tau_i$, for $1 \leq i \leq r$, induces an isomorphism $k[X_1, \dots, X_r]/(X_1^p, \dots, X_r^p) \xrightarrow{\sim} k[G]$.

Let $\mathbf{m}(G; k)$, or more simply just $\mathbf{m}$, denote the ideal in k[G] generated by $\tau_1, \dots, \tau_r$. So $\mathbf{m}$ is the one and only prime ideal in k[G]. Note that $\mathbf{m}^{(p-1)r+1} = 0$.

(2.1) **Definitions.** (1) We shall say that any element $u \in k[G]$ is a non–trivial unit if there are $\alpha_1, \dots, \alpha_r \in k$, not all zero, such that $u = 1 - \Sigma_{i=1}^r \alpha_i \tau_i$ modulo $\mathbf{m}^2$. Clearly $u^p = 1$; and so u generates a subgroup of order p in the group of units of k[G]. Denote this subgroup by $\Gamma(u)$. As follows from (3) below, $\Gamma(u)$ is a shifted subgroup of rank 1.

(2) If $u_1, \dots, u_s$ are non–trivial units in k[G], then let $\Gamma(u_1, \dots, u_s)$ denote the elementary abelian p–subgroup of the group of units of k[G] generated by $u_1, \dots, u_s$.

The inclusion of $\Gamma = \Gamma(u_1, \dots, u_s)$ in k[G] induces a homomorphism of group rings $i_\Gamma\colon k[\Gamma] \longrightarrow k[G]$.

(3) Suppose that $u_1, \dots, u_s$ are non–trivial units in k[G]. For $1 \leq j \leq s$, let $u_j = 1 - \Sigma_{i=1}^r \alpha_{ji} \tau_i$ modulo $\mathbf{m}^2$. Then $\Gamma(u_1, \dots, u_s)$ is said to be a shifted subgroup of rank s in k[G] if the vectors $\alpha_j = (\alpha_{j1}, \dots, \alpha_{jr}) \in k^r$, for $1 \leq j \leq s$, are linearly independent. It is easy to see that a shifted subgroup of rank s is indeed an elementary abelian p–group of rank s.

(4) If a nontrivial unit $u = 1 - \Sigma_{i=1}^{r} \alpha_i \tau_i$, then we also denote $\Gamma(u)$ by $\Gamma(\alpha)$, where $\alpha = (\alpha_1, ..., \alpha_r) \in k^r$.

(5) Given $u, v \in k[G]$, we shall write $u \sim v$ if $u - v \in \mathbf{m}^2$.

Some important properties of shifted subgroups are listed in the following theorem. Proofs may be found in [Carlson, 1983]; but we shall include the proof of (1), which is most often used subsequently in this paper.

(2.2) **Theorem.** (1) If $\Gamma \subseteq k[G]$ is a shifted subgroup of rank s, then $i_\Gamma : k[\Gamma] \longrightarrow k[G]$ is injective, and $k[G]$ is a free $k[\Gamma]$–module (via i_Γ). If $s = r$, then i_Γ is an isomorphism.

(2) If $u_1, ..., u_s$ are non–trivial units in $k[G]$ such that $\Gamma = \Gamma(u_1, ..., u_s)$ has order p^s, then $k[G]$ is a free $k[\Gamma]$–module if and only if $1 - u_1, ..., 1 - u_s$ are linearly independent modulo $\mathbf{m}^2$: i.e. if and only if $u_1, ..., u_s$ generate Γ as a shifted subgroup.

(3) Suppose that $u, v \in k[G]$ are non– trivial units such that $u \sim v$. Let M be a finitely generated $k[G]$–module. The M is a free $k[\Gamma(u)]$–module if and only if M is a free $k[\Gamma(v)]$–module.

(4) Dade's Lemma. If k is algebraically closed, and if M is a finitely generated $k[G]$–module, then M is a free $k[G]$–module if and only if M is a free $k[\Gamma(\alpha)]$–module for all non–zero $\alpha \in k^r$.

Proof. (1) Suppose $\Gamma = \Gamma(u_1, ..., u_s)$ where $u_j = 1 - \Sigma_{i=1}^{r} \alpha_{ji}\tau_i$ modulo $\mathbf{m}^2$ for $1 \leq j \leq s$. Let $\sigma_j = 1 - u_j \in k[\Gamma]$, and let $\sigma_j' = \Sigma_{i=1}^{r} \alpha_{ji}\tau_i$. So $\sigma_j \sim \sigma_j'$. And since Γ is a shifted subgroup of rank s we may assume that $\sigma_1', ..., \sigma_s'$ are linearly independent. Now choose $\sigma_{s+1}', ..., \sigma_r'$ so that $\{\sigma_1', ..., \sigma_r'\}$ is a basis for the k–vector subspace of $k[G]$ spanned by $\tau_1, ..., \tau_r$. Let $u_i = 1 - \sigma_i'$ for $s + 1 \leq i \leq r$ and let $\Gamma' = \Gamma(u_{s+1}, ..., u_r)$.

Now $i_\Gamma : k[\Gamma] \longrightarrow k[G]$ and $i_{\Gamma'}: k[\Gamma'] \longrightarrow k[G]$ induce, via the direct sum of commutative k–algebras, a homomorphism $\varphi : k[\Gamma] \otimes k[\Gamma'] \longrightarrow k[G]$. Since $\sigma_1', ..., \sigma_r'$ generate $k[G]$ as a k–algebra, since $\sigma_j' - \sigma_j \in \mathbf{m}^2$ for $1 \leq j \leq s$, and since $\mathbf{m}^{(p-1)r+1} = (0)$, it is clear that φ is surjective. Hence φ is an isomorphism, since its domain and codomain are finite–dimensional k–vector spaces of the same dimension.

For (2) see [Carlson, 1983], Theorem 6.2, and Corollary (1.5)(2) above. (Carlson assumes that k is algebraically closed.)

For (3) see [Carlson, 1983], Lemma 6.4, and Corollary (1.5)(2) above.

For (4) see [Carlson, 1983], Theorem 4.4, or [Dade, 1978].

(2.3) **Remarks.** (1) A subgroup Γ of the group of units of $k[G]$ is a shifted subgroup of rank s if and only if there exist $u_1, ..., u_s \in k[G]$, such that $\{1 - u_1, ..., 1 - u_s\} \subseteq \mathbf{m}$, the image of $\{1 - u_1, ..., 1 - u_s\}$ under the quotient map $\mathbf{m} \longrightarrow \mathbf{m}/\mathbf{m}^2$ is linearly independent over k, and $\Gamma = \Gamma(u_1, ..., u_s)$.

And to put it another way, a subgroup Γ of the group of units of $k[G]$ is a shifted subgroup if and only if Γ is generated by a finite number of elements of the coset $1 + \mathbf{m}$, and the homomorphism $\mathbf{m}(\Gamma; k)/\mathbf{m}(\Gamma; k)^2 \longrightarrow \mathbf{m}/\mathbf{m}^2$ induced by i_Γ is injective.

(2) Let H be a subgroup of G of rank s. Then $G \cong H \times G/H$; and so there is a homomorphism $q_H : G \to H$ such that $q_H j_H = 1_H$, where $j_H : H \to G$ is the inclusion. It follows

that $H \subsetneq k[G]$ is a shifted subgroup of rank s; and $i_H : k[H] \to k[G]$ induces an injection $m(H; k)/m(H; k)^2 \to m/m^2$.

In the following proposition we are concerned with a subgroup $H \subsetneq G$ and a shifted subgroup $\Gamma \subsetneq k[G]$; and we want to know when $k[G/H]$ is a free $k[\Gamma]$–module via the homomorphism $k[\Gamma] \to k[G/H]$ obtained by composing i_Γ with the homomorphism $k[G] \to k[G/H]$ induced by the quotient map. Let V be the k–vector space m/m^2, and let V_H be the image of $m(H; k)/m(H; k)^2$ in V. By Remarks (2.3)(2) above, $\dim_k V_H = \text{rk } H$. For a shifted subgroup $\Gamma \subsetneq k[G]$ let V_Γ be the image of $m(\Gamma; k)/m(\Gamma; k)^2$ in V under the homomorphism induced by i_Γ. By definition of a shifted subgroup $\dim_k V_\Gamma = \text{rk}\Gamma$.

(2.4) **Proposition.** In the situation described above, for a subgroup $H \subsetneq G$ and a shifted subgroup $\Gamma \subsetneq k[G]$, $k[G/H]$ is a free $k[\Gamma]$–module if and only if $V_\Gamma \cap V_H = 0$.

Proof. Suppose $V_\Gamma \cap V_H = 0$. Let $\pi : k[G] \to k[G/H]$ be induced by the quotient map, and let $\pi' : V \to m(G/H; k)/m(G/H; k)^2$ be induced by π. Clearly π' is surjective and $V_H \subsetneq \ker \pi'$. Hence $V_H = \ker \pi'$. And so $\pi' | V_\Gamma$ is injective. Hence π maps Γ isomorphically onto a shifted subgroup $\pi(\Gamma) \subsetneq k[G/H]$. Since $k[G/H]$ is a free $k[\pi(\Gamma)]$–module by Theorem (2.2)(1), $k[G/H]$ is a free $k[\Gamma]$–module.

If $V_\Gamma \cap V_H \neq 0$, then there are two possibilities: (i) π does not map Γ isomorphically onto $\pi(\Gamma)$, or (ii) π maps Γ isomorphically onto $\pi(\Gamma)$, but $\pi(\Gamma)$ is not a shifted subgroup of $k[G/H]$. In case (i) the result is clear. In case (ii) the result follows from Theorem (2.2)(2).

The following corollary is very useful in the applications of shifted subgroups to transformation groups.

(2.5) **Corollary.** Let $H_1, \ldots, H_n$ be subgroups of G such that, for $1 \leq i \leq n$, $\text{rk } H_i \leq t < r = \text{rk } G$. Then there is an extension field E of k with finite degree over k, and a shifted subgroup $\Gamma \subsetneq E[G]$ of rank $r - t$ such that $E[G/H_i]$ is a free $E[\Gamma]$–module for $1 \leq i \leq n$.

Proof. Let K be the algebraic closure of k. Using the notation of Proposition (2.4) for $K[G]$ instead of $k[G]$, since K is an infinite field, there is a subspace $W \subsetneq V$ such that $W \cap V_{H_i} = 0$ for $1 \leq i \leq n$, and $\dim_k W = r - t$. Hence there is a shifted subgroup $\Gamma \subsetneq K[G]$ of rank $r - t$ such that $K[G/H_i]$ is a free $K[\Gamma]$–module for $1 \leq i \leq n$.

But Γ is defined in terms of the elements of G using a finite number of coefficients in K. Let E be the extension field of k generated by these coefficients. So Γ may be viewed as a shifted subgroup in $E[G]$. Now it follows that $E[G/H_i]$ is a free $E[\Gamma]$–module, for $1 \leq i \leq n$, by Corollary (1.5)(2).

(2.6) **Remark.** For an important generalization of this result to $k[G]$–modules which are not necessarily permutation modules see [Kroll, 1984].

3. Topological notation and constructions

Let G be a finite group and let k be a commutative ring (with identity). Let $P_* \to k \to 0$ be a projective resolution of k viewed as a trivial $k[G]$–module; and let $\hat{P}_*$ be a complete resolution of k (see [Brown,1982], Chap. VI, Section 3). We shall assume, as we may, that P_* and $\hat{P}_*$ have finite type, and that $\hat{P}_i = P_i$ for $i \geq 0$. In particular there is an obvious map $\hat{P}_* \longrightarrow P_*$ which is the identity on $\hat{P}_i$ for $i \geq 0$ and zero on $\hat{P}_i$ for $i < 0$.

(3.1) **Definitions.** Let C^* be a cochain complex of $k[G]$–modules with $C^i = 0$ for $i < 0$.

(1) Let $\beta^*_G(C^*) = \mathrm{Hom}_{k[G]}(P_*,C^*)$. In particular $\beta^n_G(C^*) = \bigoplus_{i=0}^{n} \mathrm{Hom}_{k[G]}(P_i,C^{n-i})$. The differential d on $\beta^*_G(C^*)$ is given in terms of the differentials d_P and d_C on P_* and C^*, respectively, by the formula $df(x) = d_C(f(x)) - (-1)^n f(d_P(x))$ for $f \in \beta^n_G(C^*)$.

(2) Let $H^*_G(C^*) = H(\beta^*_G(C^*),d)$.

(3) Let $\hat{\beta}^*_G(C^*) = \mathrm{Hom}_{k[G]}(\hat{P}_*,C^*)$, where $\hat{\beta}^n_G(C^*) = \bigoplus_{i=-\infty}^{n} \mathrm{Hom}_{k[G]}(\hat{P}_i,C^{n-i})$. Note that we are taking the direct sum here not the direct product. The differential, d, on $\hat{\beta}^*_G(C^*)$ is defined in the same way as for $\beta^*_G(C^*)$.

(4) Let $\hat{H}^*_G(C^*) = H(\hat{\beta}^*_G(C^*),d)$.

(5) The first filtration on $\beta^*_G(C^*)$ is defined by $F^p\beta^n_G(C^*) = \bigoplus_{i=p}^{n}\mathrm{Hom}_{k[G]}(P_i,C^{n-i})$.

The second filtration on $\beta^*_G(C^*)$ is defined by $F^q\beta^n_G(C^*) = \bigoplus_{j=q}^{\infty}\mathrm{Hom}_{k[G]}(P_{n-j},C^j)$.

The first and second filtrations on $\hat{\beta}_G(C^*)$ are defined similarly.

The following spectral sequences are standard. (See [Brown, 1982], Chap. VII, and [Allday, Puppe], §4.6.)

(3.2) **Proposition.** The first filtrations gives rise to spectral sequences

(1) $E_2^{p,q} = H^p(G;H^q(C^*)) \Rightarrow H^*_G(C^*)$; and

(2) $E_2^{p,q} = \hat{H}^p(G;H^q(C^*)) \Rightarrow \hat{H}^*_G(C^*)$.

The second filtration gives rise to a spectral sequence

(3) $E_1^{p,q} = H^q(G;C^p) \Rightarrow H^*_G(C^*)$;

and, if C^* is bounded above, i.e. there is an integer n such that $C^i = 0$ for $i > n$, a spectral sequence

(4) $E_1^{p,q} = \hat{H}^q(G;C^p) \Rightarrow \hat{H}^*_G(C^*)$.

(3.3) **Corollary.** The map $\hat{P}_* \longrightarrow P_*$ induces a natural homomorphism $\theta^*: H^*_G(C^*) \longrightarrow \hat{H}^*_G(C^*)$. And, if $H^j(C^*) = 0$ for all $j > n$, then $\theta^*: H^j_G(C^*) \longrightarrow \hat{H}^j_G(C^*)$ is an isomophism for all $j > n$.

Proof. The existence of θ^* is immediate. So suppose $H^j(C^*) = 0$ for all $j > n$. Let C^*_t be the cochain complex with $C^i_t = C^i$ for $i < n$, $C^n_t = Z^n$, the cocycles of degree n, and $C^i_t = 0$ for

$i > n$. The inclusion $C_t^* \longrightarrow C^*$ is a weak equivalence, and hence the first spectral sequences show that $H_G^*(C_t^*) \cong H_G^*(C^*)$ and $\hat{H}_G^*(C_t^*) \cong \hat{H}_G^*(C^*)$.

Now $\beta_G^j(C_t^*) = \hat{\beta}_G^j(C_t^*)$ for $j \geq n$. Hence $H_G^j(C_t^*) \cong \hat{H}_G^j(C_t^*)$ for $j > n$.

In order to work with paracompact finitistic spaces using Alexander–Spanier or Čech cohomology we need to review some notation and terminology concerning coverings.

(3.4) **Definitions.** Let X be a paracompact G–space, and let $A \subseteq X$ be a closed invariant subspace. Let $\mathcal{U}$ be an open covering of X.

(1) Let $\mathcal{U}_A = \{U \in \mathcal{U} \mid U \cap A \neq \phi\}$.

(2) The Čech nerve of $\mathcal{U}$, denoted $\check{\mathcal{U}}$, is the abstract simplicial complex with vertices the non–empty members of $\mathcal{U}$, and $\{U_0, ..., U_n\}$ a simplex of $\check{\mathcal{U}}$, where $U_i \in \mathcal{U}$ for $0 \leq i \leq n$, if $\bigcap_{i=0}^n U_i \neq \phi$. The subcomplex $\check{\mathcal{U}}_A$ is defined by saying that a simplex $\{U_0, ..., U_n\}$ of $\check{\mathcal{U}}$ is a simplex of $\check{\mathcal{U}}_A$ if $\bigcap_{i=0}^n U_i \cap A \neq \phi$.

(3) The Vietoris nerve of $\mathcal{U}$, denoted $\overline{\mathcal{U}}$, is the abstract simplicial complex with vertices the points of X, and $\{x_0, ..., x_n\}$ a simplex of $\overline{\mathcal{U}}$ if $\{x_0, ..., x_n\} \subseteq U$ for some $U \in \mathcal{U}$. The subcomplex $\overline{\mathcal{U}}_A$ is defined by saying that a simplex $\{x_0, ..., x_n\}$ of $\overline{\mathcal{U}}$ is a simplex of $\overline{\mathcal{U}}_A$ if $\{x_0, ..., x_n\} \subseteq A$.

(4) $\mathcal{U}$ is said to be finite–dimensional if $\check{\mathcal{U}}$ is a finite–dimensional abstract simplicial complex, in which case the geometric realization $|\check{\mathcal{U}}|$ is a finite–dimensional CW–complex.

(5) X is said to be finitistic if every open covering of X has a finite–dimensional refinement: i.e. if finite–dimensional coverings are cofinal.

(6) $\mathcal{U}$ is said to be an invariant covering of X if for any $U \in \mathcal{U}$ and $g \in G$, $gU \in \mathcal{U}$.

(7) $\mathcal{U}$ is said to be a Čech–G–covering of X if $\mathcal{U}$ is invariant and if $gU \cap U \neq \phi$ implies $gU = U$ for any $U \in \mathcal{U}$ and $g \in G$. In this case, for any $U \in \mathcal{U}$, let $G_U = \{g \in G \mid gU = U\}$. (Čech–G–coverings are just called G–coverings in [Bredon, 1972]).

(8) $\mathcal{U}$ is said to be faithful if $\mathcal{U}$ is a Čech–G–covering, and if, for any $U \in \mathcal{U}$, there is a $x \in X$ such that $G_U \subseteq G_x$.

(3.5) **Lemma.** If X is a paracompact G–space, then locally finite faithful Čech–G–coverings are cofinal. If X is also finitistic, then locally finite finite–dimensional faithful Čech–G–coverings are cofinal.

Proof. Most of this is contained in [Bredon, 1972], Chap. III, Theorem 6.1. Since G is finite, X has a covering by open slices, which is a faithful Čech–G–covering. And clearly any Čech–G–covering which refines a faithful Čech–G–covering is also faithful.

(3.6) **Definitions.** Let X be a paracompact G–space, let $A \subseteq X$ be a closed invariant subspace, and let Λ be a k–module.

(1) Let $\overline{C}^*(X, A; \Lambda) = \varinjlim_{\mathcal{U}} C^*(\overline{\mathcal{U}}, \overline{\mathcal{U}}_A; \Lambda)$, where $\mathcal{U}$ ranges over the faithful Čech–G–coverings of X, and $C^*(\overline{\mathcal{U}}, \overline{\mathcal{U}}_A; \Lambda)$ is the ordered cochain complex of the pair $(\overline{\mathcal{U}}, \overline{\mathcal{U}}_A)$

with coefficients in Λ. Then $\overline{C}^*(X, A; \Lambda)$ is the Alexander–Spanier cochain complex of (X, A) with coefficients in Λ as defined, for example, in [Spanier, 1966], Chap. 6, sec. 4. Clearly $\overline{C}^*(X, A; \Lambda)$ is a cochain complex of $k[G]$–modules. If G is an elementary abelian p–group, k is a field of characteristic p, and if $\Gamma \subseteq k[G]$ is a shifted subgroup, then $\overline{C}^*(X, A; \Lambda)$ is also a cochain complex of $k[\Gamma]$–modules.

(2) Let $H_G^*(X, A; \Lambda) = H_G^*(\overline{C}^*(X, A; \Lambda))$, and let $\hat{H}_G^*(X, A; \Lambda) = \hat{H}_G^*(\overline{C}^*(X, A; \Lambda))$.

Define $H_\Gamma^*(X, A; \Lambda)$ and $\hat{H}_\Gamma^*(X, A; \Lambda)$ similarly if G is an elementary abelian p–group, k is a field of characteristic p, and $\Gamma \subseteq k[G]$ is a shifted subgroup.

(3) If X is a G–CW–complex, and A is G–CW–subcomplex, or more generally, if (X, A) is a relative G–CW–complex, then let $W_*(X, A; k)$, respectively $W^*(X, A; \Lambda)$, be the cellular chain complex of (X, A) with coefficients in k, respectively the cellular cochain complex of (X, A) with coefficients in Λ.

The following lemma is proven in detail in [Allday, Puppe].

(3.7) **Lemma.** If X is a paracompact G–space and $A \subseteq X$ is a closed invariant subspace, then

$$\hat{H}_G^*(X, A; \Lambda) \cong \varinjlim_{\mathcal{U}} \hat{H}_G^*(|\check{\mathcal{U}}|, |\check{\mathcal{U}}_A|; \Lambda)$$

$$\cong \varinjlim_{\mathcal{U}} \hat{H}_G^*(W^*(|\check{\mathcal{U}}|, |\check{\mathcal{U}}_A|; \Lambda)),$$

and similarly with H_G^* instead of $\hat{H}_G^*$.

If G is an elementary abelian p–group, k is a field of characteristic p, and if $\Gamma \subseteq k[G]$ is a shifted subgroup, then the corresponding results also hold for H_Γ^* and $\hat{H}_\Gamma^*$.

Furthermore $H_G^*(X, A; \Lambda) \cong H^*(X_G, A_G; \Lambda)$, the Alexander–Spanier cohomology of the pair (X_G, A_G) with coefficients in Λ, where X_G, for example, is the Borel construction on X; i.e. $X_G = (EG \times X)/G$.

(3.8) **Remarks.** (1) $H_G^*(X, A; \Lambda)$, respectively $\hat{H}_G^*(X, A; \Lambda)$, is called the equivariant, respectively the equivariant Tate, cohomology of (X, A) with coefficients in Λ.

(2) $\hat{H}_G^*$, like H_G^*, has natural long exact sequences for pairs, Mayer–Vietoris sequences, and tautness properties. For example, if A and B are closed invariant subspaces of X with $X = A \cup B$, then there is a long exact Mayer–Vietoris sequence

$$\dots \to \hat{H}_G^j(X; \Lambda) \to \hat{H}_G^j(A; \Lambda) \oplus \hat{H}_G^j(B; \Lambda) \to \hat{H}_G^j(A \cap B; \Lambda) \to \hat{H}_G^{j+1}(X; \Lambda) \to \dots$$ and

$$\hat{H}_G^*(A; \Lambda) \cong \varinjlim_{V} \hat{H}_G^*(V; \Lambda),$$ where V ranges over the closed invariant neighborhoods of A. The same holds for H_Γ^* and $\hat{H}_\Gamma^*$. (See [Allday, Puppe], § 4.6.)

We finish this section by recalling W.-Y. Hsiang's definition of the p–rank of a space.

(3.9) **Definitions.** Let $G \cong (\mathbb{Z}/(p))^r$, and let $\Phi : G \times X \to X$ be an action of G on a space X.

(1) The rank of Φ is $\mathrm{rk}\Phi := r - \max\{\mathrm{rk}\, G_x \mid x \in X\}$. Thus if $\mathrm{rk}\Phi = \rho$, then p^ρ is the order of the smallest orbit. $\mathrm{rk}\Phi = r$ if and only if G is acting freely; and $\mathrm{rk}\Phi = 0$ if and only if

$X^G \neq \phi$. Thus $\mathrm{rk}\Phi$ measures in a certain sense the extent to which the action is free.

(2) The p–rank of X is $\mathrm{rk}_p(X) := \sup \mathrm{rk}\Phi$, where Φ ranges over all elementary abelian p–group actions on X.

(3.10) **Remark.** We could also define the free p–rank of X to be $\mathrm{frk}_p(X) := \sup \{r \mid (\mathbb{Z}/(p))^r$ can act freely on $X\}$. There are well known examples where $\mathrm{rk}_p(X) > \mathrm{frk}_p(X)$. For example, $\mathrm{rk}_3(\mathbb{C}P^2) = 1$ but $\mathrm{frk}_3(\mathbb{C}P^2) = 0$: see Examples (5.5)(2) below.

4. Equivariant Tate cohomology.

Equivariant Tate cohomology and finitistic spaces were introduced by Swan in [Swan, 1960]. We shall recall Swan's main theorem immediately following the next definition.

(4.1) **Definition.** Let G be a compact Lie group and let X be a paracompact G–space. Then the singular set of X is defined to be $X_1 := \{x \in X \mid G_x \neq \{1\}\}$. X_1 is clearly invariant, and by the Slice Theorem it is closed.

(4.2) **Theorem.** Let G be a finite group, k a commutative ring with identity, Λ a k–module, X a paracompact finitistic G–space and $A \subseteq X$ is a closed invariant subspace. Then restriction induces an isomorphism

$$\hat{H}^*_G(X, A; \Lambda) \xrightarrow{\sim} \hat{H}^*_G(X_1, A_1; \Lambda).$$

For a proof see [Swan, 1960] or [Allday, Puppe], §4.6.

In applying Swan's Theorem the following easy lemma to be found in [Adem, 1988] and inspired by [Heller, 1959] is useful.

(4.3) **Lemma.** Let G be a finite group and let k be a field of characteristic p where p divides $|G|$. Let C^* be a cochain complex of k[G]–modules such that $C^i = 0$ for all $i < 0$ and $H^j(C^*) = 0$ for all $j > N$, where N is some integer. Then, for any integer m,

$$\dim_k \hat{H}^{m+1}(G; H^0(C^*)) \leq \dim_k \hat{H}^{m+1}_G(C^*) + \sum_{j=1}^{N} \dim_k \hat{H}^{m-j}(G; H^j(C^*)).$$

(4.4) **Corollary** ([Heller, 1959], [Adem, 1988]). Let X be a paracompact finitistic space such that $H^*(X; \mathbb{F}_p) \cong H^*(S^a \times S^b; \mathbb{F}_p)$ as graded $\mathbb{F}_p$–vector spaces where a and b are integers such that $0 < a < b$. Then $\mathrm{frk}_p(X) \leq 2$. (See Remark (3.10))

Proof. Suppose that $G = (\mathbb{Z}/(p))^3$ is acting freely on X. By Swan's Theorem, $\hat{H}^*_G(X; \mathbb{F}_p) = 0$. Now Lemma (4.3) with $m = a + b$ and $C^* = \overline{C}^*(X; \mathbb{F}_p)$ yields a contradiction.

Now we would like to prove that, under the conditions of the Corollary (4.4), $\mathrm{rk}_p(X) \leq 2$. In [Adem, 1988], Adem did this by introducing shifted subgroups. Here then, following the next definition, is a shifted version of Swan's Theorem.

(4.5) **Definition.** Let $G \cong (\mathbb{Z}/(p))^r$ and let k be a field of characteristic p. Let X be a paracompact G–space. Then, using the notation introduced immediately above Proposition (2.4), for any shifted subgroup $\Gamma \subseteq k[G]$, let $X(\Gamma;k) = \{x \in X \mid V_\Gamma \cap V_{G_x} \neq 0\}$. Note that $X(\Gamma; k)$ is invariant, and, by the Slice Theorem, it is closed. Also $X(G; k) = \{x \in X \mid V_{G_x} \neq 0\} = X_1$.

(4.6) **Theorem.** Let $G \cong (\mathbb{Z}/(p))^r$, let k be field of characteristic p, let X be a paracompact finitistic G–space, let $A \subseteq X$ be a closed invariant subspace, and let $\Gamma \subseteq k[G]$ be a shifted subgroup. Then restriction induces an isomorphism

$$\hat{H}^*_\Gamma(X, A; k) \xrightarrow{\sim} \hat{H}^*_\Gamma(X(\Gamma; k), A(\Gamma; k); k).$$

Proof. Thanks to the long exact sequences (see Remarks (3.8)(2)) it is enough to prove the result when $A = \phi$. Suppose the result has been proven in case $X(\Gamma; k) = \phi$. If $X(\Gamma; k) \neq \phi$, let W_1 be a closed invariant neighbourhood of $X(\Gamma; k)$ and let W_2 be the complement of the interior of W_1. So $W_2(\Gamma; k) = \phi$ and $(W_1 \cap W_2)(\Gamma; k) = \phi$. By the Mayer–Victoris sequence, therefore, $\hat{H}^*_\Gamma(X; k) \cong \hat{H}^*_\Gamma(W_1; k)$. The result now follows by the tautness property (Remarks (3.8(2))).

So it remains to show that $\hat{H}^*_\Gamma(X; k) = 0$ if $X(\Gamma; k) = \phi$. Let $\mathcal{U}$ be a faithful finite–dimensional Čech–G–covering of X. For any $y \in |\check{\mathcal{U}}|$, there is a $x \in X$ such that $G_y \subseteq G_x$. (Since $\mathcal{U}$ is a Čech–G–covering, the maximal isotropy groups of $|\check{\mathcal{U}}|$ occur at the vertices; and these are all contained in isotropy groups of X since $\mathcal{U}$ is faithful.) Now, by Proposition (2.4), each $k[G/G_y]$, for $y \in |\check{\mathcal{U}}|$, is a free $k[\Gamma]$–module. Thus each $W_i(|\check{\mathcal{U}}|; k)$, and hence, by Theorem (1.1)(2), also each $W^i(|\check{\mathcal{U}}|; k)$, is a free $k[\Gamma]$–module.

By the second spectral sequence (Proposition (3.2)(4)), $\hat{H}^*_\Gamma(W^*(|\check{\mathcal{U}}|; k)) = 0$. So $\hat{H}^*_\Gamma(X; k) = 0$ by Lemmas (3.5) and (3.7).

(4.7) **Corollary.** Let $G \cong (\mathbb{Z}/(p))^r$, let X be a paracompact finitistic G–space, and let $A \subseteq X$ be a closed invariant subspace. Let ρ be the rank of the action on $X - A$: i.e. $\rho = r - \max\{\mathrm{rk}\, G_x | x \in X - A\}$. Then there is a finite field k of characteristic p and a shifted subgroup $\Gamma \subseteq k[G]$ of rank ρ such that

$$\hat{H}^*_\Gamma(X, A; k) = 0.$$

Proof. Let $H_1, \ldots, H_n$ be the isotropy groups of G on $X - A$. By Corollary (2.5) and its proof, there is a field k, which is of finite degree over $\mathbb{F}_p$, and a shifted subgroup $\Gamma \subseteq k[G]$ of rank ρ, such that $V_\Gamma \cap V_{H_i} = 0$ for $1 \leq i \leq n$. So $X(\Gamma; k) = A(\Gamma; k)$; and the result follows.

Combining Corollary (4.7) with Lemma (4.3) as in the proof of Corollary (4.4) we get immediately the first part of the following corollary. The second part of the following requires a little more work with the first spectral sequence.

(4.8) **Corollary.** Let X be a paracompact finistic G–space such that $H^*(X; \mathbb{F}_p) \cong H^*(S^a \times S^b; \mathbb{F}_p)$ as graded $\mathbb{F}_p$–vector spaces where a and b are integers such that $0 < a < b$. Then

(1) $\mathrm{rk}_p(X) \leq 2$. If $X = S^a \times S^b$, then $\mathrm{rk}_2(X) = 2$. If $X = S^a \times S^b$ and a, b and p are odd, then $\mathrm{rk}_p(S^a \times S^b) = 2$.

(2) If $a + b$ and p are odd, then $\mathrm{rk}_p(X) \leq 1$. (If a and b are even, and p is odd, then $\mathrm{rk}_p(X) = 0$. This follows from Theorem (5.1) below.)

Returning to the notation established immediately before Proposition (2.4), suppose that K and H are subgroups of $G \cong (\mathbb{Z}/(p))^r$ and that $V_K \subseteq V_H$, where V_K and V_H are defined using some field k of characteristic p. Then it is clear that $V_K \subseteq V_H$ if V_K and V_H are defined using $\mathbb{F}_p$ instead of the extension k. Hence $K \subseteq H$. Thus, using any field k of characteristic p, for actual subgroups $K, H \subseteq G$, $V_K \subseteq V_H$ if and only if $K \subseteq H$.

Furthermore it is clear that, for subgroups $K, H \subseteq G$, $V_{K \cap H} = V_K \cap V_H$. Hence we have the following lemma.

(4.9) **Lemma.** Let $G \cong (\mathbb{Z}/(p))^r$ and let k be a field of characteristic p. Let $\Gamma \subseteq k[G]$ be a shifted subgroup. Then there is a unique subgroup $K(\Gamma) \subseteq G$ such that $V_\Gamma \subseteq V_{K(\Gamma)}$, and $K(\Gamma) \subseteq K$ for any subgroup $K \subseteq G$ with $V_\Gamma \subseteq V_K$.

Γ is not necessarily a shifted subgroup of $k[K(\Gamma)]$, but there is a shifted subgroup $\Gamma' \subseteq k[K(\Gamma)]$ such that $\mathrm{rk}\Gamma' = \mathrm{rk}\Gamma$ and $V_{\Gamma'} = V_\Gamma$. (And hence $K(\Gamma') = K(\Gamma)$.)

If X is a G–space, then $X(\Gamma; k)$ is contained in the singular set of the $K(\Gamma)$–action on X, although it may be properly contained therein. If, however, $\mathrm{rk}\Gamma = 1$, then $X(\Gamma; k) = X^{K(\Gamma)}$.

Furthermore, given any non–trivial subgroup $K \subseteq G$, there is a finite field k of characteristic p and a shifted subgroup $\Gamma \subseteq k[K]$ with $\mathrm{rk}\,\Gamma = 1$ and $K(\Gamma) = K$.

Proof. $K(\Gamma) = \cap \{K \subseteq G \mid V_\Gamma \subseteq V_K\}$. All is now clear except perhaps the last statement. To prove the last statement let $\{u_1, ..., u_r\}$ be a basis for V defined over $\mathbb{F}_p$ such that $\{u_1, ..., u_s\}$ is a basis for V_K where $s = \mathrm{rk}\,K$. Let k be an extension field of $\mathbb{F}_p$ of degree at least s; and let $\alpha_1, ..., \alpha_s \in k$ be linearly independent over $\mathbb{F}_p$. Now choose $\Gamma \subseteq k[K]$ so that V_Γ defined over k is the one–dimensional space spanned by $\alpha_1 u_1 + ... + \alpha_s u_s$.

Now we have the following corollary of Theorem (4.6).

(4.10) **Corollary.** Let $G \cong (\mathbb{F}/(p))^r$, let X be a paracompact finitistic G–space, let $A \subseteq X$ be a closed invariant subspace, and let $K \subseteq G$ be a non–trivial subgroup. Then there is a finite field k of characterisitc p, and a shifted subgroup $\Gamma \subseteq k[K]$ of rank one, such that restriction induces an isomorphism

$$\hat{H}^*_\Gamma(X, A; k) \xrightarrow{\sim} \hat{H}^*_\Gamma(X^K, A^K; k) \cong \hat{H}^*(\Gamma; k) \otimes_k H^*(X^K, A^K; k)$$
$$\cong \hat{H}^*(\Gamma, k) \otimes_{\mathbb{F}_p} H^*(X^K, A^K; \mathbb{F}_p).$$

(4.11) **Remarks.** (1) Since $\mathrm{rk}\,\Gamma = 1$, $\dim_k \hat{H}^i(\Gamma; k) = 1$ for all $i \in \mathbb{Z}$.

(2) If $\Gamma' \subseteq k[G]$ is a shifted subgroup with $V_{\Gamma'} = V_\Gamma$, where Γ is as in Corollary (4.10), then, using $\hat{H}^*_{\Gamma'}$ instead of $\hat{H}^*_\Gamma$, one still gets the first isomorphism of Corollary (4.10), but not in general the second, since Γ' might not be acting trivially on $\mathbb{C}^*(X^K, A^K; k)$.

(3) For a given k, Lemma (4.9) gives a function $\Gamma \longmapsto K(\Gamma)$ from the set of non–trivial shifted subgroups of k[G] to the set of non–trivial subgroups of G; and the lemma and its proof show that this function is surjective when restricted to shifted subgroups of rank one if $\dim_{\mathbb{F}_p}(k) \geq \mathrm{rk}(G)$.

(4) Let $G \cong (\mathbb{Z}/(p))^r$ be generated by $g_1, ..., g_r$, let k be a field of characteristic p, and let $\Gamma \subseteq k[G]$ be a rank one shifted subgroup. Let $\tau_i = 1 - g_i \in m$ and let $\bar{\tau}_i$ be the image of τ_i in

$V = m/m^2$ for $1 \le i \le r$. Suppose that V_Γ is spanned by $\alpha_1\bar{\tau}_1 + \dots + \alpha_r\bar{\tau}_r$, where $\alpha_i \in k$ for $1 \le i \le r$. Now, if p is odd, $H^*(G; k) \cong \Lambda(s_1, \dots, s_r) \otimes k[t_1, \dots, t_r]$, where $\deg s_i = 1$ and $\deg t_i = 2$; and $H^*(\Gamma; k) \cong \Lambda(s) \otimes k[t]$, where $\deg s = 1$ and $\deg t = 2$. $i_\Gamma : k[\Gamma] \longrightarrow k[G]$ induces a homomorphism $i_\Gamma^* : H^*(G; k) \longrightarrow H^*(\Gamma; k)$. i_Γ^* is multiplicative; and if $s_1, \dots, s_r, t_1, \dots, t_r$ correspond to $\tau_1, \dots, \tau_r$ and s, t correspond to $\alpha_1\tau_1 + \dots \alpha_r\tau_r$, then i_Γ^* is given by $i_\Gamma^*(s_i) = \alpha_i s$ and $i_\Gamma^*(t_i) = \alpha_i^p t$, for $1 \le i \le r$.

If $p = 2$, $H^*(G; k) \cong k[t_1, \dots, t_r]$ and $H^*(\Gamma; k) = k[t]$, where $\deg t_i = \deg t = 1$, for $1 \le i \le r$. And i_Γ^* is given by $i_\Gamma^*(t_i) = \alpha_i t$. (See [Carlson, 1983] or [Allday, Puppe], § 3.11.)

Next we shall describe some of Adem's results on exponents in equivariant integral Tate cohomology. Recall that if A is an abelian group and $a \in A$, then $\exp(a)$ is the order of a ($\exp(a) = \infty$ if a does not have finite order), and $\exp(A)$ is the least positive integer n such that $na = 0$ for all $a \in A$ ($\exp(A) = \infty$ if no such integer exists).

(4.12) **Definition**. Let G be a finite group, and let X be a G–space. Let P be a one–point space with trivial G–action. The map $X \longrightarrow P$ induces a homomorphism

$$\varepsilon(X) : \hat{H}^*(G; \mathbb{Z}) \cong \hat{H}_G^*(P; \mathbb{Z}) \longrightarrow \hat{H}_G^*(X; \mathbb{Z}).$$

Let $I = \mathrm{im}[\varepsilon(X) : \hat{H}^0(G; \mathbb{Z}) \longrightarrow \hat{H}_G^0(X; \mathbb{Z})]$. Since $\hat{H}^0(G; \mathbb{Z}) \cong \mathbb{Z}/(|G|)$, $\exp(I) \,\big|\, |G|$. Now the Tate G–exponent of X is defined to be

$$\hat{e}(X; G) := |G|/\exp(I).$$

In [Adem, 1989] Adem proves the following two theorems. It is not hard to deduce the second theorem from the first (see also [Allday, Puppe], §4.6), although we shall not do it here. The proof of the first theorem, however, requires a lot of background material from [Benson, Carlson, 1987] and [Carlson, 1989]. This latter material involves the theory of varieties of G–modules, and hence, at least implicitly, it involves shifted subgroups.

(4.13) **Theorem**. Let G be a finite group, and let $\Phi : G \times X \longrightarrow X$ be an action of G on a paracompact finitistic space X. Let κ be the largest integer r such that there is a prime number p and a subgroup $H \subseteq G$ with $H \cong (\mathbb{Z}/(p))^r$ and $X^H \neq \phi$. i.e. $\kappa = \max\{\mathrm{rk}_p(G_x) \,|\, x \in X, p \text{ prime}\}$. Then there are homogeneous elements $\xi_1, \dots, \xi_\kappa$ of positive degree in $H^*(G; \mathbb{Z})$ such that

$$\exp(\hat{H}_G^*(X; \mathbb{Z})) \,\Big|\, \prod_{i=1}^{\kappa} \exp(\xi_i).$$

In particular, if $G \cong (\mathbb{Z}/(p))^r$, and so $\kappa = r - \mathrm{rk}(\Phi)$, then $p^\kappa \hat{H}_G^*(X; \mathbb{Z}) = 0$. (This follows since each $\exp(\xi_i) = p$ in this case.)

(4.14) **Theorem**. Let $G \cong (\mathbb{Z}/(p))^r$, and let $\Phi : G \times X \longrightarrow X$ be an action of G on a paracompact finitistic space X. Let $\mathrm{rk}(\Phi) = \rho$. Then $\hat{e}(X; G) = p^\rho$.

(4.15) **Corollary**. With G, Φ, X and ρ as in Theorem (4.14), and I as in Definition (4.12),

$$\exp(I) = \exp(\hat{H}_G^0(X; \mathbb{Z})) = \exp(\hat{H}_G^*(X; \mathbb{Z})) = p^\kappa, \text{ where } \kappa = r - \rho.$$

(4.16) **Remark**. Theorem (4.14) should be compared to the results of [Browder, 1983], [Browder, 1988] and [Gottlieb, 1986].

If X is connected, then, in the situation of Theorem (4.14), from the first spectral sequence, it is easy to see that

$$\hat{e}(X; G) \Big| \prod_{j=1}^{\infty} \exp(\hat{H}^{-j-1}(G; H^j(X; \mathbb{Z}))).$$

Thus one has the following.

(4.17) **Corollary.** Let G, Φ, X and ρ be as in Theorem (4.14). Suppose that X is connected and that G acts trivially on $H^*(X; \mathbb{Z})$. Then $H^j(X; \mathbb{Z}) \neq 0$ for at least $\rho + 1$ different values of j.

In particular if $H^*(X; \mathbb{Z}) \cong H^*((S^n)^m; \mathbb{Z})$ as graded abelian groups, where $n > 0$, and if G is acting trivially on $H^*(X; \mathbb{Z})$, then $\rho \leq m$.

Corollary (4.17) has an interesting complement owing to C. Baumgartner ([Baumgartner]) using coefficients in $\mathbb{F}_p$. To state the theorem we need the following definition.

(4.18) **Definition.** Let $G \cong (\mathbb{Z}/(p))^r$, let k be a field of characteristic p, and, as before, let $m = m(G; k)$. Let M be a non-zero $k[G]$-module. Let $\ell(M)$ be the least positive integer n such that $m^n M = 0$. Thus $\ell(M) = 1$ if and only if M is a trivial $k[G]$-module (i.e. G acts trivially on M). If $M = 0$, then set $\ell(M) = 0$.

(4.19) **Theorem.** Let $G \cong (\mathbb{Z}/(p))^r$, and let $\Phi : G \times X \to X$ be a an action of G on a paracompact finitistic space X. Suppose that each connected component of X is open in X. Let $\operatorname{rk} \Phi = \rho$. Then

$$\sum_{j=0}^{\infty} \ell(H^j(X; \mathbb{F}_p)) \geq \rho + 1.$$

In particular, if G acts trivially on $H^*(X; \mathbb{F}_p)$, then $H^j(X; \mathbb{F}_p) \neq 0$ for at least $\rho + 1$ different values of j. Hence if $H^*(X; \mathbb{F}_p) \cong H^*((S^n)^m; \mathbb{F}_p)$ as graded $\mathbb{F}_p$-vector spaces, and if G is acting trivially on $H^*(X; \mathbb{F}_p)$, then $\rho \leq m$.

A proof of this theorem is given in [Allday, Puppe], first for free actions in §1.4, and then it is deduced for more general actions in §4.6 using shifted subgroups, mainly Corollary (4.7) above.

(4.20) **Remarks.** (1) For the situation in Corollary (4.17) where $H^*(X; \mathbb{Z}) \cong H^*((S^n)^m; \mathbb{Z})$, but G is not acting trivially on $H^*(X; \mathbb{Z})$, see [Adem, Browder, 1988].

(2) For the original proof of Theorem (4.19) in the case of free actions when $p = 2$ see [Carlsson, 1983]. See also [Adem, 1988], Proposition 5.9, for non-free actions (using shifted subgroups) when $p = 2$.

5. Smith theory

One of the classical results of Smith theory is the following. If a finite group G acts freely on a paracompact finitistic space X, and if $H^*(X; \mathbb{Z})$ is finitely generated (i.e. $H^*(X; \mathbb{Z})$ has finite type and is zero in all high degress), then $|G|$ divides the Euler characteristic $\chi(X)$. (See, e.g., [Bredon, 1972], Chap. III, Theorem 7.10.) In this section we shall prove the following version of this result by means of shifted subgroups.

(5.1) **Theorem.** Let X be a paracompact finitistic space such that $\dim_{\mathbb{F}_p} H^*(X; \mathbb{F}_p) < \infty$. Suppose that $\operatorname{rk}_p(X) \geq r$. (See Definitions (3.9).) Then $p^r \mid \chi_p(X)$, where $\chi_p(X) := \Sigma_{j=0}^{\infty}(-1)^j \dim_{\mathbb{F}_p} H^j(X; \mathbb{F}_p)$.

(5.2) **Remarks.** (1) Theorem (5.1) seems to go back at least as far as [Brown, 1974]. From the original result in Smith theory (as in [Bredon, 1972], Chap. III, Theorem 7.10) and an inductive argument, Brown proves the following (see also [Brown, 1982], Chap. IX, Theorem (10.1)).

If G is a finite p–group, if X is a paracompact finitistic G–space, if $A \subseteq X$ is a closed invariant subspace, if $\dim_{\mathbb{F}_p} H^*(X, A; \mathbb{F}_p) < \infty$, and if n is any integer which divides the order of every orbit of G on $X - A$, then n divides $\chi_p(X, A) := \sum_{j=0}^{\infty} (-1)^j \dim_{\mathbb{F}_p} H^j(X, A; \mathbb{F}_p)$.

Hence if G is any finite group, if X is a paracompact finitistic G–space, if $A \subseteq X$ is a closed invariant subspace, if $H^*(X, A; \mathbb{Z})$ is finitely generated, and if n is any integer which divides the order of every orbit of G on $X - A$, then n divides $\chi(X, A)$.

The point of the proof below is merely to illustrate how shifted subgroups can be used to obtain Theorem (5.1) quickly and directly in the manner of Smith theory. See also [Adem, 1986], Chap. III, Proposition 4.4, for a similar approach.

Note that thanks to Brown's result and Adem's theorem (Theorem (4.14) above) if $\dim_{\mathbb{F}_p} H^*(X; \mathbb{F}_p) < \infty$, then, under the conditions of Theorem (4.14), $\hat{e}(X; G)$ divides $\chi_p(X)$.

(2) Suppose that G is an elementary abelian p–group, that X is a finite G–CW–complex, and that the rank of the action is ρ. Then each cellular chain group $W_i(X; \mathbb{F}_p)$ is a direct sum of $\mathbb{F}_p[G]$–modules of the form $\mathbb{F}_p[G/G_x]$ where $|G/G_x| \geq p^\rho$. Hence $\dim_{\mathbb{F}_p} W_i(X; \mathbb{F}_p)$ is a multiple of p^ρ; and so $p^\rho | \chi(X)$.

(3) A special case of (2) appears in [Gottlieb, 1986]. Let $G = (\mathbb{Z}/(p))^r$ and let M be a smooth closed connected oriented manifold with G acting on M in a smooth orientation preserving way with rank ρ. In particular, by the results of [Illman, 1978], M is a finite G–CW–complex. Let $x \in H^n(M; \mathbb{Z})$, where $n = \dim M$, be the orientation class. Then Gottlieb shows that $p^\rho x$ is the least positive integer multiple of x in $\mathrm{im}[i^* : H^n_G(M; \mathbb{Z}) \longrightarrow H^n(M; \mathbb{Z})]$, where $i : M \longrightarrow M_G$ is the inclusion of the fibre in the bundle $M_G \longrightarrow BG$. It is easy to see that the Euler class $e(M)$ is in $\mathrm{im}\, i^*$. Hence $p^\rho | \chi(M)$. (To see that $e(M) \in \mathrm{im}\, i^*$, choose $N \gg n$ and let E^N_G be a N–connected compact free G–manifold. Let $M^N_G = (E^N_G \times M)/G$ and $B^N_G = E^N_G/G$. Let $P_N : M^N_G \longrightarrow B^N_G$ be the bundle map. Let $T(M^N_G)$ be the tangent bundle of M^N_G and let $\xi = \ker(d_{P_N}) \subseteq T(M^N_G)$. Then $i^*_N(\xi) = T(M)$, where $i_N : M \longrightarrow M^N_G$ is the inclusion of the fibre. Hence $i^*_N(e(\xi)) = e(M)$. Since N is very large, $e(M) \in \mathrm{im}\, i^*$.)

Proof of Theorem (5.1). Suppose $G \cong (\mathbb{Z}/_{(p)})^m$ is acting on X with rank r. By Corollary (4.7) and its proof, there is a finite field k of characteristic p and a shifted subgroup $\Gamma \subseteq k[G]$ such that $\mathrm{rk}\Gamma = r$, $\hat{H}^*_\Gamma(X; k) = 0$, and (by Proposition (2.4)) $k[G/_{G_x}]$ is a free $k[\Gamma]$–module for all $x \in X$. Now let $\mathcal{U}$ be a Čech–G–covering of X (see Definitions (3.4) and Lemma (3.5)). Then each ordered chain group $C_i(\overline{\mathcal{U}}; k)$ is a free $k[\Gamma]$–module; and hence each ordered cochain group $C^i(\overline{\mathcal{U}}; k)$ is a free $k[\Gamma]$–module by Theorem (1.1)(2). Thus each Alexander–Spanier cochain group $\overline{C}^i(X; k) = \underset{\mathcal{U}}{\underrightarrow{\lim}}\, C^i(\overline{\mathcal{U}}; k)$ is a free $k[\Gamma]$–module by Corollary (1.5)(1). So $C^* := \overline{C}^*(X; k)$ is a cochain complex of free $k[\Gamma]$–modules.

Let $n = \max\{j |\ H^j(X; \mathbb{F}_p) \neq 0\}$. Define C^*_t by $C^i_t = C^i$ for $i < n$, $C^n_t = Z^n$ the cocycles of degree n, and $C^i_t = 0$ for $i > n$. The inclusion $C^*_t \longrightarrow C^*$ is clearly a weak equivalence. So, by the first spectral sequence (Proposition (3.2)(2)), $\hat{H}^*_\Gamma(C^*_t) \cong \hat{H}^*_\Gamma(C^*) = \hat{H}^*_\Gamma(X; k) = 0$. Now, by the second spectral sequence (Proposition (3.2)(4)), $\hat{H}^*(\Gamma; C^n_t) = 0$. So C^n_t is a free $k[\Gamma]$–module by Theorem (1.4). Thus C^*_t is also a cochain complex of free $k[\Gamma]$–modules.

Let $(C^*_t)^\Gamma$ denote the cochain subcomplex of C^*_t consisting of all cochains fixed by Γ. We shall show that $\chi H(C^*_t) = p^r\, \chi H((C^*_t)^\Gamma)$. The theorem will then follow since $H(C^*_t) = H(C^*) = H^*(X; k) \cong H^*(X : \mathbb{F}_p) \otimes_{\mathbb{F}_p} k$.

Now we can use induction on $\mathrm{rk}\Gamma$. To see this, suppose that $\Gamma = \Gamma_1 \times \Gamma_2$, and write $k[\Gamma] \cong k[X_1, ..., X_r]/(X_1^p, ..., X_r^p)$ where $1 - X_1, ..., 1 - X_s$ generate Γ_1 (where $s = \mathrm{rk}\Gamma_1$) and $1 - X_{s+1}, ..., 1 - X_r$ generate Γ_2. Then $k[\Gamma]^{\Gamma_1}$ consists of all multiples of $X_1^{p-1} \dots X_s^{p-1}$: i.e. $k[\Gamma]^{\Gamma_1}$ is the free $k[\Gamma_2] = k[X_{s+1}, ..., X_r]/(X^p_{s+1}, ..., X^p_r)$–module generated by $X_1^{p-1} \dots X_s^{p-1}$. Hence, by induction, we are reduced to the following lemma, which is standard in Smith theory (see, e.g., [Bredon, 1972], Chap. III).

(5.3) **Lemma.** Let $G = \mathbb{Z}/_{(p)}$, let k be a field of characteristic p, and let C^* be a cochain complex of free $k[G]$–modules such that $C^i = 0$ for $i < 0$, $C^i = 0$ for $i > n$, where n is some positive integer, and $\dim_k H(C^*) < \infty$. Then $\dim_k H((C^*)^G) < \infty$; and $\chi H(C^*) = p\chi H((C^*)^G)$.

Proof. Let g generate G, let $\tau = 1 - g \in k[G]$, and let $\nu = (1-g)^{p-1}$. Then $(C^*)^G = \nu C^*$.

By the first spectral sequence $H^*_G(C^*)$ has finite type, since $\dim_k H(C^*) < \infty$. By the second spectral sequence $H^*_G(C^*) \cong H(\nu C^*)$, since C^* is free over $k[G]$. By the second spectral sequence (for equivariant Tate cohomology), $\hat{H}^*_G(C^*) = 0$. So, by Corollary (3.3), $H^j_G(C^*) = 0$ for all $j > n$. Thus $\dim_k H(\nu C^*) < \infty$.

Now for each i, $0 \le i \le p-2$, there is an exact sequence

$$0 \longrightarrow \nu C^* \longrightarrow \tau^i C^* \longrightarrow \tau^{i+1} C^* \longrightarrow 0$$

where the second map is just multiplication by τ. So, by induction, $\dim_k H(\tau^i C^*) < \infty$, for $0 \le i \le p-1$, and $\chi H(\tau^i C^*) = \chi H(\nu C^*) + \chi H(\tau^{i+1} C^*)$, for $0 \le i \le p-1$. Hence $\chi H(C^*) = p\chi H(\nu C^*)$.

(5.4) **Examples.** Let X be a paracompact finitistic space.

(1) If $H^*(X; \mathbb{F}_p) \cong H^*(S^{2n_1} \times \dots \times S^{2n_m}; \mathbb{F}_p)$ as graded $\mathbb{F}_p$–vector spaces, where $n_1, ..., n_m$ are non–negative integers, then $\chi_p(X) = 2^m$. Hence $\mathrm{rk}_p(X) = 0$ if p is odd; and $\mathrm{rk}_2(X) \le m$. Clearly $\mathrm{rk}_2(X) = \mathrm{frk}_2(X) = m$, if $X = S^{2n_1} \times \dots \times S^{2n_m}$.

(2) Suppose that $H^*(X; \mathbb{F}_p) \cong H^*(\mathbb{C}P^n; \mathbb{F}_p)$ as graded $\mathbb{F}_p$–vector spaces, where $n \ge 0$. Then $\chi_p(X) = n + 1$. So $\mathrm{rk}_p(X) \le r$, where $r = \max\{j : p^j \text{ divides } n+1\}$. If p is odd, and if $H^*(X; \mathbb{Z})$ is finitely–generated, then $\mathrm{frk}_p(X) = 0$ by the Lefschetz Fixed Point Theorem (which for

maps of prime period is a theorem of Smith theory, and hence applies to paracompact finitistic spaces). (If p is odd, and if $H^*(X; \mathbb{F}_p) \cong H^*(\mathbb{C}P^n; \mathbb{F}_p)$ as graded $\mathbb{F}_p$–algebras, then it is easy to see that $\mathrm{frk}_p(X) = 0$ by the Localization Theorem, which is stated at the beginning of Section 6 below.)

In [Skjelbred, 1975] it is shown that if p is odd, then $\mathrm{rk}_p(\mathbb{C}P^n) = r$, where $r = \max\{j : p^j \mid n + 1\}$.

Now suppose that $H^*(X; \mathbb{F}_p) \cong H^*(\mathbb{C}P^{n_1} \times \dots \times \mathbb{C}P^{n_m}; \mathbb{F}_p)$ as graded $\mathbb{F}_p$–vector spaces, where $n_1, \dots, n_m$ are non–negative integers. Then $\chi_p(X) = (n_1 + 1) \dots (n_m + 1)$. Hence $\mathrm{rk}_p(X) \le r_1 + \dots + r_m$, where $r_i = \max\{j : p^j \mid n_i + 1\}$ for $1 \le i \le m$. In particular, it follows from Skjelbred's result above, that, if p is odd, $\mathrm{rk}_p(\mathbb{C}P^{n_1} \times \dots \times \mathbb{C}P^{n_m}) = r_1 + \dots + r_m$.

6. Equivariant cohomology.

If $G \cong (\mathbb{Z}/(p))^r$ and k is a field of characteristic p, then let $R(G; k)$ denote the polynomial part of $H^*(G; k)$. Thus (see Remarks (4.11)(4)) $R(G; k) \cong k[t_1, \dots, t_r]$, where $\deg t_i = 2$ for $1 \le i \le r$, if p is odd; and $R(G; k) = H^*(G; k) \cong k[t_1, \dots, t_r]$, where $\deg t_i = 1$ for $1 \le i \le r$, if $p = 2$. A basic form of the Localization Theorem in equivariant cohomology is the following. If X is a paracompact finitistic G–space, and if $A \subseteq X$ is a closed invariant subspace, then restriction induces an isomorphism of $S^{-1}R(G; k)$–modules, $S^{-1}H^*_G(X, A; k) \xrightarrow{\sim} S^{-1}H^*_G(X^G, A^G; k)$, where $S \subseteq R(G; k)$ is the multiplicative set generated by all non–zero homogeneous linear polynomials.

In view of Theorem (4.6) above it is not surprising that one has the following shifted version of the Localization Theorem. (See also the final theorem of [Assadi, 1988].)

(6.1) **Theorem.** Let $G \cong (\mathbb{Z}/(p))^r$, let k be a field of characteristic p, let X be a paracompact finitistic G–space, let $A \subseteq X$ be a closed invariant subspace, and let $\Gamma \subseteq k[G]$ be a shifted subgroup. Then restriction induces an isomorphism of $S^{-1}R(\Gamma;k)$–modules

$$S^{-1}H^*_\Gamma(X, A; k) \xrightarrow{\sim} S^{-1}H^*_\Gamma(X(\Gamma; k), A(\Gamma; k); k),$$

where $S \subseteq R(\Gamma; k)$ is the multiplicative set generated by all non–zero homogeneous linear polynomials.

Before beginning the proof we need a remark.

(6.2) **Remark.** It can be shown (see, e.g. [Allday, Puppe], §3.11) that for any cochain complex C^* of $k[\Gamma]$–modules, $H^*_\Gamma(C^*)$ is naturally a graded $H^*(\Gamma; k)$–module. Furthermore, the long exact sequences and Mayer–Vietoris sequences for H^*_Γ (see Remarks (3.8)(2)) are long exact sequences of $H^*(\Gamma; k)$–modules. Also the first spectral sequence for $H^*_\Gamma(C^*)$ is a spectral sequence of $H^*(\Gamma; k)$–modules. And the map $H^*_G(C^*) \to H^*_\Gamma(C^*)$, induced by the inclusion of $\beta^*_G(C^*)$ into $\beta^*_\Gamma(C^*)$, and the resulting map of first spectral sequences, are homomorphisms of $H^*(G; k)$–modules, where a $H^*(\Gamma; k)$–module is viewed as a $H^*(G; k)$–module via the map $H^*(G; k) \to H^*(\Gamma; k)$.

Proof of Theorem (6.1). By the long exact sequence for the pair (X, A) and Remark (6.2), it is enough to prove the result when $A = \phi$. By the Mayer–Vietoris and tautness trick (as in the proof of Theorem (4.6)) it is enough to show that $S^{-1}H^*_\Gamma(X; k) = 0$ if $X(\Gamma; k) = \phi$.

Let $\mathcal{U}$ be a faithful finite–dimensional Čech–G–covering of X (see Definitions (3.4) and Lemma (3.5)). As in the proof of Theorem (4.6), $\hat{H}^*_\Gamma(|\check{\mathcal{U}}|; k) = 0$. So, by Corollary (3.3), $H^j_\Gamma(|\check{\mathcal{U}}|; k) = 0$ for all $j > \dim \mathcal{U}$. Hence $S^{-1}H^*_\Gamma(|\check{\mathcal{U}}|; k) = 0$. Thus $S^{-1}H^*_\Gamma(X; k) = 0$ by Lemma (3.7).

Recalling the definition of $K(\Gamma)$ from Lemma (4.9) one has the following corollary.

(6.3) **Corollary**. Let G, k, X, A, Γ and S be as in Theorem (6.1). Suppose $rk\Gamma = 1$. Then restriction induces an isomorphism

$$S^{-1}H^*_\Gamma(X, A; k) \xrightarrow{\sim} S^{-1}H^*_\Gamma(X^{K(\Gamma)}, A^{K(\Gamma)}; k).$$

If $\Gamma \subseteq k[K(\Gamma)]$, then clearly $S^{-1}H^*_\Gamma(X^{K(\Gamma)}, A^{K(\Gamma)}; k) \cong S^{-1}H^*(\Gamma; k) \otimes_k H^*(X^{K(\Gamma)}, A^{K(\Gamma)}; k)$. And, with the notation of Remarks (4.11)(4), $S^{-1}H^*(\Gamma; k) \cong \wedge(s) \otimes k[t, t^{-1}]$, respectively $k[t, t^{-1}]$, if p is odd, respectively 2.

Note also that for any topological pair (Y, B), $H^*(Y, B; k) \cong H^*(Y, B; \mathbb{F}_p) \otimes_{\mathbb{F}_p} k$ if $\dim_{\mathbb{F}_p} k < \infty$.

Lemma (4.9) associates an actual subgroup $K(\Gamma) \subseteq G$ to each shifted subgroup $\Gamma \subseteq k[G]$. In the next lemma we shall show how to find $K(\Gamma)$ from the kernel of the homomorphism $i^*_\Gamma : H^*(G; k) \longrightarrow H^*(\Gamma; k)$ when $rk\Gamma = 1$. First, however, we need some definitions and notation.

(6.4) **Definitions.** Let $G \cong (\mathbb{Z}/(p))^r$, and let $R = R(G; \mathbb{F}_p)$, the polynomial part of $H^*(G; \mathbb{F}_p)$. Let $g_1, \ldots, g_r$ generate G, and let $t_1, \ldots, t_r$ be the corresponding generators of R. That is, $R = \mathbb{F}_p[t_1, \ldots, t_r]$. If p is odd, let $s_1, \ldots, s_r$ be the corresponding exterior generators in $H^1(G; \mathbb{F}_p)$.

To save writing we shall let L denote the $\mathbb{F}_p$–vector space spanned by $t_1, \ldots, t_r$, i.e. the space of homogeneous linear polynomials in R.

(1) If $K \subseteq G$ is a subgroup, and $i^*_K : H^*(G; \mathbb{F}_p) \longrightarrow H^*(K; \mathbb{F}_p)$ is the restriction homomorphism, then let $PK = \ker(i^*_K|R)$. It follows that PK is a homogeneous prime ideal generated by homogeneous linear polynomials. And the height of PK, ht PK, is equal to $\dim_{\mathbb{F}_p}(PK \cap L)$, which in turn is equal to cork K, the corank of K (i.e. $r - rk\, K$).

(2) For any prime ideal $P \subseteq R$, let σP be the ideal generated by $P \cap L$. Thus $\sigma P = PK$ for some $K \subseteq G$.

(6.5) **Lemma.** Let G be as in Definitions (6.4), let k be a field of characteristic p, and let $\Gamma \subseteq k[G]$ be a shifted subgroup of rank one. Suppose that, in the notation of Remarks (4.11)(4), V_Γ is spanned by $\alpha_1\bar{\tau}_1 + \ldots + \alpha_r\bar{\tau}_r$. (I.e. $V_\Gamma = V_{\Gamma(\alpha)}$ in the notation of Definitions (2.1)(4).) Let $I(\alpha)$ be the kernel of the composition $R \longrightarrow R(G; k) \xrightarrow{i^*_\Gamma} R(\Gamma; k)$. Then $\sigma I(\alpha)$ is generated by $\{\sum_{i=1}^r c_i t_i \in L \mid \sum_{i=1}^r c_i\alpha_i = 0\}$; and $\sigma I(\alpha) = PK(\Gamma)$.

Proof. The statement about the generators of $\sigma I(\alpha)$ is immediate from Remarks (4.11)(4). We shall prove that $\sigma I(\alpha) = PK(\Gamma)$ assuming that p is odd: the case p = 2 is slightly more simple. Let M be the k–vector space spanned by $s_1, \ldots, s_r$; and let $\varphi : M \longrightarrow L \otimes k$ be the isomorphism given by $\varphi(s_i) = t_i$ for $1 \le i \le r$. Then, where $V = m/m^2$ as usual (see above Proposition (2.4)), M corresponds to the k–vector space dual V^*, and $\{s_1, \ldots, s_r\}$ is the dual basis for $\{\bar\tau_1, \ldots, \bar\tau_r\}$. Since V_Γ is spanned by $\alpha_1\bar\tau_1 + \ldots + \alpha_r\bar\tau_r$, the annihilator of V_Γ in V^*, $\mathrm{ann}(V_\Gamma)$, is equal to $\{\Sigma_{i=1}^r \lambda_i s_i \in M \mid \Sigma_{i=1}^r \lambda_i \alpha_i = 0\}$. Also $\mathrm{ann}(V_K) = \varphi^{-1}((L \cap PK) \otimes_{\mathbb{F}_p} k)$.

Thus $V_\Gamma \subseteq V_K \Leftrightarrow \mathrm{ann}\,(V_K) \subseteq \mathrm{ann}(V_\Gamma) \Leftrightarrow$ for all $\Sigma_{i=1}^r c_i t_i \in PK$, $\Sigma_{i=1}^r c_i \alpha_i = 0 \Leftrightarrow PK \subseteq \sigma I(\alpha)$. Hence $\sigma I(\alpha) = PK(\Gamma)$.

For some applications the next lemma is useful.

(6.6) **Lemma.** Let $G \cong (\mathbb{Z}/(p))^r$, and let $R = R(G; \mathbb{F}_p)$ as above. Let $f_1, \ldots, f_n$ be polynomials in R such that the ideal $(f_1, \ldots, f_n) \neq R$. Then there is a finite extension field k of $\mathbb{F}_p$, and an $\alpha = (\alpha_1, \ldots, \alpha_r) \in k^r$, such that $f_1(\alpha) = \ldots = f_n(\alpha) = 0$, and $\mathrm{rk}\, K(\alpha) \ge r - n$, where $K(\alpha) = K(\Gamma(\alpha))$.

Proof. Let $J = (f_1, \ldots, f_n)$ and let $F = \bar{\mathbb{F}}_p$, the algebraic closure. Let V(J) be the variety of J in F^r.

By Remarks (4.11)(4), with the notation of Lemma (6.5), for any $\alpha \in F^r$, $I(\alpha) = \{f \in R \mid f(\alpha) = 0\}$. Thus, by Hilbert's Nullstellensatz, $\sqrt{J} = \bigcap_{\alpha \in V(J)} I(\alpha)$. So, by Lemma (6.5), $\sqrt{J} \supseteq \bigcap_{\alpha \in V(J)} PK(\alpha)$, and this latter intersection is finite since there are only a finite number of subgroups of G. Hence if P is a minimal prime ideal of J, $P \supseteq PK(\alpha)$ for some $\alpha \in V(J)$. So $r - \mathrm{rk}\, K(\alpha) = \mathrm{ht}\, PK(\alpha) \le \mathrm{ht}\, P$; and, since $J = (f_1, \ldots, f_n)$, $\mathrm{ht}\, P \le n$ by the Krull Height Theorem.

Next we shall give an example to illustrate the use of Theorem (6.1) and Lemma (6.6).

(6.7) **Example.** We shall give a quick proof of the familiar fact that if X is a paracompact finitistic space such that $H^*(X; \mathbb{F}_p) \cong H^*((S^{2n-1})^3; \mathbb{F}_p)$ as graded $\mathbb{F}_p$–algebras, then the rank of any elementary abelian p–group action on X is at most 3 if the group acts trivially on $H^*(X; \mathbb{F}_p)$. (Cf. Theorem (4.19).)

Let $G \cong (\mathbb{Z}/(p))^r$ act on X. Let $x_1, x_2, x_3 \in H^{2n-1}(X; \mathbb{F}_p)$ be generators. In the Leray–Serre spectral sequence of $X_G \to BG$, let f_1, f_2, f_3 be the polynomial parts of the transgressions of x_1, x_2, x_3. By Lemma(6.6) we can choose finite k and $\alpha \in k^3$ such that $f_1(\alpha) = f_2(\alpha) = f_3(\alpha) = 0$ and $\mathrm{rk}\, K(\alpha) \ge r - 3$. From the first spectral sequence for $H^*_{\Gamma(\alpha)}(X; k)$, since x_1, x_2, x_3 transgress to zero, it is clear that $S^{-1}H^*_{\Gamma(\alpha)}(X; k) \neq 0$. (In the spectral sequence for $H^*_G(X; \mathbb{F}_p)$, $d_{2n}(x_i x_j) = f_i x_j - f_j x_i$ modulo terms involving the exterior part of $H^*(G, \mathbb{F}_p)$, if p is odd. So, in the spectral sequence for $H^*_\Gamma(X; k)$, the $x_i x_j$ classes all transgress. It follows that $S^{-1}E_\infty^{*,2n-1} \neq 0$. See Remark (6.8) below.)

Thus, by Corollary (6.3), $X^{K(\alpha)} \neq \phi$. So the rank of the action is at most 3.

(6.8) **Remark.** In general, for a shifted subgroup $\Gamma \subseteq k[G]$, the first spectral sequence for $H^*_\Gamma(X; k)$ is not multiplicative with respect to the cup product in $H^*(X; k)$. This is because the standard cup product on the cochains, $\overline{C}^*(X; k)$, is not compatible with any of the standard diagonals $k[\Gamma] \longrightarrow k[\Gamma] \otimes k[\Gamma]$. An example which shows this non–multiplicativity is the following.

The quaternionic eight–group acts freely and linearly on $\mathbb{R}^4$. Thus $G \cong (\mathbb{Z}/(2))^2$ can act freely on $X = \mathbb{R}P^3$. In the first spectral sequence for $H^*_G(X; \mathbb{F}_2)$, the generator $x \in H^1(X; \mathbb{F}_2)$ must transgress to $t_1^2 + t_1t_2 + t_2^2 \in H^2(G; \mathbb{F}_2)$, and $x^2 = Sq^1(x)$ transgresses to $Sq^1(t_1^2 + t_1t_2 + t_2^2) = t_1t_2(t_1 + t_2)$, modulo $(t_1^2 + t_1t_2 + t_2^2)$. Choose $\alpha = (\alpha_1, \alpha_2)$ in $\mathbb{F}_2$ so that $\alpha_1^2 + \alpha_1\alpha_2 + \alpha_2^2 = 0$, and $\alpha \neq 0$. Then, in the first spectral sequence for $H^*_{\Gamma(\alpha)}(X; \mathbb{F}_2)$, x transgresses to zero, but x^2 transgresses to a non–zero multiple of $t^3 \in H^3(\Gamma(\alpha); \mathbb{F}_2)$.

In Example (6.7) above, the information about the differentials in the first spectral sequence for $H^*_{\Gamma(\alpha)}(X; k)$ is obtained from the homomorphism from the first spectral sequence of $H^*_G(X; k)$ to that of $H^*_{\Gamma(\alpha)}(X; k)$. (See Remark (6.2).)

REFERENCES

[Adem, 1986] A. Adem, Finite transformation groups and their homology representations, Thesis, Princeton University, 1986.

[Adem, 1988] A. Adem, Cohomological restrictions on finite group actions, J. of Pure and Applied Algebra 54 (1988), 117–139.

[Adem, 1989] A. Adem, Torsion in equivariant cohomology, Comment. Math. Helvetici 64 (1989), 401–411.

[Adem, 1989a] A. Adem, Homology representations of finite transformation groups, Algebraic Topology, Proceedings, Arcata 1986, Edited by G.Carlsson, R. L. Cohen, H. R. Miller and D. C. Ravenel, Lecture Notes in Mathematics 1370, 15–23, Springer–Verlag, Berlin 1989.

[Adem, Browder, 1988] A. Adem and W. Browder, The free rank of symmetry of $(S^n)^k$, Invent. Math. 92 (1988), 431–440.

[Allday, Puppe, 1985] C. Allday and V. Puppe, On the localization theorem at the cochain level and free torus actions, Proc. Göttingen Conf. on Algebraic Topology 1984, Edited by . L. Smith, Lecture Notes in Mathematics 1172, 1–16, Springer–Verlag, Berlin 1985.

[Allday, Puppe] C. Allday and V. Puppe, Cohomological Methods in Transformation Groups, to appear.

[Assadi, 1988] A. Assadi, Varieties in finite transformation groups, Bull. Amer. Math. Soc. (New Series) 19 (1988), 459–463.

[Assadi, 1989a] A. Assadi, Integral representations of finite transformation groups I, J. of Pure and Applied Algebra 59 (1989), 215–226.

[Assadi, 1989 b] A. Assadi, Integral representations of finite transformation groups II: non–simply–connected spaces, J. of Pure and Applied Algebra 59 (1989), 227–236.

[Assadi] A. Assadi, An algebraic invariant for finite group actions and applications, to appear.

[Baumgartner] C. Baumgartner, On the cohomology of free p–torus actions, to appear.

[Benson, 1984] D. J. Benson, Modular Representation Theory: New Trends and Methods, Lecture Notes in Mathematics 1081, Springer–Verlag, Berlin 1984.

[Benson, Carlson, 1987] D. J. Benson and J. F. Carlson, Complexity and multiple complexes, Math. Zeit. 195 (1987), 221–238.

[Bredon, 1972] G. E. Bredon, Introduction to Compact Transformation groups, Academic Press, New York 1972.

[Browder, 1983] W. Browder, Cohomology and group actions, Invent. Math. 71 (1983), 599–607.

[Browder, 1988] W. Browder, Actions of elementary abelian p–groups, Topology 27 (1988), 459–472.

[Brown, 1974] K. S. Brown, Euler characteristics of discrete groups and G–spaces, Invent. Math 27 (1974), 229–264.

[Brown, 1982] K. S. Brown, Cohomology of Groups, Graduate Texts in Mathematics 87, Springer–Verlag, Berlin 1982.

[Carlson, 1983] J. F. Carlson, The varieties and the cohomology ring of a module, J. of Algebra 85 (1983), 104–143.

[Carlson, 1989] J. F. Carlson, Exponents of modules and maps, Invent. Math. 95 (1989), 13–24.

[Carlsson, 1983] G. Carlsson, On the homology of finite free $(\mathbb{Z}/2)^n$–complexes, Invent. Math. 74 (1983), 139–147.

[Dade, 1978] E. C. Dade, Endo–permutation modules over p–groups II, Ann. of Math. 108 (1978), 317–346.

[Fuller, 1989] K. R. Fuller, Artinian Rings, Notas de Mathematica 2, Universidad de Murcia 1989, a supplement to F. W. Anderson and K. R. Fuller, Rings and Categories of Modules, Graduate Texts in Mathematics 13, Springer–Verlag, Berlin, 1974.

[Gottlieb, 1986] D. H. Gottlieb, The trace of an action and the degree of a map, Trans. Amer. Math. Soc. 293 (1986), 381–410.

[Heller, 1959] A. Heller, A note on spaces with operators, Illinois J. Math. 3 (1959), 98–100.

[Illman, 1978] S. Illman, Smooth equivariant triangulations of G–manifolds for G a finite group, Math. Ann. 233 (1978), 199–220.

[Kroll, 1984] O. Kroll, Complexity and elementary abelian p–groups, J. of Algebra 88 (1984), 155–172.

[Skjelbred, 1975] T. Skjelbred, Actions of p–tori on projective spaces, University of Oslo, Insitute of Mathematics, Preprint Series No. 4, January 27, 1975.

[Spanier, 1966] E. H. Spanier, Algebra Topology, McGraw–Hill, New York, 1966.

[Swan, 1960] R. G. Swan, A new method in fixed point theory, Comment. Math. Helvetici 34 (1960), 1–16.

EQUIVARIANT FINITENESS OBSTRUCTION AND ITS GEOMETRIC APPLICATIONS - A SURVEY

Paweł Andrzejewski (*)

Instytut Matematyki, Uniwersytet Szczeciński
ul. Wielkopolska 15, 70-451 Szczecin 3, Poland

The purpose of this article is to present one of the topics in the theory of transformation groups. The topic deals with finiteness questions in equivariant geometric topology. As constructions below show the equivariant finiteness obstruction is closely connected with functors of algebraic K-theory and it illustrates well the increased importance and influence of these functors on group actions on manifolds and CW-complexes.

The history of the finiteness obstruction has begun with fundamental Swan's paper [53] where algebraic ideas of the obstruction have appeared and a variant of the finiteness obstruction for spaces with universal covering being homotopy sphere is considered. In a nutshell, if for a finite group G with periodic cohomology one wishes to construct a finite, free G-complex X homotopy equivalent to the sphere one proceeds as follows. We choose a finite 2-dimensional complex K with $\pi_1(K) = G$. Then by homological algebra the finitely dominated free G-complex X with $X \simeq S^n$ is established and now the finiteness obstruction decides whether X can be chosen to be finite. This is essential if one wants to replace X by a G-manifold and finally by the standard sphere with free G-action. This leads directly to the space form problem [35]. Swan's work has become a motivation both for Wall's papers [55], [56] and for a lot of activities about the space form problem (see [54], [35]). Soon after appearing Wall's paper [55] Siebenmann [46] applied the finiteness obstruction to geometric problem of constructing a boundary in an open manifold. In 1976 J. Baglivo [6] made an attempt to establish the finiteness obstruction in the equivariant context and quite shortly the equivariant finiteness obstruction has shown its usefulness in problems of transformation groups [38], [39], [40], [41], [43], [42], [17], [14], [12].

At present there are (at least) four different (but equivalent) approaches to equivariant finiteness obstruction and we wish to describe three of them in section 2. Next we show some elementary properties and among them the universal characterization given by Lück [33], [34]. Section 4 gives various applications of the obstruction to geometric problems. Recently M. Steinberger and J. West, using Chapman's and Ferry's ideas, have developed the controlled version of the equivariant finiteness obstruction. We outline their construction (which seems to be still unfinished) and show how this theory can be applied to solution of several questions in equivariant geometric topology.

Almost all of the results discussed in this paper were discovered during the past decade and even during the past five years.

I would like to express my appreciation to the referee for his suggestions, comments and complements

---

(*) Supported by Polish scientific grant RP. I. 10

which have improved the exposition of this paper. I am also indebted to W. Lück who has pointed out my attention to paper [21].

1. Preliminary remarks and constructions

We refer the reader to Bredon's and tom Dieck's books [8], [13] for background information on transformation groups used in this paper.

In the following, G denotes an arbitrary compact Lie group, unless otherwise is specifically stated.

The G-space X is said to be G-dominated by a G-space K if X is a G-homotopy retract of K. If K is a finite G-complex then X is called finitely G-dominated.

We collect some elementary properties of spaces G-dominated by finite G-complexes.

Proposition 1.1 [19], [31] *If G-space X is G-dominated by an n-dimensional finite G-complex K then X has the G-homotopy type of an (n+1)-dimensional G-complex Y. If moreover $n \geq 3$ then Y can be taken to be n-dimensional* ([34] prop. 11.10 and prop. 14.9 c).

Corollary 1.2 (of the proof) *The G-space X is finitely G-dominated iff $X \times S^1$ has the G-homotopy type of a finite G-complex.*

For any closed subgroup H of G the fixed point set X^H is an NH-space as well as a WH-space where $WH = NH/H$. Let X^H_α be a connected component of X^H and denote $(NH)_\alpha = \{ n \in NH : n(X^H_\alpha) = X^H_\alpha \}$ and $(WH)_\alpha = \{ w \in WH : w(X^H_\alpha) = X^H_\alpha \}$. Both $(NH)_\alpha$ and $(WH)_\alpha$ are compact Lie groups.

Proposition 1.3 [1], [18] *Suppose a G-complex X is G-dominated by a G-complex K and let $\phi: K \to X$ denote the domination map with the section $s: X \to K$. Let X^H_α and K^H_β be components of X^H and K^H, respectively, such that $s(X^H_\alpha) \subset K^H_\beta$. Then $(WH)_\alpha = (WH)_\beta$ and K^H_β $(WH)_\alpha$-dominates X^H_α.*

Since we are interested in equivariant homotopy type of finitely G-dominated G-spaces in view of prop. 1.1 we may suppose X is a connected G-complex.

If $p : \tilde{X} \to X$ denotes its universal covering then we can consider the lifting of the action of the group G on X to the covering action of a group G^* on $\tilde{X}$ (see §5 in [25] for details). The group G^* is a Lie group and fits into the exact sequence

$$0 \longrightarrow \pi_1(X) \longrightarrow G^* \xrightarrow{\pi} G \longrightarrow 0 .$$

Moreover $\tilde{X}$ is a G^*-complex ([25] thm. 6.6). It turns out that there is a natural isomorphism

$$\pi_0(G^*) \cong \pi_1(EG \times_G X)$$

which, in the special case when the group G acts freely on X, reduces to

$$\pi_0(G^*) \cong \pi_1(X/G).$$

Let further A be a G-invariant subcomplex of X such that the inclusion $A \subset X$ induces an isomorphism $\pi_1(A,x_0) \cong \pi_1(X,x_0)$ of fundamental groups. Then one can define an action of $\pi_0(G^*)$ on homotopy and homology groups $\pi_n(X,A,x_0)$, $H_n(X,A)$ such that it makes them into modules over the group ring $\mathbf{Z}[\pi_0(G^*)]$ (see §7 in [25]). These modules will be needed in the construction of the equivariant Wall finiteness obstruction in the next section.

We conclude with two fundamental properties of finitely G-dominated complexes. The first of them asserts that the number of orbit types of a finitely G-dominated G-complex cannot be too large.

Proposition 1.4 ([21] thm. 1.4 or [34] prop. 2.12) *Let K be a G-complex with finitely many orbit types and let X be a G-complex G-dominated by K. Then X is G-homotopy equivalent to a G- complex Y such that the number of orbit types of Y is not greater than that of K.*

Finally it turns out that making use of the equivariant cell-attaching technique one can generalize the results of Wall [55] and Baglivo [6].

Proposition 1.5 ([2] lemma 1.3) *If G-complex X is finitely G-dominated then it is G-homotopy equivalent to a G-complex Y of finite type.*

2. Three approaches to equivariant finiteness obstruction

As it was mentioned in the introduction there are four equivalent constructions of the equivariant finiteness obstruction. The fourth of them which is not discussed here has a global algebraic character and is based on the theory of modules over a category [34]. This global approach covers all our constructions listed below and is identified with them by appropriate splitting theorems (see [34] def. 14.4, thm 14.46, prop. 11.15 and thm 10.34).

The approach which follows ideas of C.T.C. Wall [55] and D. Anderson [1] seems to be the most classical. In a result of this construction one gets a family of invariants which decide whether finitely G-dominated G-complex is G-homotopy finite. In order to define this family we have to begin with the case of a relatively free action used as an inductive step.

We say that the action of G on the pair (X,A) is relatively free if G acts freely outside A. We say that the G-CW-pair (X,A) is relatively finite if $\overline{X - A}$ has finite number of G-cells.

Let now the relatively free G-CW-pair (X,A) be G-dominated by a relatively free, relatively finite G-CW-pair (K,L) via the map $\phi: (K,L) \to (X,A)$ and let $q: \hat{K} \to K$ be the pull-back of $p : \tilde{X} \to X$ by ϕ i.e.

$$\hat{K} = \{ (\tilde{x},k) \in \tilde{X} \times K : p(\tilde{x}) = \phi(k) \} .$$

The group G^* acts on $\hat{K}$ by the formula $g^*(\tilde{x},k) = (g^*(\tilde{x}), \pi(g^*)(k))$. Then $\hat{K}$ is a G^*-complex and G^* acts freely on $\hat{K} - \hat{L}$. The cellular chain complexes $C_*(\hat{K},\hat{L})$ and $C_*(\tilde{X},\tilde{A})$ are complexes of free $\mathbb{Z}[\pi_0(G^*)]$-modules and $C_*(\hat{K},\hat{L})$ is finite. Moreover the map ϕ induces the domination map $C_*(\hat{K},\hat{L}) \to C_*(\tilde{X},\tilde{A})$ of chain complexes. We define the relative equivariant Wall finiteness obstruction as

$$w_G(X,A) = w(C_*(\tilde{X},\tilde{A})) \in \tilde{K}_0(\mathbb{Z}[\pi_0(G^*)])$$

where $w(C_*)$ is the algebraic finiteness obstruction [56],[20],[45]. The following generalizes results given in [1] and [21].

Proposition 2.1 *Let a relatively free G-CW-pair (X,A) be G-dominated by a relatively free, relatively finite G-CW-pair (K,L). Suppose L is of finite type and X contains a finite G-subcomplex B. Then there exist a relatively free, relatively finite G-CW-pair (Y,A) with $B \subset Y$ and a G-homotopy equivalence*

$$h: (Y,A\cup B) \to (X,A\cup B)$$

with $h|_{A\cup B} = id_{A\cup B}$ iff $w_G(X,A) = 0$ in $\tilde{K}_0(\mathbb{Z}[\pi_0(G^*)])$. Moreover, if $\dim(K-L) = n$ then Y can be chosen in such a way that $\dim(Y-A) = \max(3,n)$.

In the case when in the relatively free G-CW-pair (X,A) the subcomplex A is empty, proposition 2.1

and the isomorphism $\pi_0(G^*) \cong \pi_1(X/G)$ gives immediately.

Corollary 2.2 *Assume that the group G acts freely on X. Then $w_G(X) = w(X/G)$ and X has the G-homotopy type of a finite G-complex iff $w(X/G) = 0$.*

Roughly speaking the family of obstructions we want to introduce is defined for each component X^H_α by means of the invariants $w_G(X,A)$. Precisely let H denote a closed subgroup of G and let X^H_α be a connected component of $X^H \neq \emptyset$. We define an equivalence relation $\approx$ in the set of such components X^H_α, by setting $X^H_\alpha \approx X^K_\beta$ iff there exists an element $n \in G$ such that $nHn^{-1} = K$ and $n(X^H_\alpha) = X^K_\beta$. We denote the set of equivalence classes of this relation by $\underline{CI}(X)$. Note that this definition is functorial i.e. a G-map $f: X \to Y$ induces a map $\underline{CI}(f): \underline{CI}(X) \to \underline{CI}(Y)$.

If now X is finitely G-dominated by a complex K and X^H_α denotes a component of X^H which represents an element of the set $\underline{CI}(X)$ then the group $(WH)_\alpha$ acts on the pairs $(X^H_\alpha, X^{>H}_\alpha)$ and $(K^H_\beta, K^{>H}_\beta)$ in such a way that $(X^H_\alpha, X^{>H}_\alpha)$ is relatively free and $(K^H_\beta, K^{>H}_\beta)$ is relatively free and relatively finite. By the relative version of prop. 1.3 we have that $(K^H_\beta, K^{>H}_\beta)$ $(WH)_\alpha$-dominates $(X^H_\alpha, X^{>H}_\alpha)$.

Definition [2] We define a Wall-type invariant $w^H_\alpha(X)$ to be

$$w^H_\alpha(X) = w_{(WH)_\alpha}(X^H_\alpha, X^{>H}_\alpha) \in \tilde{K}_0(\mathbb{Z}[\pi_0(WH)^*_\alpha]).$$

The elements $w^H_\alpha(X)$ are invariants of the equivariant homotopy type and they vanish for finite G-complexes. Moreover the invariant $w^H_\alpha(X)$ does not depend (up to canonical isomorphism) on the choice of representative X^H_α from the equivalence class $[X^H_\alpha]$ in $\underline{CI}(X)$ [2]. The fundamental property of invariants $w^H_\alpha(X)$ shows that they are really obstructions to homotopy finiteness of X.

Theorem 2.3 *Let a G-complex X be G-dominated by a finite G-complex K. Then there exist a finite G-complex Y and a G-homotopy equivalence $h: Y \to X$ iff all the invariants $w^H_\alpha(X)$'s vanish. Moreover if the complex X contains a finite G-subcomplex B and $\dim K = n$ then Y and h can be chosen in such a manner that $B \subset Y$, $\dim Y = \max(3,n)$ and $h|_B = id_B$.*

Sketch of the proof (the reader is referred to [2] and [21] for details).

Assume that $w^H_\alpha(X) = 0$ for any equivalence class $[X^H_\alpha]$ in $\underline{CI}(X)$. Note that the set $\underline{CI}(X)$ consists of one connected component from each WH-component $(WH)X^H_\alpha$. One can assume, in view of proposition 2.14 in [34] that H runs through a complete set of representatives for all the isotropy types (H) occurring in X. We may suppose, in view of prop. 1.4, the set $\underline{CI}(X)$ is finite.

If $\{ X^{H_i}_{\alpha_j} : 1 \le i \le r, 1 \le j \le p \}$ denotes a complete set of representatives of elements of $\underline{CI}(X)$ we proceed by induction on the number of these elements. Precisely for each pair (i,j) one constructs a G-complex $Y_{i,j}$ and a G-homotopy equivalence $f_{i,j} : Y_{i,j} \to X$ such that

1) $(Y_{i,j})^H$ is WH-finite for any subgroup H of G with $H \in (H_k)$ for some k, $1 \le k < i$.

2) $G(Y_{i,j})^{H_i}_{\beta_m}$ is G-finite for any component $(Y_{i,j})^{H_i}_{\beta_m}$ of $(Y_{i,j})^{H_i}$ corresponding to $X^{H_i}_{\alpha_m}$ under $f_{i,j}$ for $1 \le m \le j$.

3) $B \subset Y_{i,j}$ and $f_{i,j}|_B = id_B$.

Then the complex Y_{r,α_p} obtained as a result of the final inductive step will be a finite G-complex G-homotopy equivalent to X.

In [31] S. Kwasik has generalized the construction of the finiteness obstruction presented by S. Ferry [19] to the equivariant case. Ferry's definition was historically the second one and we briefly recall this approach following the description given in [2], §3.

Let $\phi: K \to X$ be a domination map with the section $s: X \to K$. If $A = s\phi: K \to K$ then denote by $T(A)$ the mapping torus of A obtained from the mapping cylinder $M(A)$ by identification of the top and bottom of $M(A)$ by means of the identity map. Let $B: T(A) \to X \times S^1$ denote the G-homotopy equivalence of prop.1.1. Let $u : S^1 \to S^1$ be the homeomorphism given by the complex conjugation and B^{-1} the homotopy inverse to B. Denote by $\tau(h) \in Wh_G(T(A))$ the torsion of the G-homotopy equivalence

$$h = B^{-1}\circ(id_X \times u)\circ B : T(A) \to T(A).$$

<u>Definition</u> [31],[2] An <u>equivariant Ferry obstruction to finiteness</u> is defined to be

$$\sigma_G(X) = B_*(\tau(h)) \in Wh_G(X \times S^1) .$$

One can show that this obstruction is well-defined ([19], thm. 2.3) . Furthermore making use of the product formula for equivariant Whitehead torsion [24] and carrying over Ferry's ideas to the equivariant context one can prove

<u>Theorem 2.4</u> [2] *The finitely G-dominated G-complex X has the equivariant homotopy type of a finite G-complex iff* $\sigma_G(X) = 0$.

In 1985 W. Lück [33] presented another (geometric) construction of the equivariant obstruction to finiteness. His approach has an advantage that one can derive all the formal properties of the equivariant finiteness obstruction easily from this geometric description. Also one gets a characterization of the finiteness obstruction by a universal property.

Let Y be an arbitrary G-complex. Let us consider the set of G-maps $f: X \to Y$ where X ranges through finitely G-dominated G-complexes. We define an equivalence relation as follows: $f_0: X_0 \to Y$ and $f_4 : X_4 \to Y$ are equivalent iff there exists a commutative diagram

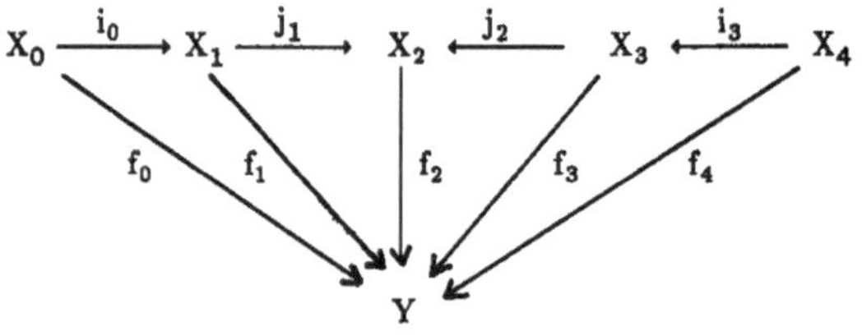

such that j_1 and j_2 are G-homotopy equivalences and i_0 , i_3 are inclusions such that the G-CW-pairs (X_1,X_0) and (X_3,X_4) are relatively finite. Let $Wa^G(Y)$ denote the set of equivalence classes. The disjoint union induces an addition on $Wa^G(Y)$ and the inclusion of the empty space defines a neutral element. One can show that this addition gives $Wa^G(Y)$ the structure of an abelian group ([33], p. 370).

<u>Definition</u> Let X be a finitely G-dominated G-complex. We define its <u>geometric obstruction to finiteness</u> as $w^G(X) = [\, id: X \to X\,] \in Wa^G(X)$.

The following theorem contains principial properties of the invariant $w^G(X)$. Its proof follows the treatment of Whitehead torsion by Cohen in [11].

<u>Theorem 2.5</u> ([33] thm. 1.1 or [34] §3) *Let X be finitely G-dominated.*

(a) $Wa^G : G\text{-}CW \to Ab$ *is a covariant functor from the category of equivariant CW-complexes*

to the category of abelian groups.

(b) $w^G(X)$ *is an invariant of the G-homotopy type.*

(c) G-complex X *is G-homotopy equivalent to a finite G-complex iff* $w^G(X) = 0$.

It turns out that there are natural relations between these definitions. More precisely, one can show that in some sense they are equivalent. We wish to discuss it shortly. The first equivalence between $\sigma_G(X)$ and $w^H_\alpha(X)$ was established in [2]. It looks as follows.

Suppose X is finitely G-dominated by K. In [25] Illman showed that there is a natural isomorphism

$$Wh_G(X \times S^1) \cong \bigoplus_{C(X)} Wh(\pi_0(WH)^*_\alpha \times \mathbb{Z}).$$

Furthermore for an arbitrary split short exact sequence of groups

$$0 \to R \to P \to \mathbb{Z} \to 0$$

we have the natural Bass-Heller-Swan decomposition of the Wh-functor [7]

$$Wh(P) = Wh(R) \oplus \tilde{K}_0(\mathbb{Z}[R]) \oplus N .$$

In particular we have an epimorphism

$$S : Wh(P) \longrightarrow \tilde{K}_0(\mathbb{Z}[R]) .$$

Hence we obtain the natural decomposition

$$Wh_G(X \times S^1) \cong \bigoplus_{C(X)} Wh(\pi_0(WH)^*_\alpha) \oplus \bigoplus_{C(X)} \tilde{K}_0(\mathbb{Z}[\pi_0(WH)^*_\alpha]) \oplus N$$

and the natural epimorphism

$$S : Wh_G(X \times S^1) \longrightarrow \bigoplus_{\underline{CI}(X)} \tilde{K}_0(\mathbb{Z}[\pi_0(WH)^*_\alpha]) .$$

<u>Theorem</u> <u>2.6</u> *The equivariant finiteness obstruction* $\sigma_G(X)$ *decomposes into the family of obstructions* $w^H_\alpha(X)$. *Precisely, the image of the (H,α)-component* $\sigma_G(X)^H_\alpha$ *of the obstruction* $\sigma_G(X)$ *under epimorphism*

$$Wh(\pi_0(WH)^*_\alpha \times \mathbb{Z}) \longrightarrow \tilde{K}_0(\mathbb{Z}[\pi_0(WH)^*_\alpha])$$

is equal to the equivariant Wall-type obstruction $w^H_\alpha(X)$.

The proof of this result is technical in nature and we refer for it to [2], thm. 4.1.

In order to relate Lück's obstruction $w^G(X)$ to $w^H_\alpha(X)$'s a sort of the realization result for the equivariant finiteness obstruction is needed.

<u>Proposition</u> <u>2.7</u> *Let* Y *be a finite G-complex and let* $\{ w^H_\alpha \}$, $w^H_\alpha \in \tilde{K}_0(\mathbb{Z}[\pi_0(WH(Y))^*_\alpha])$ *denote an arbitrary family of elements. Then there exists a finitely G-dominated G-complex* X *with* $Y \subset X$ *and a G-retraction* $r: X \to Y$ *such that* $r_*(w^H_\alpha(X)) = w^H_\alpha$.

The proof of this proposition is similar to that of theorem 2.3. It also goes by induction on the number of elements of the set $\underline{CI}(Y)$ and the inductive step is based on the analogous realization result for relatively free action. The details can be found in [4].

Let now Y be a G-complex. We define the homomorphism

$$F : Wa^G(Y) \longrightarrow \bigoplus_{\underline{CI}(Y)} \tilde{K}_0(\mathbb{Z}[\pi_0(WH(Y))^*_\alpha])$$

by the formula $F([f: X \to Y]) = \sum f_*(w^H_\alpha(X))$ where

$$f_* : \tilde{K}_0(\mathbb{Z}[\pi_0(WH(X))^*_\alpha]) \longrightarrow \tilde{K}_0(\mathbb{Z}[\pi_0(WH(Y))^*_\alpha])$$

denotes the homomorphism induced by f on $\tilde{K}_0$.

Theorem 2.8 *If X is finitely G-dominated then the homomorphism*

$$F : Wa^G(X) \longrightarrow \bigoplus_{Cl(X)} \tilde{K}_0(\mathbb{Z}[\pi_0(WH(X))^*_\alpha])$$

is an isomorphism and $F(w^G(X)) = \sum w^H_\alpha(X)$.

We sketch the proof (details can be found in [4]). This result was also independently proved by Lück [34] thm 14.12.

Let an element $[f: Y \to X]$ belongs to kerF. One can replace $f: Y \to X$ by an extension $g: Z \to X$ such that (Z,Y) is relatively finite,

$$g_* : \pi_0(Z^H) \to \pi_0(X^H)$$

is bijective and

$$g_* : \pi_1(Z^H_\alpha) \to \pi_1(X^H_\alpha)$$

is an isomorphism. Then $F([g]) = 0$ and one has $w^H_\alpha(Z) = 0$ for any component Z^H_α. It follows from theorem 2.3 that there exists a finite G-complex Z_1 G-homotopy equivalent to Z and it shows that $[f] = 0$ in $Wa^G(X)$.

To prove surjectivity of F one replaces F by

$$F_1 : Wa^G(K) \longrightarrow \bigoplus_{Cl(K)} \tilde{K}_0(\mathbb{Z}[\pi_0(WH(K))^*_\alpha])$$

with K a finite G-complex. By virtue of prop. 2.7 there exists a finitely G-dominated G-complex L and G-retraction $r: L \to K$ such that $r_*(w^H_\alpha(L)) = w^H_\alpha$. Then $F_1([r]) = \sum w^H_\alpha$.

3. Categorical characterization and other properties of the equivariant finiteness obstruction

It is clear that the equivariant finiteness obstruction has many properties similar to those of equivariant Whitehead torsion. These properties can be used to show that the equivariant finiteness obstruction is uniquely characterized by a certain universal property. In view of theorems 2.6 and 2.8 it is enough to state them for obstructions $w^H_\alpha(X)$'s.

Let us suppose G-complex X is of the form $X = X_1 \cup X_2$ with $X_0 = X_1 \cap X_2$.

Proposition 3.1 *If G-complexes X_i, $i = 0,1,2$ are finitely G-dominated then X is also finitely G-dominated and one has the following equality*

$$w^H_\alpha(X) = \sum_\beta (i_{1\beta})_* w^H_\beta(X_1) + \sum_\beta (i_{2\beta})_* w^H_\beta(X_2) - \sum_\beta (i_{0\beta})_* w^H_\beta(X_0)$$

with $i_k : X_k \to X$ being inclusions. Here β runs over the preimage of α under the induced map $Cl(i_k): Cl(X_k) \to Cl(X)$ and $i_{k\beta} : \pi_0(WH)^*_\beta \to \pi_0(WH)^*_\alpha$ is the homomorphism induced by the restriction of i_k to the β-component of the H-fixed point set.

Proposition 3.2 *If a G-complex X is finitely G-dominated and $f : X \to Y$ is a G-homotopy equivalence then Y is also finitely G-dominated and $f_*(w^H_\alpha(X)) = w^H_\alpha(Y)$.*

Now following Lück [33], [34] one can introduce the notion of a universal functorial additive invariant (UFAI). Let $\mathcal{C}$ be a small full subcategory of the category of G-spaces containing $\emptyset$ and $\{pt\}$. We assume that $\mathcal{C}$ is closed under G-homotopy equivalences and G-push-outs.

A functorial additive invariant (FAI) (B,b) for $\mathcal{C}$ consists of a functor $B : \mathcal{C} \to Ab$ from $\mathcal{C}$ to the category of abelian groups and an assignement b associating to an object X in $\mathcal{C}$ an element $b(X) \in B(X)$ such that the following conditions are fullfilled:

(a) If $f : X \to Y$ is a G-homotopy equivalence in $\mathcal{C}$ then $B(f)(b(X)) = b(Y)$.

(b) If f is G-homotopic to g then $B(f) = B(g)$.

(c) Given a G-push-out in $\mathcal{C}$

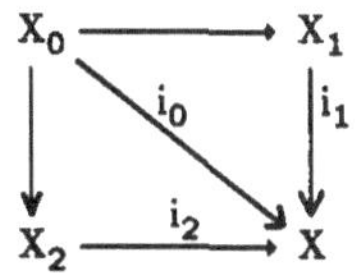

with $X_0 \to X_2$ a G-cofibration the following formula is valid:

$$b(X) = B(i_1)(b(X_1)) + B(i_2)(b(X_2)) - B(i_0)(b(X_0)) .$$

(d) $b(\emptyset) = 0$.

A functorial additive invariant (U,u) for $\mathcal{C}$ is said to be <u>universal</u> if for any FAI (B,b) of $\mathcal{C}$ there exists a unique natural transformation $F : U \to B$ such that $F(X)(u(X)) = b(X)$ holds for all objects X in $\mathcal{C}$.

One can prove that UFAI always exists ([33] prop. 2.4 or [34] thm 6.1).

We also need the notion of so-called equivariant Euler characteristic which is the UFAI for the category of finite G-complexes. First we define a functor A^G : G-Top $\to$ Ab. For G-space X we put $A^G(X)$ to be the free abelian group generated by the set $\underline{CI}(X)$. A G-map $f : X \to Y$ induces a homomorphism

$$A^G(f) : A^G(X) \to A^G(Y) .$$

Let X be a G-space that is finitely G-dominated. We define its <u>equivariant Euler characteristic</u> $\chi^G(X) \in A^G(X)$ by

$$\chi^G(X) = \sum_{\underline{CI}(X)} \chi\left(X_\alpha^H/_{(WH)_\alpha} , X_\alpha^{>H}/_{(WH)_\alpha} \right).$$

In view of theorem 2.8 one can denote

$$\tilde{K}_0^G(X) = \bigoplus_{\underline{CI}(X)} \tilde{K}_0(\mathbb{Z}[\pi_0(WH)_\alpha^*]) \quad \text{and} \quad w^G(X) = \sum_{\underline{CI}(X)} w_\alpha^H(X) .$$

<u>Theorem 3.3</u> *(a) The pair (A^G, χ^G) is the UFAI for the category of G-spaces having the G-homotopy type of a finite G-complex.*

(b) The pair $(A^G \oplus K^G, \chi^G \oplus w^G)$ is the UFAI for the category of G-spaces having the G-homotopy type of a finitely dominated G-complex.

The above characterization was showed in [33] thm 4.1 or in [34] thm 6.9.

Observe that the product formula for equivariant Whitehead torsion [24] along with theorem 2.6 gives rise to the product formula for equivariant finiteness obstruction.

<u>Proposition 3.4</u> *Let G and P denote compact Lie groups. Let X be a G-complex G-dominated by a finite G-complex K and Y be a P-complex P-dominated by a finite P-complex L. Then the product X x Y is (G x P)-dominated by K x L and*

$$w_{\alpha \times \beta}^{H \times Q}(X \times Y) = \bar{\chi}_\beta^Q(Y) \cdot i_*(w_\alpha^H(X)) + \bar{\chi}_\alpha^H(X) \cdot j_*(w_\beta^Q(Y))$$

where

$$\overline{\chi}^{Q}_{\beta}(Y) = \chi\left(Y^{Q}_{\beta}/(WQ)_{\beta}, Y^{>Q}_{\beta}/(WQ)_{\beta}\right), \quad \overline{\chi}^{H}_{\alpha}(X) = \chi\left(X^{H}_{\alpha}/(WH)_{\alpha}, X^{>H}_{\alpha}/(WH)_{\alpha}\right)$$

and

$$i : \pi_0(WH)^*_{\alpha} \longrightarrow \pi_0(WH)^*_{\alpha} \times \pi_0(WQ)^*_{\beta}, \quad j : \pi_0(WQ)^*_{\beta} \longrightarrow \pi_0(WH)^*_{\alpha} \times \pi_0(WQ)^*_{\beta}$$

denote inclusions. Moreover, each obstruction $w^S_{\gamma}(X \times Y)$ *, where* (S,γ) *is not of a product form, equals zero.*
The same formula was also obtained (algebraically) by Lück [34] thm 14.19 and cor.14.48. This product formula was used by tom Dieck and Petrie in the study of homotopy representations of finite groups ([14] § 8).

Finally a few words about the restriction formula. Let H denote a closed subgroup of G . Recently S. Illman ([27] §8) has defined (geometrically) the restriction homomorphism

$$\mathrm{res}^G_H : \mathrm{Wh}_G(X) \to \mathrm{Wh}_H(X) .$$

If f: X → Y is a G-homotopy equivalence then, by definition, $\mathrm{res}^G_H(\tau_G(f)) = \tau_H(f)$ and we obtain $\mathrm{res}^G_H(\sigma_G(X)) = \sigma_H(X)$. An algebraic description of the restriction homomorphism is given in [26] §1 and in [34] def. 14.36, thm 14.37 and prop. 14.40 and it allows to determine H-obstructions $w^K_{\alpha}(X)$ in terms of G-obstructions. Illman has also announced [27] that the restriction homomorphism satisfies the natural associativity property

$$\mathrm{res}^H_K \cdot \mathrm{res}^G_H = \mathrm{res}^G_K .$$

4. Some geometric applications

The product formula of the preceding section gives rise immediately to the following nice geometric result.

Proposition 4.1 *Let G be a finite group and X a G-complex G-dominated by a finite one. Let V be any unitary complex representation of the group G and s(V) its unit sphere. Then the product X x s(V) with the diagonal G-action has the G-homotopy type of a finite G-complex.*

Since the assertion of the proposition 4.1 depends on the fact that s(V) has well-behaved equivariant Euler characteristic it is also true for other G-spaces with vanishing G-Euler characteristic (cf [33] cor. 6.4).

In many situations when studying group actions one encounters the problem whether the G-space (or G-complex) obtained as a result of various constructions has the equivariant homotopy type of a finite G-complex. What is often guaranteed is a weaker assumption - equivariant finite domination - and we are resulting in a question when does the equivariant finiteness obstruction of a given G-complex vanish ? Such situations appear most frequently when one studies smooth, locally smooth or topological actions on manifolds. On the other hand there is an old problem posed by K. Borsuk in early fifties (and solved finally by J. West in 1977 [62]) which says whether a compact ANR-space has the homotopy type of a finite complex and it is natural to ask on the equivariant version of this question. It is well-known that any compact G-ANR is finitely G-dominated (see [37] prop. 10.1 or [30]). In particular, it applies to G-manifolds.

If M is a compact manifold with smooth G-action then the triangulation results of S. Illman [22], [23] show that M carries the structure of a finite G-complex. If we assume that the action of the group G is locally smooth then the situation changes drastically. Namely for a finite group G an n-dimensional, locally smooth G-manifold M^n has the G-homotopy type of an n-dimensional G-complex [32] but on the other hand examples given by Quinn [44], Dovermann and Rothenberg [18] §12 and Weinberger [60] p. 532 or [61] thm 16(a) show that the equivariant finiteness obstruction of M can be non-zero for certain M and a suitable finite group G . Even more, Steinberger and West [49] §8 using their equivariant topological s-cobordism theorem have constructed compact, locally smooth G-manifolds realizing elements of a subgroup of the group $\tilde{K}_0^G(X)$ for any compact G-ANR X . It is worth to note here that even in the case of a finite G-complex X not all elements of the group $\tilde{K}_0^G(X)$ can be realized as the equivariant finiteness obstruction of compact G-manifolds [50] (cf also §5). These examples show that equivariant version of the Borsuk conjecture fails in the case, when G-manifold M (or G-ANR) has at least two orbit types. However, if G acts freely on the manifold M then M/G is a compact topological manifold and in this case the equivariant finiteness obstruction of M vanishes. The following results yield an affir-mative answer to the equivariant Borsuk conjecture in some special cases.

Proposition 4.2 [3] *Let G be a compact Lie group and X a compact G-ANR-space with one orbit type. Then X has the G-homotopy type of a finite G-complex.*

Proposition 4.3 [31] *Let T^n denote a torus group and let X be a compact T^n-ANR. If for every isotropy subgroup $H \subset T^n$ occurring on X the fixed point set X^H is simply connected then X has the equivariant homotopy type of a finite T^n-complex.*

One should point out that in the case of finite G there is a very close relation between the equivariant finiteness obstruction and G-surgery on equivariant complexes [42]. The idea that originates from Milnor and Swan was successfully developed by Oliver and Petrie [38], [43], [42] and shortly looks as follows. Let $f: X \to Y$ be a G-map of finite G-complexes. First one constructs an appropriate $\mathcal{F}$-resolution $f_1: X_1 \to Y$ extending f. The resolution f_1 has "correct" homology except one dimension $H_n(M(f_1))$ which is projective $\mathbb{Z}G$-module and defines an element $(-1)^n[\, H_n(M(f_1))\,]$ of $\tilde{K}_0(\mathbb{Z}G)$ (so called Swan invariant). One can then attach infinitely many free G-n-cells to $M(f_1)$ to obtain a G-complex W which has the "right" homology and turns out to be finitely G-dominated. Its equivariant finiteness obstruction $w_G(W)$ corresponds to the projective class group invariant $(-1)^n[\, H_n(M(f_1))\,]$ above. Results similar to those of Oliver and Petrie was also obtained (in less generality) by Ku and Ku [29]. In particular they have proved the following generalization of Swan's theorem.

Theorem 4.4 ([29] thm 1.3) *Let G be a finite group of order q with periodic cohomology of period n and let $d = (q, \phi(q))$ where ϕ is the Euler ϕ-function. Suppose F is a simply connected r-dimensional integral homology r-sphere. Then there exists a finite G-complex X homotopy equivalent to S^{r+dn}. Moreover G acts semifreely on X with $X^G = F$.*

The Swan invariant detects the equivariant finiteness obstruction also in homology propagation of group actions (see e.g. [9], [59] §2).

It turns out that - at least under some restrictions - the equivariant finiteness obstruction is closely connected with the problem of existence of regular neighbourhoods. This observation was made by F. Quinn [44] prop. 2.1.2 and cor. 2.1.3 and K. H. Dovermann [16]. For instance, if a finite group G acts locally smoothly and semifreely on a compact manifold M then the fixed point set M^G has an equivariant closed regular neighbourhood in M iff all equivariant finiteness obstructions $w_\alpha^H(M)$'s vanish.

In 1965 L. C. Siebenmann in his thesis [46] applied Wall finiteness obstruction to geometric question giving necessary and sufficient conditions for attaching the boundary to an open manifold. It is natural to ask whether the equivariant finiteness obstruction can be applied to produce an equivariant boundary in a smooth open G-manifold. The answer is yes although the author did not yet verify all details. Up to the author's knowledge nobody has presented the equivariant version of Siebenmann's theorem in final written form and we only outline some constructions in the case of a smooth G-manifold.

The existence of the G-collar neighbourhood of the boundary reduces the question to the following : when does a G-end $\mathcal{E}$ of a smooth open G-manifold M admit an equivariant collar neighbourhood ?
Recall that a G-end $\mathcal{E}$ is called (equivariantly) tame if
(1) there exists a decreasing sequence of G-neighbourhoods $N_1 \supset N_2 \supset \ldots$ of $\mathcal{E}$ with $\cap N_i = \emptyset$ satisfying certain π_1-isomorphism conditions and
(2) each N_i is finitely G-dominated.
The inclusions $N_{i+1} \subset N_i$ induce isomorphisms of fundamental groups

$$\pi_1((N_{i+1})^H_\alpha) \doteq \pi_1((N_i)^H_\alpha)$$

and the sequence of obstructions $\{w^H_\alpha(N_i)\}_i$ defines an element $w^H_\alpha(M,\mathcal{E})$ of the group

$$\tilde{K}_0(\mathbb{Z}[\pi_0(WH(\mathcal{E}))^*_\alpha]) = \varprojlim_i \tilde{K}_0(\mathbb{Z}[\pi_0(WH(N_i))^*_\alpha]).$$

Both the group $\tilde{K}_0(\mathbb{Z}[\pi_0(WH(\mathcal{E}))^*_\alpha])$ and the invariant $w^H_\alpha(M,\mathcal{E})$ do not depend on the choices involved [3]. Moreover, if the G-end $\mathcal{E}$ has a G-collar then it is tame and $w^H_\alpha(M,\mathcal{E}) = 0$. Now one can formulate the following version of Siebenmann's boundary theorem for smooth G-manifolds.
Theorem 4.5 *Let G be a finite group and M a smooth G-manifold with one G-end $\mathcal{E}$. Suppose M satisfies a certain variant of the "gap hypothesis" i.e. if $M^H_\alpha \supsetneq M^K_\beta$ then $\dim M^H_\alpha - \dim M^K_\beta \geq 3$. Let $\mathcal{E}$ be tame and $w^H_\alpha(M,\mathcal{E}) = 0$ for each element $[M^H_\alpha]$ of the set Cl(M) . If $\dim M^H_\alpha \geq 6$ then there exists a G-collar around the end $\mathcal{E}$.*

The proof of this theorem goes by induction like in theorems 2.3 and 2.7. A weaker version of theorem 4.5 can be found in [3]. Another particular case of the equivariant boundary theorem for actions on Euclidean spaces was established by A. Assadi in [5] § 6.

Proposition 5.1 in [38] and above theorem yield the following characterization.
Proposition 4.6 *Let M^n be a smooth G-manifold with compact non-empty boundary and one G-end $\mathcal{E}$. Suppose M satisfies the above gap hypothesis and additionally*
1) the inclusion $\partial M \subset M$ is G-(n-2)-connected.
2) the natural map $\tilde{K}_0(\mathbb{Z}[\pi_0(WH(\mathcal{E}))^*_\alpha]) \to \tilde{K}_0(\mathbb{Z}[\pi_0(WH(M))^*_\alpha])$ is an isomorphism.
3) $\dim M^H_\alpha \geq 6$.
Then M is G-diffeomorphic to $\partial M \times \langle 0,1)$.

In conclusion one has to mention that there is an alternative approach to the equivariant Siebenmann's theorem given by Steinberger and West [49], [51] following Chapman's ideas [10]. They develop the equivariant version of the boundary theorem in controlled setting thus obtaining an equivariant analogue of Chapman's result ([51] thm. 2.4) but their end theorem is only formulated in [51] and it relies heavily on the fundamental paper [50] which is still (as far as author knows) in preparation.

5. The controlled version of the equivariant finiteness obstruction

T. Chapman and S. Ferry developed a geometric controlled simple homotopy theory from Cohen's geometric Whitehead group [11] . In particular, Chapman's exposition [10] shows many applications of this theory to various geometric questions and among other things it presents a controlled version of Ferry's finiteness obstruction. A few years ago Steinberger and West started to develop Chapman's work in an equivariant context and it turned out that their analysis yielded several interesting and surprising applications in equivariant geometric topology. In this section we sketch the outlines of an equivariant version of Chapman's ideas.

In what follows B will denote a fixed, finite-dimensional, compact, metrizable G-space. The G-space X in which we are interested is a locally finite G-complex equipped with an equivariant proper control map $p: X \rightarrow B$. We fix a positive real number ϵ and let $DR(X,p)_\epsilon$ be the set of equivariant proper $p^{-1}(\epsilon)$-strong deformation retractions $r : Y \rightarrow X$ of relatively finite G-CW-pairs (Y,X) . Now let $DR^{PL}(X,p)_\epsilon$ denote the set of equivariant classes of members of $DR(X,p)_\epsilon$ under the equivariant relation generated by equivariantly $p^{-1}(\epsilon)$-homotopy commutative diagrams

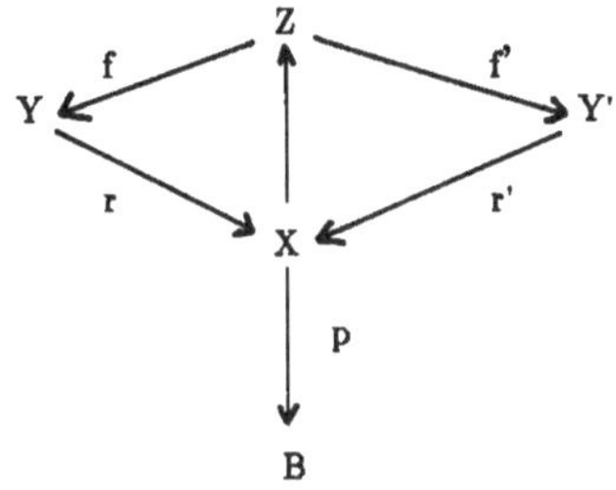

with $X \subset Z$ where f and f' are equivariant cellular surjections with each point inverse $f^{-1}(y)$ being G_y-simple-homotopy trivial and $f|_X = f'|_X = id_X$. Define $Wh_G(X,p)_\epsilon$ to be the set of invertible elements of $DR^{PL}(X,p)_\epsilon$ under the operation of union over X . If the control map p is an identity on X we simply denote $DR^{PL}(X)_\epsilon$, $Wh_G(X)_\epsilon$ etc. The group $Wh_G(X,p)_\epsilon$ has many functorial properties [10], [50], [52] and for $\epsilon < \delta$ we have a control-relaxation homomorphism

$$Wh_G(X,p)_\epsilon \rightarrow Wh_G(X,p)_\delta .$$

We set $Wh_G(X,p)_c = \varprojlim_\epsilon Wh_G(X,p)_\epsilon$, the inverse limit. It turns out, unlike the inequivariant case, that in the case when $B = X$ and p is the identity, $Wh_G(X)_c$ is usually nonzero [50] .

Following Ferry [19] and Chapman [10] one defines $\tilde{K}_0^G(X,p)_\epsilon$ to be the subgroup of $Wh_G(X \times S^1, q)_\epsilon$ comprised of those elements invariant under the standard geometric transfers. Here S^1 has the trivial G-action and the control is $X \times S^1 \rightarrow X \rightarrow B$. $\tilde{K}_0^G(X,p)_c$ is the inverse limit. The group $\tilde{K}_0^G(X,p)_\epsilon$ has also several natural properties (cf [10] §6). For instance one has

Lemma 5.1 [51] *For any commutative diagram of G-maps*

$$
\begin{array}{ccc}
X_0 & \xrightarrow{g} & X_1 \\
{\scriptstyle P_0}\downarrow & & \downarrow{\scriptstyle P_1} \\
B & \xrightarrow{f} & B
\end{array}
$$

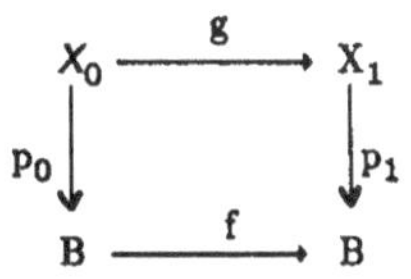

and for $\epsilon , \delta > 0$ with $\delta < f^{-1}(\epsilon)$ there is an induced map

$$g_* : \tilde{K}_0^G(X,p)_\delta \longrightarrow \tilde{K}_0^G(X,p)_\epsilon$$

which makes the controlled K-group a functor on such data. Moreover, if $g_t : X_0 \rightarrow X_1$ is a homotopy over f then $g_{0*} = g_{1*}$.

Remark. Here $\delta < f^{-1}(\epsilon)$ means that for any $x \in B$ there exists $y \in B$ with $K(y,\delta) \subset f^{-1}(K(x,\epsilon))$ where $K(x,\epsilon)$ denotes an open ball in B of radius ϵ.

Proposition 5.2 [51] *The inverse system* $\{\tilde{K}_0^G(X,p)_\epsilon\}$ *is stable i.e. there is a map,* j *, from the constant inverse system with value* $\tilde{K}_0^G(X,p)_c$ *to* $\{\tilde{K}_0^G(X,p)_\epsilon\}$ *such that for each* $\epsilon > 0$ *there is a* $\delta < \epsilon$ *and a commutative diagram*

$$\begin{array}{ccc} \tilde{K}_0^G(X,p)_c & \xrightarrow{\ j\ } & \tilde{K}_0^G(X,p)_\delta \\ \| & \swarrow & \downarrow \\ \tilde{K}_0^G(X,p)_c & \xrightarrow{\ j\ } & \tilde{K}_0^G(X,p)_\epsilon \end{array}$$

so that for sufficiently small ϵ *the inverse system is constant modulo some relaxation of control.*

Let now $X \xrightarrow{s} K \xrightarrow{\phi} X$ be a G-$p^{-1}(\epsilon)$-domination of X by a finite G-complex K. We can repeat the construction of section 2 obtaining that X is equivariantly $p^{-1}(\delta)$-homotopy equivalent to the infinite mapping cylinder $I(A)$ of the homotopy idempotent $A = s\phi$. Here $\delta = a\epsilon$, a being a structural constant of the theory. The mapping cylinder $M(h)$ of h equivariantly $p^{-1}(\gamma)$-deformation retracts to $T(A)$ yielding an element of $DR(X \times S^1, q)_\gamma$. This produces a unique element $\sigma_\epsilon(X) \in \tilde{K}_0^G(X,p)_\beta$ with $\beta = b\epsilon$, b constant.

Theorem 5.3 [51] *If* X *is G-*$p^{-1}(\epsilon)$*-homotopy finite then* $\sigma_\epsilon(X) = 0$ *. Conversely, if* $\sigma_\epsilon(X) = 0$ *then there is a finite G-complex* Y *equivariantly* $p^{-1}(d\epsilon)$*-homotopy equivalent to* X *. Here* d *is again a universal constant. Moreover, if* $\delta < \epsilon$ *then* $\sigma_\epsilon(X)$ *is the image of* $\sigma_\delta(X)$ *under the relaxation homomorphism*

$$\tilde{K}_0^G(X,p)_\delta \rightarrow \tilde{K}_0^G(X,p)_\epsilon .$$

Similarly if (X,A) is a G-CW-pair then there is a controlled relative finiteness obstruction $\sigma_\epsilon(X,A) \in \tilde{K}_0^G(X,p)_\beta$ whose vanishing implies that X is $p^{-1}(d\epsilon)$-equivalent rel A to a relatively finite relative G-complex built on A . Moreover, if $i : A \subset X$ then $\sigma_\epsilon(X) = \sigma_\epsilon(X,A) + i_*\sigma_\epsilon(A)$.

The last assertion of theorem 5.3 allows to define elements $\sigma_c(X)$ and $\sigma_c(X,A)$ in the group $\tilde{K}_0^G(X,p)_c$ induced by suitable sequences $\{\sigma_\epsilon(X)\}$ and $\{\sigma_\epsilon(X,A)\}$ in the inverse system $\{\tilde{K}_0^G(X,p)_\epsilon\}$. The element $\sigma_c(X)$ is called the controlled equivariant finiteness obstruction. By definition, for any $\epsilon > 0$ there exists a natural relaxation homomorphism

$$\Phi : \tilde{K}_0^G(X,p)_c \rightarrow \tilde{K}_0^G(X,p)_\epsilon$$

which happens to be generally neither surjective [50] nor injective [58] and of course $\Phi(\sigma_c(X)) = \sigma_\epsilon(X)$. Moreover, for any $\epsilon > 0$ we have a (forgetful) homomorphism

$$q : \tilde{K}_0^G(X,p)_\epsilon \rightarrow \tilde{K}_0^G(X)$$

such that $q(\sigma_\epsilon(X)) = \sigma_G(X)$.

As it was previously mentioned Steinberger and West has applied their controlled equivariant h-cobordism theorem ([49] p. 194 or [52] thm. 5.2) and equivariant analogue of Chapman's realization results [10] to produce examples of compact, locally smooth G-manifolds realizing elements of the group $\tilde{K}_0^G(X)_c$.

Theorem 5.4 *Let* G *be a finite group,* X *a finite G-complex and* $x \in \tilde{K}_0^G(X)_c$ *an arbitrary element. Then there is a compact, locally smooth G-manifold* M *(with boundary) containing* X *and an equivariant retraction* $r : M \rightarrow X$ *which induces an isomorphism of fundamental groups and satisfies* $r_*(\sigma_c(M)) = x$.

The proof can be found in [49] §8. Thus the image of $\Phi : \tilde{K}_0^G(X)_c \rightarrow \tilde{K}_0^G(X)$ is comprised of precisely those elements of $\tilde{K}_0^G(X)$ that are obstructions to equivariant finiteness of compact, locally smooth G-manifolds (or G-ANR's).

As results of Steinberger and West show the equivariant controlled finiteness obstruction has proved its power in solutions of many open questions in equivariant geometric topology (see [49] p. 183-184) .

The first application leads to an equivariant controlled end theorem. Since its statement is similar to that given above in section 4 we only point out essential differences. We shall hereinafter use the term G-manifold to mean an n-dimensional topological manifold with a locally smooth action of the finite group G satisfying the "codimension ≥ 3 gap hypothesis" (see theorem 4.5).

Suppose given a G-manifold M whose fixed-point components have dimension at least 6 and a G-map $p: M \to B$, with B a locally compact metrizable G-space. The sequences $\{V\}$ of neighbourhoods V of the of neighbourhoods V of the end $\mathcal{E}$ of p and $\{\epsilon\}$ of positive real numbers determine inverse systems $\{\tilde{K}_0^G(V,p)_\epsilon\}$ and $\{Wh_G(V,p)_\epsilon\}$. Taking the inverse limit of the first system and $\varprojlim^1$ of the second , we obtain groups $\tilde{K}_0^G(\mathcal{E}(p))$ and $\varprojlim^1 Wh_G(\mathcal{E}(p))$, respectively. Necesarry conditions of equivariant tameness over B and equivariant 1-movability at ∞ are defined by fully equivariant analogues of the definitions in [10] (see also [49] p. 196). The condition of tameness over B allows to define the collection $\{\sigma_\epsilon(V,FrV)\}$ and gives a well-defined element, $\sigma(\mathcal{E}(p)) \in \tilde{K}_0^G(\mathcal{E}(p))$.

Theorem 5.5 [50],[52],[49] *The obstruction $\sigma(\mathcal{E}(p))$ vanishes iff for any decreasing sequence $\epsilon_i > 0$ and for any neigh-bourhood V_0 of the end $\mathcal{E}$ there is a sequence V_i of neighbourhoods of the end $\mathcal{E}$ such that $V_i \subset V_{i-1} - Fr(V_{i-1})$ and $\overline{V_i - V_{i-1}}$ is a G-ϵ_i-h-cobordism from FrV_i to ∂V_{i+1}. Moreover, if $\sigma(\mathcal{E}(p)) = 0$ there is an obstruction $\tau(\mathcal{E}(p)) \in \varprojlim^1 Wh_G(\mathcal{E}(p))$ which vanishes iff the end $\mathcal{E}$ admits an equivariant collar over B .*

The next use of the controlled equivariant finiteness obstruction (in a relative form) is a controlled splitting obstruction for elements of $Wh_G(X,p)_\epsilon$. It strictly follows Chapman's treatment in [10] §§ 9-11 and it is summarized in [49] p. 197.

The controlled equivariant boundary theorem is closely related to the equivariant product structure theorem. Namely, let a G-manifold M be provided with a locally smooth PL structure Σ_0 near ∂M . We consider locally smooth PL structures Σ on $M \times \mathbf{R}$ which agree with $\Sigma_0 \times \mathbf{R}$ near $\partial M \times \mathbf{R}$ (so-called structures rel ∂). We say that a structure Σ on $M \times \mathbf{R}$ rel ∂ admits a product structure if Σ is G-isotopic rel ∂ to a structure of the form $\Theta \times \mathbf{R}$, where Θ is a locally smooth PL structure on M which agrees with Σ_0 near ∂M .

If now Σ is a locally smooth PL structure rel ∂ on $M \times \mathbf{R}$ one can consider the projection $p: M \times \mathbf{R} \to M$ and define the positive end obstruction $\sigma(\mathcal{E}_+(p))$ of Σ over M. Steinberger and West have observed that the product structure theorem of Kirby and Siebenmann [28], Essay III, fails equivariantly and then they have successfully identified the controlled equivariant end obstruction $\sigma(\mathcal{E}(p))$ to be the obstruction to equivariant product structure [51]. Using the equivariant end theorem 5.5 they showed in [51] the following.

Theorem 5.6 Σ *admits a product structure rel ∂ iff the positive end obstruction of Σ , $\sigma(\mathcal{E}_+(p))$ over M vanishes.*

Theorem 5.6 has very interesting consequences. We mention here three corollaries. Let V be a linear representation of the group G . Let $Top_G(V)$, $PL_G(V)$ and $O_G(V)$ be the group of equivariant self-homeomorphisms, the group of equivariant PL automorphisms and the group of equivariant linear automorphisms of V , respectively.

Corollary 5.7 *Let V be a G-representation satisfying "gap hypothesis" and with dim $V^G \geq 5$. Then stabilization maps*

$$\pi_0\left(\frac{Top_G(V)}{PL_G(V)}\right) \longrightarrow \pi_0\left(\frac{Top_G(V\oplus \mathbf{R})}{PL_G(V\oplus \mathbf{R})}\right)$$

and

$$\pi_0\left(\frac{Top_G(V)}{O_G(V)}\right) \longrightarrow \pi_0\left(\frac{Top_G(V\oplus \mathbf{R})}{O_G(V\oplus \mathbf{R})}\right)$$

are onto.

Corollary 5.8 *Let V and W be as above. Then $V \oplus \mathbf{R}$ is G-homeomorphic to $W \oplus \mathbf{R}$ iff V is G-homeomorphic to W.*

Corollary 5.9 *Let V be a G-representation with dim $V^G \geq 5$ which satisfies a certain "vanishing condition"* ([47], p. 70) *in degrees $\geq k$, where $1 \leq k \leq$ dim V^G. Then the space $Top_G(V \oplus \mathbf{R})/Top_G(V)$ is (dim V^G - k)-connected.*

Similar results were obtained independently by Madsen and Rothenberg [36] thm B and cor. C.

More interestingly yet, Steinberger and West have made an analysis of Kirby-Siebenmann program [28], Essay III to prove the existence of the handlebody decomposition on a manifold and they remarked that only the product structure theorem fails equivariantly. It leads directly to the obstruction to the existence of an equivariant handlebody decomposition. In order to exibit Steinberger-West results in this direction we need some definitions.

Let H be a subgroup of G. A G-handle of type H is $G \times_H (D^k \times D_\rho)$ where D_ρ is the unit representation disk of an orthogonal representation $\rho : H \to O(r)$ of H and D^k is a k-disk with trivial H-action. The index of the handle is k. Thus a 0-handle of type H is just $G \times_H D_\rho$. An equivariant handle decomposition of a G-manifold M is a sequence $M_0 \subset M_1 \subset \dots \subset M_m = M$ of codimension zero equivariant submanifolds with each $\overline{M_i - M_{i-1}}$ a disjoint union of equivariant k-handles of type H for some $k \leq n$ and $H \subset G$; a handle $G \times_H (D^k \times D_\rho)$ is of course attached by an (equivariant) embedding of $G \times_H (S^{k-1} \times D_\rho)$ into the previous stage.

For smooth actions of compact Lie groups equivariant Morse theory provides equivariant handle decompositions ([57] thm 4.6). Locally linear PL actions of finite groups on manifolds may be given equivariant handle decompositions by triangulating and taking barycenter stars with respect to the second derived (equivariant) subdivision as in the inequivariant case.

If we assume that the action of the group G is locally smooth then examples of compact manifolds not equivariantly homotopy equivalent to finite G-complexes do not admit equivariant handle decompositions. In other words, a compact locally smooth G-manifold M admitting an equivariant handle decomposition must have the equivariant finiteness obstruction $\sigma_G(M) = 0$. By sufficiently fine subdivision of a handle decomposition it certainly has the ϵ-homotopy type of a finite G-complex for every $\epsilon > 0$ and we conclude that $\sigma_c(M) = 0$. The fundamental result of Steinberger and West asserts that vanishing of $\sigma_c(M)$ is sufficient for M to have an equivariant handle decomposition. In order to formulate it assume that each component M^H_α has dimension at least 6. Let $N \subset \partial M$ be a codimension zero G-submanifold with equivariantly bicollared boundary.

Theorem 5.10 [51], [52] §7. *There is an equivariant handle decomposition of* M *on a closed collar of* N *iff the controlled equivariant finiteness obstruction* $\sigma_c(M,N)$ *is zero in* $\tilde{K}_0^G(M)_c$.

The idea of the proof is to adapt the Kirby-Siebenmann argument [28], Essay III to the equivariant, locally smooth context by means of the equivariant product structure theorem above. There is also an isovariant version without "gap hypothesis" [47] thm 8.

We have already mentioned that (generally) the control-relaxation map

$$\Phi : \tilde{K}_0^G(X)_c \longrightarrow \tilde{K}_0^G(X)$$

is not injective. This fact was proved by D. Webb [58] who has precisely calculated that for $X = \mathbf{RP}^2$ with trivial action of the group $\mathbf{Z}_{21}$ the control-relaxation homomorphism Φ has non-trivial kernel. This calculation along with Steinberger-West realization result (theorem 5.4) gives the following corollary.

Proposition 5.11 *For* G *cyclic of order 21 there is a compact , locally smooth G-manifold* M *(with fundamental group* $\mathbf{Z}_2$ *) which has the G-homotopy type of a finite G-complex but admits no equivariant handle structure.*

References

1. D.R. Anderson: *Torsion invariants and actions of finite groups,* Michigan Math. J. 29 (1982), 27-42.
2. P. Andrzejewski: *The equivariant Wall finiteness obstruction and Whitehead torsion,* Transformation Groups, Poznań 1985, pp. 11-25, Lecture Notes in Math. 1217, Springer Vlg 1986.
3. P. Andrzejewski: *An application of equivariant finiteness obstruction - equivariant version of Siebenmann's theorem,* (preprint, to appear).
4. P. Andrzejewski: *A complement to the theory of equivariant finiteness obstruction* (preprint 1989).
5. A.H. Assadi: *Extensions of group actions from submanifolds of disks and spheres* (preprint).
6. J.A. Baglivo: *An equivariant Wall obstruction theory,* Trans. Amer. Math. Soc. 256 (1979), 305-324.
7. H. Bass, A. Heller, R. Swan: *The Whitehead group of a polynomial extension,* Publ. Math. IHES 22 (1964), 67-79.
8. G.E. Bredon: *Introduction to compact transformation groups,* Academic Press, N.Y. 1972.
9. S.E. Cappell, S. Weinberger: *Homology propagation of group actions,* Comm. Pure and Appl. Math. 40 (1987), 723-744.
10. T.A. Chapman: *Controlled simple homotopy theory and applications,* Lecture Notes in Math. 1009, Springer Vlg 1983.
11. M.M.Cohen: *A course in simple-homotopy theory,* Graduate Texts in Math., Springer Vlg 1973.
12. T. tom Dieck: *Über projektive Moduln und Endlichkeitshindernisse bei Transformationsgruppen,* Manuscripta Math. 34 (1981), 135-155.
13. T. tom Dieck: *Transformation groups,* de Gruyter, Berlin 1987.
14. T. tom Dieck, T. Petrie: *Homotopy representations of finite groups,* Publ. Math. IHES 56 (1982), 129-170.
15. W. Dorabiała: *On the equivariant homotopy type of G-fibrations,* (preprint, to appear).
16. K.H. Dovermann: personal communication.
17. K.H. Dovermann, M. Rothenberg: *An equivariant surgery sequence and equivariant diffeomorphism and homeomorphism classification,* Topology Symposium (Siegen 1979), pp. 257-280, Lecture Notes in Math. 788, Springer Vlg 1980.

18. K.H. Dovermann, M. Rothenberg: *Equivariant surgery and classification of finite group actions on manifolds,* Memoirs Amer. Math. Soc. 379 (1988).
19. S. Ferry: *A simple-homotopy approach to the finiteness obstruction,* Shape Theory and Geometric Topology, pp. 73-81, Lecture Notes in Math. 870, Springer Vlg 1981.
20. S.M. Gersten: *A product formula for Wall's obstruction,* Amer. J. Math. 88 (1966), 337-346.
21. K. Iizuka: *Finiteness conditions for G-CW-complexes,* Japan. J. Math. 10 (1984), 55-69.
22. S. Illman: *Smooth equivariant triangulations of G-manifolds for G a finite group,* Math. Ann. 233 (1978), 199-220.
23. S. Illman: *The equivariant triangulation theorem for actions of compact Lie groups,* Math Ann. 262 (1983), 487-501.
24. S. Illman: *A product formula for equivariant Whitehead torsion and geometric applications,* Transformation Groups, Poznań 1985, pp. 123-142, Lecture Notes in Math. 1217, Springer Vlg 1986.
25. S. Illman: *Actions of compact Lie groups and the equivariant Whitehead group,* Osaka J. Math. 23 (1986), 881-927.
26. S. Illman: *On some recent questions in equivariant simple homotopy theory,* Transformation Groups and Whitehead Torsion, Proc. RIMS 633, Kyoto Univ. 1987, pp.19-33.
27. S. Illman: *The restriction homomorphism* Res_H: $Wh_G(X) \to Wh_H(X)$ *for* G *a compact Lie group,* (preprint, 1989).
28. R.C. Kirby, L.C. Siebenmann: *Foundational essays on topological manifolds, smoothings and triangulations,* Ann. Math. Studies 88 (1977), Princeton Univ. Press.
29. H.T. Ku, M.C. Ku: *Obstruction theory for finite group actions,* Osaka J. Math. 18 (1981), 509-523.
30. S. Kwasik: *On the equivariant homotopy type of G-ANR's,* Proc. Amer. Math. Soc. 267 (1981), 193-194.
31. S. Kwasik: *On equivariant finiteness,* Compositio Math. 48 (1983), 363-372.
32. S. Kwasik: *Locally smooth G-manifolds,* Amer. J. Math. 108 (1986), 27-37.
33. W. Lück: *The geometric finiteness obstruction,* Proc. London Math. Soc. 54 (1987), 367-384
34. W. Lück: *Transformation groups and algebraic K-theory,* Lecture Notes in Math. 1408, Springer Vlg 1989.
35. I. Madsen, C.B. Thomas, C.T.C. Wall: *The topological spherical space form problem-II; existence of free actions,* Topology 15 (1976), 375-382.
36. I. Madsen, M. Rothenberg: *On the classification of G-spheres III: TOP automorphism groups,* preprint 14 (1985), Aarhus Univ.
37. M. Murayama: *On G-ANR's and their G-homotopy types,* Osaka J. Math. 20 (1983), 479-512.
38. R. Oliver: *Fixed-point sets of group actions on finite acyclic complexes,* Comment. Math. Helv. 50 (1975), 155-177.
39. R. Oliver: *Smooth compact Lie group actions on disks,* Math. Zeit. 149 (1976), 79-96.
40. R. Oliver: *G-actions on disks and permutation representations II,* Math. Zeit. 157 (1977), 237-263.
41. R. Oliver: *G-actions on disks and permutation representations,* J. Algebra 50 (1978), 44-62.
42. R. Oliver, T. Petrie: *G-CW-surgery and* $K_0(\mathbb{Z}G)$, Math. Zeit. 179 (1982), 11-42.
43. T. Petrie: *G-maps and the projective class group,* Comment. Math. Helv. 51 (1976), 611-626.
44. F. Quinn: *Ends of maps, II,* Invent. Math. 68 (1982), 353-424.
45. A.A. Ranicki: *The algebraic theory of finiteness obstruction,* Math. Scand. 57 (1985), 105-126.
46. L.C. Siebenmann: *The obstruction to finding a boundary for an open manifold in dimension greater than five,* Ph. D. thesis, Princeton 1965.
47. M. Steinberger: *The equivariant topological s-cobordism theorem,* Invent. Math. 91 (1988), 61-104.
48. M. Steinberger, J.E. West: *Equivariant h-cobordisms and finiteness obstructions,* Bull. Amer. Math. Soc. 12 (1985), 217-220.
49. M. Steinberger, J.E. West: *On the geometric topology of locally linear actions of finite groups,* Geometric and Algebraic Topology, pp. 181-204, Banach Center Publ. 18, Warsaw 1986.

50. M. Steinberger, J.E. West: *Equivariant controlled simple homotopy theory* (in preparation)
51. M. Steinberger, J.E. West: *Controlled finiteness is the obstruction to equivariant handle decomposition,* (preprint).
52. M. Steinberger, J.E. West: *Equivariant handles in finite group actions,* (preprint).
53. R.G. Swan: *Periodic resolutions for finite groups,* Ann. Math. 72 (1960), 267-291.
54. C.B. Thomas, C.T.C. Wall: *The topological spherical space form problem-I,* Compositio Math. 23 (1971), 101-114.
55. C.T.C.Wall: *Finiteness conditions for CW-complexes,* Ann. Math. 81 (1965), 55-69.
56. C.T.C.Wall: *Finiteness conditions for CW-complexes, II,* Proc. Royal Soc. London, Ser. A, 295 (1966), 129-139.
57. A.G. Wasserman: *Equivariant differential topology,* Topology 8 (1969), 127-150.
58. D. Webb: *Equivariantly finite manifolds with no handle structure,* (preprint).
59. S. Weinberger: *Constructions of group actions: a survey of some recent developments,* Group actions on manifolds, pp. 269-298, Contemporary Math. 36 (1985).
60. S. Weinberger: an example in: Problems submitted to the AMS Summer Research Conference on Group Actions, Group actions on manifolds, ed. R.E. Schultz, pp. 513-568, Contemporary Math. 36 (1985).
61. S. Weinberger: *Class numbers, the Novikov conjecture and transformation groups,* Topology 27 (1988), 353-365.
62. J.E. West: *Mapping Hilbert cube manifolds to ANRs,* Ann. Math. 106 (1977), 1-18.

On Conic Spaces

GIORA DULA*§**

Abstract. In this paper, the notions of conic spaces, Thom spaces, Hopf Invariants and construction X_k are surveyed.

Introduction

The main purpose of this survey paper is to advocate the usage of *conic spaces*. Section 1, called Conic Spaces or CS for short, gives their definition (CS1), and points out that every CW complex (CS2), and every simplicial complex (CS3) are particular cases of conic spaces. It follows by examples of conic (non CW and non simplicial) presentations of three important spaces, and by some theorems concerning conic spaces.

The second section (TS) takes on one of the theorems mentioned in CS, namely, the fact that Thom spaces of bundles over conic bases are conic. Since this is a review paper, the section reviews some older results in the same direction.

The third section (HI) reviews the subject of Hopf Invariants. Those Hopf invariants come into the description of the relative attaching maps in Thom spaces so that some of the constructions in HI are mentioned in TS too, but HI discusses some older constructions.

The fourth section Xk reviews the construction Xk. Some of this work have not been published. Xk could be viewed as Thom spaces of bundles over the conic space $J_k(S^n)$, and hence the forth section is a continuation of CS and of TS.

CS1-Conic spaces

A space B is a *conic space* if B is the limit $\lim\limits_{n\longrightarrow\infty} B_n$, where B_n is defined inductively as the mapping cone of $f_n : A_{n-1} \longrightarrow B_{n-1}$, and B_0 is a discrete set. B_n is called the nth *conic skeleton* (and is called an n cone in **[FHLT]**. See also **[May2]**), and f_n is called the nth *attaching map*. It is assumed throughout that each A_i has the homotopy type of countable CW complex and that B is the limit of B_n as topological spaces. CA_{n-1} is called the nth *conic cell.*

Of particular importance is the case when B_0 is a singlton $\{*\}$. Then the first attaching map $f_1 : A_0 \longrightarrow \{*\}$ is the trivial map and B_1 is suspension of A_0. Such conic space B is called connected, and most considerations will involve connected conic spaces.

The composition map $A_{n-1} \xrightarrow{f_n} B_{n-1} \xrightarrow{p} B_{n-1}/B_{n-2}$ of the nth attaching map and the quotient map p is called the nth *relative attaching map*. It measures the way the nth conic cell attaches to the $n-1^{st}$.

CS2-CW complexes

It is easy to observe that every CW complex is a particular case of conic space where A_{n-1} is a disjoint union of spheres of dimension $n-1$. In this case the relative attaching map gives the

---

*Department of Mathematics, Purdue University, West Lafayette, Indiana 47907.

§This is a survey paper of material related to the author's Ph.D. thesis. The thesis was carried out under supervision of Prof. M.G.Barratt. I wish to thank Prof. Barratt for his help during many years. I wish to thank Prof. Mahowald for many discussions concerning this study, and for presenting me with **[CDGM]**. I wish to thank Prof. B. Gray for presenting me with his unpublished work **[Gr1]** and discussing it with me, and Prof. H. Marcum for discussing with me his work **[Mar1]** and **[Mar3]**. I wish to thank Bob Oliver for suggestions concerning the presentation of the paper. I wish to thank Greg Henderson for proof reading the paper. I wish to thank Alex Nofech for mentioning **[May2]**.

**current address: Department of Mathematics, Bar-Ilan University, 52900 Ramat-Gan, Israel.

boundary homomorphism in the chain complex calculating the homology of B. In the connected case the disjoint union becomes a bouquet.

CS3-simplicial complexes

It is easy to observe that every simplicial complex is a particular case of a CW complex in which the attaching map carries every sphere in the domain homeomorphically onto the boundary of an n simplex in B_{n-1}.

CS4-Examples of conic (non CW) spaces

The fact that every CW and simplicial presentations are also conic presentations, supplies many examples of conic presentations. The following are three examples of conic (non CW) presentations.

Example $J_n(\Sigma P)$

Given a space X, the James model for $\Omega\Sigma X$, denoted J(X), was first considered in **[J1]**, (and was denoted X_∞). Filtration one is given on X, while the basepoint $*$ of X is given filtration zero. There is an induced filtration on $X^{\times n}$, the nth cartesian power of X, and on $\lim_{n\to\infty} X^{\times n}$, denoted X^∞, such that under the induced equivalence relation $X^\infty/\sim$ denoted $J(X)$ becomes the free monoid on X, and $X^{\times n}/\sim$ denoted $J_n(X)$ gives a filtration on $J(X)$. In the particular case that X is a suspension ΣP, the points of exact filtration one in X form a cone CP. Then points of filtration k in $J_n(X)$ are of the form $(CP)^{\times k}$, the kth cartesian power of CP, and using the homotopy equivalence of pairs

$$(CP \times CQ, P \star Q) \simeq (C(P \star Q), P \star Q)$$

(where $P \star Q$ denotes the join of the spaces P and Q), it turns out that $J_n(\Sigma P)$ has the following conic presentation:

$$J_n(\Sigma P) = (...((* \cup CP) \cup C(P \star P) \cup ... \cup C(P \star P \star ... \star P)),$$

where the last join is of n copies of P. The attaching maps in this complex,

$$w_k : P \star P \star ... \star P \longrightarrow J_{k-1}(X),$$

were first studied by Porter in **[P]**, and were called generalized Whitehead Product maps. The conic structure presented in this example specializes to a CW structure if P is a sphere, but otherwise is a conic non CW structure of $J\Sigma P$, having less cells than any CW presentation of this space.

Example B(G)

Given a topological group G, whose underlying space is a countable CW complex, the following model for the classifying space of G, B(G), was given in a paper of Milnor **[Mi2]**. $E_n(G)$ denotes the iterated join of G with itself, where G is taken n+1 times. This is a G space, with diagonal G action on the G coordinates of the join, and orbit space $B_n(G)$. In the case that the underlying space G has the homotopy type of a countable CW complex, the same is true for $E_n(G)$ and $B_n(G)$. The projection map $h_n(G) : E_n(G) \longrightarrow B_n(G)$ is a fibration with fiber G. The particular cases $h_1(S^i) : S^{2i+1} \longrightarrow S^{i+1}$, for i=0,1,3 and 7, are the classical Hopf fibrations. There is an inclusion map $i_n(E) : E_n(G) \longrightarrow E_{n+1}(G)$, obtained by using the base point of the last G in the join. It induces a map $i_n(B) : B_n(G) \longrightarrow B_{n+1}(G)$ so that the following diagram commutes:

$$\begin{array}{ccc} E_n(G) & \xrightarrow{i_n(E)} & E_{n+1}(G) \\ {\scriptstyle h_n(G)}\downarrow & & \downarrow{\scriptstyle h_{n+1}(G)} \\ B_n(G) & \xrightarrow{i_n(B)} & B_{n+1}(G) \end{array}$$

The limit of the maps h_n is a fibration

$$h_\infty = h : E(G) \longrightarrow B(G)$$

with fiber G, from E(G) which is a contractible free G space to B(G), the classifying space of G. The sequence $B_0(G) \subset B_1(G) \subset ... \subset B_n(G) \subset ...B(G)$ is an ascending filtration of B(G), and it turns out that $B_{n+1}(G)$ is homotopy equivalent to the mapping cone of $h_n : E_n(G) \longrightarrow B_n(G)$, $B_n(G) \cup_{h_n} CE_n(G)$. This gives $B_n(G)$ the conic presentation

$$B_n(G) = (...((\{*\} \cup CG) \cup_{h_1} C(G \star G) \cup ... \cup_{h_{n-1}} C(G \star G \star ... \star G)),$$

where the last join has n G's and $n-1$ stars.

The conic structure presented in this example specializes to a CW structure if G is a sphere, but otherwise is a conic non CW structure of $B(G)$, having less cells than any CW presentation on this space.

Example $\Omega^2 S^3$

In **[CDGM]** a conic presentation is given to $\Omega^2 S^3$. Actually, the filtration $F_n\Omega^2 S^3$ was given in **[May1]**. In **[CDGM**, lemma 2.1], the authors show that F_n is actually a conic space with the presentation $F_n = F_{n-1} \cup_c CM_m(\partial I^m)$, where $M_m(X)$ is defined by:

$$M_m(X) = \frac{F(R^2, m) \times_{\Sigma_m} X}{F(R^2, m) \times_{\Sigma_m} \{*\}},$$

where $F(R^2, m)$ is the collection of m distinct points in the plane, and X needs to be a Σ_n set. Both $F(R^2, m)$ and ∂I^m are Σ_m sets, and thus $M_m(\partial I^m)$ is well defined.

The composition map $F_n(\Omega^2 S^3) \hookrightarrow \Omega^2 S^3 \longrightarrow S^1$ when localized at p, is denoted $\mathcal{F}_n \longrightarrow S^1_n$. A_n denotes the fiber of $\mathcal{F}_{np+1} \longrightarrow S^1_p$, and the authors show in theorem 1.3, that as a corollary to the conic structure of lemma 2.1, the fibration $\mathcal{F}_{np+1} \longrightarrow S^1_p$ is a product fibration. This gives properties of Brown-Gitler spectra.

CS5-A theorem about finite conic spaces

In **[FHLT]** the nth conic skeleton is called an n cone. The following theorem is proved in §5.

THEOREM D. *Given a simply connected space X which is a finite conic space, having a finite Postnikov tower of finite type localized at some prime p, then it follows that all the homology groups of X with coefficients in $\mathbb{Z}_p$ in positive dimensions vanish.*

In particular, if X is a p local CW complex, it is contractible.

This theorem D is derived from theorem A.

THEOREM A. *A simply connected space X with all the mod p homology groups finite dimensional has the property that the depth of the algebra $H_*(\Omega X, \mathbb{Z}_p)$ is not bigger than the Lusternik-Schnirelmann category of X.*

Every finite conic space composed of n cones has Lusternik-Schnirelmann category lesser than n, and thus theorem A applies to finite simply connected conic spaces with finite dimensional *mod p* homology groups.

CS6-A theorem about conic spaces

THEOREM. *A Thom space of a bundle over a conic space is conic.*

In this last theorem the 'bundle' can be a fiber bundle. The Thom space of the fiber bundle is defined [HS] as the mapping cone of the projection map. In the usual case of vector bundles, the associated sphere bundle has a Thom space in the new definition which is the same as the classical Thom complex of the original vector bundle.

This theorem follows from the work of Held and Sjerve [HS], which followed that of Wall [**Wll**](mentioned in TS6 below). In §3 they prove that given a cofiber sequence $A \xrightarrow{f} X \xrightarrow{i} C_f$, and a fiber bundle ξ over C_f, the Thom space over C_f is a mapping cone $T(X) \cup_g C(A \star Y)$, where Y is the fiber of the bundle, $\star$ denotes the join, T(X) is the Thom complex of the bundle over X, $i^*(\xi)$, induced from ξ by i, and $(A \star Y) \xrightarrow{g} T(X)$, the attaching map, is given a description in [**HS**, theorem 3.4]. The description is a composition of four maps, one of which uses a trivialization of the induced bundle $(f \circ i)^*(\xi)$ over A. [**HS**, theorem 3.5], states that in the particular case that X is a point and C_f is homotopic to ΣA, $(A \star Y) \xrightarrow{g} \Sigma Y$ is homotopic to the Hopf construction of the clutching map $A \times Y \longrightarrow A$ after both maps are suspended once.

The following obvious extension of [**HS**, 3.4] appears in [**D I**, §1.17]. B the base of ξ has a conic presentation $B_0 \subset B_1 \subset ... \subset B_n \subset ...B$, where

$$B_n = (...((* \cup CA_1) \cup C(A_2)) \cup ... \cup C(A_{n-1})),$$

then it follows that the Thom space over B is conic with a conic presentation $T_0 \subset T_1 \subset ... \subset T_n \subset ...T$, where

$$T_n = (...((T(*) \cup C(A_1 \wedge T(*))) \cup C(A_2 \wedge T(*))) \cup ... \cup C(A_{n-1} \wedge T(*))),$$

where $T(*)$ is the Thom space of the fibration induced from ξ by the inclusion of the basepoint $\{*\}$ in B.

CS7-Work of H.Marcum

In [**Mar1**] (see also [**Mar2**]), the main theorem 1.3 generalizes the theorem of [**HS**] about the structure of Thom spaces over conic spaces.

Given a diagram:

$$(*) \qquad \begin{array}{ccc} E_1 & \xrightarrow{=} & E_1 \\ {\scriptstyle j}\uparrow & & \uparrow{\scriptstyle p_1} \\ E_B & \xrightarrow{j_B} & E \\ {\scriptstyle p_B}\downarrow & & \downarrow{\scriptstyle p} \\ B & \xrightarrow{\beta} & M, \end{array}$$

in which p is a fibration with fiber F, E_B is the pull back of p and β and given a homotopy pushout

$$(**) \qquad \begin{array}{ccc} C & \xrightarrow{f} & A \\ {\scriptstyle g}\downarrow & & \downarrow{\scriptstyle \alpha} \\ B & \xrightarrow{\beta} & M, \end{array}$$

then it follows that there exist a map W and a space $E(f \star p_A)$ such that the following diagram is a homotopy pushout:

$$\begin{array}{ccc} E(f \star p_A) & \xrightarrow{W} & \mathcal{Z}(p_B, j) \\ \downarrow & & \downarrow \mu \\ A & \longrightarrow & \mathcal{Z}(p, p_1), \end{array}$$

where $\mathcal{Z}(p, p_1)$ is the double mapping cylinder of p and p_1, and μ is the obvious map between the double mapping cylinders.

The theorem of Held and Sjerve is obtained as the particular case when both A and E_1 equal a point. Then $**$ is a presentation of M as a mapping cone $M = B \cup_g CC$, and $*$ says that there exists a fibration $p : E \longrightarrow B$ which restricts to a fibration $p_B : E_B \longrightarrow B$. Then $\mathcal{Z}(p, p_1)$ is the Thom space over M, $T(M)$, $\mathcal{Z}(p_B, j)$ is $T(B)$ and $E(f \star p_A)$ specializes to be $C \star F$. Thus the conclusion of theorem 1.3 presents $T(M)$ as a mapping cone $T(B) \cup_W C(C \star F)$. Iterated usage of $**$ gives the result stated in CS6.

§2. Attaching maps in Thom spaces over Conic Spaces

TS1-Introduction

Finding the cells of the Thom space over the conic space B is not enough information for building up the space. The attaching maps are needed too. **[HS**, theorem 3.4] gives a description of the attaching maps in the mapping cone, but this description is not homotopy invariant, as it uses a choice of trivialization of the bundle over the cone. This section reviews the literature describing the homotopy type of the attaching maps. It turns out that the two key concepts appearing are 'J-homomorphism' and 'Hopf invariants'.

TS2-The structure of T_1-The J homomorphism

In the particular case that the base space B equals B_1, which is by definition ΣA_0, the Thom space of any bundle over B has a conic presentation as $T(*) \cup_{T(f_1)} CA_0 \wedge T(*)$, and is totally determined by the attaching map $T(f_1)$. As $T(*)$ equals the suspension of the fiber, $T(f_1)$ is a map $A_0 \star F \longrightarrow \Sigma F$.

In the case that both $A_0 = S^i$ and $F = S^j$ are spheres, and ξ is a spherical fibration with $O(j)$ action, classified by a homotopy class $\alpha : S^{i+1} \longrightarrow BO(j)$, $T(f_1)$ equals $J(\alpha)$, the image under the J homomorphism of α, discussed first in the paper **[WG1]** by G.W.Whitehead.

In the case that ξ has a more general structure group, still $T(f_1)$ equals $J(\alpha)$, where this J was defined in the paper **[At]** by Atiyah.

Held and Sjerve in **[HS**, theorem 3.5] discuss this map for fiber bundles over suspensions without assuming that the base or the fiber are spheres. They prove that the map $T(f_1)$ and the Hopf construction of the clutching function are homotopic, after both are suspended once (as mentioned in CS6). They do not call this map a 'J homomorphism'.

The same construction appears in the thesis of C.H.Hanks **[Ha1**, Ha2], to which he gives the name J Homomorphism. As an application of his main theorem, Hanks deduces that this J Homomorphism is *not* a homomorphism in general. In **[Ha1**, theorem 4.8] he calculates the obstruction for the J 'Homomorphism' from indeed being a homomorphism. He gives an example of a case where the obstruction is non zero. This example is repeated in **[D**, I, §3.13b]. This obstruction vanishes in the case that the base of ξ is a double suspension. In all the classical cases, the base is a highly connected sphere and the obstruction vanishes.

TS3-The structure of T_2

Given a presentation

$$B = B_2 = ((* \cup CA_0) \cup CA_1),$$

the following presentation is obtained:

$$T = T_2 = ((T(*) \cup C(A_0 \wedge T(*))) \cup C(A_1 \wedge T(*))),$$

with attaching maps $T(f_1)$ and $T(f_2)$. Thus, as $T(f_1)$ is known, only $T(f_2)$ is needed in order to give a full description of T_2. The main theorem of **[Hal]**, was shown in **[D**, I, §6], to describe the homotopy class of the relative attaching map $p_0 \circ T(f_2)$, (in the case that A_1 is a suspension) which is an element in the set $[A_1 \wedge T(*), \Sigma A_0 \wedge T(*)]$. This is a group as A_1 is a suspension, and in that group the following statement can be derived from **[Hal]**:

$$p_0 \circ T(f_2) = \sum_{n=1}^{\infty} J_n(\alpha) \circ s_n \circ HI_n(f_2),$$

where $HI_n(f_2) : A_1 \wedge T(*) \longrightarrow \Sigma A_0^{\wedge n} \wedge T(*)$ is the identity map on $T(*)$ smashed with a higher Hopf Invariant of the map f_1, s_n is a combinatorial type map permuting the order of the smash factors in $\Sigma A_0^{\wedge(n)} \wedge T(*)$, and for $\alpha : \Sigma A_1 \longrightarrow B(G)$ classifying the fibration ξ. $J(\alpha) : A_0 \wedge T(*) \longrightarrow T(*)$ is the J map defined in **[Hal]** while $J_n(\alpha) : \Sigma A_0^{\wedge(n)} \wedge T(*) \longrightarrow T(*)$ is a map defined in **[Hal]** by iterating $J(\alpha)$.

TS4-The map $T(f_i)$

It is desirable to describe the attaching map $T(f_i) : A_{i-1} \wedge T(*) \longrightarrow T_{i-1}$. There is the description of **[HS**, theorem 3.4]. However in general, this description is not a homotopy invariant, for example one of the composition factors uses a trivialization map for the bundle ξ restricted to the cone CA_{i-1}. The main theorem of **[D]**,**[D**, I, §1.32], gives a primary homotopy invariant of $T(f_i)$, namely the relative attaching map $p_{i-2} \circ T(f_i) : A_{i-1} \wedge T(*) \longrightarrow \Sigma(A_{i-2} \wedge T(*))$, in the case that A_{i-1} is a suspension. This is an element in the group $[A_{i-1} \wedge T(*), \Sigma A_{i-2} \wedge T(*)]$, and in it the following statement holds

$$p_{i-2} \circ T(f_i) = ((\epsilon_A) \wedge \underline{T(*)}) \circ J(\epsilon_B) \circ (GHI(f_i) \wedge \underline{T(*)}) + (p_{i-2} \circ f_i) \wedge \underline{T(*)},$$

where the second summand (which is simpler than the first) is the smash of the relative attaching map $p_{i-2} \circ f_i : A_{i-1} \longrightarrow B_{i-1} \longrightarrow B_{i-1}/B_{i-2} = \Sigma A_{i-2}$, and the identity map of $T(*), \underline{T(*)}$. It is the same for all bundles ξ over B, that have the same $T(*)$. In particular, it comes from the trivial bundle, while the first summand vanishes on the trivial bundle.

The first summand is more complicated than the second and is a composition of three maps. The first and the last of those three, are the same for all bundles as they are smashes of a map from the base with $\underline{T(*)}$. Only the middle factor does depend substantially on ξ. The first factor is the smash of $\underline{T(*)}$ with a version of the Hopf Invariant of f_i, discovered first by T.Ganea in **[Ga]**, which is called the *Ganea Hopf Invariant of* f_i. The second map is essentially a J homomorphism as defined by Hanks **[Hal]**, applied to the map $\alpha \circ \epsilon_B : \Sigma\Omega B_n \longrightarrow B(G)$, which is the composition of the evaluation map $\epsilon_B : \Sigma\Omega B_n \longrightarrow B_n$, with the map $\alpha : B_n \longrightarrow B(G)$ classifying ξ. The last factor is $\underline{T(*)}$ smashed with the evaluation map $\epsilon_A : \Sigma\Omega\Sigma A_n \longrightarrow \Sigma A_n$.

TS5-Examples

The following are examples for the application of the theorem mentioned in TS4.

Example $J_n(\Sigma P)$

In the case that $B = B_n = J_n(\Sigma P)$, the $n^{\underline{th}}$ stage in the James model for $\Omega\Sigma^2 P$, as defined in CS4, it is shown in [**D**, II, §4] that there is an element η in the image of J (as defined by Hanks) such that all $p_{i-2} \circ T(w_i)$, for $1 \leq i \leq n$, can be presented as a sum of elements, each of which is a composition $\lambda \circ \mu$, where μ is obtained by smashing η with an identity map of the space $P \star P \star ... \star P$, and λ is a combinatorial type map permuting the smash factors. This construction generalizes some aspects of the Barratt-Mahowald construction X_k. ([**B4**],[**Mah**], [**Gr**] and [**BJM**]).

Example $B_n(G)$

In the case that $B = B_n = B_n(G)$, the $n^{\underline{th}}$ stage in the Milnor model for $B(G)$, as defined in CS4, it is shown in [**D**, II, §3] that that there exist three elements $h_1 : G \star G \longrightarrow \Sigma G$ the attaching map of $B_2(G)$, $p_0 \circ h_2 : G \star G \star G \longrightarrow \Sigma G \star G$ the relative map of the attaching map $h_2 : G \star G \star G \longrightarrow B_2(G)$ in $B_3(G)$, and η which is a certain element in the image of J (as defined by Hanks) such that all $p_{i-2} \circ T(h_i)$, for $1 \leq i \leq n$, can be built from those three elements . The way of building $p_{i-2} \circ T(h_i)$ is slightly more complicated than in the $J_n(\Sigma P)$ case. In some particular cases the general result simplifies. One such case is when the fibration ξ is the 'line bundle' with projection map h_n discussed in CS4. Then $T(h_i)$ is h_{i+1} and the element η equals h_1. Thus the result expresses $p_{i-2} \circ h_{i+1}$ in terms of h_1 and $p_0 \circ h_2$. In the more special case that G is a sphere, there is an easy way to express $p_0 \circ h_2$ in terms of h_1, and the classical result about relative attaching maps in projective spaces is obtained, expressing $p_{i-2} \circ h_{i+1}$ in terms of h_1.

TS6-Work of C.T.C.Wall

In his paper [**Wll**], C.T.C. Wall discussed the structure of Poincaré complexes. In chapter 3 he discussed Thom complexes of vector bundles over those spaces. He called a Thom complex *reducible*, if the top cell homotopy splits. This is equivalent to the fact that the attaching map of the top cell is null homotopic. He proved that every Poincaré complex M has a stable spherical bundle ν over it with reducible Thom space M^ν. ν is unique, and M^ν is Spanier Whitehead dual to M^{trivial}.

In the next part of his work (which predated the work of Held and Sjerve mentioned in CS6 above), he discussed the structure of the Thom complex over a suspension ΣK (proposition 3.7). He gave the cell structure of $(\Sigma K)^\alpha$ for every spherical fibration α and showed that the attaching map is stably equal to the characteristic map (also called clutching map) $K \longrightarrow G$ where G is the limit of G_n, the monoid of self homotopy equivalences on $\mathbf{S}^{n-1}$.

§3. Hopf Invariants

HI1-Introduction

It turns out that relative attaching maps in the Thom spaces of bundles over conic spaces. are expressed in terms of Hopf Invariants. This section reviews some of the works concerning with Hopf Invariants. The very classical works of H.Hopf [**Ho1**], [**Ho2**], Steenrod's definition of functional cup products [**S**], the solution of the Hopf invariant one problem [**Ade**] and [**Ada**] and the works of G.Whitehead [**WG2,3**], are skipped. This survey is by no mean claimed to be complete. Another source is the book by Baues [**Bau**]. This is the only book devoted entirely to material related to Hopf invariants. We can only describe it very briefly. The Hilton-Milnor process takes place in a free group, and in the collection process commutators are produced. The book discusses commutator calculus in general. Then it derives results for distributivity laws and homotopy operations on spheres, including the higher Hopf invariants. Then it uses the previous theory with different coefficient rings to deduce results about homotopy groups.

HI2-The work of I.James

In his work,[**J1**,2], I.M.James had a contribution to the subject of Hopf Invariants. In [**J1**] he established his model JX for $\Omega\Sigma X$, discussed before in CS10. Using this model he defined in [**J2**] for each k the maps $C_k : JX \longrightarrow J(X^{\wedge k})$, where $A^{\wedge k}$ is the kth smash power of A, and called those maps combinatorial extensions. In particular C_2 is a map $JX \longrightarrow J(X^2)$. Then the kth Hopf invariant of a class $f : \Sigma A \longrightarrow \Sigma X$, $H_k(f)$, is obtained by first taking the adjoint of f, $a(f) : A \longrightarrow \Omega\Sigma X$, composing it with the kth combinatorial extension map, giving the composition $A \longrightarrow \Omega\Sigma X \longrightarrow \Omega\Sigma X^{\wedge n}$ and then adjointing back again giving $H_k(f) : \Sigma A \longrightarrow \Sigma X^{\wedge k}$. Thus not only that James' H_2 extends Whiteheads' definition (up to sign) of the operator H for any dimension q and for every space X (giving Whiteheads' case as a sphere), also the other H_k, called nowadays the ***higher James Hopf Invariants*** were introduced by James. Thus there are many Hopf Invariants indexed by N, and using them it is clear that the EHP sequence stops being exact exactly when H_3 becomes non zero. Thus the exactness of the EHP sequence is related to the vanishing of the higher James Hopf Invariants.

HI3-The work of J.P.Hilton

The work of Whitehead was generalized in another form by J.P.Hilton [**Hi1**]. He maintained the original line of Whitehead by starting with a map $f : S^q \longrightarrow S^p$, and applying the coproduct $\rho : S^p \longrightarrow S^p \vee S^p$. Using an analysis of the space $\Omega(S^i \vee S^j)$, he was able to show that there exists an infinite sum $\sum_{\tau} w_\tau \circ H_\tau$ which equals $\rho \circ f$, where τ varies over a Hall basis of the free lie algebra generated by two variables x_1, x_2, of dimensions $i-1$ and $j-1$ respectively, X_τ is a space of the form $S^{k_1} \wedge S^{k_2} \wedge \cdots \wedge S^{k_l}$, where the generator τ in the free lie algebra has coordinates $\{k_1, k_2, \ldots, k_l\}$, the map $H_\tau : S^q \longrightarrow \Sigma X_\tau$ is a generalization of the Hopf Invariant, and the map $w_\tau : \Sigma X_\tau \longrightarrow S^p \vee S^p$ is a successive composition of whitehead product maps.

HI4-The work of J.Milnor

The works of James and Hilton were generalized by that of Milnor [**Mi1**]. He was able to produce all the results obtained by Hilton (and mentioned in HI3) to the case when S^q is replaced by a space ΣP and S^p is replaced by a space ΣA. His result (theorem 4) is that $\rho \circ f$ is an infinite sum as before, indexed by τ in the Hall basis of a free Lie algebra on two variables. In order to obtain his result he defined a model for the loop space over the suspension of a space, using free group functor $\mathcal{F}$, rather than the free monoid functor used by James. His theorem 3, which is a lemma for theorem 4, specializes to give an alternative definition of James' invariants. Theorem 3 states that $\mathcal{F}(B \wedge \mathcal{F}(A))$ is homotopy equivalent to the space $\mathcal{F}(\bigvee_{i=1,\infty} B \wedge A^{\wedge i})$. Applying the classifying space operator B on the last equality gives a homotopy equivalence between $\Sigma(B \wedge \mathcal{F}(A))$ and the space $\Sigma(\bigvee_{i=1,\infty} B \wedge A^{\wedge i})$. The particular choice of B being S^0, a sphere of dimension zero, gives that $\Sigma\mathcal{F}(A)$ is homotopy equivalent to the space $\Sigma(\bigvee_{i=1,\infty} A^{\wedge i})$. Thus given a class $f : \Sigma P \longrightarrow \Sigma A$, it has an adjoint $a(f) : P \longrightarrow \Omega\Sigma A$, whose suspension is a map $\Sigma P \longrightarrow \Sigma(\mathcal{F}(A))$, using the fact that $\mathcal{F}(A)$ is Milnor's model for $\Omega\Sigma A$. Then using the above corollary of theorem 3 in [**Mi1**], there is a composition map $\Sigma P \longrightarrow \Sigma(\mathcal{F}(A)) \longrightarrow \Sigma A^{\wedge i}$ whose adjoint is a map $\Sigma P \longrightarrow \Sigma A^{\wedge i}$, which equals James' i^{th} invariant up to a sign. The next two paragraphs discuss other relations between the different generalization of the Hopf invariants.

HI5-The work of Barcus and Barratt

In the paper [**BB**] Barcus and Barratt considered the following problem: Given a map $u : K \longrightarrow X$ and an inclusion $K \hookrightarrow L$, they wanted to enumerate the homotopy classes of the extensions of u, $u' : L \longrightarrow X$. Given a class α in $\pi_q(X,*)$, they define a map $\alpha_u : \pi_1(\mathcal{F},*) \longrightarrow \pi_{q+1}(X,*)$ where $\mathcal{F}$ is the function space of basepoint preserving maps $K \longrightarrow X$. In the case that L is the space $K \cup_\alpha e^{q+1}$, obtained as a mapping cone of any representative of α, then they prove (theorem

3.3) that the homotopy classes of extensions are in one to one correspondence with the cokernel of α_u.

Then they go on to present α_u explicitly. For the particular case that a representative of α has a presentation as a composition $S^q \xrightarrow{\Phi} S^n \xrightarrow{\beta} K$, they present (corollary 4.8) $\alpha_u(\xi)$ as an infinite sum of elements using α, its Hopf invariants, the homotopy groups of X, and the operations of compositions suspensions and formations of Whitehead products.

The higher Hopf invariants they use, are part of the Hilton invariants. They choose a subfamily of the Hall basis of the free Lie algebra, in which one of the variables has degree one. In [**B2**] it is claimed that this subfamily determines the James' invariants and all the Hilton invariants. The proof of the first assertion is sketched in [**B3**]. The completion of the proof is sketched in [**B7**], as well as the procedures for finding the explicit relations. It uses twisted Lie algebras ([**B6**]) and generalized signs.

HI6-The work of Boardman and Steer

The fact that there are so many generalizations of the Hopf invariants, could mean that this notion is important, but is certainly discouraging because it seems to be that the different generalizations are not related ([**B7**] was not yet published). Maybe this is the reason that instigated Boardman and Steer to make the study appearing in ([**BS**]). They were ready to suspend the Hopf invariants certain number of times in order to obtain simplifications. Thus they define the nth Hopf invariant as an operator $\lambda_n : [\Sigma A, \Sigma B] \longrightarrow [\Sigma^n A, (\Sigma B)^n]$. It should be noted that the James' n^{th} invariant is an operator $[\Sigma A, \Sigma B] \longrightarrow [\Sigma A, \Sigma B^n]$, so that λ_n has the same domain and range as that of the $n - 1^{st}$ suspension of the James invariants. They defined a 'cup product' $\alpha \wedge \beta$ of a class α in $[\Sigma^n A, B]$ and of a class β in $[\Sigma^m A, C]$ as the composition class $\Sigma^{n+m} A \xrightarrow{\Sigma^{n+m}\Delta} \Sigma^{n+m} A \wedge A \xrightarrow{\simeq} \Sigma^n A \wedge \Sigma^m A \xrightarrow{\alpha\wedge\beta} B \wedge C$, where Δ is the reduced diagonal map, and $\simeq$ is a generalized sign homeomorphism, induced by switching the order of some of the smash factors. A ladder of Hopf invariants is a family of maps $\lambda_n : [\Sigma A, \Sigma B] \longrightarrow [\Sigma^n A, (\Sigma B)^n]$ with the following properties:

(i) λ_1 equals the identity operator.

(ii) for every n, the composition operator $\lambda_n \circ \Sigma$ is the zero operator $[A, B] \longrightarrow [\Sigma A, \Sigma B] \longrightarrow [\Sigma^n A, (\Sigma B)^n]$.

(iii) for every α and β in $[\Sigma A, \Sigma B]$, there is an equality of $\lambda_n(\alpha + \beta)$ and of $\lambda_n(\alpha) + \lambda_{n-1}(\alpha) \wedge \lambda_1(\beta) + \cdots + \lambda_1(\alpha) \wedge \lambda_{n-1}(\beta) + \lambda_n(\beta)$.

They first proved that a ladder, if it exists, must be unique. Then they went on to show that the nth James' invariants, when suspended $n-1$ times, form a ladder. Then they showed that the Barcus Barratt [**BB**] family of the Hilton-Milnor invariants, suspended $n - 1$ times form a ladder too, establishing the fact that the two invariants are the same, when sufficiently suspended. Then they went on to prove that any other Hilton-Milnor invariant, when suspended, can be expressed in terms of the ladder and of generalized signs. Finally they related the ladders to construction in framed cobordism.

HI7-Two steps in the Milnor process

The Hilton-Milnor process, as described in HI4 above, had actually two conceptual steps.

(i) presenting the free group on a wedge as the product of the free group on one the wedge summands, times a fiber.

(ii) presenting the above fiber as a free group on some wedge.

The Hilton-Milnor process is obtained by applying iteratively the two steps above. Theorem 3 mentioned in HI4 is part of the second step, of presenting the fiber as a free group on some wedge, which turns out to be the longest part in the proof. Sometimes the fiber resulting from a projection of the first step, has a wedge summand in it before the second step is applied, and thus one could apply the first step twice, and project into two wedge summands, before the need to apply the second step. This point of view leads to the next work.

HI8-The work of Berstein and Hilton

In [**BH**], I.Berstein and P.Hilton defined the flat product as follows: Given a wedge of spaces $X \vee Y$, there is the inclusion of the wedge into the product $j : X \vee Y \hookrightarrow X \times Y$. The homotopy fiber of j was denoted by $X \flat Y$ and called the flat product of X and Y. There is the map $w : X \flat Y \to X \vee Y$, and [**BH**] showed that the fibration $\Omega(X \flat Y) \longrightarrow \Omega(X \vee Y) \longrightarrow \Omega(X \times Y)$ homotopy splits and gave a particular choice of a splitting map $\tau : \Omega(X \vee Y) \longrightarrow \Omega(X \flat Y)$. Thus there is a homotopy presentation of $\Omega(X \vee Y)$ as the product $\Omega X \times \Omega Y \times \Omega(X \flat Y)$. The particular case of X being ΣA and of Y being ΣB specializes to be an intermediate step in the Hilton-Milnor process, in which $\Omega\Sigma A$ and $\Omega\Sigma B$ have been projected, and thus $X \flat Y$ generalizes the fiber in the Hilton-Milnor process denoted in [**Mil**] by R_3, which is the result of splitting $\Omega\Sigma A \times \Omega\Sigma B$ off $\Omega(\Sigma A \vee \Sigma B)$. Berstein and Hilton defined a Hopf invariant of a class $f : \Sigma P \longrightarrow \Sigma A$, as the following construction: f is composed with the coproduct map $\rho : \Sigma A \longrightarrow \Sigma A \vee \Sigma A$. The adjoint $a(\rho \circ f) : P \longrightarrow \Omega(\Sigma A \vee \Sigma A)$ is composed with $\tau : \Omega(\Sigma A \vee \Sigma A) \longrightarrow \Omega(\Sigma A \flat \Sigma A)$, and the adjoint of this composite is a class $\Sigma P \longrightarrow \Sigma A \flat \Sigma A$ which is the Hopf invariant of f as defined in [**BH**]. It is clear from the definitions that this invariant contains all the higher Hilton-Milnor invariants combined.

HI9-The work of T.Ganea

In [**Ga**] Ganea had a contribution to the subject of Hopf Invariants. Given an inclusion $d : A \hookrightarrow X$, then $f : X \hookrightarrow B$ is the inclusion of X into the reduced mapping cone of d. There is a natural lifting $\epsilon : A \longrightarrow F$ into the homotopy fiber of f. ϵ specializes to be the map $A \hookrightarrow \Omega\Sigma A$ in the case that B is CA. In theorem I, Ganea showed that the mapping cone of ϵ is homotopy equivalent to $\Omega B * A$. This implies the fact that $\Sigma\Omega\Sigma A$ is homotopy equivalent to $\Sigma A \vee (\Sigma\Omega\Sigma A) \wedge A$, and gives another proof to the fact (theorem 3 of [**Mil**]) that $\Sigma\Omega\Sigma A$ is homotopy equivalent to the wedge $\bigvee_{i=1,\infty} \Sigma A^i$.

Ganea went on defining a Hopf invariant. Using the coaction $\rho : B \longrightarrow \Sigma A \vee B$ and the map $\tau : \Omega(\Sigma A \vee B) \longrightarrow \Omega(\Sigma A \flat B)$ as defined in [**BH**], he defined a version of the Hopf invariant which is the composition map $\tau \circ \Omega\rho : \Omega B \longrightarrow \Omega(\Sigma A \flat B)$.

This map specializes to the invariant defined by Berstein and Hilton in [**BH**] in the case that X is a point, the mapping cone $X \cup CA$ becomes a suspension ΣA and the coaction ρ becomes a coproduct. In his paper Ganea mentions other versions of the Hopf invariant, all derived ¿from the one presented here. Given a class $f : \Sigma P \longrightarrow B$ it has an adjoint $a(f) : P \longrightarrow \Omega B$ and the composition $\tau \circ \Omega\rho \circ a(f) : P \longrightarrow \Omega(\Sigma A \flat B)$ has an adjoint $\Sigma P \longrightarrow (\Sigma A \flat B)$ which is the Ganea Hopf invariant of f used in [**D**].

HI10-The Ganea Hopf Invariant of certain maps

The following three Ganea Hopf invariants were calculated in [**D**].

Case I-Projective spaces

The Hopf map $h_n : E_n(G) \twoheadrightarrow B_n(G)$ was discussed in CS10. In the particular case that G is a sphere group, $G = \mathbf{S}^d$, for $d, d = 0, 1, 3$, h_n becomes the attaching map in projective spaces $\mathbf{S}^{n(d+1)-1} \twoheadrightarrow \mathsf{FP}^n$, where F is the corresponding field. This is a map from a suspension $\mathbf{S}^{n(d+1)-1}$ to a mapping cone FP^n, and [**D**, II, §2] calculates the Ganea Hopf Invariant of this map using cohomology methods. $GHI(h_n)$ is a map $\mathbf{S}^{n(d+1)-1} \longrightarrow \Sigma\Omega S^{nd} \wedge \Omega\mathsf{FP}^n$. There is an inclusion map inc of $S^{n(d+1)-1}$ into the latter space defined as the composition of the maps $S^{n(d+1)-1} \longrightarrow S^{nd} \wedge S^{d-1} \hookrightarrow \Sigma\Omega S^{nd} \wedge \Omega\mathsf{FP}^n$, where the last inclusion in the composition is the smash product of the obvious inclusion of S^{nd} into $\Sigma\Omega S^{nd}$ and of S^{d-1} into $\Omega\mathsf{FP}^n$.

THEOREM. *$GHI(h_n)$ factors as $inc \circ gs$, where $gs : S^{n(d+1)-1} \longrightarrow S^{n(d+1)-1}$ is a self map of degree $(-1)^{n(d+1)}$.*

It should be observed that this statement specializes to the famous theorem stating that h_1 has Hopf invariant one, proved by Hopf [**Ho1**], [**Ho2**].

Case II-Loop space over suspensions

The higher Whitehead product map $w_n : P \star P \star \cdots \star P \longrightarrow J_{k-1}(\Sigma P)$ was discussed in CS10, following [**P**]. In its most general form (appearing in [**P**] and in [**D**]), this is a map $w_n : P_1 \star P_2 \star \cdots \star P_n \longrightarrow FW(\Sigma P_1 \times \Sigma P_2 \times \cdots \times \Sigma P_n)$, where each P_i is a countable CW complex, the higher Whitehead product of CS10 is obtained from the most general one in the case that $P_1, P_2, \ldots, P_n$ are all equal to P, by composing it with a folding map $FW(\Sigma P \times \Sigma P \times \cdots \times \Sigma P) \longrightarrow J_{n-1}(\Sigma P)$, and FW of a product space is the *fat wedge*, consisting of all points such that at least one of the coordinates is the basepoint. Thus w_n is a map from a (homotopy) suspension $P_1 \star P_2 \star \cdots \star P_n$ to the mapping cone $FW(\Sigma P_1 \times \Sigma P_2 \times \cdots \times \Sigma P_n)$, and therefore it has a Ganea Hopf invariant $P_1 \star P_2 \star \cdots \star P_n \longrightarrow \Sigma\Omega\Sigma(Q_1 \vee Q_2 \vee \cdots \vee Q_n) \wedge \Omega FW(\Sigma P_1 \times \Sigma P_2 \times \cdots \times \Sigma P_n)$, where the quotient $FW(\Sigma P_1 \times \Sigma P_2 \times \cdots \times \Sigma P_n)/FW_{n-2}(\Sigma P_1 \times \Sigma P_2 \times \cdots \times \Sigma P_n)$ of FW by the $n-2^{nd}$ conic skeleton of FW, is of the form $\Sigma(Q_1 \vee Q_2 \vee \cdots \vee Q_n)$, and Q_i is the join of all the spaces $P_1, P_2, \ldots, P_n$ except of P_i. It can be observed that there is an inclusion map $inc_i : \Sigma Q_i \wedge P_i \hookrightarrow \Sigma\Omega\Sigma(Q_1 \vee Q_2 \vee \cdots \vee Q_n) \wedge \Omega FW(\Sigma P_1 \times \Sigma P_2 \times \cdots \times \Sigma P_n)$, which is the smash product of the obvious inclusion maps $\Sigma Q_i \hookrightarrow \Sigma\Omega\Sigma(Q_1 \vee Q_2 \vee \cdots \vee Q_n)$ and $P_i \hookrightarrow \Omega\Sigma P_i \hookrightarrow \Omega FW(\Sigma P_1 \times \Sigma P_2 \times \cdots \times \Sigma P_n)$. The following theorem was proved in [**D**, II, §4].

THEOREM. *$GHI(w_n)$ factors as $\sum_{i=1}^{n} inc_i \circ gs_i$, where $gs_i : P_1 \star P_2 \star \cdots \star P_n \longrightarrow \Sigma Q_i \wedge P_i$ is a generalized sign homeomorphism permuting the join factor P_i to become last, and the suspension coordinate after P_i to become before the last.*

Case III-$B(G)$

The Hopf map $h_n : E_n(G) \twoheadrightarrow B_n(G)$ was discussed in CS10. It has the (homotopy) suspension $E_n(G)$ as a domain, and the mapping cone $B_n(G)$ as a range. Therefore it has a Ganea Hopf invariant map $E_n(G) \longrightarrow \Sigma\Omega\Sigma E_{n-1}(G) \wedge \Omega B_n(G)$. There is an inclusion map $inc : \Sigma E_{n-1}(G) \wedge G \hookrightarrow \Sigma\Omega\Sigma E_{n-1}(G) \wedge \Omega B_n(G)$, a smash product of the obvious inclusion $\Sigma E_{n-1}(G) \hookrightarrow \Sigma\Omega\Sigma E_{n-1}(G)$ and the negative of the inclusion $G \hookrightarrow \Omega\Sigma G \hookrightarrow \Omega B_n(G)$. The following theorem (proved in [**D**, II, §3]) describes the GHI of h_n:

THEOREM. *Given that n is bigger than or equal to 2, $GHI(h_n)$ factors as $inc \circ (\alpha + \beta)$, where both α and β are the following maps of $E_n(G) \longrightarrow \Sigma E_{n-1}(G) \wedge G$:*

(i) *α is the homeomorphism $E_n(G) \longrightarrow \Sigma E_{n-1}(G) \wedge G$.*

(ii) *β is the composition $E_n(G) \xrightarrow{\alpha} \Sigma E_{n-1}(G) \wedge G \xrightarrow{-\Sigma E_{n-1}(G) \wedge \Delta_G} \Sigma E_{n-1}(G) \wedge G \wedge G \xrightarrow{r_n \wedge G} \Sigma E_{n-1}(G) \wedge G$, where r_n is the composition $\Sigma E_{n-1}(G) \wedge G \xrightarrow{\alpha^{-1}} E_n(G) \xrightarrow{h_n} B_n(G) \xrightarrow{p} B_n(G)/B_{n-1}(G) \longrightarrow \Sigma E_{n-1}(G)$ and Δ_G is the reduced diagonal map on G.*

HI11-The work of M.Walker

In his paper [**Wlk**], M.Walker had a contribution to the subject of Hopf invariants. He defined a Hopf invariant of a given diagram:

(HI11a)
$$\begin{array}{ccc} A & \xrightarrow{f} & B \\ {\scriptstyle g}\downarrow & & \\ C & & \end{array},$$

as the following construction: he denoted by $\mathbf{Z}(f,g)$ the double mapping cylinder of f and g, that is the space $B \cup_f (A \times I) \cup_g C$ obtained from the disjoint union $B \coprod (A \times I) \coprod C$ by the identifications $(a,0) \sim f(a)$ and $(a,1) \sim g(a)$. Thus the following is a homotopy pushout diagram extending $HI11a$:

(HI11b)
$$\begin{array}{ccc} A & \xrightarrow{f} & B \\ {\scriptstyle g}\downarrow & & \downarrow \\ C & \longrightarrow & \mathbf{Z}(f,g). \end{array}$$

There is another homotopy pushout diagram

(HI11c)
$$\begin{array}{ccc} \{*\} & \longrightarrow & C_f \\ \downarrow & & \downarrow \\ C_g & \longrightarrow & C_f \vee C_g, \end{array}$$

where C_h denotes the mapping cone of h. There is a map from $HI11b$ to $HI11c$ forming a commutative box:

(2.4)

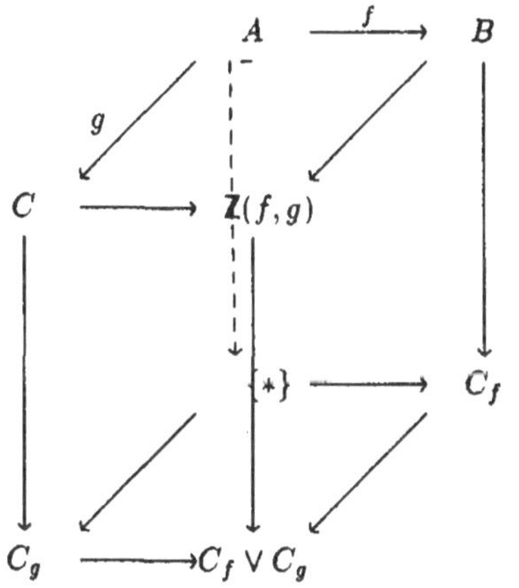

where the vertical map from $\mathbf{Z}(f,g)$ to $C_f \vee C_g$ is obtained by pinching the subset $A \times \{1/2\}$ in $\mathbf{Z}(f,g)$ to the common basepoint of the two reduced cones.

The box $HI11d$ induces a map from the pull back of diagram $HI11b$ to that of $HI11c$. The diagram $HI11c$ has a pull back denoted in **[Hi2]** by $E(C_f \vee C_g; C_f, C_g)$. The induced map of homotopy pull backs, $P \longrightarrow E(C_f \vee C_g; C_f, C_g)$, was Walker definition for the Hopf invariant of the diagram $HI11a$. It is stated in [**Hi2**, p. 122] and proved again in [**D**, p. 91] that the range of the Hopf invariant map $E(C_f \vee C_g; C_f, C_g)$ is homotopy equivalent to $\Omega(C_f \flat C_g)$, and thus the Walker generalization of the Hopf invariant can be thought of as a map $P \longrightarrow \Omega(C_f \flat C_g)$. The space P has description as the space of paths in $\mathbf{Z}(f,g)$ starting at the image of B and ending at the image of C. As the basepoint of $\mathbf{Z}(f,g)$ is the image of both basepoints of B and C, it follows that there is an inclusion of $\Omega\mathbf{Z}(f,g)$ into P. The composition map $\vartheta : \Omega\mathbf{Z}(f,g) \longrightarrow P \longrightarrow \Omega(C_f \flat C_g)$ is an invariant derived from Walker's invariant.

The GHI as a particular case

In the particular case that C equals a space with one point in diagram $HI11a$, then the definition simplifies. $Z(f,g)$ specializes to be the mapping cone C_f, C_g specializes to be the suspension of A, ΣA. Then Walker's definition specializes to be a map $P \longrightarrow \Omega(C_f \flat \Sigma A)$, and the map ϑ above specializes to be a map $\Omega C_f \longrightarrow \Omega(C_f \flat \Sigma A)$, which is exactly the Ganea Hopf invariant as defined in [G].

HI12-The work of H.Marcum

I wish to thank Prof. H. Marcum for explaining his work to me on several occasions. Also it should be remarked that **[Mar3]** is a source for more works in the area of Hopf Invariants, that were not covered in this paper in order to keep it within certain length.

In **[Mar3]** H.Marcum gave the following definition of the Hopf Invariant, given a homotopy commutative diagram

(HI12a)
$$\begin{array}{ccc} C & \xrightarrow{g} & B \\ {\scriptstyle f}\downarrow & & \downarrow{\scriptstyle \beta} \\ A & \xrightarrow{\alpha} & X, \end{array}$$

where H is a homotopy between $\alpha \circ f$ and $\beta \circ g$.

This gives rise to a homotopy commutative diagram:

(HI12b)
$$\begin{array}{ccc} B & \xrightarrow{\beta \Delta \underline{B}} & X \times B \\ {\scriptstyle g}\uparrow & & \uparrow{\scriptstyle \alpha \times \underline{B}} \\ C & \xrightarrow{f \Delta g} & A \times B \\ {\scriptstyle f}\downarrow & & \downarrow{\scriptstyle \underline{A} \times \beta} \\ A & \xrightarrow{\underline{A} \Delta \alpha} & A \times X, \end{array}$$

where $h \Delta k$ stands for the composition $(h \times k) \circ \Delta$. This last diagram induces a map of the double mapping cylinders $M(f,g) \longrightarrow E(\alpha \star \beta)$. There is a fibration $F_\alpha \star F_\beta \longrightarrow E(\alpha \star \beta) \longrightarrow X \times X$ where F_α and F_β stand for the homotopy fibers of α and β respectively. There is an embedding of $X \vee X$ into $E(\alpha \star \beta)$ as $X \vee \{*\}$ embeds into $X \times B$ which embeds into $E(\alpha \star \beta)$ and $\{*\} \vee X$ embeds into $A \times X$ which embeds into $E(\alpha \star \beta)$. This induces homotopy sections $\Omega(X \times X) \longrightarrow \Omega(E(\alpha \star \beta))$ and $\Omega(E(\alpha \star \beta)) \longrightarrow \Omega(F_\alpha \star F_\beta)$ in the fibration $\Omega(F_\alpha \star F_\beta) \longrightarrow \Omega(E(\alpha \star \beta)) \longrightarrow \Omega(X \times X)$. Given a map $h : \Sigma P \longrightarrow M(f,g)$ its adjoint $P \longrightarrow \Omega M(f,g)$ is part of the composition $P \longrightarrow \Omega M(f,g) \longrightarrow \Omega(E(\alpha \star \beta)) \longrightarrow \Omega(F_\alpha \star F_\beta)$ whose adjoint $\Sigma P \longrightarrow (F_\alpha \star F_\beta)$ is the Hopf invariant of h as defined by Marcum.

The GHI as a particular case

In the particular case that B equals the one point space $\{*\}$ then X in $HI12a$ and $M(f,g)$ both become the mapping cone of f, C_f. The space $E(\alpha \star \beta)$ becomes the union of the mapping cylinder of f, M_f, and $A \times C_f$, where the union is over A of the mapping cylinder and over $A \times \{*\}$ of the product. F_β becomes $\Omega(C_f)$, while the inclusion map $\Sigma F_\alpha \hookrightarrow C_\alpha$ has ΣC as the range and has an adjoint $F_\alpha \longrightarrow \Omega \Sigma C$. Thus there is a map $F_\alpha \star F_\beta \longrightarrow \Omega \Sigma C \star \Omega(C_f)$. Given a map $h : \Sigma P \longrightarrow C_f$ its Hopf invariant as defined by Marcum, is a map $\Sigma P \longrightarrow (F_\alpha \star F_\beta)$ whose composition with the map $F_\alpha \star F_\beta \longrightarrow \Omega \Sigma C \star \Omega(C_f)$ is the Ganea Hopf invariant.

HI13-Hopf Invariant as defined in a thesis by Dula

This section will review the definition of the Hopf invariant given in [D].

The first step- The GHI as a homotopy obstruction

Given a class $f : Q \longrightarrow B$ from a cogroup (Q, μ) to a space with a coaction (B, ρ), the following diagram can be formed

$$\text{(HI13a)} \qquad \begin{array}{ccc} Q & \xrightarrow{\quad f \quad} & B \\ {\scriptstyle \mu}\big\downarrow & & \big\downarrow{\scriptstyle \rho} \\ Q \vee Q & \xrightarrow{(i_2 \circ f) \vee (i_1 \circ \pi_1 \circ \rho \circ f)} & A \vee B, \end{array}$$

in which $\pi_1 : A \vee B \twoheadrightarrow A$ is the projection induced by pinching B to the basepoint, and $i_1 : A \hookrightarrow A \vee B$ is the inclusion. It is not necessarily true that diagram $(HI13a)$ homotopy commutes and the element $\Delta(f) = \rho \circ f - (i_1 \circ \pi_1 \circ \rho \circ f) - (i_2 \circ f)$, called in [D, I, §2] the image of f under the *deviation operator*, measures the obstruction from commutativity. There is an asymmetry in the last presentation, which can be removed by plugging the class $\pi_2 \circ \rho \circ f$ instead of that of f. Then $\Delta(f)$ becomes equal to $\rho \circ f - (i_1 \circ \pi_1 \circ \rho \circ f) - (i_2 \circ \pi_2 \circ \rho \circ f)$. $\Delta(f)$ is a class $Q \longrightarrow A \vee B$ which can be seen to become the zero class after composing with the inclusion class $A \vee B \hookrightarrow A \times B$. Therefore there is a factorization, $HI(f)$, of $\Delta(f)$ (that can be shown to be unique) through the inclusion of the homotopy fiber $w : A \flat B \longrightarrow A \vee B$. $w_\sharp$ can be seen to be a monomorphic operator, and hence the equality between $\Delta(f)$ and $w \circ HI(f)$ implies that $(HI13a)$ commutes if and only if $HI(f)$ is the zero class. $HI(f)$ is a particular case of the Hopf invariant of f as defined in [D, I, §2].

For spaces A and B that are of the homotopy type of countable CW complexes, it can be shown that the homotopy fiber $A \flat B$ is homotopy equivalent to $\Sigma\Omega A \wedge \Omega B$. Given that Q is a suspension ΣP that B is the mapping cone of a class $g : A \longrightarrow X$ and that $\rho : X \cup_g CA \longrightarrow (\Sigma A) \vee (X \cup_g CA)$ is induced by pinching the subspace $A \times \{1/2\}$ to the basepoint of the reduced cone, then it follows that the definition of $HI(f)$ of [D, I, §2] specializes to that of Ganea [Ga].

Given the category of spaces with coactions $\mathcal{CA}$, and the subcategory of cogroups $\mathcal{CG}$, there are two functors $\mathcal{F}_1$ and $\mathcal{F}_2$ from $\mathcal{CG} \times \mathcal{CA}$ to the category of groups, $\mathcal{G}$, where $\mathcal{F}_1 : \mathcal{CG} \times \mathcal{CA} \longrightarrow \mathcal{G}$ maps a pair (Q, B) to the homotopy group $[Q, B]$, and $\mathcal{F}_2 : \mathcal{CG} \times \mathcal{CA} \longrightarrow \mathcal{G}$ maps a pair (Q, B) to the homotopy group $[Q, A \flat B]$. Then it is shown that HI is a natural transformation $\mathcal{F}_1 \longrightarrow \mathcal{F}_2$.

The second step- Most general definition

The most general definition of the Hopf invariant in [D] comes ¿from the observation that the role played by the coaction ρ in the previous analysis was minor. A similar construction can be carried out, when a coaction is replaced by a *split map*, which is any map $\rho : R \longrightarrow R_1 \vee R_2 \vee \cdots \vee R_n$ into a finite bouquet of spaces. Then given a class $f : Q \longrightarrow R$, the following diagram can be formed:

$$\text{(HI13b)} \qquad \begin{array}{ccc} Q & \xrightarrow{\quad f \quad} & R \\ {\scriptstyle \mu}\big\downarrow & & \big\downarrow{\scriptstyle \rho} \\ Q \vee Q \vee \cdots \vee Q & \xrightarrow{(i_n \circ \pi_n \circ \rho \circ f) \vee \cdots \vee (i_1 \circ \pi_1 \circ \rho \circ f)} & R_1 \vee R_2 \vee \cdots \vee R_n, \end{array}$$

and a *deviation operator* can be formed, whose value on f is the class $\rho \circ f - (i_1 \circ \pi_1 \circ \rho \circ f) - (i_2 \circ \pi_2 \circ \rho \circ f) - \cdots - (i_n \circ \pi_n \circ \rho \circ f)$. $\Delta(f)$ is a class $Q \longrightarrow R_1 \vee R_2 \vee \cdots \vee R_n$ which can be seen to become the zero class after composing with the inclusion class $R_1 \vee R_2 \vee \cdots \vee R_n \hookrightarrow R_1 \times R_2 \times \cdots \times R_n$. Therefore there is a factorization, $HI(f)$, of $\Delta(f)$ (that can be shown to be unique) through the inclusion of the homotopy fiber $w : F \longrightarrow R_1 \vee R_2 \vee \cdots \vee R_n$. $w_\sharp$ can be seen to be a monomorphic operator, and hence the equality between $\Delta(f)$ and $w \circ HI(f)$ implies that $(HI13b)$ commutes if and only if $HI(f)$ is the zero class. $HI(f)$ is the Hopf invariant of f as defined in [**D**, I, §2]. It can be seen to be a natural transformation between two functors. In the case that the split map is a coaction map, then HI specializes to the definition given above, and also to the GHI.

Relation with the Walker's Hopf invariant

The generalization of the Hopf invariant related to diagram $(HI13b)$ has the following two applications.

(i) The calculation of $GHI(h_n)$, mentioned in [**HI10**, III] simplifies using a split (non coaction) map.

(ii) There is a relation between a particular case of the Hopf invariant as defined by Walker [**Wlk**], and a particular case of the Hopf invariant defined above.

HI14-Hopf Invariant as defined by Koschorke and Sanderson

In the paper of Koschorke and Sanderson [**KS**], there is a relation between Hopf Invariants and bordism sets of imersions of manifolds. Namely, for a given manifold without boundary V, and a given vector bundle ξ over B, there is a set $\mathfrak{J}^k_m(V,\xi)$ defined as follows: $B_{m,k}$ is the space of maps ϕ from subsets A of $\mathbf{R}^m$ of k elements into B. There is a map from $B_{m,k}$ into the kth cartesian power of $B \times \mathbf{R}^m$, and this map pulls the direct sum of k copies of $\xi \oplus t$, where t is a trivial m bundle, back to $B_{m,k}$ producing a bundle $\xi_{m,k}$ over $B_{n,k}$. There is a forgetful map from $B_{m,k}$ to $C_{m,k}$, where the latter is the space of all subsets of $\mathbf{R}^m$ with k elements, sending a map ϕ to its domain. Koschorke and Sanderson consider the space of all tiples $(M, g, \overline{g})$, such that M is a closed manifold, $g = (g_1, g_2)$ is an embedding of M into $V \times C_{m,k}$ such that $g_1 : M \longrightarrow V$ is an imersion, ν is the normal bundle of the imersion g_1, $\overline{g}$ is a map of bundles from ν to $\xi_{m,k}$, and the following diagram commutes:

$$\begin{array}{ccccc}
 & & \nu & \xrightarrow{\overline{g}} & \xi_{m,k} \\
 & & \downarrow & & \downarrow \\
V & \xleftarrow{g_1} & M & \longrightarrow & B_{m,k} \\
 & & {\scriptstyle g_2}\downarrow & & \downarrow \\
 & & C_{m,k} & \xrightarrow{=} & C_{m,k}.
\end{array}$$

Two such diagrams are considered bordant, if the images under g_1 are bordant over $V \times I$. The set of bordism class is denoted $\mathfrak{J}^k_m(V,\xi)$. This is the 'geometry' side.

On the other hand there is an 'algebra' side. Let V_c denote the one point compactification of V, and let $T(\xi)$ be the Thom space of ξ. $C^k_m(X)$ denotes the space of all Φ, such that Φ is a finite subset of functions ϕ as above, with the extra property that each ϕ is into $X \setminus \{*\}$. The following is the main theorem of [**KS**]:

THEOREM. *There is a bijection* $\beta : \mathfrak{J}^k_m(V,\xi) \longrightarrow [V_c, C^k_m(T(\xi))]$.

In the particular case $k = 1$ theorem 2 states that β fits into the following diagram:

$$\begin{array}{ccc} \mathfrak{d}_m(V,\xi) & \longrightarrow & \mathfrak{d}_0(V\times \mathbf{R}^m, \xi\oplus t) \\ {\scriptstyle\beta}\downarrow & & {\scriptstyle\beta}\downarrow \\ [V_c, C_m(T(\xi))] & \longrightarrow & [\Sigma^m V_c, \Sigma^m(T(\xi))]. \end{array}$$

Using the geometrical side the authors produce operations $\psi^k_m : \mathfrak{d}_m(V,\xi) \longrightarrow \mathfrak{d}^k_m(V,\xi)$. Those have the properties that ψ^1_m is the identity, all ψ's are zero on a 'suspension' elements, and the suspensions of the ψ obey a Cartan furmula. Specializing to certain cases and moving to the algebra side, gives various generalizations of the Hopf invariants discussed by Snaith [**Sn**], Segal [**Se**] and James [**J2**].

§4. Construction X_k

Xk1-Introduction

Construction Xk, reviewed in this section, can be viewed (at least in some cases) as the Thom space of a spherical fibration over the conic space $J_k(S^q)$, which is a finite filtration of the James model mentioned in HI2 above. The review is more important as some of the works reviewed are unpublished.

Xk2-The work of M.G.Barratt

In an unpublished paper, [**B5**], M.G.Barratt had a construction on a mapping cone X of a map $f : S^p \longrightarrow S^q$. The construction uses $X\wedge X$ which is described by the schematic diagram:

$$\begin{array}{ccc} e^{p+1}\wedge e^{p+1} & \xrightarrow{\partial f\wedge \underline{e^{p+1}}} & S^q\wedge e^{p+1} \\ {\scriptstyle \underline{e^{p+1}}\wedge\partial f}\downarrow & & \downarrow{\scriptstyle \underline{S^q}\wedge\partial f} \\ e^{p+1}\wedge S^q & \xrightarrow{\partial f\wedge \underline{S^q}} & S^q\wedge S^q, \end{array}$$

in which the four spaces present the four cells of $X\wedge X$, and the four maps present the relative attaching maps from the boundary of a cell to the lower skeleton. This is denoted in the diagram by having the sign ∂ in front of a map. It can be observed that there exists a map $\Delta : S^q\wedge e^{p+1} \longrightarrow e^{p+1}\wedge S^q$, commuting the smash factors, that makes the following diagram homotopy commute, up to the sign $(-1)^{pq}$:

$$\begin{array}{ccc} e^{p+1}\wedge e^{p+1} & \xrightarrow{\partial f\wedge \underline{e^{p+1}}} & S^q\wedge e^{p+1} \\ {\scriptstyle \underline{e^{p+1}}\wedge\partial f}\downarrow & \Delta\swarrow & \downarrow{\scriptstyle \underline{S^q}\wedge\partial f} \\ e^{p+1}\wedge S^q & \xrightarrow{\partial f\wedge \underline{S^q}} & S^q\wedge S^q, \end{array}$$

using the map Δ, X_2 can be defined as a quotient of $X\wedge X$ in the following way: points of $S^q\wedge e^{p+1}$ are identified with their images under Δ in $e^{p+1}\wedge S^q$. Points in the cells $S^q\wedge S^q$ and $e^{p+1}\wedge e^{p+1}$ identify with their image under a map of degree $(-1)^{pq}$. In the particular case that $(-1)^{pq} = 1$, the top and bottom cells are not altered by identifications, and X_2 has three cells, $e^{p+1}\wedge e^{p+1}, e^{p+1}\wedge S^q$and $S^q\wedge S^q$. Similar constructions give X_k as a quotient of $X^{\wedge k}$. X_k can be defined for more complicated spaces X.

Xk3-The work of B.Gray

In [**Gr2,3**] the previous construction was applied for the following question. Given a map $f : S^p \longrightarrow S^q$, and its mapping cone X, is X an 'q-ringed space'? This is a space X, an inclusion map $S^q \hookrightarrow X$, and a projection map $\mu : X \wedge X \longrightarrow S^q \wedge X$ such that the following diagrams commute:

$$\begin{array}{ccc} S^q \wedge X & = & S^q \wedge X \\ {\scriptstyle i\wedge \underline{X}}\downarrow & & \| \\ X \wedge X & \xrightarrow{\mu} & S^q \wedge X, \end{array}$$

$$\begin{array}{ccc} X \wedge S^q & \xrightarrow{(-1)^q T} & S^q \wedge X \\ {\scriptstyle i\wedge \underline{X}}\downarrow & & \| \\ X \wedge X & \xrightarrow{\mu} & S^q \wedge X, \end{array}$$

where T is the twisting map.

The existence of that structure on X guarantees that the suspension spectrum of X is a ring spectrum. There is an element r(f) in a quotient of the group $\pi_{2p+1}(S^q)$ with the property that $0 \in$ r(f) if and only if X is a q ringed space. This element appeared in **[B4]** and in **[T, page 30]**. Other elements are defined in **[Gr2,3]**, which measure the obstruction of the q-ringed space from being associative and commutative.

The relation between the work of Gray and of Barratt can be seen in the following form: for a given $f : S^p \longrightarrow S^q$, and its mapping cone X, (X_2 of Xk2) is constructed. Then the element r(f) measures the obstruction to the fact that the top cell of X_2 attaches trivially to the lower skeleton. The Lower skeleton is homotopic to $S^q \wedge X$, if $(-1)^{pq} = 1$. If indeed r(f)=0, there is a pinch map

$$\pi : X_2 \sim (S^q \wedge X) \vee (\text{top cell}) \longrightarrow S^q \wedge X,$$

pinching the top cell. In this case μ in the definition of q-ringed space can be obtained as the composition

$$X \wedge X \xrightarrow{\text{quotient map}} X_2 \xrightarrow{\pi} S^q \wedge X.$$

Xk4-Work of Komornicki

Komornicki in **[K]** used the work of Gray to consider the following question: $\alpha : S^{n+k} \longrightarrow S^n$ is given with mapping cone $C = S^n \cup_\alpha e^{n+k+1}$. E is a given ring spectrum. Then $C \wedge E$ is a spectrum with $(C \wedge E)_m = C \wedge (E_{m-n})$. It is a left and right module over E, with unit map $E \longrightarrow C \wedge E$, induced by the inclusion of S^n into C. What is the obstruction for making $C \wedge E$ a ring spectrum which is an algebra over E? **[K]** describes an element $\tilde{\alpha}_*$ in $E_{n+2k+1}(C)$, the $n + 2k + 1^{st}$ E homology of the spectrum C, which carries the obstruction. The construction of $\tilde{\alpha}_*$ uses the Barratt-Mahowald construction X_k. **[K]** also gives obstructions for the ring $C \wedge E$ from being associative and commutative, following the work of Gray.

Xk5-The work of M.Mahowald

In **[Mah]**, M.Mahowald considered ring spectra which are Thom complexes. L is a space and ξ is a spherical fibration over L classified by a map $f : L \longrightarrow BF$, where F is $\lim_{n\to\infty} F_n$, and F_n is the monoid of self maps of S^{n-1}. The Thom spectrum of f, T(f), can be made into a suspension spectrum of the Thom space of $f_n : L^n \longrightarrow BF_n$, f restricted to the n skeleton. Theorem 1.1 states that T(f) is a ring spectrum, when given that L is an H space and f is an H map. Theorem 1.2 states that $T(f) \wedge T(f) = L_+ \wedge T(f)$, where L_+ is the disjoint union of L and a point, when given that f is a loop map from the loop space L. In later sections some examples are reviewed, and theorem 1.2 is used for T(f) resolutions.

The relation between [**Mah**] and [**B5**] is as follows: Let $\alpha : S^m \longrightarrow BF(n)$ be given with J map $f = J(\alpha) : S^{m+n} \longrightarrow S^n$. Then the mapping cone of f, X, is the Thom complex of the fibration induced by α. Assume further that a map $\beta : S^{m+1} \longrightarrow B^2F(n)$ is given such that the composition $S^m \overset{i}{\hookrightarrow} \Omega\Sigma S^m \overset{\Omega\beta}{\longrightarrow} \Omega B^2F(n) \sim BF(n)$ is homotopic to α. Then the Thom spaces of the induced fibrations $\alpha^*(\xi)$ over S^n and $(\Omega\beta)^*(\xi)$ over $\Omega\Sigma S^m$, the spaces $T(\alpha)$ (which is X) and $T(\Omega\beta)$, are related in the following form: $T(\Omega\beta)$ is the X_∞ construction (as in Xk2) of $T(\alpha)$ (which is X). The last statement appears in another work of B.Gray.

Xk6-More work of B.Gray

In [**Gr1**], B.Gray had taken a functorial point of view. Let $\mathcal{C}, \mathcal{C}_1$ be two categories, such that $\mathcal{C}$ is closed under pushouts. There is an associative bifunctor $\times : \mathcal{C} \times \mathcal{C} \longrightarrow \mathcal{C}$ called product (although it need not be the categorial product). There is a natural transformation $\sqcup : \times \longrightarrow \times$, sending A×B to B×A. There are functors, $\mathcal{T} : \mathcal{C}_1 \longrightarrow \mathcal{C}$ called the terminal functor, and $\mathcal{U} : \mathcal{C}_1 \longrightarrow \mathcal{C}$ called the unit functor, and a natural transformation $\eta : \mathcal{U} \longrightarrow \mathcal{T}$, such that the following diagram commutes:

$$\begin{array}{ccc} \mathcal{U} \times \mathcal{U} & = & \mathcal{U} \times \mathcal{U} \\ {\scriptstyle id\times\eta}\downarrow & & \downarrow{\scriptstyle \eta\times id} \\ \mathcal{U} \times \mathcal{T} & \overset{\sqcup}{\longrightarrow} & \mathcal{T} \times \mathcal{U}. \end{array}$$

Then [**Gr1**] proves that there exists a functor $\mathcal{T}_\infty : \mathcal{C}_1 \longrightarrow \mathcal{C}$, such that $\mathcal{T}_\infty(X)$ is a free object in $\mathcal{C}$, for every object X in $\mathcal{C}_1$, and an 'inclusion' transformation $inc : \mathcal{T} \longrightarrow \mathcal{T}_\infty$ such that every morphism $\lambda : \mathcal{T}(X) \longrightarrow F$ into a free object factors through inc. $\mathcal{T}_\infty$ can be given as $lim_{n\to\infty}\mathcal{T}_n$, where $\mathcal{T}_n(X)$ is a 'quotient' of $X^{\times n}$, the nth cartesian power of X under $\times$.

James' model is the particular case of this construction of Gray, in which $\mathcal{C}_1$ is the category of based topological spaces which are homotopy equivalent to countable CW complexes, $\mathcal{C}$ equals $\mathcal{C}_1$, $\mathcal{T}$ is the identity functor, $\mathcal{U}$ is the functor sending each space to its basepoint, $\eta(X)$ is the transformation including the one point space $\{*\}$ in X, $\times$ is the cartesian product and $\sqcup$ is the natural homeomorphism of $A \times B$ with $B \times A$.

Barratt-Mahowald construction X_k is another particular case carried out in the category of inclusion maps $S^n \hookrightarrow X$, where n is any natural number and X has the homotopy type of a countable CW complex, $\mathcal{T}$ sends an arrow to its range, $\mathcal{U}$ sends an element to its domain, and $\eta(\xi)$ for $\xi : S^n \overset{i}{\hookrightarrow} X$ is the map i, ¿from the domain $\mathcal{U}(\xi)$ to the range $\mathcal{T}(\xi)$. The product $\times$ is the smash product $\wedge$, and $\sqcup$ is the obvious twisting map.

The last important result of [**Gr1**] is the following; given a functor

$$\mathcal{F} : (\mathcal{C}_1, \mathcal{C}, \times, \mathcal{T}, \mathcal{U}, \eta, \sqcup) \longrightarrow (\mathcal{C}'_1, \mathcal{C}', \times', \mathcal{T}', \mathcal{U}', \eta', \sqcup')$$

that commutes with the structure, it induces a commutative diagram:

$$\begin{array}{ccc} \mathcal{T} & \overset{inc}{\longrightarrow} & \mathcal{T}_\infty \\ {\scriptstyle \mathcal{F}}\downarrow & & \downarrow{\scriptstyle \mathcal{F}} \\ \mathcal{T}' & \overset{inc'}{\longrightarrow} & \mathcal{T}'_\infty. \end{array}$$

The following exemplifies the last theorem: let $\mathcal{C}_1$ be the category of diagrams $\xi : \{*\} \overset{i}{\hookrightarrow} X \overset{f}{\longrightarrow} BO$, of stable orthogonal bundles over the based space X. Then $\mathcal{T}_\infty(\xi)$ is the diagram $\{*\} \overset{i}{\hookrightarrow} JX \overset{f}{\longrightarrow} BO$, an element in the space consisting of stable orthogonal bundles over the

James model for $\Omega\Sigma X$. Let $\mathcal{F}$ be the functor sending a stable bundle to its Thom spectra. Then the Thom spectrum over JX, $\mathcal{F}(T_\infty(\xi))$, is obtained from the one over X, $T'(\mathcal{F}(\xi))$, by means of the Barratt-Mahowald construction X_k, where the role of the sphere is taken by the Thom class, the sphere spectrum over a point, $\mathcal{F}(\mathcal{U}(\xi))$.

Xk7-Ravenel's $X(n)$

In **[R]**, the ring spectra $X(n)$ were defined. Let $\Omega SU(n)$ map to BU by the composition $\vartheta : \Omega SU(n) \longrightarrow \Omega SU \longrightarrow BU$, where the first map is the inclusion and the second exists by Bott periodicity theorem. ϑ pulls the stable bundle back to $\Omega SU(n)$ and the Thom spectrum of this bundle is defined to be $X(n)$. **[R]** studied some properties of $X(n)$ and so did **[Hop]**. These spectra play an important role in the proof of the nilpotency theorem **[DHS]**, and so do the spectra $X(n,k)$.

$X(n,k)$ are defined as follows: there is a fibration $\xi : \Omega SU(n-1) \longrightarrow \Omega SU(n) \longrightarrow \Omega S^{2n-1}$, with base homotopy equivalent to the James model JS^{2n-2}. The inclusion of the kth filtration $J_kS^{2n-2} \longrightarrow JS^{2n-2}$ pulls ξ back, giving a commutative diagram:

$$\begin{array}{ccccc} \Omega SU(n-1) & \longrightarrow & \Omega SU(n) & \longrightarrow & \Omega S^{2n-1} \\ \| & & \uparrow{\scriptstyle\alpha} & & \uparrow \\ \Omega SU(n-1) & \longrightarrow & E_{n,k} & \longrightarrow & J_kS^{2n-2}, \end{array}$$

and the composition map $\vartheta \circ \alpha : E_{n,k} \longrightarrow \Omega SU(n) \longrightarrow BU$ pulls the stable bundle back to $E_{n,k}$. The Thom spectrum of the pull back is by definition $X(n,k)$.

Xk8-Relation to the thesis of Dula

$X(n,k)$ is easily seen to be a double suspension spectrum. The lth space in the spectrum $X(n,k)$, $X(n,k)_l$, is obtained using the following diagram:

$$\begin{array}{ccccc} E_{n,k,l} & \longrightarrow & \Omega SU(n)_l & \longrightarrow & BU(l) \\ \downarrow & & \downarrow & & \downarrow \\ E_{n,k} & \xrightarrow{\alpha} & \Omega SU(n) & \xrightarrow{\vartheta} & BU, \end{array}$$

in which the inclusion $BU(l) \hookrightarrow BU$ pulls back the fibration $\vartheta \circ \alpha$. The universal $l-$ plane bundle over $BU(l)$ pulls back to a bundle η over $E_{n,k,l}$ that has fiber R^{2l}. The Thom space of η is $X(n,k)_l$. The base of $\eta, E_{n,k,l}$, is a total space of a fibration λ with projection map $E_{n,k,l} \longrightarrow E_{n,k} \longrightarrow J_kS^{2n-2}$, such that η followed by λ is an iterated fibration in the sense of [D] with a conic subbase J_kS^{2n-2}. Thus by [D], $X(n,k)_l$ has a conic presentation as follows:

$$(\ldots(X(n-1)_l \cup C(S^{2n-3} \wedge X(n-1)_l) \cup \cdots \cup C(S^{2k(n-2)-1} \wedge X(n-1)_l).$$

The map $BU(l) \longrightarrow BU(l+1)$ induces on the Thom space level, a map $\Sigma^2 X(n,k)_l \longrightarrow X(n,k)_{l+1}$. The limit on l is the spectrum $X(n,k)$ that has the following structure in the stable category:

$$(\ldots(X(n-1) \cup C(S^{2n-3} \wedge X(n-1)) \cup \cdots \cup C(S^{2k(n-2)-1} \wedge X(n-1)).$$

The properties $X(n,0) = X(n-1)$ and $X(n,\infty) = X(n)$ follow from the definitions or from the last conic presentation. Thus $X(n,k)$ form an increasing (conic in the stable category) filtration of the pair $(X(n), X(n-1))$.

It also follows from [D, II, §4] that in the conic space $X(n,k)_l$, the relative attaching map of the top cell

$$S^{2k(n-2)} \wedge X(n-1)_l \longrightarrow S^{2(k-1)(n-2)} \wedge X(n-1)_l$$

is a k fold sum of an appropriate suspension of the map attaching the second cell to the Thom class

$$S^{2(n-2)} \wedge X(n-1)_l \longrightarrow X(n-1)_l.$$

As the property of being k fold sum is preserved in direct limits, the statement holds in $X(n,k)$ and in $X(n)$.

Xk9-Relation between the works of Gray and Ravenel

It follows that $X(n,k)$ is the Barratt-Mahowald construction X_k in the stable category of $X(n,1)$. In particular $X(n)$ is $X_\infty(X(n,1))$.

References

Ada. J.F.Adams, *On the Non-Existence of Elements with Hopf Invariant One*, . Ann. of Math. **72** (1960), 20–104.

Ade. J.Adem, *The Iteration of Steenrod Squares in Algebraic Topology*, Proc. Natl. Acad. Sci. **38** (1952), 720–726.

At. M.F.Atiyah, *Thom Complexes*, Proc. Lond. Math. Soc. **11** (1961), 291–310.

BB. W.D.Barcus and M.G.Barratt, *On the Homotopy Classification of the Extensions of a Fixed Map*, Trans. Amer Math. Soc. **88** (1957), 57–74.

B1. M.G.Barratt, *Track Groups I,II*, Proc. Lond. Math. Soc. **5** (1955), 71–106, 285–329.

B2. ———, *Higher Hopf Invariants*, Mimeographed notes, University of Chicago.

B3. ———, *Remark on James' invariants*, "Conference on Algebraic Topology," Aarhus University 1962, pp. 102–103.

B4. ———, *Spaces of Finite Characteristic*, Quat. Jour. of Math. **11** (1960), 124–136.

B5. ———, *On the Construction X_k*, Unpublished.

B6. ———, *Twisted Lie Algebras*, "Geometric applications of homotopy theory II (Northwestern 1977) M.G.Barratt and M.E.Mahowald edtrs," Lect. Notes in Math. 658, pp. 9–15.

B7. ———, *Taming Hopf Invariants*, "Combinatorial methods in topology and algebraic geometry (Rochester 1978) J.R.Harper and R.Mandelbaum edtrs," Contemporary mathematics, Amer. Math. Soc. vol 44, pp. 201–205.

BJM. M.G.Barratt, J.D.S.Jones and M.E.Mahowald, *The Kervaire Invariant and the Hopf Invariant*, "Algebraic Topology Proceedings(Seattle 1985)H.Miller and D.C.Ravenel edtrs," Lect. Notes in Math. 1286, pp. 135–173.

BH. I.Berstein and P.J.Hilton, *On Suspensions and Comultiplications*, Topology **2** (1963), 73–82.

BS. J.M.Boardman and B.Steer, *On Hopf Invariants*, Comm. Math. Helv. **42** (1967), 180–221.

CDGM. F.R.Cohen D.M.Davis P.G.Goerss and M.E.Mahowald, *Integral Brown-Gitler spectra*, Proc. Amer. Math. Soc. **103** (1988), 1299–1304.

DHS. E.S.Devinatz M.J.Hopkins and J.H.Smith, *Nilpotence and Stable Homotopy theory I*, Ann. of Math. **128** (1988), 207–241.

D. G.Dula, *The Conic Structure of Thom Spaces*, Thesis, Northwestern University, Aug. 1988.

FHLT. Y.Félix, S.Halperin, J.M.Lemaire and J.C.Thomas, *Mod p Loop Spaces Homology*, Inv. Math. **95** (1989), 247–262.

Ga. T.Ganea, *On the Homotopy Suspension*, Comm. Math. Helv. **43** (1968), 225–234.

Gr1. B.Gray, *Ring Spectra and the James Construction*, Mimeographed Notes, 1968.

Gr2. ———, *Operations and Two Cell Complexes*, "Conference on Algebraic Topology, V.Gugenheim edtr," University of Illinois at Chicago Circle,1968, pp. 61–68.

Gr3. ———, *Ring Spectra*, "Conference on Algebraic Topology," Aarhus University 1970, pp. 143–153.

Ha1. C.H.Hanks, *On the Hopf Construction Associated to a Composition*, Thesis, Northwestern University, Aug. 1975.

Ha2. ———, *On the Hopf Construction Associated with a Composition*, "Geometric Applications of Homotopy Theory II, M.G.Barratt and M.E.Mahowald edtrs," Lect. Notes in Math. 658, pp. 191–205.

HS. R.P.Held and D.K.Sjerve, *On the Homotopy Properties of Thom Complexes*, Math. Zeit. **135** (1974), 315–323.

Hi1. P.J.Hilton, *On the Homotopy Groups of the Union of Spheres*, Jour. Lond. Math. Soc. **30** (1955), 154–172.

Hi2. ———, "Homotopy Theory and Duality," Notes in Mathematics and its Applications, Gordon and Breach, 1965.

Ho1. H.Hopf, *Über die Abbildungen der Dreidimensionalen Sphare auf die Kugelflache*, Math. Ann. **104** (1931), 639–665

Ho2. ———, *Über die Abbildungen von Spharen auf Spharen Niedriger dimension*, Fund. Math. **25** (1935), 427–440.

Hop. M.J.Hopkins, Thesis, Northwestern University, Aug. 1985.

J1. I.M.James, *Reduced Product Spaces*, Ann. of Math. **62** (1955), 170–197.

J2. ———, *On the Suspension Triad*, Ann. of Math. **63** (1956), 191–247.

K. W.Komornicki, *Multiplication in two-cell spectra*, "Geometric Applications of Homotopy Theory II, M.G.Barratt and M.E.Mahowald edtrs," Lect. Notes in Math. 658, pp. 215–223.

KS. U. Koschorke and B. Sanderson, *Self intersections and higher Hopf Invariants*, Topology (1978), 283–290.

Mah. M.E.Mahowald, *Ring Spectra which are Thom Complexes*, Duke Math. Jour. **46** (1979), 549–559.

Mar1. H.J.Marcum, *Fibrations over double mapping cylinders*, Ill. jour. of Math. **24** (1980), 344–358.

Mar2. ———, *Homotopy decompositions for product spaces*, "Atas do Décimo Primero Colóquio Brasileiro de Matemática," IMPA, Rio de Janeiro, 1978, pp. 665–680.

Mar3. ———, *Functional Properties of Hopf Invariant*, to appear in Quaestiones Mathematicae.

May1. J.P. May, "The Geometry of Iterated Loop Spaces," Springer Verlag, Lecture Notes in Mathematics, 1972.

May2. ———, *The dual Whitehead theorems*, "Topological topics, article presented to P.J.Hilton on his sixtieth birthday," London Mathematical Society, Lecture Notes in Mathematics, I.M.James edtr., 1983, pp. 46–64.

Mi1. J.W.Milnor, *On the Construction FK*, Mimeographed Lecture Notes, Princeton University, 1956.

Mi2. ———, *Construction of Universal Bundles II*, Ann. of Math. **63** (1956), 430–436.

P. G.J.Porter, *Higher Order Whitehead Products*, Topology **3** (1965), 123–165.

R. D.C.Ravenel, *Localization with Respect to Certain Periodic Homology Theories*, Amer. Jour. of Math. **106** (1984) 351–414.

S. N.E.Steenrod, *Cohomology Invariants of Mappings*, Ann. of Math. **50** (1949), 954–988.

T. H.Toda, "Composition Methods in the Homotopy Groups of Spheres," Annalls of Math. Studies, Princeton University, 1962.

Wlk. M.Walker, *Homotopy Pullbacks and the Hopf Invariant*, Jour. Lond. Math. Soc. **19** (1979), 153–158.

Wll. C.T.C. Wall, *Poincaré Complexes I*, Ann. of Math. **86** (1967), 213–245.

WG1. G.W. Whitehead, *On the Homotopy Groups of Spheres and Rotation Groups*, Ann. of Math. **43** (1942), 634–640.

WG2. ———, *A Generalization of the Hopf Invariant*, Ann. of Math. **51** (1950), 192–237.

WG3. ———, *On the Freüdenthal Theorems*, Ann. of Math. **57** (1953), 209–228.

Keywords. conic spaces, Thom spaces, Hopf Invariants, construction X_k

1980 *Mathematics subject classifications*: 55-02,55Q25,55Q50,55R65

Computations of Stable Pseudoisotopy Spaces for Aspherical Manifolds

by

F. T. Farrell and L. E. Jones

Let $\mathcal{P}(X)$ denote the semi-simplical space of stable pseudoisotopies having compact support in the topological space X. Let $\mathcal{P}_*(S^1)$ denote the stable pseudoisotopy spectrum of the circle. Results obtained by the authors during the last four years show that for many spaces X the homotopy type of $\mathcal{P}(X)$ can be computed in a simple way from $\mathcal{P}_*(S^1)$. The purpose of this paper is to give a survey of such results, and to formulate a general conjecture along the same lines.

Here is an outline of the paper.

In §1 we review some basic definitions and results from pseudoisotopy theory.

In §2 we give a simple formula for the homotopy type of $\mathcal{P}(M)$ in terms of the stable pseudoisotopy spectrum $\mathcal{P}_*(S^1)$, for M equal any compact closed Riemannian manifold with sectional curvature $K \leq 0$.

In §3 we use the formula of §2 to obtain information about the homotopy groups $\pi_j(\mathcal{P}_k(M))$, the algebraic K-groups $K_i(Z\pi_1 M)$, and the Whitehead groups $Wh_i(\pi_1 M)$, for M as in §2.

In §4 we conjecture that a suitable homotopy version of the formula given for $\mathcal{P}(M)$ in §2 should give the homotopy type of $\mathcal{P}(X)$ for any connected aspherical space X having torsion free fundamental group. (Note that M of §2 is aspherical with torsion free $\pi_1 M$.)

§1. Pseudoisotopies and Homology Theory

Let M denote a compact manifold. The <u>space of pseudoisotopies</u> of M, denoted P(M), consists of all homomorphisms $h : M\times[0,1] \to M\times[0,1]$ which are the identity on $M \times 0$. Note, if I^q denotes the q-fold Cartesian product of [0,1] with itself, then there is an inclusion map $P(M\times I^q) \to P(M\times I^{q+1})$ gotten by forming the product of $h : (M\times I^q) \times [0,1] \to (M\times I^q) \times [0,1]$ with the identity map $I \to I$. The direct limit space $\lim_{q \to \infty} P(M\times I^q)$ is called the <u>space of stable pseudoisotopies</u> of M and is denoted by $\mathcal{P}(M)$.

There is the following simple method for delooping the space $\mathcal{P}(M)$ infinitely often. Let R^k be k-dimensional Euclidean space, and let

$P^b(M\times R^k)$ denote the space of bounded pseudoisotopies on $M \times R^k$. (Recall that a pseudoisotopy $h : (M\times R^k) \times [0,1] \to (M\times R^k) \times [0,1]$ is called <u>bounded</u> if there is a number $N > 0$ such that for any $(x,y,t) \in M \times R^k \times [0,1]$ we have that $|y-y'| < N$ where $(x',y',t') = h(x,y,t)$.) There are the usual inclusions $P^b(M\times I^q\times R^k) \to P^b(M\times I^{q+1}\times R^k)$, permiting the definition of $P^b(M\times R^k)$ as the direct limit space $\lim_{q\to\infty} P^b(M\times I^q\times R^k)$. The following lemma is due to A. Hatcher [11].

<u>Lemma 1.1</u>. For any integer $k \geq 1$ there is a natural homotopy equivalence $\Omega P^b(M\times R^k) \simeq P^b(M\times R^{k-1})$.

<u>Definition 1.2</u>. The <u>stable pseudoisotopy spectrum</u> for M, denoted $P_*(M)$, consists of the collection of spaces $\{P_k(M) : k \in Z\}$ defined by $P_k(M) = P^b(M\times R^k)$ for $k \geq 0$ and by $P_k(M) = \Omega^{|k|}P(M)$ for $k < 0$.

Note that Lemma 1.1 assures the $P_*(M)$ is a Ω-spectrum. We remark that by appealing to semi-simplical constructions, the space $P(X)$ and the Ω-spectrum $P_*(X)$ can be constructed for any topological space X (c.f. [11; §1], [15; §5]). In the event that X is not compact, we insist that every n-simplex in each $P_k(X)$, $k \geq 0$, consists of an n-parameter family of stable pseudoisotopies which have compact support in X.

We remind the reader that Ω-spectra are the "coefficients" for generalized homology theories. Let X be a given topological space and let S_* denote a given Ω-spectrum. For each integer j define a space $\mathbb{H}_j(X,S_*)$ to be the direct limit space $\lim_{i\to\infty} \Omega^i(X\times S_{j+i}/X\times s_{j+i})$ where s_{j+i} is the base point of the space S_{j+i}. Note that the collection of spaces $\{\mathbb{H}_j(X,S_*) : j \in Z\}$ is a Ω-spectrum, which we denote by $\mathbb{H}_*(X,S_*)$ and call the <u>homology spectrum</u> for X with coefficients in S_*. Recall that the <u>homology groups</u> $H_j(X,S_*)$ for X with coefficients in S_* are defined to be the homotopy groups $\pi_j(\mathbb{H}_*(X,S_*))$ of the Ω-spectrum $\mathbb{H}_*(X,S_*)$.

There are more complicated versions of generalized homology theories where the Ω-spectrum of coefficients S_* is "stratified and twisted" over the space X (c.f. [13], [15; appendix]). The following version is taken from F. Quinn's paper [15].

<u>Definition 1.3</u>. A mapping $f : E \to X$ is a <u>simplicially stratified fibration</u> if there is a triangulation K for the space X such that the following hold.

(a) For each simplex $e \in K$ the mapping $f : f^{-1}(e-\partial e) \to e - \partial e$

is a fibration.

(b) For each simplex $e \in K$ there is a neighborhood for ∂e in e, denoted by U, and there are deformation retracts $r : U \times [0,1] \to U$ of U onto ∂e, and $r' : f^{-1}(U) \times [0,1] \to f^{-1}(U)$ of $f^{-1}(U)$ onto $f^{-1}(\partial e)$, such that $f(r'(p,t)) = r(f(p),t)$ holds for all $p \in f^{-1}(U)$ and $t \in [0,1]$.

Let K' denote the first barycentric subdivision of the triangulation K given in 1.3. For each integer j define the space $P_j(f)$ to be the quotient space $(\bigcup_{e \in K'} P_j(f^{-1}(e)) \times e)/\approx$, where the equivalence relation $\approx$ simply identifies $P_j(f^{-1}(e')) \times e'$ with its image in $P_j(f^{-1}(e)) \times e$ under the map induced by the inclusion $e' \subset e$, for every pair $e,e' \in K'$ satisfying $e' \subset e$. Note that X may be identified with the subspace $(\bigcup_{e \in K'} I_e \times e)/\approx$ of $P_j(f)$, where I_e is the identity pseudoisotopy in $P_j(f^{-1}(e))$. Note that the collection $P_*(f) = \{P_j(f) : j \in Z\}$ is not in general a Ω-spectrum.

<u>Definition 1.4</u>. The <u>homology spectrum</u> $\mathbb{H}_*(X,P_*(f))$ for X with coefficients in $P_*(f)$ is the collection of spaces $\{\mathbb{H}_j(X,P_*(f)) : j \in Z\}$ defined by

$$\mathbb{H}_j(X,P_*(f)) = \lim_{i \to \infty} \Omega^i(P_{j+i}(f)/X).$$

The <u>k'th homology group</u> $H_k(X,P_*(f))$ for X with coefficients in $P_*(f)$ is defined to be the k'th homotopy group $\pi_k(\mathbb{H}_*(X,P_*(f)))$ of the Ω-spectrum $\mathbb{H}_*(X,P_*(f))$.

We note that the homology spectrum $\mathbb{H}_*(X,P_*(f))$ can be constructed for any map $f : E \to X$ whose range space can be triangulated. However, the following lemma holds only for maps $f : E \to X$ which satisfy 1.3(a)(b) (c.f. [15; appendix]).

<u>Lemma 1.5</u>. Let $f : E \to X$ be a map satisfying 1.3(a)(b). Then the following hold.

(a) The homology spectrum $\mathbb{H}_*(X,P_*(f))$ is independent of the triangulation K.

(b) There is a spectral sequence with $E^2_{i,j} = H_i(X,\pi_j(P_*(f)))$ which abuts to $H_{i+j}(X,P_*(f)))$. Here $\pi_i(P_*(f))$ denotes the stratified and twisted system of groups $\{\pi_i(P_*(f^{-1}(x))) : x \in X\}$ over X, where $\pi_i(P_*(f^{-1}(x)))$ is the i'th homotopy group of the Ω-spectrum $P_*(f^{-1}(x))$.

The inclusion maps $f^{-1}(e) \subset E$, $e \in K'$, induce maps

$P_j(f^{-1}(e)) \to P_j(E)$, $e \in K'$, which inturn induce maps $\phi_j : P_j(f) \to P_j(E)$. Define $A_j : \mathbb{H}_j(X, P_*(f)) \to P_j(E)$ to be the direct limit as $i \to \infty$ of the composite maps

$$\Omega^i(P_{j+i}(f)/X) \xrightarrow{\Omega^i(\phi_{j+i})} \Omega^i(P_{j+i}(E)) \simeq P_j(E) .$$

Note that the collection of all such maps $A_* : \mathbb{H}_*(X, P_*(f)) \to P_*(E)$ is a mapping of Ω-spectra, which is called the assembly map. We note that the assembly map can be defined for any map $f : E \to X$ whose range can be triangulated. However, the following lemma holds only if $f : E \to X$ satisfies 1.3(a)(b) (c.f. [15; appendix]).

Lemma 1.6. If $f : E \to X$ satisfies 1.3(a)(b) then the assembly map $A_* : \mathbb{H}_*(X, P_*(f)) \to P_*(E)$ is independent of the triangulation K .

Remark 1.7. If the map $f : E \to X$ equals the first factor projection map $X \times Y \to X$ for some space Y , then there is a weak equivalence of Ω-spectra $\mathbb{H}_*(X, P_*(f)) \simeq H_*(X, P_*(Y))$. Thus there is an assembly map $A_* : \mathbb{H}_*(X, P_*(Y)) \to P_*(X \times Y)$.

§2. A Simple Formula for $P_*(M)$ In Terms of $P_*(S^1)$

Our goal in this section is to give a simple method for computing the homotopy type of $P(M)$ in terms of the stable pseudoisotopy spectrum $P_*(S^1)$ of the circle, provided that M is a compact closed Riemannian manifold with sectional curvature $K \le 0$ everywhere. We describe the method in the following three examples.

In each of the following examples we will need the following notation and terminology. Recall that a closed geodesic in M is a smooth map $g : S^1 \to M$ from the circle of constant speed such that for all sufficiently small arcs $A \subset S^1$ the distance in M from $g(\partial_- A)$ to $g(\partial_+ M)$ is equal the length of $g : A \to M$. Note that if a closed geodesic $g : S^1 \to M$ is essential (not homotopic to a constant map) then g must be an immersion. A closed geodesic $g : S^1 \to M$ is simple if for any rotation $r : S^1 \to S^1$ (other than the identity) there is a point $y \in S^1$ such that $g(y) \ne g(r(y))$. Two closed geodesics $f, g : S^1 \to M$ are said to be equivalent if there is a diffeomorphism $h : S^1 \to S^1$ such that $g \circ h = f$. We let G denote all equivalence classes of the closed essential simple geodesics in M . We can equip G with a topology by insisting that a sequence $[g_i]$, $i = 1,2,\ldots$, of equivalence classes in G converges to another equivalence class $[g]$ if and only if there is a number $N > 0$ and there are points $p_i \in S^1$, $i = 1,2\ldots$, such that $\lim_{i \to \infty} g_i(p_i) = g(1)$, and such that $\lim_{i \to \infty} dg_i(T(S^1)_{p_i}) = dg(T(S^1)_1)$, and such that length $(g_i) < N$ for all i .

Example 2.1. M is a compact closed Riemannian manifold with $K < 0$.

In this case G is a countable infinite set, in fact for any number $N > 0$ there are only a finite number of equivalance classes in G which have length less than or equal to N (c.f. [3]). Thus, we have the following lemma.

Lemma 2.1. G is a countable infinite discrete topological space.

Let $[g_1]$, $[g_2]$, $[g_3]$, ... be an enumeration of all the elements of G , where each $g_i : S^1 \to M$ is a closed essential simple geodesic in M . For $g \in G$ let S^1_g be a copy of the circle associated to g . Since there is an obvious inclusion $P_*(S^1_{[g_1]}) \times \ldots \times P_*(S^1_{[g_n]}) \subset P_*(S^1_{[g_1]}) \times \ldots \times P_*(S^1_{[g_{n+1}]})$, we may define $\bigoplus_{g \in G} P_*(S^1_g)$ to be the direct limit space $\lim_{n \to \infty} P_*(S^1_{[g_1]}) \times \ldots \times P_*(S^1_{[g_n]})$. Each map $g_i : S^1 \to M$ induces a map $P_*(g_i) : P_*(S^1_{[g_i]}) \to P_*(M)$. Now by summing the $P_*(g_1), \ldots, P_*(g_n)$, and then taking the direct limit as $n \to \infty$, we get a map $\bigoplus_{g \in G} P_*(g) : \bigoplus_{g \in G} P_*(S^1_g) \to P_*(M)$.

The authors have proven the following theorem in [6] and [8; appendix].

Theorem 1.1. The mapping $\bigoplus_{g \in G} P_*(g) : \bigoplus_{g \in G} P_*(S^1_g) \to P_*(M)$ is a weak equivalence of Ω-spectra. (A mapping $F_* : S_* \to S'_*$ between Ω-spectra is a weak equivalence if each $F_j : S_j \to S'_j$ induces an isomorphism on all homotopy groups.)

Example 2.2. M is isometric to the m-torus .

We note that this is a special case where the sectional curvature $K = 0$ everywhere. We have the following structure lemma for G .

Lemma 2.2. G has a countable infinite number of components G_α , $\alpha \in H$. Each component G_α is homeomorphic to the $(m-1)$-torus.

Proof of Lemma 2.2. Let $H \subset H_1(M,Z)$ denote a maximal subset of the first homology group of M which satisfies the following two properties: if $\alpha \in H$, then $-\alpha \notin H$; if $\alpha \in H$, then α is divisible only by ± 1 . For each $\alpha \in H$ we may choose an integral basis for the integral lattice $Z^m \subset R^m$, which inturn yields a splitting

$M = \underbrace{S^1 \times S^1 \times \ldots \times S^1}_{m\text{-fold}}$ (because $M = R^m/Z^m$) , such that the first factor of this splitting represents the homology class α . Note that each of the paths $\{S^1 \times p_2 \times p_3 \times \ldots \times p_m : p_i \in S^1\}$ represents a distinct equivalence class in G , and any two such geodesics are parallel (i.e: there liftings to R^m are parallel straight lines). Let $G_\alpha \subset G$ denote the subset $\{S^1 \times p_2 \times p_3 \times \ldots \times p_m : p_i \in S^1\} \subset G$.

This completes the proof of Lemma 2.2.

Now we want to state an analogue of Theorem 2.1 for M equal the m-torus. Since G contains an uncountable number of points, the "addition" construction $\bigoplus_{g \in G} P_*(S^1_g)$ of 1.1 no longer makes sense. Roughly speaking, what we should do is first "integrate" over each continuous family of closed geodesics G_α , $\alpha \in H$, and then "add" over the countable set H . The functor which accomplishes the "integration' followed by the "addition" is the homology spectrum $\mathbb{H}_*(\ ,P_*(S^1))$ with coefficients in $P_*(S^1)$.

For each component G_α of G there is a homeomorphism $h_\alpha : G_\alpha \times S^1 \to M$ such that each map $h_\alpha : p \times S^1 \to M$, $p \in G_\alpha$, is a simple closed geodesic which represents the homology class α (see the proof of Lemma 2.2). Let $h : G \times S^1 \to M$ be the map with $h|G_\alpha \times S^1 = h_\alpha$ for all $\alpha \in H$, and let $P_*(h) : P_*(G \times S^1) \to P_*(M)$ be the map induced by h . We define a map $F_* : \mathbb{H}_*(G,P_*(S^1)) \to P_*(M)$ to be the composite of $P_*(h) : P_*(G \times S^1) \to P_*(M)$ with the assembly map A_* : $\mathbb{H}_*(G,P_*(S^1)) \to P_*(G \times S^1)$ of 1.7.

The following theorem is a special case of a result recently obtained by the authors [9].

<u>Theorem 2.2</u>. The mapping $F_* : \mathbb{H}_*(G,P_*(S^1)) \to P_*(M)$ is a weak equivalence of Ω-spectra.

The following corollary to Theorem 2.2 was pointed out to the authors by Dieter Puppe. In this corollary we shall need the following notation. Let I denote a countable collection of non-negative integers (not necessarily distinct), and for each $k \in I$ let $P_{*+k}(S^1)$ denote the Ω-spectrum having for its j'th level space the space $P_{j+k}(S^1)$. Let $\bigoplus_{k \in I} P_{*+k}(S^1)$ denote the direct limit as $i \to \infty$ of the finite products $\chi_{k \in I_i} P_{*+k}(S^1)$, where I_i denotes the first i integers in I with respect to some fixed ordering of I .

Corollary 2.2. There is a weak equivalence $F'_* : \bigoplus_{k \in I} P_{*+k}(S^1) \to P_*(M)$ of Ω-spectra. Here I is the countable set of integers which satisfies the following properties: if $k \in I$ then $0 \leq k \leq m-1$, where $m = \dim(M)$; any integer k which satisfies $0 \leq k \leq m-1$ must occur an infinite number of times in I.

Proof of Corollary 2.2. Let X be any space and let S_* be any Ω-spectrum. Let $\bar{\mathbb{H}}_*(X,S_*)$ denote Ω-spectrum cofiber for the map $\mathbb{H}_*(pt.,S_*) \to \mathbb{H}_*(X,S_*)$ which is induced by a map $pt. \to X$ from the single point space. The following weak equivalences of Ω-spectra are well known.

(a) $\mathbb{H}_*(X,S_*) \simeq \bar{\mathbb{H}}_*(X,S_*) \times \mathbb{H}_*(pt.,S_*)$.

(b) $\mathbb{H}_*(pt.,S_*) \simeq S_*$.

(c) $\bar{\mathbb{H}}_*(\Sigma X,S_*) \simeq \bar{\mathbb{H}}_{*+1}(X,S_*)$.

(d) $\bar{\mathbb{H}}_*(\vee_i S^{m_i}) \simeq \times_i S_{*+m_i}$, where $\vee_i S^{m_i}$ denotes a finite wedge product of spheres.

Now since each component G_α of G is an $(m-1)$-torus (by Lemma 2.2), it follows that each suspension ΣG_α is a finite wedge of spheres. So by applying (a)-(d) (with $S_* = P_*(S^1)$ and $X = G_\alpha$) we get:

(e) $\mathbb{H}_*(G_\alpha,P_*(S^1)) \simeq P_*(S^1) \times (\times_i P_{*+m_i-1}(S^1)$.

We also have the following:

(f) $\mathbb{H}_*(G,P_*(S^1)) \simeq \bigoplus_{\alpha \in H} \mathbb{H}_*(G_\alpha,P_*(S^1))$.

The conclusion to Corollary 2.2 follows from (e)(f) and Theorem 2.2.

Example 2.3. M is a compact Riemannian manifold with $K \leq 0$.

This example contains both examples 2.1 and 2.2. We are therefore looking for a formula for $P_*(M)$ in terms of $P_*(S^1)$ which reduces to the formulae of Theorems 2.1 and 2.2 when M satisfies the hypotheses of these theorems.

There is the following alternate description of the space G, which is useful because it also describes a simplicially stratified fibration $p : E \to G$. Let $S(M)$, $RP(M)$ denote the unit sphere bundle and the real projective bundle associated to the tangent bundle of M. There is a geodesic flow $g^t : S(M) \to S(M)$, $t \in (-\infty,\infty)$, on $S(M)$, and there is a smooth one-dimensional foliation $\mathcal{G}$ of $RP(M)$ whose leaves are covered by the orbits of g^t under the canonical 2-fold covering map $S(M) \to RP(M)$. Fix a Riemannian metric on $RP(M)$, and for any positive number s let E_s denote the union of all closed leaves of G which have length less than or equal to s. Let $p_s : E_s \to G_s$ denote the quotient map obtained by collapsing each closed leaf of G (contained in E_s) to a point. Finally, let $p : E \to G$ denote the direct limit as $s \to \infty$ of the maps $p_s : E_s \to G_s$. Let $h : E \to M$ denote the direct limit as $s \to \infty$ of the composite maps

$$E_s \xrightarrow{\text{inclusion}} RP(M) \xrightarrow{\text{proj.}} M .$$

It is not obvious that the mapping $p : E \to G$ just constructed is a simplically stratified fibration (see Def. 1.3); however, this can easily be deduced from the following lemma.

<u>Lemma 2.3</u>. Let G_α be a given connected component of G. There are compact smooth manifolds, A,B and a smooth fiber bundle projection $q : A \to B$ having a circle for fiber. There are smooth group actions $A \times H \to A$, $B \times H \to B$ by a finite group H which commute with the projection may $q : A \to B$. The mapping $p : p^{-1}(G_\alpha) \to G_\alpha$ is topologically equivalent to $\bar{q} : A/H \to B/H$, where $\bar{q}$ is the quotient of the map q under the H-action. The action $A \times H \to A$ is free.

Since the proof of Lemma 2.3 is fairly involved it is not presented here.

It follows from Lemma 2.3 that $p : E \to G$ is a simplically stratified fibration. Thus, by 1.5(a) and 1.6, there is a well defined assembly map $A_* : H_*(G, \mathcal{P}_*(p)) \to \mathcal{P}_*(E)$. Let $\mathcal{P}_*(h) : \mathcal{P}_*(E) \to \mathcal{P}_*(M)$ denote the map induced by $h : E \to M$, and define $F_* : \mathbb{H}_*(G, \mathcal{P}_*(p)) \to \mathcal{P}_*(M)$ to be the composition of A_* with $\mathcal{P}_*(h)$.

The following theorem has recently been obtained by the authors [9].

<u>Theorem 2.3</u>. The mapping $F_* : \mathbb{H}_*(G, \mathcal{P}_*(p)) \to \mathcal{P}_*(M)$ is a weak equivalance of Ω-spectra.

Recall that $H_j(G, \mathcal{P}_*(p))$ is equal to the homotopy group $\pi_j(\mathbb{H}_*(G, \mathcal{P}_*(p)))$. Thus we can apply Theorem 2.3 and Lemma 1.5(b) to deduce the following corollary to Theorem 2.3.

<u>Corollary 2.3</u>. There is a spectral sequence with $E^2_{i,j} = H_i(X, \pi_j(\mathcal{P}_*(p)))$ which abuts to $\pi_{i+j}(\mathcal{P}_*(M))$. Here $\pi_j(\mathcal{P}_*(p))$ denotes the stratified and twisted system of groups $\{\pi_j(\mathcal{P}_*(f^{-1}(x)) : x \in G\}$ over G.

<u>Remark 2.3</u>. We leave as an excercise for the reader to check that the mapping $F_* : \mathbb{H}_*(G, \mathcal{P}_*(p)) \to \mathcal{P}_*(M)$ of Theorem 2.3 is equivalent to the

mappings $F_* : \mathbb{H}_*(G\,P_*(S^1)) \to P_*(M)$ of Theorem 2.2 or $\bigoplus_{g \in G} P_*(g) : \bigoplus_{g \in G} P_*(S^1_g) \to P_*(M)$ of Theorem 2.1 when M is either the m-torus or a Riemannian manifold with strictly negative sectional curvature. Note that every fiber of $p : E \to G$ is a circle (c.f. Lemma 2.3). Thus the formula $P_*(M) \simeq H_*(G, P_*(p))$ of Theorem 2.3 should be thought of as a "stratified and twisted version" of the formula $P_*(M) \simeq H_*(M, P_*(S^1))$ of Theorem 2.2 (c.f. Remark 1.7). Note that information about the homotopy groups $\{\pi_j(P_*(S^1))\}$ can be fed directly into Corollary 2.3 to get information about the homotopy groups $\{\pi_{i+j}(P_*(M))\}$ because $p^{-1}(x) = S^1$ for every $x \in G$ in Corollary 2.3.

§3. Computations of $\pi_j(P_*(M))$, $K_i(Z\pi_1 M)$, $Wh_i(\pi_1 M)$

Let M denote a compact closed Riemannian manifold with sectional curvature $K \leq 0$ everywhere. Our goal in this section is to compute the homotopy groups $\pi_j(P_*(M))$ for $-\infty < j < \infty$, the algebraic K-groups $K_i(Z\pi_1 M)$ for $-\infty < i < \infty$, and the Whitehead groups $Wh_i(\pi_1 M)$ for $i \geq 1$. To do this we use Corollary 2.3 and the following two theorems.

Theorem 3.1. Let X denote a path connected topological space.

(a)

$$\pi_j(P_*(X)) = \begin{cases} K_{j+2}(Z\pi_1 X) & \text{if } j \leq -3 ; \\ \tilde{K}_0(Z\pi_1 X) & \text{if } j = -2 ; \\ Wh(\pi_1 X) & \text{if } j = -1 . \end{cases}$$

Here $\tilde{K}_0(Z\pi_1 X)$ is the projective class group for $Z\pi_1 X$.

(b) If X is an aspherical space then

$\pi_j(P_*(X)) \otimes Z(1/N) = Wh_{j+2}(\pi_1 X) \otimes Z(1/N)$ for all $j \geq 0$, where $N = [(j+4)/2]!$.

Theorem 3.2. Let X denote the circle.

(a) $Wh_i(\pi_1 X) = 0$ for all $i \geq 2$.

(b) $Wh(\pi_1 X) = 0$, $\tilde{K}_0(Z\pi_1 X) = 0$, and $K_i(Z\pi_1 X) = 0$ for all $i \leq -1$.

Theorem 3.1(a) is due to D. Anderson and W.-c. Hsiang [2]. Theorem 3.1(b) is due mainly to F. Waldhausen [19] who proved it rationally; A. Nicas [14] then sharpened the result to its present form. Theorem 3.2 is certainly true in much greater generality than we have stated it. For example, if X is a space such that $\pi_1 X$ is a torsion free poly-(finite or cyclic) group then F. T. Farrell and W.-c. Hsiang [5] have proven that X satisfies 3.2(b), and F. Quinn [16] has announced the proof that X satisfies 3.2(a). F. Waldhausen [18] has shown that a large class of spaces X (including the n-torus) satisfy 3.2(a), and A. Nicas [14] has proven a rational version of 3.2(a) for X equal any compact flat Riemannian manifold. It is a possibility that Theorem 3.2 even holds true for any space X with torsion free fundamental group!

We can now formulate and verify the calculations promised at the beginning of this section. The following result was proven by the authors in [6] for the special case that M has strictly negative sectional curvature. The more general case, when M has non-positive sectional curvature, was only recently obtained by the authors [9].

<u>Theorem 3.3</u>. Let M be a compact closed and connected Riemannian manifold with sectional curvature $K \leq 0$. Then

$$\pi_j(P_*(M)) = 0 \quad \text{if} \quad j < 0 ,$$

and

$$\pi_j(P_*(M)) \otimes Z(1/N) = 0 \quad \text{if} \quad j \geq 0 , \text{ where}$$

$$N = [(j+4)/2] ! .$$

<u>Proof of Theorem 3.3</u>. First note that it follows from 3.1 and 3.2 that the conclusion of Theorem 3.3 holds true for M equal the circle. Now apply Corollary 2.3.

Note that Theorems 3.1 and 3.3 now immediately imply the following corollary.

<u>Corollary 3.4</u>. For M as in 3.3 we have that $Wh(\pi_1 M) = 0$, $\tilde{K}_0(Z\pi_1 M) = 0$, $K_i(Z\pi_1 M) = 0$ for all $i \leq -1$, and $Wh_i(\pi_1 M) \otimes Z(1/N) = 0$ for all $i \geq 2$, where $N = [(i+2)/2]!$.

Since $Wh_i(\pi_1 M)$ is defined in terms of $K_i(Z\pi_1 M)$ (c.f. [18]), Corollary 3.4 implies the following calculation for $K_i(Z\pi_1 M)$.

<u>Corollary 3.5</u>. For M as in 3.3 we have, for all integers n, that

$$K_n(Z\pi_1 M) \otimes Q = H_n(M,Q) \oplus \left(\bigoplus_{i=1}^{\infty} H_{n-1-4i}(M,Q) \right).$$

<u>Remark 3.6</u>. The computation for $Wh_i(\pi_1 M)$ given in 3.4 can be improved to the following: $Wh_i(\pi_1 M) \otimes Z(1/N) = 0$ for any $i \geq 2$ where $N = [(i+1)/2]!$. In particular, we have that $Wh_2(\pi_1 M) = 0$ for any compact Riemannian manifold M with sectional curvature $K \leq 0$. To verify this improved calculation, we note that Theorem 3.1(b) is slightly

weaker than A. Nicas' result [14; 2.4]. We use [14; 2.4] in place of 3.1(b), and we use a spectral sequence comparison argument (where one of the spectral sequences occurs in Corollary 2.3 and the other is related to the Whitehead spectrum $Wh_*(\pi_1 M))$, but otherwise argue as before.

Remark 3.7. Hatcher and Wagoner [12], and Volodin [17], have derived a very simple formula for the group $\pi_0(P_*(X))$, for X equal any compact manifold (c.f. [11; 3.1]). If we assume that X is a compact closed and connected Riemannian manifold with sectional curvature $K \leq 0$, then we can use the calculation $Wh_2(\pi_1 M) = 0$ of 3.6, and we can use the fact that M is aspherical, to simplify the Hatcher-Wagoner-Volodin formula to

$$\pi_0(P_*(M)) = H_0(\pi_1 M, Z_2[\pi_1 M]/Z_2[1]) ,$$

where in computing the homology group $H_0(\ )$ we let $\pi_1 M$ act on $Z_2[\pi_1 M]$ by conjugation. Computing from this formula, and using the fact that $\pi_1(M)$ has an infinite number of non-conjugate classes (c.f. [3]), we conclude that

$$\pi_0(P_*(M)) = Z_2^\infty .$$

This is an improvement over the calculation $\pi_0(P_*(M)) \otimes Z(1/2) = 0$ given in 3.3.

§4. A Conjectured Formula for $P_*(X)$ In Terms of $P_*(S^1)$.

Let X denote an arbitrary path connected topological space. In this section we associate to X a simplically stratified fibration $\bar{p} : \bar{E} \to \bar{G}$, having the circle for fiber, by a construction which is the homotopy analogue of the construction of $p : E \to G$ given in Example 3.3. We conjecture that the homology spectrum $\mathbb{H}_*(\bar{G}, P_*(\bar{p}))$ and the stable pseudoisotopy space $P_*(X)$ are weakly equivalent for any space X which is both aspherical and has torsion free fundamental group. This conjecture is verified for X equal to a compact closed and connected Riemannian manifold with sectional curvature $K \leq 0$, by showing that the Ω-spectra $\mathbb{H}_*(G, P_*(p))$ and $\mathbb{H}_*(\bar{G}, P_*(\bar{p}))$ are weakly equivalent.

To motivate the construction of $\bar{p} : \bar{E} \to \bar{G}$, we consider the case when X is a compact Riemannian manifold with sectional curvature $K \leq 0$, and first give an alternate construction for the map $p : E \to G$ in Example 3.3. Let S denote the space of all maps $\{S^1 \to S^1\}$ which have constant non-zero speed. Note that S is a semi-group with respect to the composition operation. Let G denote the space of all essential closed geodesics $\{g : S^1 \to X\}$.. Note that S acts on the right of G by $(g,h) \to g\circ h$ for $g \in G$, $h \in S$, and S acts on the right of $G \times S^1$ by $(g,x,h) \to (g\circ h, h^{-1}(x))$ for $g \in G$, $x \in S^1$, $h \in S$. It is not difficult to see that the map $p : E \to G$ is equal to the quotient of the first factor projection map $G \times S^1 \to G$ under the right action of S.

Now it seems clear what to do to get $\overline{p} : \overline{E} \to \overline{G}$ by a homotopy theoretic version of the preceeding construction. Let C denote the space of all continuous maps $\{g : S^1 \to X\}$, and let C^*, S^* denote the singular complexes of the spaces C, S. Thus a k-simplex $\Delta_g \in C^*$ is represented by a continuous map $g : \Delta^k \times S^1 \to X$ where Δ^k denotes the standard k-simplex, and a k-simplex $\Delta_h \in S^*$ is represented by a continuous map $h : \Delta^k \times S^1 \to \Delta^k \times S^1$ such that for each $y \in \Delta^k$ we have that $h(y \times S^1) = y \times S^1$ and the composite map

$$\begin{array}{ccc} S^1 = y \times S^1 & \xrightarrow{\ h\ } & y \times S^1 = S^1 \\ & \underset{h_y}{\longrightarrow} & \end{array}$$

is an element of S. Note that S^* is a simplical semi-group which acts on the right of C^* and $C^* \times S^1$ as follows: $(\Delta_g, \Delta_h) \to \Delta_{g\circ h}$ defines the right action of S^* on C^*; $(y_1, q, y_2) \to (y_3, h_y^{-1}(q))$ defines the right action of S^* on $C^* \times S^1$, where y_1, y_2, y_3 are points of $\Delta_g, \Delta_h, \Delta_{g\circ h}$ which correspond to the same point $y \in \Delta^k$, and $q \in S^1$. Now define $\overline{p} : \overline{E} \to \overline{G}$ to be the quotient of the first factor projection map $C^* \times S^1 \to C^*$ under the right action of S^*. Note that $C^* \times S^1$ comes equipped with a map $f : C^* \times S^1 \to X$ defined as follows: for any k-simplex $\Delta_g \in C^*$, represented by $g : \Delta^k \times S^1 \to X$, and for any $y' \in \Delta_g$ corresponding to the point $y \in \Delta^k$, set $f(y',q) = g(y,q)$ for any $q \in S^1$. Since f is invariant under the right action by S^* on $C^* \times S^1$, it induces a map $\overline{f} : \overline{E} \to X$.

The space $\overline{G}$ comes equipped with a cell structure via the equality $\overline{G} = C^*/S^*$. The first barycentric subdivision of this cell structure, denoted by K, is a triangulation for $\overline{G}$ with respect to which the map $\overline{p} : \overline{E} \to \overline{G}$ is a simplically stratified fibration as in 1.3. Thus there is a well defined assembly map $\overline{A}_* : \mathbb{H}_*(\overline{G}, P_*(\overline{p})) \to P_*(E)$ from the homology spectrum $\mathbb{H}_*(\overline{G}, P_*(\overline{p}))$ (c.f. 1.5(a), 1.6). Let $P_*(\overline{f}) : P_*(\overline{E}) \to P_*(X)$ denote the map induced by $\overline{f} : \overline{E} \to X$, and define $\overline{F}_* : \mathbb{H}_*(\overline{G}, P_*(\overline{p})) \to P_*(X)$ to be the composition of $\overline{A}_*$ with $P_*(\overline{f})$.

Conjecture 4.1. Suppose that X is a path connected aspherical space which has a torsion free fundamental group. Then the map

$$\bar{F}_* : \mathbb{H}_*(\bar{G}, P_*(\bar{p})) \to P_*(X)$$

is an equivalence of Ω-spectra.

We note that 4.1 would have many consequences by which its veracity could be checked. There is a spectral sequence with $E^2_{i,j} = H_i(\bar{G}, \pi_j(P_*(\bar{p})))$ which abuts to $\pi_{i+j}(P_*(X))$ (c.f. 1.5(b)). Using this spectral sequence we can argue as in §3 that Theorem 3.3 and Corollaries 3.4 and 3.5 would hold true more generally for any path connected aspherical space X with torsion free fundamental group, if 4.1 were true. In particular the truth of 4.1 would imply that each of the following open problems has a positive solution.

Problem I. Show $Wh(\Gamma) = 0$ for any torsion free group Γ.

Problem II. Show that $Wh_i(\Gamma) \otimes Q = 0$, $i > 1$, for any torsion free group Γ.

Conjecture 4.1 is motivated by Theorem 2.3, however, it is not yet clear that 4.1 is essentially a restatement of Theorem 2.3 in the special case that X is a compact closed and connected Riemannian manifold with sectional curvature $K \leq 0$. The following theorem clarifies this situation.

Theorem 4.2. Suppose that X is a compact closed and connected Riemannian manifold with sectional curvature $K \leq 0$. Then there is a commutative diagram of mappings of Ω-spectra

$$\begin{array}{ccc} \mathbb{H}_*(\bar{G}, P_*(\bar{p})) & \xrightarrow{\bar{F}_*} & P_*(X) \\ {\scriptstyle I_*}\uparrow & \nearrow{\scriptstyle F_*} & \\ \mathbb{H}_*(G, P_*(p)) & & \end{array}$$

where I_* is a weak equivalence of Ω-spectra and where $F_*, \bar{F}_*$ are the mappings of Theorem 2.3 and Conjecture 4.1. Thus it follows from Theorem 2.3 that Conjecture 4.1 is true for such an X.

Proof of Theorem 4.2. Let G^* denote the singular complex for the space G of all essential closed geodesics $\{g : S^1 \to X\}$. Note that S^* acts on the right of G^* and $G^* \times S^1$. So we may define another simplicially stratified fibration $p' : E' \to G'$ to be the quotient of the first factor projection map $G^* \times S^1 \to G^*$ under the right action by

S^* . We let $I^1_* : \mathbb{H}_*(G',P_*(p')) \to \mathbb{H}_*(\bar{G},P_*(\bar{p}))$ denote the mapping of Ω-spectra which is induced by the inclusion map $(E',G') \subset (\bar{E},\bar{G})$.

Note that Lemma 2.3 allows us to choose a triangulation L for G with respect to which $p : E \to G$ is a simplically stratified fibration. There is a natural inclusion of L into the semi-simplical complex G' which is covered by an inclusion $E \subset E'$ of stratified fibrations. We let $I^2_* : \mathbb{H}_*(G,P_*(p)) \to \mathbb{H}_*(G',P_*(p'))$ denote the mapping of Ω-spectra which is induced by the inclusion $(E,G) \subset (E',G')$.

We define $I_* : \mathbb{H}_*(G,P_*(p)) \to \mathbb{H}_*(\bar{G},P_*(\bar{p}))$ to be the composition of I^2_* with I^1_* . It is straightforward to check that the diagram in 4.2 commutes. Thus to complete the proof of 4.2 it remains to show that both I^1_* , I^2_* are weak equivalences of Ω-spectra.

The argument that I^2_* is a weak equivalence is based on the fact that $p' : E' \to G'$ is just the singular complex version of $p : E \to G$. Details are left to the reader.

To prove that I^1_* is a weak equivalence it will suffice to verify the following claim.

<u>Claim</u>. There is a homotopy $\bar{f}_t : \bar{E} \to X$, $t \in [0,1]$, of the map $\bar{f} : \bar{E} \to X$ which satisfies the following properties.

(a) $\bar{f}_t|E' = \bar{f}|E'$ for all $t \in [0,1]$.

(b) For any fiber $\bar{p}^{-1}(x)$ of $\bar{p} : \bar{E} \to \bar{G}$ the composite map $S^1 = \bar{p}^{-1}(x) \xrightarrow{\bar{f}_1} X$ is a closed geodesic of X .

To get such a homotopy of $\bar{f} : \bar{E} \to X$ we first use the theory of approximating continuous functions by smooth functions to get a homotopy $\bar{f}_t : \bar{E} \to X$, $t \in [0,1/2]$, of $\bar{f}$ which satisfies (a) and the following property.

(c) For any fiber $\bar{p}^{-1}(x)$ of $\bar{p} : \bar{E} \to \bar{G}$ the composite map $S^1 = \bar{p}^{-1}(x) \xrightarrow{\bar{f}_{1/2}} X$ is a smooth map.

Next we use the theory of deforming smooth functions to harmonic functions, due to S. I. Alber [1], J. Eells and J. H. Sampson [4], and P. Hartman [10], to get the rest of the homotopy $\bar{f}_t : \bar{E} \to X$, $t \in [1/2,1]$. In more detail, we use the following facts concerning deformations of smooth maps $s : S^1 \to X$ to harmonic maps $h : S^1 \to X$ (c.f. [10]).

(1) A map $h : S^1 \to X$ is harmonic if and only if it is a closed geodesic.

(2) Let C^∞ denote the space of all smooth maps $\{g : S^1 \to X\}$ equipped with the metric $D : C^\infty \times C^\infty \to R$ given by $D(s,r) = \text{l.u.b.} \{d(s(x),r(x)) : x \in S^1\}$, where $d(\ ,\ )$ is the Riemannian metric on X. Then there is a semi-flow $\phi : C^\infty \times [0,\infty) \to C^\infty$ on C^∞ (i.e: each $\phi : C^\infty \times t \to C^\infty$, $t \in [0,\infty)$, is a homeomorphism which is the identity map if $t = 0$, and $\phi(s,t+t') = \phi(\phi(s,t),t')$ holds for all $t,t' \in [0,\infty)$, $s \in C^\infty$). This semi-flow satisfies (3)-(6) below.

(3) For each $s \in C^\infty$, the $\lim_{t\to\infty} \phi(s,t)$ converges in $(C^\infty, D(\ ,\ ))$ to a harmonic map.

(4) There is $\varepsilon > 0$ such that for any $s,r \in C^\infty$ with $D(s,r) \leq \varepsilon$ we have that $D(\phi(s,t),\phi(r,t)) \leq D(r,s)$ holds for any $t \in [0,\infty)$.

(5) For any $h \in S$ let $|h|$ denote the constant speed of h. For any $t \in [0,\infty)$, $h \in S$, $s \in C^\infty$ we have that $\phi(s \circ h,t) = \phi(s,|h|^2 t) \circ h$.

(6) If $s \in G$ then $\phi(s,t) = s$ for all $t \in [0,\infty)$.

Now to get $\bar{f}_t : \bar{E} \to X$, for $t \in [1/2,1]$, we set (for each $t \in [1/2,1)$, $x \in \bar{G}$) $\bar{f}_t : \bar{p}^{-1}(x) \to X$ equal to $\phi(\bar{f}_{1/2}|\bar{p}^{-1}(x), \beta_x(t))$, where $\beta_x : [1/2,1) \to [0,\infty)$ is a suitable homeomorphism depending on x, and we set $\bar{f}_1 : \bar{p}^{-1}(x) \to X$ equal to $\lim_{t\to\infty} \phi(f_{1/2}\, \bar{p}^{-1}(x),t)$. Note that properties (1),(6) imply that $\bar{f}_t : \bar{E} \to X$, $t \in [0,1)$, satisfies properties (a),(b) of the claim. The other properties (2)-(5) are needed to show that the $\beta_x : [1/2,1) \to [0,\infty)$, $x \in \bar{G}$, can be chosen in such a way that $\bar{f}_t : \bar{E} \to X$, $t \in [1/2,1]$, is a continuous homotopy.

This completes the proof of Theorem 4.2.

References

(1) S. I. Al'ber, Spaces of mappings into a manifold with negative curvature, Dokl. Akad. Nauk SSSR, Tom 178 (1968), No. 1.

(2) D. R. Anderson and W.-c. Hsiang, The functors K_{-i} and pseudoisotopies of polyhedra, Ann. of Math. 105 (1977), 201-223.

(3) W. Ballman, M. Gromov, and V. Schroeder, Manifolds of nonpositive curvature, Birkhauser (1985).

(4) J. Eells and J. H. Sampson, Harmonic mappings of Riemannian manifolds, Amer. J. Math. 86 (1964), 109-160.

(5) F. T. Farrell and W.-c. Hsiang, The Whitehead groups of poly-(finite or cyclic) groups, J. London Math. Soc. 24 (1981), 308-324.

(6) F. T. Farrell and L. E. Jones, K-theory and dynamics, II, Ann. of Math. 126 (1987), 451-493.

(7) F. T. Farrell and L. E. Jones, Algebraic K-theory of spaces stratified fibered over hyperbolic orbifolds, Proc. Nat. Acad. Sci. U. S. A. 83 (1986), 5364-5366.

(8) F. T. Farrell and L. E. Jones, A topological analogue of Mostow's rigidity theorem, J. Amer. Math. Soc. 2 (1989), 257-370.

(9) F. T. Farrell and L. E. Jones, Rigidity and other topological aspects of compact non-positively curved manifolds, Bull. Amer. Math. Soc., in press.

(10) P. Hartman, On homotopic harmonic maps, Canadian J. Math. 19 (1967), 673-687.

(11) A. E. Hatcher, Concordance spaces, higher simple homotopy theory, and applications, Proc. Symp. Pure Math. 32 (1978), 3-21.

(12) A. E. Hatcher and J. B. Wagoner, Pseudoisotopies of compact manifolds, Asterisque 6 (1973).

(13) I. M. James, Ex-homotopy theory, Illinois J. Math. 15 (1971), 324-337.

(14) A. J. Nicas, On the higher Whitehead groups of a Bieberbach group, Trans. Amer. Math. Soc. 287 (1985), 853-859.

(15) F. Quinn, Ends of maps, II, Invent. Math. 68 (1982), 353-424.

(16) F. Quinn, Algebraic K-theory of poly-(finite or cyclic) groups, Bull. Amer. Math. Soc. 12 (1985), 221-226.

(17) I. A. Volodin, Generalized Whitehead groups and pseudoisotopies, Uspehi Mat. Nauk. 27 (1972), 229-230.

(18) F. Waldhausen, Algebraic K-theory of generalized free products, Ann. of Math. 108 (1978), 135-256.

(19) F. Waldhausen, Algebraic K-theory of topological spaces, I. Proc. Symp. Pure Math. 32 (1978), 35-60.

Professor Lowell Jones
Department of Mathematics
State University of New York at StonyBrook
Stony Brook, New York 11794 USA

Professor F. T. Farrell
Department of Mathematics
Columbia University
New York, New York 10027 USA

THE FUNDAMENTAL GROUPS OF ALGEBRAIC VARIETIES

F.E.A. Johnson
Department of Mathematics, University College,
Gower Street, London WC1E 6BT.

and

E.G. Rees
Department of Mathematics, University of Edinburgh,
James Clerk Maxwell Building, The King's Buildings,
Mayfield Road, Edinburgh EH9 3JZ.

§1. If M is a compact, closed manifold then its fundamental group is finitely presented; if $n \geq 4$ then every such group is the fundamental group of some n-dimensional M. By the theorems of Nash, Tognoli ([17] is a good reference) every compact closed manifold is diffeomorphic to a real, projective, algebraic variety, so every finitely presented group is the fundamental group of such a variety. In this paper we study the classes of groups that are isomorphic to the fundamental groups of compact closed manifolds that are either projective, Kähler or complex of (complex) dimension n; we denote these classes of groups by $\mathcal{P}_n$, $\mathcal{K}_n$ and $\mathcal{C}_n$ respectively. Since each structure is less restrictive than its predecessor one has inclusions $\mathcal{P}_n \subset \mathcal{K}_n \subset \mathcal{C}_n$. We let $\mathcal{S}_n$ stand for any one of these and $\mathcal{S} = \bigcup_n \mathcal{S}_n$. Since the product of two manifolds carrying one of these structures also has the structure one sees that $\Gamma_1 \in \mathcal{S}_m$, $\Gamma_2 \in \mathcal{S}_n$ implies $\Gamma_1 \times \Gamma_2 \in \mathcal{S}_{m+n}$. The fundamental group of the complex projective line, $\mathbb{P}_1\mathbb{C}$, is trivial so $\{1\} \in \mathcal{P}_1$; hence by taking the product with $\mathbb{P}_1\mathbb{C}$ one sees that $\Gamma \in \mathcal{S}_n$ implies $\Gamma \in \mathcal{S}_{n+1}$, that is $\mathcal{S}_n \subset \mathcal{S}_{n+1}$.

A compact, complex 1-dimensional manifold is a Riemann surface and every Riemann surface is an algebraic curve, hence

(1.1) $$\mathcal{P}_1 = \mathcal{K}_1 = \mathcal{C}_1 .$$

J-P Serre [15] showed that every finite group is in $\mathcal{P}_2$ so clearly $\mathcal{P}_1 \neq \mathcal{P}_2$. One can also see that $\mathcal{P}_1 \neq \mathcal{P}_2$ because the direct product of two non-trivial surface groups is not a surface group. Recall [6] that a complex manifold is Kähler if it has a Hermitian metric such that the 2-form given by its imaginary part is closed. The induced metric on a smooth projective variety has this property. The cohomology of a Kähler manifold has a Hodge decomposition

$$H^r(M;\mathbb{C}) = \bigoplus_{p+q=r} H^{p,q}$$

and since $H^{p,q} = \bar{H}^{q,p}$ one has that $b_1(M)$ is even.

As a partial converse to this result, K. Kodaira [11, Theorem 25] proved that a complex surface whose first Betti number is even is a deformation of an algebraic surface. This shows that it is diffeomorphic to a smooth projective variety. Hence, if $\Gamma \in \mathscr{C}_2$ and $\dim H_1(\Gamma;\mathbb{R})$ is even then $\Gamma \in \mathscr{P}_2$. Since $\Gamma \in \mathscr{K}_2$ implies that $\dim H_1(\Gamma;\mathbb{R})$ is even, one obtains

(1.2) $$\mathscr{P}_2 = \mathscr{K}_2 .$$

The Hopf surface [9] is a complex surface diffeomorphic to $S^1 \times S^3$ and so its fundamental group is $\mathbb{Z}$. Now $H_1(\mathbb{Z};\mathbb{R})$ has dimension one, so $\mathbb{Z} \notin \mathscr{K}_2$, and therefore

(1.3) $$\mathscr{K}_2 \neq \mathscr{C}_2 .$$

Somewhat analogously to the case of smooth manifolds one has

(1.4) $$\mathscr{P}_2 = \mathscr{P}_n \quad \text{for all } n \geq 2 .$$

This result is a consequence of the Lefschetz hyperplane theorem [12, p.41] and Bertini's theorem. Let X_n be a smooth projective manifold, by Bertini's theorem there is a smooth hyperplane section Y_{n-1} say. By Lefschetz's theorem, $\pi_k(X,Y) = 0$ for $k < n$, so if $n \geq 3$, X and Y have isomorphic fundamental groups, proving (1.4). We do not know if $\mathscr{K}_n = \mathscr{K}_{n+1}$ for $n \geq 2$, one cannot use a sectioning argument like that above since the existence of such sections characterizes projective manifolds amongst Kähler manifolds. However, one does have

(1.5) $$\mathscr{C}_2 \neq \mathscr{C}_3 .$$

Let Γ be the nilpotent group of upper triangular 3×3 matrices with 1 for each diagonal entry and Gaussian integers above the diagonal. It is a discrete subgroup of the nilpotent Lie Group $G_{\mathbb{C}}$ which has complex entries above the diagonal. The quotient $G_{\mathbb{C}}/\Gamma$ is a complex manifold called the Iwasawa manifold and its fundamental group is Γ. So $\Gamma \in \mathscr{C}_3$. Now, suppose $\Gamma \in \mathscr{C}_2$, say $\pi_1 X_2 = \Gamma$ then, by the result [11, Theorem 2.5] of Kodaira already mentioned, since $H_1(\Gamma;\mathbb{R})$ has dimension four, X is diffeomorphic to a projective manifold and this would imply that $\Gamma \in \mathscr{K}_2$. However by (2.9) below this is not the case.

Note that (1.5) shows that, unlike the projective case, the complex case is not analogous to what happens for smooth manifolds.

A number of related questions that would be interesting to answer are

(a) Is $\mathscr{K}_n = \mathscr{K}_{n+1}$ for $n \geq 2$?

(b) Is $\mathscr{K}_n = \mathscr{P}_n$ for $n \geq 3$?

(c) Is $\mathscr{C}_n = \mathscr{C}_{n+1}$ for $n \geq 3$?

(d) Does every finitely present group belong to $\mathscr{C}$?

§2. In this section we use rational homotopy theory methods to show that certain groups cannot belong to the class $\mathscr{K}$. We start by recalling some basic terminology, there are several relevant sources such as [4], [7] and [16].

If Γ is a (discrete) group, we denote its lower central series by

$L_1(\Gamma) = \Gamma$; $L_{i+1}(\Gamma) = [\Gamma, L_i(\Gamma)]$ for $i \geq 1$. Define $\mathcal{L}_i(\Gamma) = L_i(\Gamma)/L_{i+1}(\Gamma) \otimes \mathbb{R}$ for $i \geq 1$, then, using the commutator in Γ , $\mathcal{L}(\Gamma) = \bigoplus_{i\geq 1} \mathcal{L}_i(\Gamma)$ is a Lie algebra. Define $\mathcal{M}(\Gamma)$ to be the free graded algebra generated (in degree 1) by the dual $\mathcal{L}(\Gamma)^*$. The dual of the Lie bracket $[\ ,\ ] : \Lambda^2\mathcal{L}(\Gamma) \to \mathcal{L}(\Gamma)$ is denoted $d : \mathcal{M}_1(\Gamma) \to \Lambda^2\mathcal{M}_1(\Gamma) = \mathcal{M}_2(\Gamma)$, so

$$dx(y \wedge z) = x[y,z] .$$

One extends d to be a derivation on the whole of $\mathcal{M}(\Gamma)$. It is well known that as a consequence of the Jacobi identity in $\mathcal{L}(\Gamma)$, one obtains the relation $d^2 = 0$. By their construction, $\mathcal{L}(\Gamma)$ and $\mathcal{M}(\Gamma)$ depend only on the nilpotent completion $\hat{\Gamma} = \varprojlim \Gamma/L_i(\Gamma)$.

Using the polynomial forms on the simplices of a complex X , D. Sullivan [15] constructs a differential graded algebra $\mathcal{E}(X)$ which represents the real homotopy type of X and has a minimal model $\mathcal{M}(X) \to \mathcal{E}(X)$ inducing an isomorphism on cohomology. When X is a smooth manifold the d.g.a. of smooth differential forms $\mathfrak{A}(X)$ can be used instead of $\mathcal{E}(X)$. The space X is called formal if there is a homomorphism $\lambda : \mathcal{M}(X) \to H^*(X)$ inducing an isomorphism on cohomology. It is proved in [4, p.270] that compact Kähler manifolds are formal. When X is a $K(\hat{\Gamma},1)$ space, one has $\mathcal{M}(X) \cong \mathcal{M}(\Gamma)$ and so $\mathcal{M}(\Gamma)$ can be regarded as a minimal model for Γ (or $\hat{\Gamma}$) . When X is a simplicial complex with fundamental group Γ , one has an inclusion $\mathcal{M}(\Gamma) \to \mathcal{M}(X)$ and the composition $\mathcal{M}(\Gamma) \to \mathcal{M}(X) \to \mathcal{E}(X)$ is a 1-minimal model for X ; that is, the induced map $H^i(\Gamma) \to H^i(X)$ is an isomorphism for $i = 0, 1$ and a monomorphism for $i = 2$. In the case where Γ is a discrete, cocompact subgroup of a nilpotent Lie group G then $\mathcal{M}(\Gamma)$ is isomorphic to the d.g.a. of (left) invariant differential forms on G , see [14] for related arguments.

To illustrate these ideas we prove the following

(2.1) <u>Theorem.</u> If G is a nilpotent Lie group and $\Gamma \subset G$ is a discrete cocompact subgroup then G/Γ is formal if and only if G is abelian.

<u>Proof</u>. The minimal model $\mathcal{M}(G/\Gamma)$ is given by $\Lambda^*(\mathfrak{g}^*)$ where $\mathfrak{g}^*$ is the dual of the Lie algebra $\mathfrak{g}$ of G .

By the definition of d , it vanishes if and only if $\mathfrak{g}$ is abelian which is true if and only if G is abelian.

If G is abelian then $\Lambda^*\mathfrak{g}^* \cong H^*(G/\Gamma)$ and so G/Γ is formal.

Conversely, if G/Γ is formal then there is a map of d.g.a.'s $\lambda : \Lambda^*\mathfrak{g}^* \to H^*(G/\Gamma)$ that induces an isomorphism on cohomology. In particular if $m = \dim G$ then this map is an isomorphism $\mathbb{R} \cong \Lambda^m\mathfrak{g}^* \to H^m(G/\Gamma) \cong \mathbb{R}$. However, if $d \neq 0$ then $H^1(\Lambda^*\mathfrak{g}^*)$ has dimension less than m and hence so does $H^1(G/\Gamma)$;

the kernel of λ is a non-trivial ideal and hence λ must be zero in dimension m. This contradiction shows that $d = 0$.

(2.2) Corollary. [2, Theorems A and B]. Under the above hypotheses, if G/Γ admits a Kähler structure then G is abelian.

Proof. By [4] every compact Kähler manifold is formal.

Now, we return to consider what these ideas imply about the class $\mathcal{K}$.

(2.3) Definition. A group Γ is formal in dimension one if there is a differential graded algebra $\mathfrak{A}$ with trivial differential and a homomorphism

$$\lambda : \mathcal{M}(\Gamma) \to \mathfrak{A}$$

that induces an isomorphism on H^0 and H^1 and a monomorphism on H^2.

(2.4) Proposition. If X is a formal space then $\pi_1(X)$ is formal in dimension one.

Proof. Take $\mathfrak{A}$ to be $H^*(X)$ and λ to be the composite of the map $\mathcal{M}(X) \to H^*(X)$, given by the formality condition, and the map $\mathcal{M}(\Gamma) \to \mathcal{M}(X)$.

(2.5) Corollary. If $\Gamma \in \mathcal{K}$ then Γ is formal in dimension one.

Proof. Kähler manifolds are formal [4].

If Γ_1 and Γ_2 are both formal in dimension one, so is $\Gamma_1 \times \Gamma_2$ since

$$\mathcal{M}(\Gamma_1 \times \Gamma_2) = \mathcal{M}(\Gamma_1) \otimes \mathcal{M}(\Gamma_2) .$$

It is easy to see that the free abelian groups and free groups are formal in dimension one.

(2.6) Proposition. If Γ is formal in dimension one then $H^2(\Gamma)$ is generated by decomposable elements.

Proof. Let $x \in H^2(\Gamma)$ be represented by $\xi \in \mathcal{M}_2(\Gamma)$ so $\xi = \sum_{i=1}^{r} \eta_i \wedge \zeta_i$ for some η_i, $\zeta_i \in \mathcal{M}_1(\Gamma)$. Now $\lambda(\eta_i)$, $\lambda(\zeta_i)$ are in $H^1(\Gamma)$ from (2.3); this proves the result.

If $x, y, z \in H^1(\Gamma)$ are such that $xy = yz = 0$ in $H^2(\Gamma)$, their Massey product $\langle x,y,z\rangle$ is defined as follows: Choose representatives $\tilde{x}, \tilde{y}, \tilde{z} \in \mathcal{M}(\Gamma)$ for x, y, z; then $\tilde{x}\tilde{y} = da$ and $\tilde{y}\tilde{z} = db$ for some $a, b \in M^1(\Gamma)$ and $d(a\tilde{z} + \tilde{x}b) = 0$. Define

$$\langle x,y,z\rangle = \{a\tilde{z} + \tilde{x}b \mid da = \tilde{x}\tilde{y} , db = \tilde{y}\tilde{z}\}$$

modulo the boundaries $d\mathcal{M}^1(\Gamma)$. Clearly a and b can be varied by any closed class so the set $\langle x,y,z\rangle$ is a coset of $xH^1(\Gamma) + H^1(\Gamma)z$.

(2.7) Proposition. If Γ is formal in dimension one, and $x, y, z \in H^1(\Gamma)$ are such that $\langle x,y,z,\rangle$ is defined, then $0 \in \langle x,y,z\rangle$.

Proof. One can choose reprsentatives for x, y, z in the d.g.a. $\mathfrak{A}$ of (2.3).

Since $d = 0$ in $\mathfrak{A}$ the result follows.

Now we consider some examples.

(2.8) Let Γ be the Heisenberg group of upper triangular 3×3 matrices with integer entries and ones on the diagonal. This group, Γ, is not formal in dimension one and so $\Gamma \notin \mathcal{K}$.

Proof. The algebra of Maurer-Cartan forms on G, the real 3×3 Heisenberg group

$$G = \left\{A = \begin{bmatrix} 1 & x & z \\ 0 & 1 & y \\ 0 & 0 & 1 \end{bmatrix} : x, y, z \in \mathbb{R}\right\}$$

is generated by the entries of $A^{-1}dA$ i.e. by dx, $-dy$ and $dz - xdy$. Call these θ_1, θ_2 and θ_3 respectively.

So $\mathcal{M}^1(\Gamma) = \mathrm{Sp}\{\theta_1, \theta_2, \theta_3\}$

$d\theta_1 = d\theta_2 = 0$, $d\theta_3 = \theta_1 \wedge \theta_2$.

So $H^1(\Gamma) = \mathrm{Sp}\{\theta_1, \theta_2\}$,

$H^2(\Gamma) = \mathrm{Sp}\{\theta_1\theta_3, \theta_2\theta_3\}$.

Since cup products from $H^1(\Gamma)$ to $H^2(\Gamma)$ are all zero, it follows from (2.6) that $\Gamma \notin \mathcal{K}$. This can also be seen using Massey products. Since the cup products from $H^1(\Gamma)$ to $H^2(\Gamma)$ are zero, Massey products of one dimensional classes have zero indeterminancy. However $-\theta_1\theta_3 \in \langle\theta_1,\theta_1,\theta_2\rangle$ and this is non-zero.

Since all cup products of 1-dimensional classes vanish, the theorem of [10] (see (3.1) below) also gives $\Gamma \notin \mathcal{K}$.

The referee remarked that (2.8) also follows from the results of [5]. If $\mathcal{R}(\Gamma,G)$ is the variety of representations $\Gamma \to G$ where G is a compact real algebraic group and Γ is the fundamental group of a compact Kähler manifold, then Theorem 1 of [5] asserts that $\mathcal{R}(\Gamma,G)$ has quadratic singularities at worst. However, in §9.1 of [5] it is shown that $\mathcal{R}(\Gamma,G)$ can have a cubic singularity at $\rho = 1$ when Γ is the 3×3 Heisenberg group.

(2.9) Let Γ be the Heisenberg group of upper triangular 3×3 matrices with Gaussian integer entries. This group is not formal in dimension one and so $\Gamma \notin \mathcal{K}$.

Proof. Let $G_{\mathbb{C}}$ be the complexification of the group G of the previous examples. Then the Maurer-Cartan forms are generated by

$$dx_1,\ dx_2,\ dy_1,\ dy_2,\ dz_1 - x_1dy_1 + x_2dy_2 \text{ and } dz_2 - x_2dy_1 - x_1dy_2 .$$

Call these θ_1, θ_2, θ_3, θ_4, θ_5 and θ_6 respectively, so

$d\theta_i = 0$, $1 \leq i \leq 4$; $d\theta_5 = -\theta_1\theta_3 + \theta_2\theta_4$, $d\theta_6 = -\theta_2\theta_3 - \theta_1\theta_4$.

So
$$H^1(\Gamma) = \mathrm{Sp}\{\theta_1, \theta_2, \theta_3, \theta_4\}$$
$$H^2(\Gamma) = \mathrm{Sp}\{\theta_1\theta_2,\ \theta_1\theta_3 = \theta_2\theta_4,\ \theta_1\theta_4 = -\theta_2\theta_3,\ \theta_3\theta_4,\ \theta_3\theta_5 - \theta_4\theta_6,\ \theta_3\theta_6 + \theta_4\theta_5,\ \theta_1\theta_6 + \theta_2\theta_5,\ \theta_1\theta_5 - \theta_2\theta_6\} .$$

The span of the decomposable elements in $H^2(\Gamma)$ can be at most six dimensional since $\dim H^1(\Gamma) = 4$. Hence, by (2.6), $\Gamma \notin \mathcal{K}$. Explicitly, the class $\theta_3\theta_5 - \theta_4\theta_6$ is not a linear combination of decomposables. Theorem (3.1) does not apply to this example. With real coefficients there are no non-trivial Massey products, however with complex coefficients there are non-trivial Massey products and so this would also show $\Gamma \notin \mathcal{K}$ since formality is independent of the field of coefficients. With the notation above one sees that with real coefficients the only cohomologically trivial products of one-dimensional classes are the squares. However with complex coefficients, let
$\theta = \theta_1 + i\theta_2$, $\varphi = \theta_3 + i\theta_4$, $\psi = \theta_5 + i\theta_6$ then $\theta \wedge \varphi = -d\psi$ so
$-\theta \wedge \psi \in \langle\theta,\theta,\varphi\rangle$ and $0 \notin \langle\theta,\theta,\varphi\rangle$.

(2.10) Let Γ_n be the Heisenberg group of $(n+2) \times (n+2)$ matrices with ones on the diagonal and integer entries in the first row and last column. For $n > 1$, Γ_n is formal in dimension one.

[We do not know if Γ_n is in $\mathcal{K}$ for $n > 1$.]

Proof. $\mathcal{M}(\Gamma_n)$ is generated by $\theta_1, \ldots, \theta_n, \varphi_1, \ldots, \varphi_n$ and ψ (all in dimension one) with $d\theta_i = d\varphi_i = 0$; $d\psi = \sum_{i=1}^{n} \theta_i\varphi_i$. Since $H^2(\Gamma_n)$ is a quotient of $\Lambda^2 H^1(\Gamma_n)$, one can take $\mathfrak{A}$ to be $H^2(\Gamma_n)$ in degree 2 and zero in higher degrees.

§3. Some other methods have been used to study these problems and we review them briefly.

In [10] the Hard Lefschetz Theorem is used to refine the condition that $\Gamma \in \mathcal{K} \Rightarrow b_1(\Gamma)$ is even.

(3.1) If $\Gamma \in \mathcal{K}$ then there is a linear map $\alpha : H^2(\Gamma;\mathbb{R}) \to \mathbb{R}$ such that the skew pairing given by the composition with the cup product

$$H^1(\Gamma;\mathbb{R}) \otimes H^1(\Gamma;\mathbb{R}) \xrightarrow{\cup} H^2(\Gamma;\mathbb{R}) \xrightarrow{\alpha} \mathbb{R}$$

is non-degenerate.

If M_n is a compact Kähler manifold of complex dimension n and Kähler class $\omega \in H^2(M;\mathbb{R})$ then the Hard Lefschetz Theorem states that the map induced by iterated cup product with ω is an isomorphism

$$\cup\, \omega^{n-r} : H^r(M) \longrightarrow H^{2n-r}(M) \quad \text{for} \quad 0 \leqslant r \leqslant n.$$

Now consider the map

$$H^1(M;\mathbb{R}) \otimes H^1(M;\mathbb{R}) \longrightarrow H^{2n}(M;\mathbb{R}) \cong \mathbb{R}$$

given by $x \otimes y \longrightarrow x \cup y \cup \omega^{n-1}$. We first show this is non-degenerate. By Poincaré duality, given x, there is a $y' \in H^{2n-1}(M)$ such that $x \cup y'$ is non-zero. By the Hard Lefschetz Theorem, there is a $y \in H^1(M)$ such that

$y' = y \cup \omega^{n-1}$. This is the required class y .

Consider the map $i : M \longrightarrow K(\Gamma,1)$ induced by $\pi_1(M) \cong \Gamma$ and the corresponding diagram on cohomology

$$\begin{array}{ccccc} H^1(\Gamma) \otimes H^1(\Gamma) & \xrightarrow{\cup} & H^2(\Gamma) & & \\ \downarrow i^* & & \downarrow & & \\ H^1(M) \otimes H^1(M) & \xrightarrow{\cup} & H^2(M) & \xrightarrow[\cup\, \omega^{n-1}]{} & H^{2n}(M) \cong \mathbb{R} . \end{array}$$

Since i^* is an isomorphism, the result (3.1) follows.

Since a covering of a Kähler manifold is also Kähler one can see that (3.1) also applies to any subgroup of finite index in Γ . Using this one can prove for example that if $\Gamma = G_1 * G_2$ is the free product of two groups both having non-trivial finite images then $\Gamma \notin \mathcal{K}$ [10]. This result has also been announced by D. Arapura [1] who uses mixed Hodge structures. One can also deduce several other such results, for example that $G * \mathbb{Z} \notin \mathcal{K}$ since if $H_1(G;\mathbb{Z}) \neq 0$ then G has a finite image and if $H_1(G;\mathbb{Z}) = 0$ then $b_1(G*\mathbb{Z}) = 1$.

More recently, M. Gromov [8] has proved, using L^2 cohomology, that no non-trivial free product lies in $\mathcal{K}$.

Using results on harmonic mappings from Kähler manifolds to locally symmetric spaces, J.A. Carlson and D. Toledo have proved that if Γ is cocompact in $SO(1,n)$ for $n > 2$ then $\Gamma \notin \mathcal{K}$. Of course for $n = 2$ such groups are the fundamental groups of Riemann surfaces and are in $\mathcal{K}$. They conjecture that if $\Gamma \subset SO(p,q)$ is cocompact and $p, q > 2$ then $\Gamma \notin \mathcal{K}$.

In [1], D. Arapura, using mixed Hodge structures on cohomology with coefficients, studies which groups can be the fundamental group of compact normal varieties. He shows that the class of fundamental groups arising from these varieties equals the class $\mathcal{P}$. A similar remark can be made about the fundamental groups of normal analytic spaces whose resolution is Kähler.

If one allows non-normal singularities then the class of fundamental groups is much larger, for example, the fundamental group of a nodal cubic is $\mathbb{Z}$ and non-trivial free products can occur as fundamental groups of singular curves.

References

[1] D. ARAPURA
'Hodge theory with local coefficients and fundamental groups of varieties'
Bull. Amer. Math. Soc. 20 (1989) 169-172.

[2] C. BENSON and C.S. GORDON
'Kähler and symplectic structures on nilmanifolds'
Topology 27 (1988) 513-518.

[3] J.A. CARLSON and D. TOLEDO
'Harmonic mappings of Kähler manifolds to locally symmetric spaces'
Publ. Math. I.H.E.S. (to appear).

[4] P. DELIGNE, P. GRIFFITHS, J. MORGAN and D. SULLIVAN
'Real homotopy theory of Kähler manifolds'
Invent. Math. 29 (1975) 245-274.

[5] W.M. GOLDMAN and J.J. MILLSON
'The deformation theory of representations of fundamental groups of compact Kähler manifolds'
Publ. Math. I.H.E.S. 67 (1988) 43-96.

[6] P. GRIFFITHS and J. HARRIS
'Principles of algebraic geometry'
Wiley, New York (1978).

[7] P.A. GRIFFITHS and J.W. MORGAN
'Rational homotopy theory and differential forms'
Birkäuser (1981).

[8] M. GROMOV
'Sur le groupe fondamental d'une variété kählérienne'
C.R. Acad. Sci. Paris 308 (1989) 67-70.

[9] H. HOPF
'Zur topologie der komplexen mannigfaltigkeiten' in 'Studies and essays presented to R. Courant'
Interscience (1948) 167-185.

[10] F.E.A. JOHNSON and E.G. REES
On the fundamental group of a complex algebraic manifold'
Bull. Lond. Math. Soc. 19 (1987) 463-466.

[11] K. KODAIRA
'On the structure of compact complex analytic surfaces I'
Amer. J. Math 86 (1964) 751-798.

[12] J. MILNOR
'Morse theory'
Annals of Math. Studies 51, Princeton Univ. Press (1963).

[13] J.W. MORGAN
'The algebraic topology of smooth algebraic varieties'
Pub. Math. I.H.E.S. 48(1978) 137-204.

[14] K. NOMIZU
'On the cohomology of compact homogeneous spaces of nilpotent Lie groups'
Ann. of Math. 59 (1954) 531-538.

[15] J-P. SERRE
'Sur la topologies des variétés algébriques en caractéristique p'
Symp. Int. de Topologia Algebraica, Mexico, UNESCO (1958) 24-53.

[16] D. SULLIVAN
'Infinitesimal computations in algebraic topology'
Publ. Math. I.H.E.S. 47 (1977) 269-331.

[17] A. TOGNOLI
'Algebraic approximation of manifolds and spaces'
Springer Lecture Notes in Math. 842 (1981) 73-94.

Invariants of graphs and their applications to knot theory

Kunio Murasugi,* University of Toronto
Toronto, Ontario M5S 1A1

Introduction

In the early years of knot theory, the graph was one of the main tools used to study knots in 3-space $\mathbf{R}^3$ (or 3-sphere S^3). The progress of algebraic topology since the early 1920's, however, helped to establish knot theory as one of the major branches of low dimensional manifold topology. As a result, the topic of research changed from the knot K itself to the knot complement $S^3 - K$ (or knot manifold). The knot complement in fact determines the knot, as proven very recently [GLu].

In 1984, V.F.R. Jones defined a new polynomial invariant for knots or links. This discovery opened a new era in knot theory. The invariant was unexpectedly defined through operator algebras, but its combinatorial description indicated that through graph theory knot theory could most benefit from this new invariant.

In this survey we will consider two types of graphs associated with each link diagram and define a few of their invariants.

In §§1-3, we consider the first type of graph and discuss an application of the polynomial invariant of a graph to links and, in particular, to alternating links which are related to T.G. Tait's long standing conjectures.

In §§4-7, the second type of graph will be considered and another invariant of graphs will be defined. Applications including a partial determination of the braid index and the amphicheirality problem of alternating links will be discussed in these sections.

We will use many standard terminologies and notations of graph theory. We restrict ourselves to finite graphs, but most of our graphs are signed, i.e. either + 1 or -1 is assigned to each edge in a graph.

For a graph G, $V(G)$ and $E(G)$ denote the set of vertices and edges, respectively. $|X|$ denotes the cardinality of a set X.

Throughout this paper , a graph frequently represents the geometric realization of a finite 1-dim CW-complex in $\mathbf{R}^3$. A vertex and an edge correspond to a 0-simplex and 1-simplex, respectively. Therefore, we are free to use many terms from algebraic topology. Denote $\beta_i(X)$ the ith Betti number of X.

A graph G is *separable* if there is a vertex v_0, called a cut vertex, such that $\beta_0(G) < \beta_0(G - \{v_0\})$. Otherwise, G is non-separable. A *block* is a maximal non-separable connected subgraph of G. A connected graph G is decomposed into finitely many blocks $G_1, G_2, \cdots, G_n$, and we say that G is the *block sum* of $G_1, G_2, \cdots, G_n$.

An alternate sequence of vertices v_i and edges e_i, $C = \{v_0, e_1, v_1, e_2, \ldots, v_{k-1}, e_k, v_k\}$ is called a cycle if v_{i-1} and v_i are ends of the edge e_i for $i = 1, 2, \ldots, k$, and $v_k = v_0$. The length of C is k. A cylce is called *simple* if $e_i \neq e_j$ and $v_i \neq v_j$ for any i and j, $i \neq j$, except $v_k = v_0$.

---

* Partially supported by NSERC No. A 4034.
Most of the material in §§5-7 is from the joint work with J.H. Przytycki [MP]

§1 Jones polynomial of a graph.

Given a knot or link L in $S^3 = \mathbf{R}^3 \cup \{\infty\}$, we project it into $\mathbf{R}^2$. A link diagram D of L is an image of L in which we specify the arc running over the other. We assume throughout this paper that L is oriented and D has the orientation induced from that of L.

A link diagram is called *alternating* if when one travels along each component, over crossing and under crossing appear alternatively. Fig 1.1(a) represents an alternating diagram, but Fig 1.1(b) does not. A link L is called an *alternating link* if L has an alternating diagram.

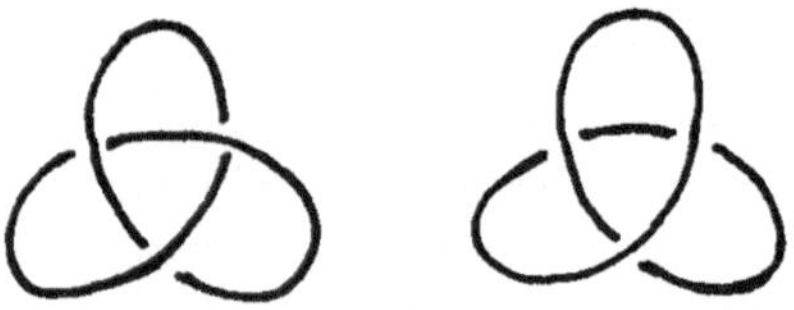

(a) (b)

Fig. 1.1

Now there are several ways to associate a graph with a link. In this paper, we will use two different types of graph. The first graph, defined below, was used by C. Bankwitz [B] in 1930 to study alternating knots. We call it the *graph* of a link L (or more precisely the *graph of a diagram* D of L).

Let D be a link diagram of a link L. D divides $\mathbf{R}^2$ into a finite number of domains $R_1, R_2, \cdots, R_n$ which will be classified as shaded or unshaded. There is no common edge between two shaded or unshaded domains. With D we associate a planar signed graph G_D as follows. Take a point v_i from each shaded domain R_i. These points form a set of vertices of G_D. Suppose that two shaded domains R_i and R_j meet at a crossing c_k. If R_i is different from R_j, then we join v_i and v_j by an edge e_k passing through c_k. If R_i and R_j are the same domain, then we form a loop e_k passing through c_k. Se Fig. 1.2.

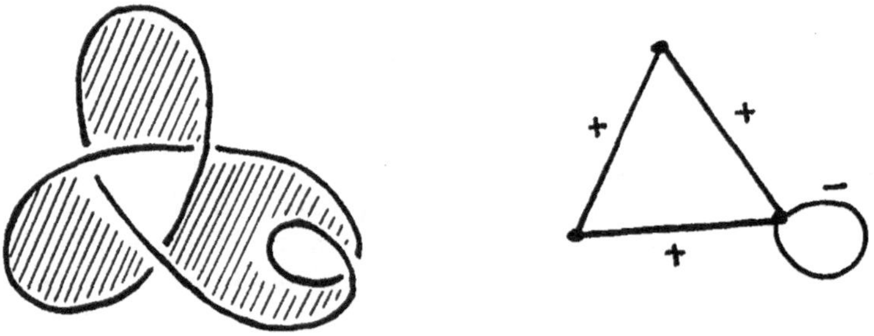

Fig. 1.2

Each edge of G_D, therefore, corresponds to a crossing of D. Furthermore, each edge e_i is signed with + 1 or -1 according to whether the twist at the crossing is positive or negative. (See Fig. 1.3)

$sign(e) = +1$ $sign(e) = -1$

Fig 1.3

Now, we fix some notations, before we define an integer polynomial $F_G(x, y, z)$ for any finite signed graph G.

Let H be a subgraph of G. Denote $p(H)$ and $n(H)$, respectively, the number of positive and negative edges in H. A (not necessarily connected) subgraph H is called a *spanning subgraph* of G if H contains all vertices of G. In particular, the *maximal positive spanning subgraph* P is the spanning subgraph that contains all positive edges but no negative edges. Analogously, the *maximal* negative spanning subgraph N is defined. We reserve P and N for these subgraphs.

Let $\mathcal{S}_G(r, s)$ be the set of all spanning subgraphs H of G such that $\beta_0(H) = r + 1$ and $\beta_1(H) = s$. Therefore $\mathcal{S}_G(0, 0)$ is, in particular, the set of all spanning trees in G.

Definition 1.1 We define

$$F_G(x, y, z) = \sum_{r,s}\{\sum_{H\in\mathcal{S}_G(r,s)} x^{p(H)-n(H)}\}y^r z^s \tag{1.1}$$

where the second summation runs over all spanning subgraphs H in $\mathcal{S}_G(r, s)$. $F_G(x, y, z)$ will be called the *Jones polynomial* of a graph G.

From the definition, we have

Proposition 1.2 [Mu 4]

(1) *If G has n connected components $G_1, \cdots, G_n$, then*

$$F_G(x, y, z) = y^{n-1}\prod_{i=1}^{n} F_{G_i}(x, y, z)$$

(2) *If G is connected and is the block sum of m blocks $G_1, \cdots, G_m$, then*

$$F_G(x, y, z) = \prod_{i=1}^{m} F_{G_i}(x, y, z).$$

If G is a positive graph, i.e., $n(G) = 0$, then $F_G(x, y, z)$ is of a much simpler form.

Proposition 1.3 *If G is a positive (connected) graph, then*

$$F_G(x, y, z) = x^{v-1}\sum_{r,s}|\mathcal{S}_G(r, s)|\left(\frac{y}{x}\right)^r (xz)^s. \tag{1.2}$$

where $v = |V(G)|$.

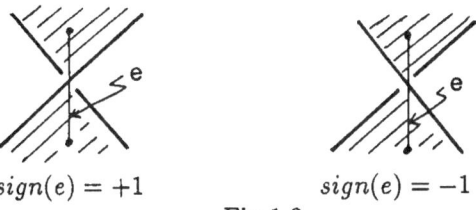

Proof. For $H \in S_G(r,s)$, $p(H) = v - 1 - r + s$ and $n(H) = 0$. Therefore, (1.1) is reduced to

$$F_G(x,y,z) = \sum_{r,s}\{\sum_{H \in S_G(r,s)} \left(\frac{y}{x}\right)^r (xz)^s\} x^{v-1} = x^{v-1}\sum_{r,s} |S_G(r,s)| \left(\frac{y}{x}\right)^r (xz)^s.$$

Remark 1.4 It is possible to define a similar polynomial $F_M(x,y,z)$ for a (circuit) matroid M.

$F_G(x,y,z)$ is invariant under 2-isomorphism. Two graphs G_1 and G_2 are said to be 2-isomorphic if one is obtained from the other by applying the following two operations Ω_1 and Ω_2 finitely many times. Let G be the one-point union of two subgraphs H and K which meet at a vertex v. Then $\Omega_1(G)$ is another one-point union of H and K which meet at a different vertex v'. To define $\Omega_2(G)$, suppose that G is obtained from two disjoint graphs H and K by identifying vertices u_1 and u_2 of H with v_1 and v_2 of K, respectively. $\Omega_2(G)$ is a new graph obtained from H and K by modifying the identification so that $u_1 = v_2$ and $u_2 = v_1$. (Cf [W].) See Fig. 1.4.

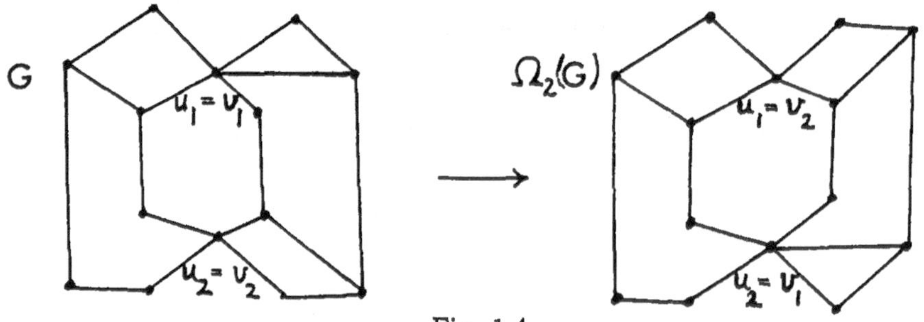

Fig. 1.4

Example 1.5 (1) If G consists of only one vertex, then $F_G(x,y,z) = 1$.
(2) If $G =$  , the $F_G(x,y,z) = x^3 z + 3x^2 + 3xy + y^2$.

§2 First Tait Conjecture

In late last century, P.G. Tait proposed three marvellous conjectures on alternating knots. Two of them have been completely solved [K , Mu 2,3, Th 1,2]. (See also Theorems 2.4 and 3.3.) The third (unsolved) conjecture is restated in the simple terms used in graph theory in §3. (See Conjecture A.). In this section, we consider the first conjecture.

For a Laurent polynomial $f = \sum_{i_1 \cdots i_n} a_{i_1 \cdots i_n} x_1^{i_1} \cdots x_n^{i_n} \in \mathbf{Z}[x_1^{\pm 1}, \cdots, x_n^{\pm 1}]$, we define $max\ deg_{x_j} f = max\{i_j | a_{i_1 \cdots i_n} \neq 0\}$ and $maxdeg f = max\{i_1 + \cdots + i_n | a_{i_1 \cdots i_n} \neq 0\}$ $min\ deg_{x_j} f$ and $min\ deg f$ are defined analogously. Furthermore, we denote $x_j - span\ f = max\ deg_{x_j} f - min\ deg_{x_j} f$ and $span\ f = max\ deg\ f - min\ deg\ f$.

Now it is easy to evaluate $max\ deg\ P_L(x,y,z)$ and $min\ deg\ P_L(x, y^{-1}, z^{-1})$. In fact, using a simple combinatorial argument on spanning subgraphs, we have the following theorem.

Theorem 2.1 [Mu 4] *For any signed graph G,*

$$\begin{aligned} max\ deg F_G(x,y,z) &= p(G) + \beta_0(P) + \beta_1(P) - 1 \\ min\ deg F_G(x,y^{-1},z^{-1}) &= -\{n(G) + \beta_0(N) + \beta_1(N) - 1\}, \end{aligned}$$

where P and N denote, respectively, the maximal positive and negative spanning subgraphs.

If G is a planar graph, we can define the dual graph G^* associated with a planar imbedding $\hat{G}(\subset \mathbf{R}^2)$ of G. G^* is a graph imbedded in $\mathbf{R}^2$. The set of vertices $V(G^*)$ of G^* and the set of domains in $\mathbf{R}^2 - \hat{G}$ are in one-to-one correspondence, and moreover $E(G^*)$ and $E(\hat{G})$ are in one-to-one correspondence in such a way that $e^* \in E(G^*)$ and its corresponding edge e in $E(\hat{G})$ have exactly one point, not a vertex, in common. G^* is also signed by assigning -1 (or +1) to an edge of G^* if the sign of the edge in $\hat{G}$ is +1 (or -1). See Fig. 2.1, where edges of G^* are depicted by broken lines.

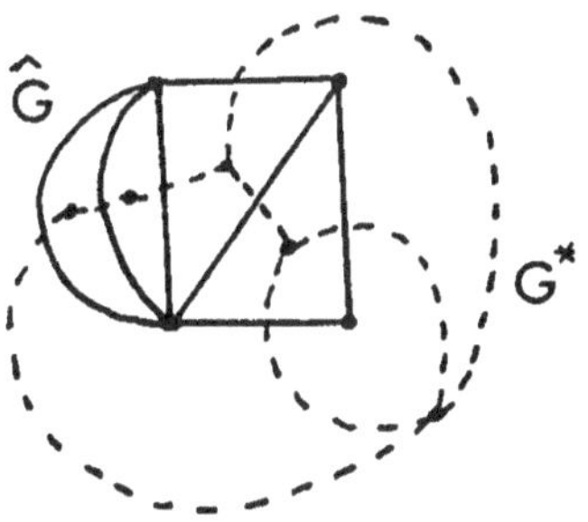

Fig. 2.1

Although the dual graph depends on an imbedding of G in $\mathbf{R}^2$, the Jones polynomial of the dual graph is uniquely determined no matter how G imbeds in $\mathbf{R}^2$ and we have
Proposition 2.2 [Mu 4] *If G^* is a dual graph of a planar graph G, then*

(2.1) $$x^{p(G)-n(G)} F_{G^*}(x,z,y) = F_G(x,y,z)$$

Warning: Variables y and z are interchanged in the left-hand side.

A graph G is called *reduced* if G has no isthmuses and no loops, where an isthmus is an edge e such that $\beta_0(G) < \beta_0(G-e)$. A link diagram D is called *reduced* if the graph G_D associated with D is reduced. Therefore, D is reduced iff D has no nugatory crossings i.e. D does not have crossings like 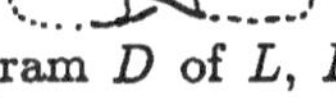 or  .

Since $F_{G_D}(x,y,z)$ depends on a link diagram D of L, $F_{G_D}(x,y,z)$ need not be an invariant of the link type. However, simple substitutions $y = z = -(x+x^{-1})$ in $F_{G_D}(x,y,z)$ suffices to obtain a link type invariant: $\hat{F}_{G_D}(x) = F_{G_D}(x, -(x+x^{-1}), -(x+x^{-1}))$. In fact, $\hat{F}_G(x)$ is *equivalent* to the Jones polynomial $V_L(t)$ of L and we have
Proposition 2.3 [Mu 4, Theorem 12.1] *For some integer k,*

(2.2) $$\hat{F}_G(x) = \pm x^k V_L(x^2)$$

If G is a positive (connected) graph, then (1.2) is reduced to

$$\hat{F}_G(x) = x^{v-1} \sum_{r,s} (-1)^{r+s} |S_G(r,s)| (1+x^{-2})^r (1+x^2)^s \tag{2.3}$$

Furthermore, it is shown that if G is positive and reduced, then for any spanning subgraph H of G,

(2.4) (1) $\beta_1(H) < \beta_1(P)$

(2) $2\beta_0(H) + 1 < 2\beta_0(N)$.

Using (2.4) we can conclude that

$$\begin{aligned} &max\ deg \hat{F}_G(x) = v - 1 + 2\beta_1(P), \text{ and} \\ &min\ deg \hat{F}_G(x) = v - 1 - 2\{\beta_0(N) - 1\}. \end{aligned}$$

If G is positive and reduced, then $P = G$ and N consists of only vertices of G, and hence $\beta_1(P) = \beta_1(G)$ and $\beta_0(N) = |V(G)|$. Therefore, we have

$$\begin{aligned} span\ \hat{F}_G(x) &= 2\beta_1(P) + 2\beta_0(N) - 2 \\ &= 2\beta_1(G) + 2|V(G)| - 2 \\ &= 2|E(G)|. \end{aligned}$$

We should note that an alternating (non-split) link is the very link that admits a diagram whose graph is positive (with a proper shading).

If a link diagram D is not alternating, then it follows from Theorem 2.1 that

$$\begin{aligned} span \hat{F}_G(x) \leq\ & p(G) + \beta_0(P) + \beta_1(P) - 1 \\ & + n(G) + \beta_0(N) + \beta_1(N) - 1. \end{aligned}$$

A simple homological algebraic argument will prove that

$$\beta_0(P) + \beta_1(P) + \beta_0(N) + \beta_1(N) \leq |E(G)| + 2,$$

and hence we have

$$span\ \hat{F}_G(x) \leq 2|E(G)|.$$

Since each edge of G_D corresponds to a crossing, we have proved the following theorem

Theorem 2.4 [Mu 2] *If D_1 and D_2 are reduced (connected) alternating diagrams of an alternating link L, then D_1 and D_2 have exactly the same number of crossings. Furthermore, a non-alternating diagram of an alternating link L cannot have fewer crossings than those possessed by a reduced alternating diagram of L.*

Theorem 2.4 is exactly what the first Tait conjecture claims.

§3. Tait's second and third conjectures

The second conjecture of Tait was more subtle than the first. A proof requires a more precise evaluation of the degrees of $V_L(x^2)$.

To each crossing of a link diagram D, we assign +1 or -1 as depicted in Fig. 3.1.

$w(c) = +1$ $w(c) = -1$

Fig. 3.1.

A crossing c is called *positive* (or *negative*) if $w(c) = +1$ (or $w(c) = -1$). Let $p(D)$ and $n(D)$ denote the number of positive and negative crossings in D. The integer $w(D) = p(D) - n(D)$ is called the *Tait number* of D (or the writhe of D).

The second Tait conjecture claims that $w(D)$ is the same for any reduced alternating diagrams D of an alternating link.

To prove this conjecture, first we note that

$$V_K(x^2) = (-1)^a x^b \hat{F}_{G_D}(x), \tag{3.1}$$

where $a = p(G_D) - n(G_D)$ and $b = -\frac{1}{2}\{a - 3w(D)\}$ [K].

Now to evaluate $max\ deg V_L(x^2)$ and $min\ deg V_L(x^2)$, we introduce a new invariant of a graph which has not been previously studied.

For a signed graph G, the adjacent matrix A_G is a $|V(G)| \times |V(G)|$ integer matrix whose rows (and columns) correspond to the vertices of G. Suppose the ith row (and the ith column) corresponds to a vertex v_i. Then for $i \neq j$, the (i,j) entry a_{ij} of A_G is given by the number of negative edges minus the number of positive edges connecting v_i and v_j. For $i = j$, we set $a_{ii} = -\sum_{\substack{i=1 \\ i \neq j}}^{|V(G)|} a_{ij}$. Since A_G is symmetric, the signature of A_G is well-defined. It is called the *signature* of G and is denoted by $\sigma(G)$. $\sigma(G)$ is invariant under 2-isomorphism [Mu 5].

If G is a positive (or negative) connected graph, then $\sigma(G) = |V(G)| - 1$ (or $\sigma(G) = -(|V(G)| - 1)$). More generally, we have the following theorem

Theorem 3.1 [Mu 4]

$$\beta_0(N) + \beta_1(N) - n(G) - 1 \leq \sigma(G) \leq p(G) - (\beta_0(P) + \beta_1(P) - 1). \tag{3.2}$$

Equalities hold simultaneously in (3.2) iff each block of G is either a positive or a negative graph.

Let $\sigma(L)$ be the signature of a link [Mu 1]. Theorem 3.1 combined with a theorem in [*GLi*] implies

Proposition 3.2 [Mu 4] *For a link diagram D of a link L,*

$$-2n(D) - \sigma(L) \leq min\ deg\ V_L(x^2) \tag{3.3}$$
$$max\ deg V_L(x^2) \leq 2p(D) - \sigma(L).$$

Equalities hold simultaneously in (3.3) iff D is a reduced alternating diagram or the connected sum of reduced alternating diagrams.

Therefore, if D is a reduced alternating diagram, then $p(D)$ and $n(D)$ are both link type invariants and so is $w(D)$. In other words, we have proved

Theorem 3.3 [Mu 3] *Let D_1 and D_2 be reduced alternating diagrams of an alternating link L. Then*

$$(3.4)\qquad \begin{aligned} p(D_1) &= p(D_2) = max\ deg V_L(t) + \frac{\sigma(L)}{2}, \text{ and} \\ n(D_1) &= n(D_2) = -min\ deg V_L(t) - \frac{\sigma(L)}{2} \end{aligned}$$

and hence

$$(3.5)\qquad w(D_1) = w(D_2).$$

Theorem 3.5 proves Tait's second conjecture.

The third conjecture by Tait claims that two reduced alternating diagrams of the same link can be transformed into each other by a finite number of applications of a single operation, now called the *Conway flyping.* This conjecture is equivalent to the following conjecture.

Conjecture A. *Let G_1 and G_2 be two positive reduced graphs associated with two reduced alternating diagrams D_1 and D_2 of an alternating link L. Then G_2 is obtained from G_1 by a finite sequence of two operations depicted in Fig. 3.2.*

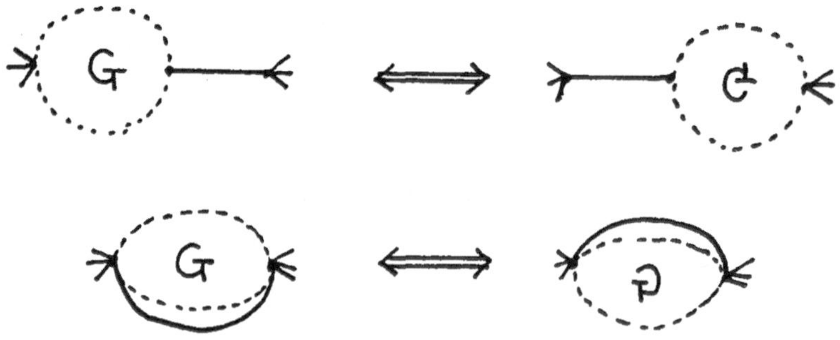

Fig. 3.2.

Conjecture A implies Theorems 2.4 and 3.3. Conjecture A is far from being solved at present, although some serious attempts have been made during the last several years [Mu 4, Th 3]. Conjecture A may not be true, but a slightly weaker conjecture B seems plausible.

Conjecture B *(Weak Tait Conjecture) Under the same assumption as in Conjecture A, G_1 and G_2 are 2-isomorphic.*

Many invariants of a graph are in fact invariant under 2-isomorphism. Since the Jones polynomial $F_G(x, y, z)$ of a graph G is invariant under 2-isomorphism, Conjecture B at least implies

Conjecture C. *Let D_1 and D_2 by reduced alternating diagrams of an alternating link L. Let G_1 and G_2 be positive graphs associated with D_1 and D_2, respectively. Then*

$$(3.6)\qquad F_{G_1}(x, y, z) = F_{G_2}(x, y, z).$$

If G is positive, we may assume without loss of generality that $x = 1$. Some simplified versions of Conjecture C are in fact proved.

Theorem 3.4 *Under the same assumption in Conjecture C,*

$$(3.7) \quad \begin{aligned} F_{G_1}(1,t,t^{-1}) &= F_{G_2}(1,t,t^{-1}) \\ F_{G_1}(1,t^{-1},t) &= F_{G_2}(1,t^{-1},t) \quad \text{(See [Mu 4])} \end{aligned}$$

$$(3.8) \quad \begin{aligned} F_{G_1}(1,t,-1) &= F_{G_2}(1,t,-1) \\ F_{G_1}(1,-1,t) &= F_{G_2}(1,-1,t) \quad \text{(See [Th 2]).} \end{aligned}$$

We still do not know, however, whether or not $F_{G_1}(1,t,t) = F_{G_2}(1,t,t)$.

§4 Seifert graphs

The second type of graph used in knot theory is the so-called *Seifert graph*.

Let L be a link and D a diagram of L. If we split D at each crossing (Fig. 4.1), D

Fig. 4.1

is decomposed into finitely many circles on a plane, called *Seifert circles*.

Let $s(D)$ denote the number of Seifert circles in D and $c(D)$ the number of crossings in D. The *Seifert graph* Γ_D (associated with D) is a graph with $s(D)$ vertices $v_1, \cdots, v_{s(D)}$ and $c(D)$ edges $e_1, \cdots, e_{c(D)}$. Each vertex corresponds to a Seifert circle and each edge corresponds to a crossing. Two distinct vertices v_i and v_j are connected by e_k iff two Seifert circles S_i and S_j (corresponding to v_i and v_j) are joined by a crossing c_k (corresponding to e_k). Furthermore, each edge is given the same sign as that of the corresponding crossing in D. (See Fig. 3.1) Therefore, the Seifert graph is a signed (planar) graph. A Seifert graph has no loops. In contrast to graphs discussed in §§1-2, a Seifert graph does not represent a link type. In other words, the exact same graph may represent different links. In fact, the figure eight knot (Fig. 4.2 (a)) and a 3-component link (Fig. 4.2 (b)) have the same Seifert graph Γ (Fig. 4.2 (c)).

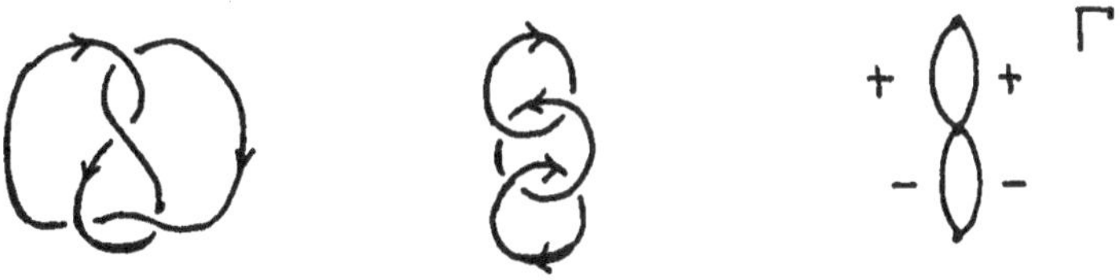

(a) (b) (c)

Fig. 4.2.

If Γ_D is not separable, D is called a *special* diagram, and Γ_D uniquely represents a link. There is no ambiguity in recovering the original link diagram from a non-spearable Seifert graph. A Seifert graph is a spine of some orientable spanning surface of L, and hence it is *bipartite*, i.e. every cycle of Γ has an even length. Particularly interesting examples of Seifert graph arise from *braids*.

A Seifert graph of a closed braid is the block sum of multiple-edge graphs (See Fig. 4.3).

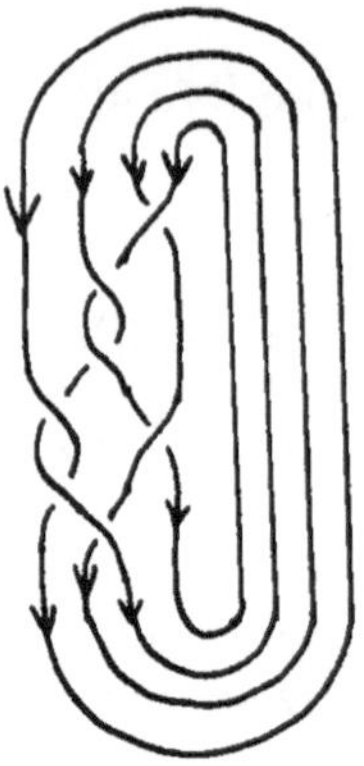

Fig. 4.3

The theory of braids has played a fundamental role in the discovery of the new polynomials [J]. The original Jones' theory depends on the fact that every oriented link is represented as a closure of an n-braid for some n [A]. The minimum number of strings n needed to represent L as a closed n-braid, is called the ***braid index*** of L, denoted by $\mathrm{b}(L)$. Before the new polynomial was discovered, it was almost impossible to evaluate $\mathrm{b}(L)$ for L, except for a few limited types of links. The recent development of the new polynomials, however, has revealed a strong connection between these polynomials and the braid index of a link. And now the evaluation of the braid index is possible at least for certain type of links. In the next section, we will discuss the recent progress on this problem.

§5 The braid index

We begin with a definition of a new integer polynomial $P_L(v, z)$ in variables v and z [FYHLMO,J, LM, PT].

Let L_+, L_- and L_0 be links which are identical except in the neighborhood of a crossing, where they look like

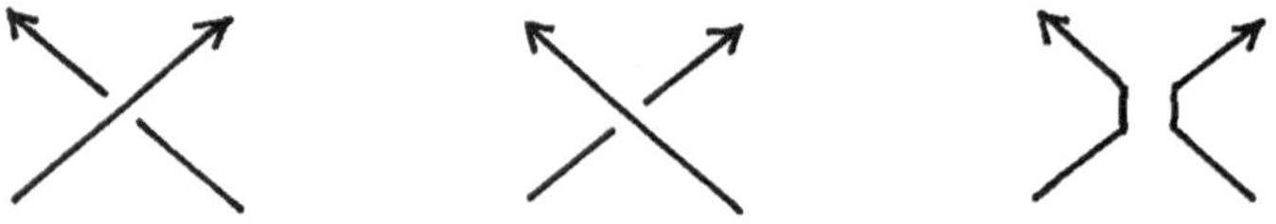

L_+ L_- L_0

Fig. 5.1

Then $P_L(v, z)$ satisfies the following formula

$$\frac{1}{v}P_{L_+}(v,z) - vP_{L_-}(v,z) = zP_{L_0}(v,z). \tag{5.1}$$

If L is a trivial knot, then

$$P_L(v,z) = 1. \tag{5.2}$$

The integer polynomial uniquely defined by (5.1) and (5.2) will be called the ***skein polynomial*** of a link L. The skein polynomial is a generalization of the Jones polynomial.

Example 5.1

P$(v,z) = (v^{-1} - v)z^{-1}$

P$(v,z) = v^{-2}z^{-1} - vz + z^{-1}$

Since $P_L(v,z)$ involves two variables v and z, we can define the $v - span\ P_L(v,z)$ and the $z - span\ P_L(v,z)$. As is suggested by Theorem 5.1 below, however, $v - span\ P_L(v,z)$ is more interesting and important.

Theorem 5.1 [Mo, FW] *For any link L,*

$$v - span\ P_L(v,z) \leq 2(\mathbf{b}(L) - 1). \tag{5.3}$$

Surprisingly, equality holds in (5.3) for many knots, although inequality is sharp for some knots up to 10 crossing knots. One of the earliest conjecture on $\mathbf{b}(L)$ can be found in [FW].

Conjecture [FW] *If L is the closure of a positive braid, then*

$$v - span\ P_L(v,z) = 2(\mathbf{b}(L) - 1), \tag{5.4}$$

where a positive braid is a braid in which every Artin's generator σ_i appears with a non-negative exponent.

Although this conjecture was recently disproved in [MS], equality (5.4) holds for many alternating links, including 2-bridge links, alternating fibred links and alternating pretzel links [MP]. Unfortunately, there is an alternating link for which the equality is false [MP]. The simplest example is the alternating link depicted in Fig. 5.2.

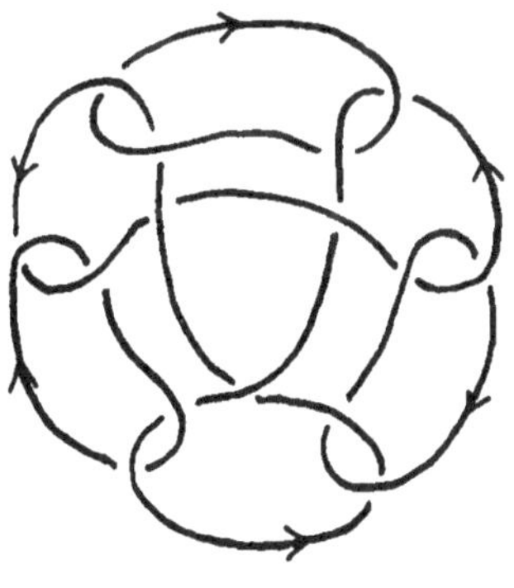

Fig. 5.2

We have seen before that a Seifert graph of a closed n-braid is the block sum of $n-1$ multiple edge graphs and hence the natural diagram of a closed n-braid has exactly n Seifert circles. (See Fig. 4.3.) Therefore, any link has at least one diagram D_0 for which $s(D_0) = \mathbf{b}(L)$, and hence, we have $min\ s(D) \leq \mathbf{b}(L)$, where the minimum is taken over all diagrams of L. In 1987, Yamada proved the reverse inequality. In fact, he proved

Theorem 5.2 [Y] *For any diagram D of L, $s(D) \geq \mathrm{b}(L)$.*

This theorem suggests that for any link diagram D of L, the study of the *surplus* $s(D) - \mathrm{b}(L)$ would eventually lead to the determination of $\mathrm{b}(L)$. We may ask, for example, for what diagram D, is $s(D) - \mathrm{b}(L)$ equal to 0? If D is a reduced alternating diagram, $s(D) - 1$ equals the degree of the reduced Alexander polynomial of L and hence $s(D) - \mathrm{b}(L)$ is a link type invariant. Our study of the surplus $s(D) - \mathrm{b}(L)$ leads to a new invariant of a graph G, called the *index* of G, which is a topic of the next section.

§6 Index of graphs

Definition 6.1 An edge e of a graph G is called *singular* if e is not a loop and no other edges ($\neq e$) have the same ends as e.

Let $\mathcal{F} = \{e_1, \cdots, e_n\}$ be a set of singular edges of G. $\mathcal{F}$ is said to be *independent* if there exists an edge e_i in $\mathcal{F}$ and a vertex v that is one end of e_i, such that $\mathcal{F} - \{e_i\}$ is independent in $G/star\ v$, where $star\ v$ is the smallest subgraph of G containing all edges incident to v, and $G/star\ v$ is the graph obtained from G by contracting $star\ v$ to a point. We define the empty set to be independent. The index of G, $ind\ G$, is the maximal number of independent edges in G.

Example 6.2 For the graph G depicted in Fig. 6.1, $ind\ G = 2$.

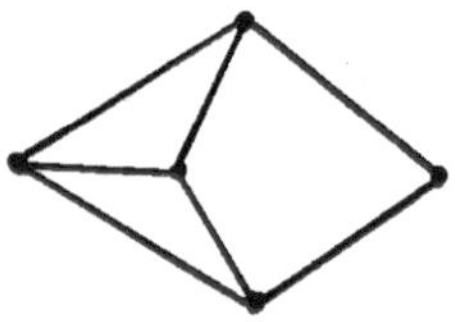

Fig. 6.1

We note that $ind\ G = 0$ iff G has no singular edges.

We can define a slightly different, but almost equivalent invariant of G, called the *cycle* index of G.

Definition 6.3 A set $\mathcal{F} = \{e_1, \cdots, e_n\}$ of singular edges in G is called *cyclically* independent if no k edges ($1 \leq k \leq n$) occur on a simple cycle of length at most $2k$. The cycle index, $\alpha(G)$, of G is the maximal number of cyclically independent edges in G.

Generally $ind(G) \leq \alpha(G)$. Very recently, however, P. Traczyk proved the reverse inequality for bipartite graphs.

Theorem 6.4 [Tr] *Let G be a bipartite graph. Then $ind(G) \geq \alpha(G)$ and hence, $ind\ G = \alpha(G)$.*

Theorem 6.4 is false for a non-bipartite graph. Using Theorem 6.4, it is easy to show the following

Theorem 6.5

(1) *For a bipartite graph, $ind(G)$ is an invariant under 2-isomorphism.*

(2) *If G_1 and G_2 are both bipartite, then $ind(G_1 \vee G_2) = ind\ G_1 + ind\ G_2$ where $G_1 \vee G_2$ denotes the one point union of G_1 and G_2.*

For a link diagram D of a link L, the index of D, $ind\ D$, is defined as the index of the Seifert graph Γ_D associated with D. Recall that Γ_D is bipartite. Now, given a link diagram D of L, it is possible to find another diagram D_0 of L with $s(D_0) = s(D) - ind\ D$, and therefore, Theorem 5.2 implies

Theorem 6.6 [MP] *For any link diagram D of a link L,*

$$\mathrm{b}(L) \leq s(D) - ind\ D. \tag{6.1}$$

For many alternating links, the equality holds in (6.1). We conjecture that this is always the case.

Conjecture D. *For an alternating link diagram D of an alternating link L,*

$$\mathrm{b}(L) \ = \ s(D) - ind\ D.$$

This conjecture is true if $ind\ D = 0$ and we have

Theorem 6.7 [MP] *Let D be an alternating link diagram of an alternating link L. If $ind\ D = 0$, then $\mathrm{b}(L) = s(D)$ and, furthermore,*

$$v - span\ P_L(v, z) = 2(\mathrm{b}(L) - 1).$$

Since an alternating fibred link L admits a reduced alternating diagram D with $ind\ D = 0$, the braid index of L is completely determined by $v - span\ P_L(v, z)$. (cf [Mu 6].)

§7 Amphicheirality

A Seifert graph Γ_D of an alternating diagram D of an alternating link L is the block sum of blocks $\Gamma_1, \cdots, \Gamma_k$. Let D_i be the alternating diagram of a (special) alternating link L_i recovered from Γ_i. We write then $L = L_1 * L_2 * \cdots * L_k$ and L is called a *$*$-product* (or *Murasugi sum or product*) of L_i. Many important numerical invariants of L are obtained from those of L_i. For example, it is known that (1) the degree of the (reduced) Alexander polynomial $\Delta_L(t)$ of L is the sum of those of L_i [Mu 1] (2) The leading coefficient $c_0(L)$ of $\Delta_L(t)$ is the product of those of L_i [Mu 1] (3) the span of the Jones polynomial $V_L(t)$ is the sum of those of L_i [Mu 2]. These observations would suggest a certain kind of uniqueness in the decomposition of L into its $*$-product for an alternating link. For example, the mirror image $\hat{L}$ of L is the $*$-product of the mirror images $\hat{L}_i$ of L_i, i.e. $\hat{L} = \hat{L}_1 * \cdots * \hat{L}_k$. Since each L_i is special alternating, it is either a positive or a negative link. An (oriented) link is called positive (or negative) if all of its crossings are positive (or negative). (See §3). If L_i is a special alternating positive link, then $\hat{L}_i$ is a special alternating negative link. Therefore the uniqueness of $*$-product decomposition would lead to the following

Conjecture E *Let $L = L_1 * \cdots * L_k$ be a $*$-product decomposition of an alternating link L. If L is amphicheiral, then*

(1) k is even, and

(2) $L_1, \cdots, L_k$ is grouped into pairs $\{L_i, L_j\}$ $1 \leq i \neq j \leq k$, in such a way that L_j is ambient isotopic to the mirror image of L_i.

As a consequence of Conjecture E, we have

Conjecture F *Suppose that L is an alternating link and is amphicheiral. Then $c_0(L)$, the leading coefficient of the reduced Alexander polynomial $\Delta_L(t)$ of L, is a square of some integer (up to sign).*

The conclusion of Conjecture F might hold for even non-alternating links. As supporting evidence for Conjecture F, we can prove Theorem 7.1 below, which is a surprising application of the theorems related to the braid index.

Theorem 7.1 [MP] *Suppose that L is an alternating link. If $c_0(L)$ is a prime, L is never amphicheiral.*

In contrast to the Alexander polynomial, the Jones polynomial is generally unsymmetric: $V_L(t) \neq V_L(t^{-1})$. Since $V_L(t)$ must be symmetric for an amphicheiral link, the Jones polynomial detects the amphicheirality for many (but not all) links. Similarly, $P_L(v,z)$ is *partially* symmetric for an amphicheiral link L: $P_L(v,z) = P_L(v^{-1},-z)$. Although $P_L(v,z)$ cannot provide a complete solution to the amphicheirality problem, it can be used to prove the non-amphicheirality of many links. In fact, the proof of Theorem 7.1 depends on the detailed evaluation of $max\ deg_v\ P(v,z)$ and $min\ deg_v P(v,z)$.

References

[A] J.W. Alexander, A lemma on systems of knotted curves, *Proc. Nat. Acad. Sci.*, U.S.A., 9 (1923) 93-95.

[B] C. Bankwitz, Über die Torsionzahlen der alternierenden Knoten, *Math. Ann.* **103** (1930) 145-161.

[FW] J. Frank-R.F. Williams, Braids and the Jones polynomial, *Trans. Amer. Math. Soc.* **303** (1987) 97-108.

[FYHLMO] P. Freyd, et al., A new polynomial invariant of knots and links, *Bull. Amer. Math. Soc.* **12** (1985) 103-111.

[GLi] C. McA Gordon - R.A. Litherland, On the signature of a link, *Invent. Math.* 47 (1978) 53-69

[GLu] C. McA Gordon - J. Luecke, Knots are determined by their complements, *J. Amer. Math. Soc.* 3 (1989), 371-415.

[J] V. F.R. Jones, Hecke algebra representations of braid groups and link polynomials, *Ann. of Math.* **126** (1987) 335-388.

[K] L.H. Kauffman, State models and the Jones polynomial, *Topology* **26** (1987) 395-407.

[LM] W.B.R. Lickorish-K.C. Millett, A polynomial invariant of oriented links, *Topology* **26** (1987) 107-141.

[Mo] H.R. Morton, Seifert circles and knot polynomials, *Math. Proc. Cambridge. Phil. Soc.* 99 (1986) 107-109

[MS] H.R. Morton-H.B. Short, The 2-variable polynomial of cable knots, *Math. Proc. Cambridge Phil. Soc.* **101** (1987), 267-278.

[Mu 1] K. Murasugi, On a certain numerical invariant of link types, *Trans. Amer. Math. Soc.* **117** (1965) 387-422.

[Mu2] — , Jones polynomials and classical conjecture in knot theory, *Topology* **26** (1987) 187-194.

[Mu3] — , Jones polynomials and classical conjecture in knot theory (II), *Math. Proc. Cambridge Phil. Soc.* **102** (1987) 317-318.

[Mu4] — , On invariants of graphs with applications to knot theory, *Trans. Amer. Math. Soc.* 314 (1989) 1-49.

[Mu5] — , On the signature of a graph. *C.R. Math. Rep. Acad. Sci. Canada* 10 (1988) 107-111.

[Mu6] — , On the braid index of alternating links, (to appear in Trans. Amer. Math. Soc.).

[MP] K. Murasugi-J.H.Przytycki, The index of a graph with applications to knot theory (preprint)

[PT] J.H. Przytycki-P. Traczyk, Invariants of links of Conway type, *Kobe J. Math.* 4 (1987) 115-139.

[R] D. Rolfsen, *Knots and links*, Publish or Perish Inc (1976).

[Th1] M. Thistlethwaite, A spanning tree expansion of the Jones polynomial, *Topology* **26** (1987) 287-309.

[Th2] — , Kauffman's polynomial and alternating links, *Topology* 27 (1988) 311-318.

[Th3] — , On flypes and alternating tangles, (preprint).

[Tr] P. Traczyk, On the index of graphs: Index versus cycle index (preprint)

[W] H. Whitney, 2-isomorphic graphs, *Amer. J. Math.* 55 (1933) 236-244.

[Y] S. Yamada, The minimal number of Seifert circles equals the braid index of a link, *Inv. math.* **89** (1987) 347-356.

University of Toronto
Toronto, Canada M5S 1A1

MORSE THEORY OF CLOSED 1-FORMS

A. V. Pazhitnov
Institute of Chemical Physics, Kosygin str.
Moscov, 117977, USSR

1. Introduction. Let M^n be a smooth manifold and ω be a real-valued 1-form on M. We say that ω is Morse form if it is closed and if locally $\omega = dh$, where h is a Morse function.

In this paper we consider the analogue of usual Morse theory for such 1-forms. The first step in this direction was undertaken by S. P. Novikov [1,2], motivated by his and I. Smeltzer's research on periodic orbit of some hamiltonian systems [3]. We present here the results of the theory as developed by S. P. Novikov, M. Sh. Farber and the author.

Usual Morse theory corresponds to the case $\omega = df$, $f : M \to \mathbb{R}^1$. Denote by $m_p(f)$ the number of zeros of index p of the Morse function f. The classical Morse inequalities state that $m_p(f) \geq b_p(M)$ (where $b_p(M)$ stands for p-th Betti number of the manifold M). One also obtains the stronger inequalities which involve torsion in integral homology : $m_p(f) \geq b_p(M) + q_p(M) + q_{p-1}(M)$, where q_p is the least possible number of generators in the group $\text{Tors}H_p(M, \mathbf{Z})$. These stronger inequalities are *sharp* in the following sense: if a manifold M^n is 1-connected and $n \geq 6$ there exists a Morse function f on M for which the mentioned inequalities turn into equalities (Smale [4]).

Now we try to find corresponding generalizations of the results for the case of closed 1-forms. First of all note that for any zero c of a Morse form ω the notion of Morse index of c is well defined; the number of zeros of index p will be called the p-th Morse number of ω and denoted by $m_p(\omega)$.

There arise three natural problems:

1) find the suitable analogues of Betti numbers, i.e. find the nontrivial numerical homotopy invariants of (M, ω), providing lower bounds for $m_p(\omega)$.
2) give (if possible) a method for calculating them in terms of usual homology of M.
3) find homotopy invariants, providing sharp lower bounds for $m_p(\omega)$.

Three subsequent sections of the present paper correspond more or less to these three problems, and contain partial solutions for them. We present formulations of theorems and the outlines of proofs sometimes omitting the technical details.

The presentation of the material is not in historical succession so we make now some historical remarks. Three problems above were partly formulated and solved by Novikov in 1981–82 (cf. [1,2]). He was concerned mainly with the case of the forms ω with rational cohomology class $[\omega]$, and introduced for them the analogues of strong Morse inequalities (see §3). He conjectured that these inequalities are sharp for $\pi_1 M^n = \mathbf{Z}$,

$n \geq 6$. This was settled by Farber [5] in 1985. The numerical estimates of Morse type for the case of arbitrary cohomology class $[\omega]$ (see §1) were obtained simultaneously in 1985 by S. P. Novikov [6] and the author [7]. In the papers [6–8] we gave methods of calculation of corresponding numerical invariants (see §2). In 1987 J. C. Sikorav proved an algebraic lemma (§4, lemma 4.3) which enables one to give the better estimates for $m_p(\omega)$, than discussed in §1. The same year the author [9,10] proved the sharpness of the inequalities for the case $\pi_1 M^n = \mathbf{Z}^m$, $n \geq 6$ (and some restrictions, for precise formulation see th. 4.2, 4.3 of §4).

2. Analogues of Betti numbers. In this section we give the simplest lower estimates for $m_p(\omega)$ which are analogues of Morse estimates $m_p(f) \geq b_p(M)$ for Morse functions f. We need some notation.

From now on M^n will denote connected compact smooth manifold without boundary. Denote by ξ the deRham cohomology class of the Morse form ω. For any $t \in \mathbf{C}$ define the 1-dimensional representation $\rho_t : \pi_1 M \to GL(1, \mathbf{C}) = \mathbf{C}^*$ as following:

$$\rho_t(\gamma) = \exp\left(t \int_\gamma \omega\right) \in \mathbf{C}^*.$$

Clearly ρ_t depends only on $\xi = [\omega]$. Consider the corresponding local systems of coefficients on M and denote by $H^*(M, \rho_t)$ the cohomologies with coefficients in it. The vector spaces $H^*(M, \rho_t)$ are finite dimensional; let $\beta_p(\xi, t) = \dim H^p(M, \rho_t)$. We need a simple lemma. The proof we postpone until §2.

LEMMA 2.1. *The numbers $\beta_p(\xi, t)$ do not depend on t for $Re(t)$ large enough.*

Thus we introduce the notation $B_p(M, \xi) = \lim_{\mathrm{Re}(t) \to \infty} \beta_p(\xi, t)$ and now we can state and prove our estimates.

THEOREM 2.2. $m_p(\omega) \geq B_p(M, [\omega]), \quad 0 \leq p \leq n.$

PROOF. First of all note that standard deRham type argument shows that cohomology with coefficients in the local system ρ_t equals the cohomology of the deRham complex $\Omega^* M$ with perturbed differential d_t where $d_t \lambda = d\lambda + t\xi \wedge \lambda$, $(\lambda \in \Omega^* M)$. The latter is equal to the kernel of corresponding Laplacian $\Delta_t = d_t^* d_t + d_t d_t^*$ (for any given Riemannian metric).

It is enough to prove that

$$m_p(\omega) \geq \beta_p(\xi, t) = \dim \ker(\Delta_t : \Omega^p(M) \to \Omega^p(M))$$

for t real and large enough.

We proceed further just as in [11], where the case $\omega = df$, $f : M \to \mathbb{R}^1$ is treated. All the proofs work equally well in the case of general ω. We reproduce the main lines here for the sake of completeness. Let $\lambda \in \Omega^p(M)$; the explicit computation shows that $\Delta_t \lambda = \Delta\lambda + |t|^2 |\omega|^2 \lambda + tG(\lambda)$, where Δ is a usual Laplacian, $\Delta = d^* d + dd^*$, G is a tensor field of type (p, p), not depending on t. Now fix any real $A > 0$ and let $t \to \infty$. Then the eigen p-forms of Δ_t concentrate in the neighbourhood of the zeros of ω.

Next we turn to to local considerations. Let c be any zero of ω of index p. We choose the local coordinates x_i in the small neighbourhood U_c of the point c in such a way that the point c represent the origin of the coordinate system and in U we have

$$\omega = \frac{1}{2}(-x_1^2 - \cdots - x_p^2 + x_{p+1}^2 + \cdots + x_n^2).$$

We can suppose that the Riemannian metric in coordinates x_i is given by $g^{ij} = \delta_{ij}$. Then one can check that the p-form

$$\lambda_c = \Big(\prod_{k=1}^{n} e^{\frac{-tx_k^2}{2}}\Big) dx_1 \wedge \ldots \wedge dx_p$$

(defined only inside U) satisfies $\Delta_t \lambda_c = 0$. We can extend the form λ_c to a global form $\widetilde{\lambda_c}$ which vanishes outside U_c. Moreover we can do it in such a way that the quadratic form $\langle \Delta_t \cdot, \cdot \rangle$ on the space generated by all the $\widetilde{\lambda_c}$ is (for sufficiently large t) less than $A\langle \cdot, \cdot \rangle$. Thus we conclude (from minimax principle) that the vector space $A^p(M)$, generated by p-eigenforms of Δ_t, having eigenvalues less than A has the dimension at least $m_p(\omega)$ (t is sufficiently large). One can also prove that the dimension of $A^p(M)$ is exactly $m_p(\omega)$ (this is more difficult and I will not reproduce it here). The spaces $A_p(M)$ form a finite-dimensional subcomplex $A^*(M) \subset (\Omega^*(M), d_t)$, and its cohomology is equal to $H^*(\Omega^*(M), d_t) = H^*(M, \rho_t)$. Now we have $m_p(\omega) = \dim A^p(M) \geq \dim H^p(M, \rho_t)$. ∎

3. Cohomology with local coefficients and Massey products. Now we have the analogue $B_p(M, \xi)$ of Betti numbers for $\xi \in H^1(M, \mathbb{R})$. The subject of this section is to compute them in terms of usual cohomology of M and Massey products of the form $\langle x, \xi, \ldots, \xi \rangle$.

First we need some generalities about the (co)homology with local coefficients.

Let k be a field. For a 1-dimensional representation $\rho : \pi_1 M \to GL(1, k)$ (i.e. a homomorphism $\rho : \pi_1 M \to k^*$) we denote by $H_*(M, \rho)$ the homology with local coefficients in k, determined by ρ. Every 1-dimensional representation factors through some epimorphism $E : \pi_1 M \to \mathbb{Z}^l$. Fix now E and denote by $R(E)$ the space of all representations, factoring through E; $R(E) = (k^*)^l \subset k^l$. Consider the covering $\widehat{E} : \widehat{M} \to M$, corresponding to the subgroup $\ker E \subset \pi_1 M$. The structure group of the covering is $\mathbb{Z}^l$. Denote by Λ the ring $k[\mathbb{Z}^l] = k[t_1^{\pm 1}, \ldots, t_l^{\pm 1}]$. The homology $H_*(M, \rho)$ is the homology of the complex $C_*(M, k)$ with differential d_ρ, which is a regular function on $R(E)$, $d_\rho \in \Lambda$. Suppose now that k is infinite. The standard argument shows that for all $\rho \notin V$ for some algebraic variety $V \subset (k^*)^l$ we have

$$\dim_k H_*(M, \rho) = \dim_{\{\Lambda\}} H_*(M, k). \tag{3.1}$$

For $\rho \notin V$ we call $H_*(M, \rho)$ general position homology (g.p. homology) and denote it by $H_*^{\text{g.p.}}(M, \rho)$ and $\dim H_p(M, \rho)$ — by $b_p^{\text{g.p.}}(M, E)$.

The same result holds for cohomology: for some algebraic variety $V' \subset (k^*)^l$ we have $\dim_k H^*(M, \rho) = \dim_{\{\Lambda\}} H^*(\widehat{M}, k)$. We denote general position numbers by $b_{\text{g.p.}}(M, E)$. Consider now some examples and specifications.

1) The local system ρ is given by a 1-form ω (as in section 2). It factors through $E : \pi_1 M \to \mathbf{Z}^l$ where l is maximal number of $\mathbf{Q}$-linearly independent periods of ω. This number is the rank of $\mathrm{im}\,(\xi : H_1(M,\mathbf{Z}) \to \mathbf{R})$ where $\xi = [\omega]$, and we will call it the degree of irrationality of ξ (or ω). Denote these periods by $a_1, \ldots, a_l$. Then the local system ρ_t is given by a curve in the space $(\mathbf{C}^*)^l$ with the coordinates $\rho(t) = (\exp(a_1 t), \ldots, \exp(a_l t))$. One checks immediately (using $a_i \in \mathbf{R}$) that for $\mathrm{Re}(t)$ sufficiently large this curve does not meet the algebraic manifold V' and so for $\mathrm{Re}(t) \to \infty$ we have $\beta_p(\xi, t) = b^p_{\mathrm{g.p.}}(M, E)$. (In particular, we have proved Lemma 2.1).
2) Suppose that $l = 1$. Then the algebraic variety V consists of a finite number of points in complex plane and they can be explicitly computed in terms of homology of corresponding cyclic cover $\widehat{M}$. Namely, consider the module $H_p(\widehat{M}, k)$ over the principal ideal domain $\Lambda = k[t, t^{-1}]$, and decompose it as a sum

$$(\Lambda)^{l_p} \oplus \Big(\bigoplus_{i=1}^{q_p} \Lambda / a_i^{(p)} \Lambda\Big)$$

where a_i are the polynomials in $k[t]$. To compute the cohomology $H^p(M, \rho)$ which is the same as the homology of the complex $\mathrm{Hom}\,(C_*(\widehat{M}), k)$ ($k[t]$ is a module over Λ induced by ρ) we apply the universal coefficient theorem:

$$0 \longrightarrow (\mathrm{Ext}^1(H_{p-1}(\widehat{M}), k) \longrightarrow H^p(\widehat{M}, \rho) \longrightarrow \mathrm{Hom}\,(H_p(\widehat{M}), k) \longrightarrow 0.$$

Now one easily sees that for general ρ the left term is zero, the right one has dimension b_p. The dimension of left term jumps up when $\rho(t)$ is the root of some $a_i^{(p-1)}$ and the dimension of right term — when $\rho(t)$ is the root of some $a_i^{(p)}$.
In particular we see that the jumps of the $b_*(\xi, t)$, where $\xi = [\omega]$ has the degree of irrationality 1, take place in these points t for which e^t is a root of some polynomial P_i in the decomposition

$$H_*(M, \mathbf{Q}) = \Big(\bigoplus \Lambda\Big) \oplus \Big(\bigoplus_{i=1}^{N} \Lambda / P_i \Lambda\Big).$$

3) Using the generic cohomology one can get the estimation of §2 type. Namely suppose that the cohomology class $\xi = [\omega]$ of the Morse form ω is the linear combination $\sum \lambda_i e_i$ of the integral classes $e_i \in H^1(M, \mathbf{Z}) = \mathrm{Hom}\,(\pi_1(M), \mathbf{Z})$. Choose the epimorphism $E : \pi_1(M) \to \mathbf{Z}$ such that all e_i are $\mathbf{Z}$-linear combinations of the coordinate functions of the projections. Then $m_p(\omega) \geq b_p^{\mathrm{g.p.}}(M, E)$.

Now we turn to computation of cohomology with g.p. local coefficients in deRham case, following [6]. Consider as in section 2 the differential in $\Omega^*(M)$ of the form $d_t = d + t\xi\,\wedge$. We are interested in general t, so we treat it as a parameter and consider the space of $\Omega^*(M)[[t]]$ consisting of the power series, converging somewhere near $t = 0$, and differential d_t on it. Suppose $\omega(t) = \omega_0 + \omega_1 t + \omega_2 t^2 + \ldots$ is d_t-cycle. After simple calculation we get $d\omega_0 = 0$, $\xi \wedge \omega_0 = -d\omega_1, \ldots$ it means that all Massey products of the type $\langle \xi, \xi, \ldots, \omega_0 \rangle$ vanish in ordinary cohomology. Factoring d_t-cocycles by d_t-coboundaries correspond to factoring by images of Massey products.

We can generalize the procedure as follows. Consider the case of m-dimensional representations of $\pi_1(M)$ (generally nonabelian) and let d_t be a I-parameter analytic deformation of d of the kind $d_t = d + \Theta(t)\wedge$, $\Theta(t) \in \Omega^1(M) \bigotimes \mathrm{End}(\mathbf{C}^m)$, $d_t^2 = 0$. (The corresponding deformation of trivial representation I is given by $\rho_t(\gamma) = \exp(\int_\gamma \Theta(t))$.

Write $\Theta(t) = \Theta_1 t + \Theta_2 t^2 + \cdots$, then integrability condition $0 = d\Theta(t) + \Theta(t) \wedge \Theta(t)$ takes form $d\Theta_n + \sum_{i=1}^{n-1} \Theta_i \wedge \Theta_{n-i} = 0$. The condition for $\omega(t) = \omega_0 + \omega_1 t + \omega_2 t^2 + \cdots$ to be a d_t-cocycle is $d\omega_0 = 0, \ldots, (\sum_{i=1}^{n} \Theta_i \omega_{n-1}) + d\omega_n = 0$. One sees at once that these conditions mean exactly that all Massey products $\langle \omega_0, \Theta_1, \ldots, \Theta_1 \rangle$ exist and vanish (precise statements of this kind we give in next section).

Now we intend to generalize this procedure to the infinite field k of any characteristic. We can not use the deRham framework any more and we must work directly with chains and cochains. The main difficulty here is to find the suitable formalism, which certainly must be the spectral sequence, converging to $H_*^{\mathrm{g.p.}}(M,k)$, with differentials expressible in terms of Massey products. The existence of such spectral sequence was conjectured by Novikov (private communication). Actually he conjectured more general thing, involving multidimensional representations:

CONJECTURE 3.1. (S. P. Novikov). *Let k be any algebraically closed field and suppose we have a representation $\rho : \pi_1 M \to GL(m,k)$, which is close enough to trivial one and is in general position.*

Then(co)homology of M with local coefficients, generated by ρ can be computed explicitly in terms of ordinary homology and for $m = 1$ it can be done by means of Massey products.

In [8] we proved this conjecture for $m = 1$ and now we will formulate the result and give main lines of proof. It appears to be more convenient to work with homology. We will assume for simplicity that the fundamental group is equal to $\mathbf{Z}^n$ although this restriction can be easily eliminated.

Suppose that X is a CW-complex $\pi_1 X = \mathbf{Z}^n$, k-infinite field. Consider the space R of representations $R = (k^*)^n \subset k^n$ and any algebraic curve $\gamma(t)$ in it such that $\gamma(0) = I = (1, \ldots, 1)$ with polynomial coordinate functions $P_i(t) = 1 + a_{1i} t + \cdots + a_{Ni} t^N$; $1 \leq i \leq n$, $a_{ij} \in k$. First of all we construct a spectral sequence (E_*^r, ∂_r), begining from $H_*(X,k)$ and converging to $H_*^{(\alpha)}(X,k)$, where (α) is a generic point of γ. Namely, consider the ring $W = S^{-1}k[t]$, where $S = \{P_1, \ldots, P_n\}$, and the short exact sequence $0 \longrightarrow W \xrightarrow{t} W \xrightarrow{\varepsilon} k \longrightarrow 0$. Now, tensoring $C_*(\widetilde{X}, k)$ with it over $\Lambda = \mathbf{Z}[t_1^{\pm 1}, \ldots, t_n^{\pm 1}]$ and passing to homology, we obtain an exact couple

$$\begin{array}{ccc} H_*(C_*(\widetilde{X},k) \otimes_\Lambda W) & \xrightarrow{t} & H_*(C_*(\widetilde{X},k) \otimes_\Lambda W) \\ \nwarrow & & \swarrow \varepsilon \\ & H_*(X,k) & \end{array}$$

and one easily shows, that the term E_∞ of the corresponding spectral sequence equals to $H_*^{(\alpha)}(X,k)$. For $n = 1$ and $\gamma(t) = 1 + t$ this spectral sequence is exactly the same as considered by Milnor in [12].

Now we proceed as to compute differentials in this spectral sequence. Denote $\xi \in H^1(X,k)$ the cohomology class, given by a tangent vector $\{dP_i/dt\}$ to a curve γ at $t=0$. We will prove that $d_r(x) = \langle x, \xi, \dots, \xi \rangle$, where indeterminacy of Massey product is reduced suitably with respect to γ. Now pass to the precise formulation.

We will need a very special kind of Massey products and I reproduce here from [13] only the definitions for this case. Let R be any commutative ring, $\xi_1, \xi_2, \dots, \xi_r \in C^*(X,R)$. Denote $(-1)^{|x|}x$ by $\overline{x}$. We say, that $\xi_1, \xi_2, \dots, \xi_r$ form an incomplete symmetric Massey triangle (i.s.t) if $d\xi_1 = 0$, $d\xi_k = \sum_{j=1}^{k-1} \overline{\xi}_{k-j}\xi_j$, $k > 1$ (ab means here $a \cup b$); if so, then $\mu = \sum_{j=1}^{r} \overline{\xi}_{r-j+1}\xi_j$ is a cocycle, called symmetric Massey product $\langle \xi_l \rangle^r$, and if $\mu = d\xi_{r+1}$, we say, than $\xi_1, \xi_2, \dots, \xi_{r+1}$ form a complete symmetric Massey triangle (c.s.t). If $r = \infty$ we call this an infinite complete symmetric triangle (∞.s.t). If $x_1, \dots, x_r \in C_*(X,R)$ and $(\xi_1, \dots, \xi_r)$ form an (c.s.t) we say, that (x_i, ξ_j) form an incomplete quasisymmetric Massey triangle (i.q.t) if $dx_1 = 0$, $dx_k = \sum_{i=1}^{k} x_i \cap \xi_{k-i} (k > 1)$ (then $\mu = \sum_{i=1}^{r} x_i \cap \xi_{r-i+1}$ is a cocycle , denote by $\langle x_1, \xi_1 \rangle^r$ etc.). We arrange this usually as follows

$$\begin{array}{ccc} \xi_1, \dots\dots, \xi_1 & & x_1, \xi_1, \dots\dots, \xi_1 \\ \xi_2, \dots, \xi_2 & & x_2, \xi_2, \dots, \xi_2 \\ \dots\dots & & \dots\dots \\ \xi_r, \xi_r & & x_r, \xi_r \end{array}$$

Now I introduce the spectral sequence (E_r, d_r).

LEMMA 3.1. *For every infinite symmetric triangle $\Delta = (\xi_1, \xi_2, \dots)$ there exist the spectral sequence (E_r, d_r), satisfying i) and ii)*

i) $E_1 = H_*(X,R)$, $d_1 x = x \cap \xi_1$

ii) if x is a cycle and there exist $x_1 = x, x_2, \dots, x_r$, forming together with ξ_j an (i.q.t) and $y = dz + \sum \overline{x}_i \cap \xi_{r-i}$ then x survives up to E_r and $d_r x = y$.

It appears that the above spectral sequence (E_r, ∂_r) coincides with $(E_r, -d_r)(\Delta)$, if we choose Δ suitably, according to the curve γ.

I shall now explain how to choose Δ. Fix a map $f: X \to T^n$ inducing iso in π_1 and H_1. Consider some (∞.s.t.) $(\xi_1, \xi_2, \dots)$ on T^n and lift it to $\mathbb{R}^n$. Since $\mathbb{R}^n$ is contractible, every 1-cocycle is canonically cohomologous to zero, and we have an (∞.q.t):

$$\begin{array}{l} h_1 = 1, \xi_1, \xi_1, \dots \\ \qquad\quad h_2, \xi_2, \dots \\ \qquad\qquad \dots\dots \end{array}$$

where $h_1 = 1 \in C^0(\mathbb{R}^n), h_i \in C^0(\mathbb{R}^n)$. Denote by e_i orthonormal basis in $\mathbb{R}^n$, by μ_i-corresponding Massey cocycle, and by $A_{ij}(\Delta)$ the value $\langle \mu_i, e_j \rangle$.

THEOREM 3.2. *For any polynomial curve $\gamma(t)$ in a space k^n given by $\gamma_i(t) = P_i(t) = 1 + a_{1i}t + \dots + a_{Ni}t^N$ and for any symmetric Massey triangle $\Delta = (\xi_1, \xi_2, \dots)$, satisfying $\xi_1 = \frac{d\gamma}{dt}|_{t=0}$ and $A_{ij}(\Delta) = a_{ij}$, the spectral sequences (E_r, ∂_r)(converging to homology with coefficients in generic point of γ) and $(E_r, -d_r)$("Massey spectral sequence") coincide.*

For the proof see [8].

Now we must be sure that infinite symmetric triangles $(\xi_1, \ldots)$ on T^n with prescribed values of A_{ij} exist. This is the contents of the following lemma, which should have been known long since, but I couldn't find it in literature.

LEMMA 3.3. *Let Y be a space and suppose $\varphi : R \to K$ is a monomorphism, where K is a field of characteristic zero, and $\varphi^* : H^*(Y, R) \to H^*(Y, K)$ is mono. Then every incomplete symmetric Massey triangle $\Delta = (\xi_1, \ldots, \xi_r)$, consisting of odd-dimensional cochains, can be completed.*

I'd like to emphasize that the proof relies heavily on the vanishing of torsion in integral homology of $T^n = B(\mathbf{Z}^n) = B\pi_1(X)$.

The idea of the proof is as follows. It suffices to complete over K. We pass to the deRham algebra $A^*(Y, K)$, which is strictly commutative, so for odd-dimensional cochains z we have $\langle z\rangle^r = 0$. Alas, the morphism $\rho_1 : A^*(Y, K) \to C^*(Y, K)$ is not multiplicative, but there exist corresponding higher homotopies (cf. [14]) ρ_i, and, finally, the completing vertex is written down explicitly using $\rho_1^{-1}(\xi_i), \rho_i$ (for detailed proof see [8]).

4. The sharp inequalities. In this section we discuss the problem 3) of the Introduction. The general remark to this section is that all the numerical invariants providing lower bounds for $m_p(\omega)$, which are known up-to-date come from homology of free abelian covers and the sharpness theorems concern correspondingly the manifolds with free abelian fundamental groups.

We begin our exposition of sharpness results with the case of Morse functions, i.e. the forms of irrationality degree zero. We cited already the classical Morse inequalities

$$m_p(f) \geq b_p(M) + q_p(M) + q_{p-1}(M) \tag{4.1}$$

where $b_p(M)$ is the rank of the group $H_p(M)$, and $q_p(M)$ is the minimal number of generators of the subgroup $\operatorname{Tors} H_p(M)$. These inequalities are sharp in the following sense: For any simply-connected manifold M^n, one can find a Morse function f on M, for which $m_p(f)$ equals the righthand side of (4.1). The proof of these inequaities goes as follows. For a given Morse function f one constructs a free chain comlex $C_*(f)$ over $\mathbf{Z}$ for which the number of generators $\mu(C_p(f))$ is equal to $m_p(f)$, and the homology $H_p(C_*(f))$ is isomorphic to $H_p(M, \mathbf{Z})$. When $C_*(f)$ is constructed the proof of (4.1) becomes an easy algebraic exercise. To construct the complex $C_*(f)$ we define $C_p(f)$ to be the free abelian group generated by critical points of f of index p, and for such point x define $dx = \sum \lambda(x, y)y$, where the sum is taken over all y of index $(p-1)$ and the incidence coefficient $\lambda(x, y)$ is defined as follows. We choose the Riemannian metric on M and consider all the paths of steepest descent (with respect to f) going from x to y. To each of them one assigns the number (+1) or (-1) according to some rule which I'll not discuss here (see [15]) and then $\lambda(x, y)$ is by definition the sum of these numbers. One checks that $(C_p(f), d)$ is the chain complex and that $H_p(C_*(f)) \approx H_p(M, \mathbf{Z})$.

The inequalities (4.1) are not sharp for the non-simply-connected manifolds. The reason is that for any regular cover $\overline{M} \longrightarrow M$ with the structure group G the Morse function f determines the complex $C_*(f)$ of free $\mathbf{Z}[G]$-modules such that $\mu_{\mathbf{Z}[G]}(C_p(F)) = m_p(f)$ and the complex $C_*(f)$ is simply homotopy equivalent to the chain complex of M, given by some triangulation of M. So the Morse number $m_p(f)$ must be not less then minimal possible number $\mu(C_p)$, where C_* runs through the free based chain complexes over $\mathbf{Z}[G]$ simply homotopy equivalent to $C_*(M)$. This number $\min_{C_*}\mu(C_p)$ is generally greater then righthand side of (4.1). The best possible bounds got in this way come from the universal cover $\widetilde{M} \longrightarrow M$. Sometimes the inequalities obtained this way are sharp. This holds e.g. for the case of Morse functions f on cobordisms $(W^n; V_0, V_1), \partial W = V_0 \bigcup V_1, f$ is constant on both components of boundary, $n \geq 6, \pi_1 V_0 \longrightarrow \pi_1 W \longleftarrow \pi_1 V_1$ are isomorphisms, $Wh(\pi_1 W) = 0$ [16].

Now we pass to the forms ω of degree of irrationality 1. To obtain the inequalities of type (4.1) Novikov ([1, 2]) constructs the analogue of the Morse complex. Namely, the form becomes exact on some infinite cyclic cover $p: \overline{M} \rightarrow M$, $p^*\omega = d\overline{f}$. We have an analog of Morse complex of the function f with an essential difference: for a given critical point of f of index p there can exist an infinite number of critical point y of index $(p-1)$ connected with x by the paths of steepest descent. To overcome the difficulty Novikov considers the completion $\widehat{\Lambda} = (\mathbf{Z}[[t]])[t^{-1}]$ of the group ring $\Lambda = \mathbf{Z}[\mathbf{Z}] = \mathbf{Z}[t^{\pm 1}]$, and constructs the complex $C_*(\overline{M}, \omega)$ with the following properties:

1) $C_*(\overline{M}, \omega)$ is a free chain complex over $\widehat{\Lambda}$, $\mu_{\widehat{\Lambda}}(C_p(\overline{M}, \omega)) = m_p(\omega)$,
2) $H_p(C_*(\overline{M}, \omega)) \approx H_p(M) \otimes_\Lambda \widehat{\Lambda}$.

Since the ring $\widehat{\Lambda}$ is the principal ideal domain we obtain

$$m_p(\omega) \geq b_p(M, [\omega]) + q_p(M, [\omega]) + q_{p-1}(M, [\omega]), \tag{4.2}$$

where $b_p(M, [\omega])$ denotes the rank of the module $H_p(M) \otimes_\Lambda \widehat{\Lambda}$ over $\widehat{\Lambda}$ and $q_p(M, [\omega])$ — the minimal number of generators of the module $\mathrm{Tors}_{\widehat{\Lambda}}(H_p(M) \otimes_\Lambda \widehat{\Lambda})$. If no ambiguity is possible we denote these numbers just $b_p(\xi), q_p(\xi)$, where $\xi = [\omega]$. Note that the cyclic covering $\overline{M} \longrightarrow M$ is uniquely determined by the cohomology class $[\omega] \in H^1(M, \mathbf{R})$ (recall that $[\omega]$ is a multiple of the integer class) so these numbers b_p, q_p really depend only on $[\omega]$.

These inequalities are exact [5] in the following sense.

THEOREM 4.1. [5]. *For any manifold M^n, $n \geq 6, \pi_1 M^n = \mathbf{Z}$, there exists a Morse form ω, representing a generator in $H^1(M, \mathbf{Z})$, such that $m_p(\omega)$ is equal to righthand side of* (4.2).

Now I will say few words about the proof. One particular case of the problem was considered in 60's already by Browder–Levine [17].They ask when the manifold M^n, $\pi_1 M^n = \mathbf{Z}$ can be smoothly fibered over the circle. (Note that the forms of degree of irrationality 1 are up to the constant just the Morse maps into the circle, so in our language the problem is following: when does there exist a smooth Morse form ω, $[\omega] \neq 0$, without critical points, i.e. $m_*(\omega) = 0$?) The answer is that for $n \geq 6$ the necessary and sufficient condition for existence of the smooth fibration is that the fiber of the map $M^n \longrightarrow S^1$, inducing iso in π_1, is homotopy equivalent to finite CW-complex. This is easily checked to be equivalent to the condition $b_p(M, [\omega]) = q_p(M, [\omega]) = 0$ for all p.

The Browder-Levine's argument is as follows. Consider the arbitrary smooth map $f: M \rightarrow S^1$, representing the generator $\xi \in H^1(M, \mathbf{Z})$ and denote by V a regular inverse image of $c \in S^1$. The manifold $W = M \setminus \mathrm{Tub}(V)$, where $\mathrm{Tub}(V)$ stands for a small open tubular neighbourhood of V in M, has the boundary ∂W consisting of two components:

$\partial W = \partial_0 W \bigcup \partial_1 W, \partial_i W \approx V$. The infinite cyclic covering M is obtained as follows: take an infinite number of copies of W (denoted by $W_n, n \in \mathbf{Z}$), and glue them together, identifying the components of the boundary: $\partial_0(W_n) \approx V \approx \partial(W_{n-1})$. Note that $\mathbf{Z}$ acts freely on M and if we denote the generator of $\mathbf{Z}$ by t then $tW_{n-1} = W_n, t(\partial_0 W_{n-1}) = \partial_0 W_n$. We can suppose that $\xi(t) = -1$. Now attaching handles to V inside W we can modify V in such a way that $\pi_1 V = 0 = \pi_1 W$, $H_*(W, V) = 0$. (We need a finite number of handles since M has the type of finite CW-complex.) Afterwards the application of Smale's theorem finishes the proof.

Farber's proof of his sharpness theorem goes the same line, but is more technically complicated. He shows that after attaching to V the finite number of handles inside W we construct a manifold V, for which $\pi_1 V = 0 = \pi_1 W$ and Betti numbers $b_p(W, V)$ and torsion numbers $q_p(W, V)$ are equal to corresponding Novikov numbers $b_p(W, [\omega])$, $q_p(W, [\omega])$. He finishes applying Smale's theorem to the cobordism $(W; \partial_0 W, \partial_1 W)$. We will not reproduce here the details of the proof of theorem 4.1, since many of them appear again in the proof of the theorem 4.2 below.

Suppose now that rk $H_1 M$ is greater than 1. There is no reason to expect (4.2) to be sharp for all classes $\xi \in H^1(M, \mathbf{R})$ of irrationality degree 1 (see the above discussion of Morse functions). Still, this holds for the classes of general position. That is the subject of the following theorem.

THEOREM 4.2. [9,10]. *Suppose $\pi_1 M^n = \mathbf{Z}^m, n \geq 6$, universal covering M^n is 4-connected. Then there exist a finite number of integer hyperplanes $\Gamma_i \subset H^1(M, \mathbf{Z})$ such that any nonzero integer cohomology class $\xi \in H^1(M, \mathbf{Z}) \setminus \bigcup_i \Gamma_i$ is represented by a Morse form ω with $m_p(\omega)$ equal to the righthand side of (4.2).*

Although the formulation says nothing about the forms of degree of irrationality > 1, the proof requires them urgently, and, in turn, gives an sharpness result for them. So I pass now to these forms.

First of all, we need the analog of the ring $\widehat{\Lambda}$ above. Denote Λ the ring $\mathbf{Z}[\mathbf{Z}^k] = \mathbf{Z}[t_1^{\pm 1}, \dots, t_k^{\pm 1}]$. Consider the abelian group $\mathbf{Z}[[t_1^{\pm 1}, \dots, t_k^{\pm 1}]]$ of all formal power series in $t_i^{\pm 1}$. For each element $\lambda = \sum \lambda_I t^I$ of this group we define supp$\lambda \subset \mathbf{Z}^k$ to be a set of indices I with $\lambda_I \neq 0$. Now let ξ be a homomorphism $\mathbf{Z}^k \to \mathbf{R}$. Novikov ring Λ_ξ^- consists of all power series λ, for which the set supp$\lambda \bigcap (\xi \leq c)$ is finite for any $c \in \mathbf{R}$.

Return now back to the forms. Let ω be a Morse form on a manifold M, $\xi = [\omega]$. Suppose $p : \widehat{M} \to M$ is a free abelian covering with the structure group $\mathbf{Z}^k$, such that $p^*\omega$ is exact: $p^*\omega = d\widehat{f}$. In this case ξ determines a homomorphism $\mathbf{Z}^k \longrightarrow \mathbf{R}$ which we'll denote by the same letter ξ. The consideration, similar to above leads us to a definition of the Novikov complex $C_*(\widehat{M}, \omega)$ which has the following properties:

1) It is a free chain complex over Λ_ξ^- and $\mu(C_p(\widehat{M}, \omega)) = m_p(\omega)$.
2) Its homology is isomorphic to $H_*(C_*(\widehat{M}) \otimes_\Lambda \Lambda_\xi^-)$.

We'll be particulary interested in the case when degree of irrationality of $\xi = [\omega]$ is equal to rk$H_1(M, \mathbf{Z})$, which we denote by m. These cohomology classes (and corresponding forms) will be called totally irrational. The corresponding homomorphism $\xi : \mathbf{Z}^m \to \mathbf{R}$ is then a monomorphism. The following lemma enables us to define in this case the corresponding numerical invariants – Betti and torsion numbers. It is due to

J. C. Sikorav (autumn 1987, private communication; the full proof can be found in [9, 10]).

LEMMA 4.3. *If $\xi : \mathbf{Z}^k \to \mathbf{R}$ is a monomorphism, the ring Λ_ξ^- is a principal ideal domain.*

Using this lemma one defines for a totally irrational class $\xi \in H^1(M, \mathbf{R})$ the numbers $b_p(\xi)$, $q_p(\xi)$ as follows. Take a maximal free abelian cover $\overline{M}^n \longrightarrow M$. Homology $H_*(\overline{M}^n)$ is a $\Lambda = \mathbf{Z}[\mathbf{Z}^m]$-module; each module $H_p(M^n) \otimes_\Lambda \Lambda_\xi^-$ is the finitely generated over the principal ideal domain, and we define $b_p(\xi)$ and $q_p(\xi)$ as its rank and torsion number.

Considering the Novikov complex we get again the inequalities (4.2). Note that the full proof of (4.2) using the Novikov complex presents technical difficulties. See [10] for the complete proof appealing only to Morse theory for compact manifolds with boundary.

Now we need a little more algebra.

Let C_* be any free finitely generated complex over $\Lambda = \mathbf{Z}[\mathbf{Z}^k]$. Then for totally irrational $\xi : \mathbf{Z}^k \to \mathbf{R}$ the ranks $b_p(C_*, \xi)$ and the torsion numbers $q_p(C_*, \xi)$ of the homology are defined. One can show that $H_p(C_* \otimes_\Lambda \Lambda_\xi^-) = H_p(C_*) \otimes_\Lambda \Lambda_\xi^-$. The next lemma was also communicated to the author in autumn 1987, for the complete proof see [9, 10].

LEMMA 4.4.(J. C. Sikorav). *The number $b_p(C_*, \xi)$ does not depend on ξ. There exists a finite set of integer hyperplanes $\Gamma_i \subset Hom(\mathbf{Z}^k, \mathbf{R}) = \mathbf{R}^k$, such that numbers $q_p(C_*, \xi)$ are constant in each connected component of the complement $Hom(\mathbf{Z}^k, \mathbf{R}) \setminus \bigcup_i \Gamma_i$.*

Using this lemma we can correctly define the numbers $b_p(C_*, \xi)$, $q_p(C_*, \xi)$ for <u>any</u> $\xi \in \mathrm{Hom}(\mathbf{Z}^k, \mathbf{R}) \setminus \bigcup_i \Gamma_i$, setting $b_*(C_*, \xi) = b_*(C_*, \xi')$, where ξ' is totally irrational and sufficiently close to ξ (the same definition for q_*). For a manifold M in consideration we will denote these numbers $B_p(\xi)$, $Q_p(\xi)$.

For our further purposes we replace the ring Λ_ξ^- by a suitable localization $S_\xi^{-1}\Lambda$. Namely, let

$$S_\xi = \{1 + \lambda | \lambda \in \Lambda,\ \mathrm{supp}\lambda \subset (\xi < 0)\}.$$

We need to know that $S_\xi^{-1}\Lambda$ is also a principal ideal domain for ξ totally irrational (see [9]). For the case $m = 1$ the corresponding localization was introduced and used by Farber [5].

The algebraic part of the Theorem 4.2 is given by

THEOREM A. *For any manifold M, $\pi_1 M = \mathbf{Z}^m$ there exists a finite set of integer hyperplanes $H_i \subset H^1(M, \mathbf{R})$(including Γ_i from lemma 4.4 and maybe something else), such that*

1) for any nonzero $\xi \in H^1(M, \mathbf{R}) \setminus \bigcup_i H_i$, the module $S^{-1}H_p(M, \mathbf{Z})$ is isomorphic to

$$(S_\xi^{-1}\Lambda)^{B_p(\xi)} \oplus \Big(\bigoplus_{j=1}^{Q_p(\xi)} S_\xi^{-1}\Lambda / a_j^{(p)} S_\xi^{-1}\Lambda\Big), \quad 0 \le p \le n, \tag{4.3}$$

where $a_j^{(p)}$ are the nonzero elements of Λ, noninvertible in $S_\xi^{-1}\Lambda$,

2) For ξ rational we have $B_p(\xi) = b_p(\xi)$, $\quad Q_p(\xi) = q_p(\xi)$.

Note that the ring $S_\xi^{-1}\Lambda$ for ξ rational is very far from being a principal ideal domain (as well as ring Λ_ξ^-). The theorem A states that inspite of this the modules $S^{-1}H_p(M,\mathbf{Z})$, which "control" the number of critical points of $\omega, [\omega] = \xi$, are just like of the "principal ideal domain" type, and the corresponding "ranks" and "torsion numbers" are precisely $b_p(\xi)$ and $q_p(\xi)$.

The geometric part of the proof of theorem 4.2 proceeds on the lines of [5, 17, 18]. To explain it in more details we fix the notations. We will assume that ξ is a projection of $\mathbf{Z}^m$ on its first summand $\mathbf{Z}$, so that there exists a system of free generators $t, t_1, \dots, t_{m-1}$ of $\mathbf{Z}^m$, such that $\xi(t) = -1, \xi(t_i) = 0$. Let $R = \mathbf{Z}[\mathbf{Z}^{m-1}] = \mathbf{Z}[t_1^{\pm 1}, \dots, t_{m-1}^{\pm 1}]$, $P = R[t]$, $\Lambda = R[t, t^{-1}]$, $S = \{1 + tQ(t) | Q(t) \in R[t]\}$, $K = S^{-1}P$, $\Gamma = S_\xi^{-1}\Lambda = t^{-1}S^{-1}P$.

The framed submanifold (N^{n-1}, ν) of M^n will be called splitting (cf. [18]) if it represents $\xi \in H^1(M,\mathbf{Z}) = [M, S^1]$ by Pontrjagin-Thom construction. We will adopt the notations similar to those introduced when we explained Browder-Levine's theorem. Namely, let W denote $M \setminus \mathrm{Tub}(V)$, where $\mathrm{Tub}(V)$ is a small open tubular neighbourhood of V. Just as above the infinite cyclic covering M is the union of infinite number of copies of W, $\widehat{M} = \bigcup_{n\in\mathbf{Z}} t^n W$, where t is the generator of the deck transformation group of the covering. We suppose that $\xi(t) = -1$. Denote by V^- the union $\bigcup_{n\geq 0} t^n W$, by V^+ the union $\bigcup_{n<0} t^n W$. We have $tV^- \subset V^-$.

We call the splitting V regular if V is connected and $\pi_1 V \longrightarrow \pi_1 M$ is the isomorphrism on the subgroup $\ker\xi$. The regular splittings exist by the argument of Farrell [18, p. 325-326] which works equally well in our situation. For the regular splitting the universal covers of W, V, V^+, V^- are just the inverse images of these subsets in universal cover of M. The homology $H_*(\widetilde{V}^-)$ is a finitely generated P-module.

THEOREM B. *Let $\pi_1 M^n = \mathbf{Z}^m$, $n \geq 6$, $\pi_2(M) = \pi_3(M) = \pi_4(M) = 0$, and for the class $\xi \in H^1(M,\mathbf{Z})$ the condition(4.3) is fulfilled. Then there exists a splitting mainfold V such that V is regular and*

$$H_p(\widetilde{V}^-)\bigotimes_P K \approx (K)^{b_p(\xi)} \oplus \Big(\bigoplus_{j=1}^{q_p(\xi)} K/a_j^{(p)}K\Big) \tag{4.5}$$

Now we'll show how to deduce Theorem 4.2 from Theorem B. Suppose that (4.5) holds. Then the homology of the pair $(\widetilde{V}^-, t\widetilde{V}^-)$ is easily computed using the exact sequence of the pair and the fact that for any P-module N we have $N/tN \approx S^{-1}N/tS^{-1}N$. We have (by excision)

$$H_*(\widetilde{W}, t\widetilde{V}) \approx (R)^{b_p(\xi)} \oplus \Big(\bigoplus_{j=1}^{q_p(\xi)} R/a_j^{(p)}R\Big) \tag{4.6}$$

Consider now the cell complex of the pair (W, tV). It is a finitely generated free R-complex. A purely algebraic argument using the representation (4.6) shows that it is homotopy to a complex C_*, having in each dimension p exactly $b_p(\xi) + q_p(\xi) + q_{p-1}(\xi)$ generators. Since $b_i(\xi) = q_i(\xi) = 0$ for $i = 0, 1, n, n-1$ we can apply results of [16] and realize the complex C_* as a Morse complex of some Morse function f on the cobordism

(W, tV), constant on the upper and lower boundaries. Glueing together V and tV we get the map $M \longrightarrow S^1$.

The proof of Theorem B goes by induction, with the help of procedure similar to that of [5]. Instead of Smale's theory of minimal Morse function on a simply-connected manifolds we use the Sharko's theory [16].

There arises the obstruction in dimension $n-3$. The situation here is similar to considered by Farrell [18]. At present we can cope with it only when the situation is essentially the same as in [18]. Namely we need $m_p(\xi) = m_{p+1}(\xi) = m_{p+2}(\xi) = 0$ for some p with $2 \leq p \leq n-4$ (where $m_*(\xi)$ stands for $b_*(\xi) + q_*(\xi) + q_{*-1}(\xi)$). (It means that we don't expect any critical points of these indices.) This holds for example if $p = 2$ and the condition of theorem B is fulfilled. Then the Farrell's arguments work and (recall that the corresponding obstruction sits in $C(\mathbf{Z}[\mathbf{Z}^{m-1}]) = 0$) we are done.

Now I'll formulate the exactness result for the cohomology classes ξ of any degree of irrationality. We say that the set $U \subset \mathbb{R}^N$ is conical if $x \in U \Rightarrow tx \in U$ for $t > 0$.

THEOREM 4.5. *Let $\pi_1 M^n = \mathbf{Z}^m$, $n \geq 6$, $\pi_2(M) = \pi_3(M) = \pi_4(M) = 0$. Then there exists a open conical subset $U \subset H^1(M, \mathbb{R}) = \mathbb{R}^m$, such that any nozero element $\xi \in U$ is represented by a Morse form ω, having the minimal number of zeros of all indices, namely, $m_p(\omega) = B_p(\xi) + Q_p(\xi) + Q_{p-1}(\xi)$.*

This theorem follows form theorem 4.2 immediately. Indeed, take any rational $\xi \notin \bigcup_i H_i$ and a Morse form ω : $[\omega] = \xi$. Then one easily shows that in sufficiently small neighbourhood of ξ in $H^1(M, \mathbb{R})$ any class ξ' is represented by a small perturbation ω' of ω, so that ω' is Morse form and $m_p(\omega') = m_p(\omega)$. Passing if necessary to a smaller neighbourhood we have $B_p(\xi') = B_p(\xi)$, $Q_p(\xi') = Q_p(\xi)$, but recall $\xi \notin \bigcup_i H_i$, hence $B_p(\xi) = b_p(\xi)$, $Q_p(\xi) = q_p(\xi)$ and since $m_p(\xi)$ is given by the righthand side of (2.2), we arrive at the conclusion.

REFERENCES

[1] Novikov S. P. *Multivalued functions and functionals analogue of Morse theory.* Dokl. AN SSSR, 1981, v.270, N 1, p. 31–35 (in Russ.).

[2] Novikov S. P. *Hamiltonian formalism and multivalued analogue of Morse theory.* Russ. Math. Surveys, 1982, v.37, N 5, p. 3–49 (in Russ.).

[3] Novikov S. P., Smeltzer I. *Periodic solutions of the Kirchhof type equations for the free motion of the solid body in the liquid and the extended Lusternik-Schnirelman-Morse theory (L-Sch-M)* I. Funkz. anal. i pril. 1981, v.15, N 3, p .54–66 (in Russ.).

[4] Smale S. *On the structure of mainfolds.* Amer. J. Math. 1962, v.84, p. 387–399.

[5] Farber M. Sh. *The exactness of Novikov inequalities.* Funcz. anal. i pril. 1985, v.19, N 1, p. 49–59 (in Russ.).

[6] Novikov S. P. *Bloch homology. Critical points of functions and closed* 1*-forms.* Dokl. AN SSSR, 1987, v.287, N 6, p. 1321–1324.

[7] Pazhitnov A. V. *An analytic proof of the real part of Novikov's inequalities.* Dokl. AN SSSR 1987, v.293, N 6, p. 1305–1307.

[8] Pazhitnov A. V. *Proof of Novikov's coniecture on homology with local coefficients over a field of finite characteristic.* Dokl. AN SSSR. 1988, v.300, N 6, p. 1316–1320.

[9] Pazhitnov A. V. *On the exactness of Novikov type inequalities for $\pi_1 M = \mathbf{Z}^m$ and Morse forms within the generic cohomology classes.* Dokl. AN SSSR. 1989, v.306, N 4, p. 544–548.

[10] Pazhitnov A. V. *On the exactness of Novikov inequalities for the manifolds with free abelian fundamental group.* Mat. Sbornik 1989, v.180, N 11, p. 1486–1523.

[11] Witten E. *Supersymmetry and Morse theory.* Journal of differential geometry, 1982, v.17, p. 661–692.

[12] Milnor J. W. *Infinite cyclic coverings.* In: *Conference on the topology of Manifolds* (edited by J.G. Hocking) Prindle Weber & Schmidt 1968, p. 115–133.

[13] Kraines D. *Higher order Massey products.* Transactions of American mathematical society 1966, v.124, N 5, p. 431–439.

[14] Bousfield A. K., Gugenheim V. K. A. M. *On PL deRham theory and rational homotopy type.* Memoirs of the American Mathematical Society, 1976, v.8, number 179.

[15] Milnor J. W. *Lectures on the h-cobordism theorem.* Princeton 1965.

[16] Sharko V. V. *K-Theory and Morse theory 1.* Preprint Kiev Inst. of Math. AN SSSR 1986, N 86.39.

[17] Browder W., Levine J. *Fibering manifolds over a circle.* Comment. Math. Helv. v.40, 1966, p. 153–160.

[18] Farrell F. T. *The obstruction to fibering a manifold over a circle.* Indiana Univ. Math. Journ. 1971, v.21, N 4, p. 315–346.

Morava K-Theories: A survey

Urs Würgler
Mathematisches Institut der Universität Bern
CH 3012 Bern

For any prime p, the Morava K-theories $K(n)^*(-)$, n a positive integer, form a family of $2(p^n-1)$-periodic cohomology theories with coefficient objects

$$K(n)^* = \pi_{-*}(K(n)) = \mathbf{F}_p[v_n, v_n^{-1}],$$

where $|v_n| = -2(p^n-1)$.They were invented in the early seventies by J. Morava in an attempt to get a better understanding of complex cobordism theory. Morava's work used rather complicated tools from algebraic geometry and, unfortunately, it seems that no published version of it exists. So topologists interested in this subject were very pleased to see the paper [18] of Johnson and Wilson where a construction of these theories together with many of their basic properties were carried out in more conventional terms.

In the period after the appearance of [18] the importance of the Morava K-theories for algebraic topology and homotopy theory became more and more obvious. First, in the work of Miller, Ravenel and Wilson (see [33]) it was shown that making use of a theorem of Morava, the cohomology of the automorphism groups of these K-theories is strongly related -via the chromatic spectral sequence- to the stable homotopy groups of the sphere. Then, in their paper [50], Ravenel and Wilson demonstrated the computability of the $K(n)$'s by calculating $K(n)^*(-)$ for Eilenberg-MacLane spaces. From this paper it also became clear that the $K(n)$ constitute a useful tool for the problem of describing the structure of $BP_*(X)$, an idea, which has found further applications in the papers of Wilson and Johnson-Wilson [60],[19]. More recently, from the work of Devinatz, Hopkins and Smith (see [11], [15]) it becomes appearent that the Morava K-theories also play a very important rôle in stable homotopy theory.

The purpose of this paper is to give a brief survey of some of the basic properties of the $K(n)$'s with the aim to help a non-specialist to get quickly informed about some important aspects of this topic. Clearly, the choice of the material we are presenting here is mostly dictated by personal taste and we pretend by no means to be complete.

In the first section we indicate where the Morava K-theories come from and sketch a method how they can be constructed. Section 2 contains a description of the stable operations in $K(n)^*(-)$ and in the third section we study some connections with other BP-related cohomology theories. In 4. some $K(n)$-computations are reviewed . Section 5 contains some properties of the connected cover $k(n)$ of $K(n)$ and in 6. we treat uniqueness questions. Finally, in section 7 we make some comments concerning the significance of Morava K-theories for certain topics of stable homotopy theory.

1 The origins of Morava K-theories

One of the key motivations which led J. Morava to the construction of his K-theories was certainly a remarkable theorem of Quillen [43] relating the theory of formal groups with complex cobordism theory. Let $MU^*(-)$ denote complex cobordism theory. Then

$$MU^* \cong \mathbf{Z}[x_1, x_2, \ldots],\ x_i \in MU^{-2i}$$

and $MU^*(-)$ is a complex-oriented theory, i.e. there is an element $y \in MU^2(\mathbf{C}P_\infty)$ such that

$$MU^*(\mathbf{C}P_\infty) \cong MU^*[[y]],\ MU^*(\mathbf{C}P_\infty \times \mathbf{C}P_\infty) \cong MU^*[[y \otimes 1, 1 \otimes y]].$$

The classifying map $m : \mathbf{C}P_\infty \times \mathbf{C}P_\infty \to \mathbf{C}P_\infty$ induces a power series

$$F_{MU}(y_1, y_2) = m^*(y) = \sum_{i,j} a_{i,j}\, y_1^i \hat{\otimes} y_2^j$$

with the three properties

$$\begin{array}{rcll} F_{MU}(x,y) & = & F_{MU}(y,x) & \text{commutativity} \\ F_{MU}(F_{MU}(x,y),z) & = & F_{MU}(x, F_{MU}(y,z)) & \text{associativity} \\ F_{MU}(x,0) & = & x & \text{identity} \end{array}$$

We define a formal group law G over a commutative ring A to be a formal power series $G(x,y) \in A[[x,y]]$ having these three properties of F_{MU}. Quillen's observation was

Theorem 1.1 *The formal group law F_{MU} over MU^* is universal in the sense that for any formal group law G over any commutative ring A, there is a unique ring homomorphism $\theta : MU^* \to A$ such that $G(x,y) = \sum \theta(a_{i,j}) x^i y^j = \theta_* F_{MU}$.*

The universal group law F_{MU} may be described rather explicitly: For any formal group F over a torsion free ring A define its *logarithm* $log_F(x) \in A \otimes \mathbf{Q}[[x]]$ by

$$log_F(x) = \int_0^x \frac{dt}{\frac{\partial F}{\partial y}(t,0)}.$$

Then $log_F(F(x,y)) = log_F(x) + log_F(y)$, i.e. log_F is an isomorphism over $A \otimes \mathbf{Q}$ between F and the additive formal group law and F is determined by its logarithm. A theorem of Mischenko [40] asserts that

$$log_{MU}(x) = \sum_{n \geq 0} [\mathbf{C}P_n] \frac{x^{n+1}}{n+1},$$

where $[\mathbf{C}P_n]$ denotes the element of MU^* determined by the complex manifold $\mathbf{C}P_n$.

A formal group law over a torsion free ring is called *p-typical* with respect to the prime p, if its logarithm is of the form $log_F(x) = \sum_{i \geq 0} l_i x^{p^i}$. This definition may be extended to rings with torsion, see e.g. [14]. A theorem of Cartier [10] asserts that every formal group law F over a torsion free $\mathbf{Z}_{(p)}$-algebra is canonically isomorphic to a p-typical formal group law F^{typ} in the sense that if $log_F(x) = \sum_{i\geq 0} a_i x^i$, then $log_{F^{typ}}(x) = \sum_{i\geq 0} a_{p^i} x^{p^i}$. Applying this result to F_{MU} over $MU^* \otimes \mathbf{Z}_{(p)}$, Quillen was able to construct a multiplicative and idempotent natural transformation

$$\epsilon_p : MU\mathbf{Z}^*_{(p)}(-) \to MU\mathbf{Z}^*_{(p)}(-)$$

whose image is represented by a ring spectrum BP, which is called the Brown-Peterson spectrum (see [9] for the original approach). On homotopy, ϵ_p is determined by

$$\epsilon_p([\mathbf{C}P_n]) = \begin{cases} [\mathbf{C}P_n] & \text{if } n = p^i - 1 \\ 0 & \text{otherwise} \end{cases}$$

This implies that the logarithm of $F_{BP} = F^{typ}_{MU} = (\epsilon_p)_* F_{MU}$ is given by

$$log_{BP}(x) = \sum_{i\geq 0} \frac{[\mathbf{C}P^{p^i-1}]}{p^i} x^{p^i} = \sum_{i\geq 0} l_i x^{p^i} \in BP^* \otimes \mathbf{Q}[[x]].$$

Moreover, F_{BP} is universal for p-typical formal group laws over $\mathbf{Z}_{(p)}$-algebras.

The BP-spectrum bears as much of informations as $MU\mathbf{Z}_{(p)}$, and, because homotopy theory is essentially a local subject, homotopy theorists concern themselves mostly with the smaller spectrum BP. If G is a formal group law over A and if $f, g \in A[[x]]$ are power series without constant term, we define $f +_G g = G(f(x), g(x))$ and for any positive integer n we set

$$[n]_G(x) = \underbrace{x +_G \cdots +_G x}_{n}.$$

The following theorem of Araki [2] is very useful and shows that it is possible to find generators of BP^* which behave well with respect to the formal group law F_{BP}. Another (and equally useful) set of generators was earlier found by Hazewinkel, see [14].

Theorem 1.2 *Let p be any prime. There is an isomorphism of* $\mathbf{Z}_{(p)}$*−algebras*

$$BP_* \cong \mathbf{Z}_{(p)}[v_1, v_2, ...]$$

where the generators $v_i \in BP_{2(p^i-1)}$ *may be chosen to be the coefficients of* x^{p^i} *in the series*

$$[p]_{F_{BP}}(x) = \sum_{i>0}{}^{F_{BP}} v_i x^{p^i}.$$

Now the construction which leads to the formal group law F_{MU} applies to every complex-oriented cohomology theory: For example, the formal group law associated to $H^*(-;R)$ is $G_a(x,y) = x + y$, the additive formal group law, and the group law associated to complex K-theory $K^*(-)$ is the multiplicative formal group law $G_m(x,y) = x + y + txy$ where $t \in K^* \cong \mathbf{Z}[t, t^{-1}]$. In general, one may ask if given a (graded) commutative ring A and a formal group law G defined over A there exists a complex-oriented cohomology theory which realises (A, G) in the sense indicated above. In this generality, an answer to this question is not known today. However, one may try to realise special types of formal groups.

A formal group law F over a commutative $\mathbf{F}_p$-algebra A is *of height* n $(n > 0)$ if the series $[p]_F(x)$ has leading term ax^{p^n} with $a \neq 0$. If $[p]_F(x) = 0$, F *is of height* ∞. Consider the ring homomorphism $\theta_n : BP^* \to A$ defined by $\theta_n(v_n) = 1$ and $\theta_n(v_i) = 0$ if $i \neq n$, and put $F_n(x,y) = (\theta)_* F_{BP}$. From theorem 1.2. we see that F_n is of height n. Now a theorem of Lazard [30] (see also [13],[14]) asserts that over a separably closed field K of characteristic $p > 0$ any formal group law G of height n is isomorphic to F_n. In view of this theorem it is certainly interesting to try to realise the formal groups F_n resp. the graded versions of them.

Theorem 1.3 *Let p be any prime. For all integers $n \geq 1$ there is a multiplicative, $2(p^n-1)$-periodic and complex-oriented cohomology theory $K(n)^*(-)$ with coefficient ring*

$$K(n)^* = \mathbf{F}_p[v_n, v_n^{-1}]$$

where v_n is of degree $|v_n| = -2(p^n-1)$ and whose associated formal group law $F_n(x,y)$ satisfies the relation

$$[p]_{F_n}(x) = v_n x^{p^n}.$$

If p is odd, the product on $K(n)^*(-)$ is commutative, for $p = 2$ it is non-commutative.

The theories $K(n)^*(-)$ of this theorem are named after Jack Morava who proved a version of 1.3. (he did not know the $K(n)$'s to be multiplicative) in the early seventies in a paper which never appeared in print. The first published reference concerning the $K(n)'s$ is the paper [18] of Johnson and Wilson. It may be interesting to notice that $K(1)$ has a rather familiar interpretation: Let $K^*(-)$ denote complex K-theory. As Adams showed (see, e.g. [2]), $K^*(-)_{(p)}$ decomposes into a direct sum of copies of a cohomology theory $G^*(-)$ which is periodic with period $2(p-1)$. Then there is an isomorphism $K^*(-) \cong G^*(-;\mathbf{F}_p)$.

To construct the $K(n)$'s one uses (co)bordism theories of stably almost-complex manifolds with singularities, see [3]. Very briefly, the idea behind the construction of these theories is as follows. By a *singularity type* Σ we mean a sequence $\{P_0, P_1, ..., P_n\}$ of closed stably almost-complex manifolds P_i of dimension p_i and with $P_0 = *$. A *n-decomposed manifold* is a manifold M together with a sequence $\{\partial_0 M, ..., \partial_n M\}$ of submanifolds of codimension 0 of the boundary ∂M of M such that $\partial M = \partial_0 M \cup \cdots \cup \partial_n M$. Baas defines a *manifold of singularity type* Σ (a Σ-manifold) to be a family $V = \{V(\omega) | \omega \subset \{0.1....,n\}\}$ of n-decomposed manifolds $V(\omega)$ with $\partial_i V(\omega) = \emptyset$ for $i \in \omega$ together with a system of diffeomorphisms (the structure maps)

$$\beta(\omega, i) : \partial_i V(\omega) \xrightarrow{\cong} V(\omega, i) \times P_i,\ i \notin \omega$$

which satisfy certain compatibility conditions (see [3]). The Σ-boundary $\delta_\Sigma V$ of a Σ-manifold V is defined by $\delta_\Sigma V = \{\delta_\Sigma V(\omega)\}$ where $\delta_\Sigma V(\omega) = \partial_0 V(\omega) = V(\omega, 0)$. $\delta_\Sigma V$ is a Σ-manifold with structure maps

$$\partial_i \delta_\Sigma V(\omega) = \partial_i V(\omega, 0) \xrightarrow{\beta(\omega,0,i)} V(\omega, i, 0) \times P_i = \delta_\Sigma V(\omega, i) \times P_i$$

for $i \notin \omega \cup \{0\}$.Notice that $dim(\delta_\Sigma V) = dim(V) - 1$ and that $\delta_\Sigma^2 V = \emptyset$.

Using this concept of manifolds, Baas was able to mimick the usual construction of a bordism theory to get for any singularity type Σ a homology theory $MU(\Sigma)_*(-)$

(this is also known as the Baas-Sullivan construction). These theories are representable by spectra $MU(\Sigma)$ which are module spectra over the ring spectrum MU. If Σ is a singularity type, we denote by Σ_i the singularity type which results from Σ by deleting the ith entry of Σ. The following theorem relates bordism theories based on manifolds of different singularity types:

Theorem 1.4 *([3]) For each i there is a natural exact sequence*

$$\cdots \to MU(\Sigma_i)_*(X) \xrightarrow{\theta_i} MU(\Sigma_i)_*(X) \xrightarrow{\eta_i} MU(\Sigma)_*(X) \xrightarrow{\delta_i} MU(\Sigma_i)_*(X) \to \cdots$$

where the natural transformations θ_i,η_i and δ_i are of degree p_i, 0 and $-(p_i+1)$, respectively. θ_i is given by multiplication with $[P_i]$.

If the sequence $\{[P_1], ..., [P_n]\}$ is regular, i.e. if for all $i = 1, ..., n$, $[P_i]$ is not a zero-divisor in $MU_*/([P_1], ..., [P_{i-1}])$, this implies that

$$MU(\Sigma)_* \cong MU_*/([P_1], ..., [P_n]).$$

In this way one can kill off any regular ideal in MU_*, and, by passing to the limit, even ideals with infinitely many generators. For example, one can kill the kernel of the map $MU_* \to BP_*$. After localizing at p this produces Brown-Peterson theory. One may continue this process by killing generators of BP_* to obtain ,for example, theories $P(n)_*(-)$, $k(n)_*(-)$ or $BP\langle n\rangle_*(-)$ with coefficients

$$\begin{aligned} BP\langle n\rangle_* &\cong \mathbf{Z}_{(p)}[v_1, ..., v_n] \\ P(n)_* &\cong \mathbf{F}_p[v_n, v_{n+1}, ...] \\ k(n)_* &\cong \mathbf{F}_p[v_n]. \end{aligned}$$

The spectrum $k(n)$ is the (-1)-connected version of the spectrum $K(n)$ of Morava K-theory. Using $k(n)$ one defines $K(n)$ by

$$K(n) = holim\{\Sigma^{-2i(p^n-1)}k(n) \xrightarrow{v_n} k(n)\}.$$

Similarly, one defines (periodic) spectra $E(n) = holim\{\Sigma^{-2i(p^n-1)}BP\langle n\rangle \xrightarrow{v_n} BP\langle n\rangle\}$ resp. $B(n) = holim\{\Sigma^{-2i(p^n-1)}P(n) \xrightarrow{v_n} P(n)\}$ with coefficients $E(n)_* = \mathbf{Z}_{(p)}[v_1, ..., v_n, v_n^{-1}]$ resp. $B(n)_* = v_n^{-1}P(n)_*$. By the construction of these spectra, one has canonical morphisms $BP \to P(n) \to K(n)$ etc.. Moreover, for different n, the $P(n)'s$ are related by stable cofibrations

$$\Sigma^{2(p^n-1)}P(n) \xrightarrow{v_n} P(n) \xrightarrow{\eta_n} P(n+1) \xrightarrow{\partial_n} \Sigma^{2p^n-1}P(n).$$

The question whether (co)bordism theories of manifolds with singularities are multiplicative is a delicate one. Using geometric constructions on Σ-manifolds, Mironov [40], Shimada-Yagita [57] and later Morava [36] constructed good products for a large class of such theories. Using purely homotopy theoretic methods, products for theories like $P(n)$, $K(n)$ etc. were constructed in [62], see also [66] for the case $p = 2$. Where they apply, these homotopy theoretic methods also give uniqueness results. In this context it is interesting to remark that the methods of Sanders [56] and unpublished work of Margolis show that for example the spectra $k(n)$ and $K(n)$ may themselves be constructed by homotopy theoretic methods, so many of the questions we are discussing here are in fact independent of the theory of manifolds with singularities.

Let $F(n)$ denote one of the spectra $P(n)$, $k(n)$ or $K(n)$. By their construction, the $F(n)$ are canonically module spectra over the ring spectrum BP and the natural map $\mu_n : BP \to F(n)$ is a map of BP-module spectra.

Theorem 1.5 *1. Suppose p is an odd prime. There is exactly one product $m_n : F(n) \wedge F(n) \to F(n)$ which makes $F(n)$ a BP-algebra spectrum compatible with the given BP-module structure . This product is associative, commutative and has a two-sided unit.*

2. Suppose p=2. There are exactly two products $m_n, \overline{m}_n : F(n) \wedge F(n) \to F(n)$ which make $F(n)$ a BP-algebra spectrum compatible with the given BP-module structure . Both are associative and have a two-sided unit. m_n and $\overline{m}_n$ are related by the formula

$$\overline{m}_n = m_n \circ T = m_n + v_n m_n(Q_{n-1} \wedge Q_{n-1})$$

where Q_{n-1} is a stable $F(n)$-operation of degree $2^n - 1$ satisfying the relation $Q_{n-1}^2 = 0$ (a Bockstein operation).

In particular, this theorem settles the question about products in the $K(n)$'s in a satisfactory manner.

2 Operations and cooperations

To apply the $K(n)$'s in concrete situations it is clearly important to know something about (stable) operations. There is a duality isomorphism

$$K(n)^*(K(n)) \cong Hom_{K(n)_*}(K(n)_*(K(n)), K(n)_*),$$

so one may consider as well the algebra $K(n)_*(K(n))$. Now from Adams [1] we know that if E is a ring spectrum such that $E_*(E)$ is a flat E_*-module, $E_*(E)$ is a Hopf algebroid and $E_*(-)$ takes values in the category of $E_*(E)$-comodules. This assumption is true for the spectra $P(n)$ and $K(n)$, so one should try to describe the structure of their cooperation Hopf algebroids. The basic information needed to compute them is contained in the following theorem [1], [43]:

Theorem 2.1 *There are elements $t_i \in BP_{2(p^i-1)}(BP)$, $t_0 = 1$, such that*

$$BP_*(BP) \cong BP_*[t_1, t_2, ...]$$

as a BP_*-algebra. The counit ϵ satisfies $\epsilon(1) = 1, \epsilon(t_i) = 0, i > 0$, and the conjugation c resp. the coproduct ψ are given by the formulas

$$\sum_{n,j \geq 0}{}^{F_{BP}} t_n c(t_j)^{p^n} = 1,$$

$$\sum_{i \geq 0}{}^{F_{BP}} \psi(t_i) = \sum_{i,j \geq 0}{}^{F_{BP}} t_i \otimes t_j^{p^i}.$$

The behaviour of the right unit η_R on the generators of BP_* is defined by

$$\sum_{i,j\geq 0}{}^{F_{BP}}\, t_i\eta_R(v_j)^{p^i} = \sum_{i,j\geq 0}{}^{F_{BP}}\, v_i t_j^{p^i}.$$

The last formula concerning the action of η_R on the v_i is due to Ravenel [45], it is extremely useful, especially for computational purposes. Combining the above theorem with work of Baas-Madsen [4] concerning $H_*(P(n);\mathbf{Z}_p)$, the fact that the ideals $I_n = (v_0, ..., v_{n-1}), n \geq 1, v_0 = p$, are invariant with respect to stable BP-operations and the stable cofibrations

$$\Sigma^{2(p^n-1)}P(n) \xrightarrow{v_n} P(n) \xrightarrow{\eta_n} P(n+1) \xrightarrow{\partial_n} \Sigma^{2p^n-1}P(n)$$

one can prove (see [62] for the case p odd and [26] for the case $p=2$)

Theorem 2.2 *For any prime p, $P(n)_*(P(n))$ is a (commutative) Hopf-algebroid over $P(n)_*$. If p is odd, there is an isomorphism of left $P(n)_*$-algebras*

$$P(n)_*(P(n)) \cong P(n)_* \otimes_{BP_*} BP_*(BP) \otimes E(a_0, a_1, ..., a_{n-1})$$

where $E(a_0, a_1, ..., a_{n-1})$ is an exterior algebra in generators a_i of degree $2p^i - 1$ and for $p = 2$,

$$P(n)_*(P(n)) \cong P(n)_*[a_0, ..., a_{n-1}, t_1, t_2, ...]/J_n$$

where $J_n = (a_i^2 - t_{i+1} : 0 \leq i \leq n-1)$. Modulo the generators a_i, $P(n)_*(P(n))$ is for all primes isomorphic to the Hopf-algebroid $BP_*(BP)/I_n$ and the coproduct resp. the conjugation are given on the generators a_i by the formulas

$$\psi_n(a_k) = \sum_{i=0}^{k} a_i \otimes a_{k-i-1}^{2^{i+1}} + 1 \otimes a_k$$

$$c_n(a_k) = -a_k - \sum_{i=0}^{k-1} c_n(a_i) a_{k-i-1}^{2^{i+1}}$$

for $p = 2$, with the obvious changes for p odd.

Observe that there is again a duality isomorphism

$$P(n)^*(P(n)) \cong Hom^*_{P(n)_*}(P(n)_*(P(n)), P(n)^*).$$

Under this isomorphism, the generators a_i correspond to Bockstein operations Q_i of degree $2p^i - 1$. In particular, $Q_{n-1} = \eta_n \circ \partial_n$.

To get from theorem 2.2. to the structure of $K(n)_*(K(n))$ one may use Landweber's exact functor theorem [29]. Let $\mathcal{BP}_n$ denote the category of $P(n)_*(P(n))$-comodules which are finitely presented as $P(n)_*$-modules (we set $P(0) = BP$ and $v_0 = p$). Then

Theorem 2.3 *Let G be a $P(n)_*$-module. The functor*

$$M \mapsto M \otimes_{P(n)_*} G$$

is exact on the category $\mathcal{BP}_n$ if and only if multiplication by v_n on G and for each $k > n$, multiplication by v_k on $G/(v_n, ..., v_{k-1})$ is monic.

For $n > 0$, this theorem has first been proved by Yagita [68]. The canonical map $\lambda_n : P(n) \to K(n)$ makes $K(n)_*$ a $P(n)_*$-module for which Landweber's theorem clearly applies. One then gets a natural multiplicative equivalence

$$P(n)_*(X) \otimes_{P(n)_*} K(n)_* \xrightarrow{\simeq} K(n)_*(X).$$

This equivalence is the mod I_n version of the theorem of Conner-Floyd. In particular, it produces an isomorphism of Hopf algebroids

$$K(n)_*(K(n)) \cong K(n)_* \otimes_{P(n)_*} P(n)_*(P(n)) \otimes_{P(n)_*} K(n)_*.$$

Combining this with theorem 2.2. and Ravenel's formula of theorem 2.1. one then obtains (see [70], [63])

Theorem 2.4 *Let p be any prime. There is an isomorphism of left $K(n)_*$-algebras*

$$\begin{aligned} K(n)_*(K(n)) &\cong K(n)_*[t_1, t_2, ...]/(v_n t_i^{p^n} - v_n^{p^i} t_i) \\ &\otimes E(a_0, a_1, ..., a_{n-1}) \end{aligned}$$

for p odd and

$$K(n)_*(K(n)) \cong K(n)_*[a_0, ..., a_{n-1}, t_1, t_2, ...]/J_n$$

for $p = 2$, where $J_n = (v_n t_i^{2^n} - v_n^{2^i} t_i, a_i^2 - t_{i+1})$. Right and left unit agree in $K(n)_*(K(n))$ and the coaction map ψ_n resp. the conjugation c_n may be described on the t_i by the formulas

$$\sum_{n,j \geq 0}{}^{F_n}\, t_n c(t_j)^{p^n} = 1,$$

$$\sum_{i \geq 0}{}^{F_n}\, \psi(t_i) = \sum_{i,j \geq 0}{}^{F_n}\, t_i \otimes t_j^{p^i},$$

and on the generators a_j as in theorem 2.2..

The intimate relation between the structure of the Hopf algebroids considered above and the respective formal group laws may be expressed in a slightly different manner. Recall that a groupoid is a small category in which every morphism is an isomorphism. Let k be a commutative ring and let $\mathcal{A}_k$ be the category of k-algebras. By a ***groupoid-scheme over k*** we mean a representable functor $G : \mathcal{A}_k \to \mathcal{G}$ from $\mathcal{A}_k$ to the category of groupoids . Here representable means that the two set-valued functors $A \mapsto ob(G(A))$ and $A \mapsto mor(G(A))$ are representable. For all A we have morphisms (natural in A)

$$mor(G(A)) \cong Hom_{\mathcal{A}_k}(C, A) \rightrightarrows Hom_{\mathcal{A}_k}(B, A) \cong ob(G(A))$$

which are induced by the maps source, target and identity of the category $G(A)$. These morphisms give rise to homomorphisms of k-algebras $\eta_R, \eta_L : B \to C$ and $\epsilon : C \to B$. Furthermore, the composition of morphisms in $G(A)$ is represented by a map $\psi : C \to C \otimes_B C$ and all these data together make (B, C) a Hopf algebroid.

Let $n \geq 0$. For any $\mathbf{F}_p$- algebra ($\mathbf{Z}_{(p)}$-algebra if $n = 0$) A consider the set $TI_n(A)$ of triples (F, G, ϕ) where F,G are p-typical formal groups of height $\geq n$ over A and $\phi : G \to F$ is a strict isomorphism. $TI_n(A)$ is a groupoid in an obvious sense and we get a functor

$$TI_n(-) : \mathcal{A}_k \to \mathcal{G}.$$

One then has the following theorem of Landweber [28]:

Theorem 2.5 *$TI_n(-)$ is a groupoidscheme over $\mathbf{F}_p$ (resp. over $\mathbf{Z}_{(p)}$ if $n=0$) which is represented by the Hopf algebroid $(BP_*/I_n, BP_*(BP)/I_n)$.*

Using theorem 2.5. it is easy to describe the group of multiplicative automorphisms of $K(n)$. In this context it is important to consider also the $\mathbf{Z}_2$-graded version of $K(n)^*(-)$ which we define by

$$K(n)^\bullet(X) = \begin{cases} \oplus_{i=0}^{q-1} K(n)^{2i}(X) & \text{if } \bullet = 0 \\ \oplus_{i=0}^{q-1} K(n)^{2i+1}(X) & \text{if } \bullet = 1 \end{cases}$$

where $q = p^n$. Let $Mult(K(n)^*(-))$ resp. $Mult(K(n)^\bullet(-))$ denote the groups of multiplicative automorphisms of $K(n)^*(-)$ resp. of $K(n)^\bullet(-)$. Let $SAut_{F_n}(\mathbf{F}_p)$ resp. $SAut^{gr}_{F_n}(K(n)_*)$ denote the groups of strict automorphisms of the formal group law F_n considered as an ungraded power series over $\mathbf{F}_p$ resp. as a graded power series over $\mathbf{F}_p[v_n, v_n^{-1}]$. Then

Theorem 2.6 *For all primes p and all $n>0$ there are isomorphisms*

$$Mult(K(n)^*(-)) \cong SAut^{gr}_{F_n}(K(n)_*)$$

$$Mult(K(n)^\bullet(-)) \cong SAut_{F_n}(\mathbf{F}_p).$$

This theorem was first proved by Morava (unpublished), see also [44], [67], [65]. Now in fact, for each n there is an isomorphism

$$SAut^{gr}_{F_n}(K(n)_*) \cong S_1 \subset \widehat{\mathbf{Z}}_p^*,$$

where S_1 denotes the group of p-adic units congruent to 1 mod (p), (see [67]), and so the elements of $Mult(K(n)^*(-))$ may be considered as some sort of (stable) Adams operations.

In the $\mathbf{Z}_2$-graded case the situation is more interesting. A theorem of Lubin and Dieudonné (see [14], [13]) asserts that if k is a field of characteristic p containing $\mathbf{F}_q$ where $q=p^n$, then the endomorphism ring of F_n over k is isomorphic to the maximal order E_n of the division algebra D_n with center $\mathbf{Q}_p$ and invariant $\frac{1}{n}$. More explicitely, E_n may be obtained from the Witt ring $W(\mathbf{F}_q)$ by adjoining an indeterminate S and setting $S^n = p$ and $Sw = w^\sigma S$ for $w \in W(\mathbf{F}_q)$, where σ denotes the lift of the Frobenius automorphism of $\mathbf{F}_q$ to $W(\mathbf{F}_q)$. Let

$$S_n = \{1 + \sum_{i\geq 1} w_i S^i \mid w_i \in W(\mathbf{F}_q)\}$$

be the group of strict units of E_n. Then there are isomorphisms

$$S_n \cong SAut_{F_n}(\mathbf{F}_q) \cong SAut_{F_n}(\overline{\mathbf{F}}_p)$$

where $\overline{\mathbf{F}}_p$ denotes the algebraic closure of $\mathbf{F}_p$. In [5], A. Baker showed that the element $1+S \in S_n$ determines a multiplicative operation

$$[1+S] : K(n) \longrightarrow \bigvee_{a\in\mathbf{Z}/(p^n-1)} \Sigma^{2a} K(n)$$

which satisfies the relation

$$[1 + S](y) = y +_{F_n} y^p \in K(n)^*(CP_\infty).$$

Putting $r_n = (p^n - 1)/(p - 1)$ one can in fact decompose $[1 + S]$ as

$$[1 + S] - 1 = \sum_{a \in \mathbf{Z}/r_n} \theta^a$$

where the $\theta^a : K(n) \to \Sigma^{2a(p-1)} K(n)$ are stable operations. The θ^a satisfy the product formula

$$m_n^*(\theta^a) = 1 \otimes \theta^a + \sum_{b \in \mathbf{Z}/r_n} \theta^b \otimes \theta^{a-b} + \theta^a \otimes 1$$

and one has

$$\langle \theta^a, t_1^k \rangle = (-1)^k \delta_{a,k}; \ 1 \leq k \leq p^n - 1.$$

Baker then obtains the following theorem:

Theorem 2.7 *The indecomposables of $K(n)^*(K(n))$ have a basis*

$$Q^0, \theta^0, \theta^1, \theta^p, ..., \theta^{p^{n-1}}$$

over $K(n)^*$, where $Q^0 \in K(n)^1(K(n)$ is the 0th Bockstein.

In [5], this theorem is stated for odd primes, but in fact it also holds for $p = 2$. Using Ravenel's calculation for the 2-line of $K(n)_*(K(n))$ [44] it is possible to describe the relations amongst these indecomposables.

An interesting family of stable operations arises also by considering the duals Q_i of the elements a_i of theorem 2.4.. We will make some comments on these Bockstein operations at the end of the next section. Let us also remark that in [59], Steve Wilson determines the unstable $K(n)$-operations by computing their dual $K(n)_*(\mathbf{K}(n))*)$ as a Hopf ring where $\mathbf{K}(n)_* = \{\mathbf{K}(n)_i\}$ denotes the Ω-spectrum representing $K(n)$.

3 Relations with other cohomology theories

A very important aspect of the Morava K-theories is the fact that they are strongly related to BP-theory and complex cobordism via several types of intermediate spectra. For example, consider the diagram

$$\begin{array}{ccc} P(n) & \xrightarrow{v_n} & \\ \downarrow{\eta_n} & \searrow & P(n) \xrightarrow{l_n} v_n^{-1}P(n) = B(n) \\ & \nearrow{\partial_n} & \\ P(n+1) & & \end{array}$$

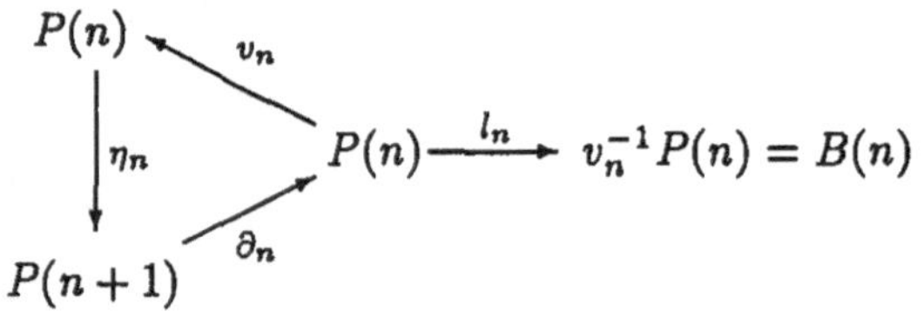

where l_n means localization with respect to v_n. The triangle is exact and determines a Bockstein spectral sequence. Assuming that we know $P(n+1)_*(X)$ for some X, then the v_n-torsion of $P(n)_*$ is determined by $P(n+1)_*(X)$ and the behaviour of this spectral sequence, whereas the v_n torsion-free part of $P(n)_*(X)$ passes monomorphically to $B(n)_*(X)$. If X is finite, this is a finite process: There is an n such that if $m > n$, then $P(m)_*(X) \cong H_*(X;\mathbf{F}_p) \otimes P(m)_*$ and the $m-th$ Bockstein spectral sequence collapses. Now the point is that in fact $B(n)_*(X)$ is determined by $K(n)_*(X)$: There is a natural isomorphism

$$B(n)_*(X) \cong K(n)_*(X) \otimes \mathbf{F}_p[v_{n+1}, v_{n+2}, \ldots]$$

(see [18] for the existence of such an isomorphism and [61] for the fact that it is natural), so in particular $B(n)_*(X)$ is a free $B(n)_*$- module whose rank equals the rank of $K(n)_*(X)$ as a $K(n)_*$-module. Because $K(n)_*(X)$ is in many cases computable and $P(0) = BP$, this process can be used to get information about $BP_*(X)$ in terms of the $K(n)_*(X)$. A beautiful example how this works in a concrete case is the Ravenel-Wilson proof of the Conner-Floyd-conjecture (see [50],[58]).

In fact, the relation between the two homology theories $B(n)_*(-)$ and $K(n)_*(-)$ is even more close as indicated above. $B(n)_*(K(n))$ may be considered as a left $B(n)_*(B(n))$-and a right $K(n)_*(K(n))$-comodule and using results of [32] one can prove the following (see [63], $\Box$ denotes the cotensor product)

Theorem 3.1 *There is a natural equivalence*

$$B(n)_*(X) \cong B(n)_*(K(n)) \Box_{K(n)_*(K(n))} K(n)_*(X)$$

of homology theories with values in the category of $B(n)_*(B(n))$-comodules.

This is of some importance if one observes that the Bockstein spectral sequences considered above are in fact spectral sequences of comodules.

In analogy to the splitting of $MU\mathbf{Z}_{(p)}$ into a wedge of suspensions of the Brown-Peterson spectrum BP one may ask if there is a similar splitting of $B(n)$ into a wedge of suspensions of $K(n)$. Unfortunately, because the formal group laws F_n and $F_{B(n)}$ are not isomorphic over $B(n)_*$, this is not the case (see [64]). However, such a splitting is possible if one completes $B(n)$ suitably. This problem was studied in [64] and, in a more general way, in [7].

First, we should explain what we mean by a "suitable completion". Let R be a commutative ring and let $\mathbf{m} \lhd R$ be a maximal ideal. We define the $\mathbf{m}$-*artinian topology* on R to be the R-linear topology on R for which the open neighbourhoods of 0 are the ideals $J \lhd R$ with $J \subset \mathbf{m}$ and R/J Artinian (the $\mathbf{m}$-co-Artinian ideals). Then the $\mathbf{m}$-artinian completion of an R-module M is defined as

$$\widehat{M} = invlim_J\,(R/J \otimes_R M).$$

If $h^*(-)$ is a multiplicative cohomology theory defined on the category $\mathcal{CW}_f$ of finite spectra, we consider in particular the functor on $\mathcal{CW}_f$

$$X \mapsto \widehat{h}^*(X) = invlim_J\,(h^*/J \otimes_{h^*} h^*(X)),$$

where J ranges over the co-Artinian ideals with respect to some (fixed) maximal ideal of h^*. Now in general, the functor $M \mapsto \widehat{M}$ is not exact, so $\widehat{h}^*(-)$ needs not be a cohomology theory. However, in certain interesting cases, this difficulty does not occur. For example, let $E(m,n)$ denote the ring spectrum obtained by Baas-Sullivan theory with coefficient ring $E(m,n)_* = \mathbf{F}_p[v_m, ..., v_n, v_n^{-1}]$, $1 \leq m \leq n$. Then $E(n,n) = K(n)$ and $E(1,n) = E(n)\mathbf{F}_p$. We define $E(0,n) = E(n)$ and $P(0) = BP$. Then we have [7]:

Theorem 3.2 *Suppose m,n are integers with $0 \leq m \leq n$. Then the functors $X \mapsto \widehat{v_n^{-1}P(m)}^*(X)$ and $X \mapsto \widehat{E(m,n)}^*(X)$ are multiplicative cohomology theories over the category $\mathcal{CW}_f$ where in both cases the co-Artinian idelas J are taken with respect to the maximal ideal $\mathbf{m} = (v_i : 0 \leq i, i \neq n)$. Moreover, these theories extend uniquely to representable ring theories over the category of all spectra .*

The proof of this theorem uses in an essential manner Landweber's exact functor theorem. The representing ring spectra of the theories constructed in the theorem are denoted $\widehat{v_n^{-1}P(m)}$ and $\widehat{E(m,n)}$ respectively and are called the *Artinian completions* of $v_n^{-1}P(m)$ resp. $E(m,n)$.

In order to obtain splittings of the spectra $\widehat{v_n^{-1}P(m)}$ one needs some facts about formal group laws. Let G_n and H_n denote the formal group laws of $\widehat{v_n^{-1}BP}$ resp. of $\widehat{E(n)}$.

Let $\mathcal{A}_p$ denote the category of Artinian local rings A with residue field $A/\mathbf{m}$ of characteristic p. If A is such a ring let $\mathbf{lift}_n(A)$ denote the groupoid whose objects are p-typical lifts of height n Lubin-Tate formal groups over $A/\mathbf{m}$ (where by a Lubin-Tate formal group over a field of characteristic p we mean a formal group whose classifying homomorphism factors through $\mathbf{F}_p[v_n, v_n^{-1}]$), and similarly for morphisms. Then $A \mapsto \mathbf{lift}_n(A)$ is a groupoid-valued functor on $\mathcal{A}_p$. Now one can show (see [7]) that there is an idempotent natural equivalence

$$\mathrm{e} : \mathbf{lift}_n(A) \xrightarrow{\simeq} \mathbf{lift}_n(A)$$

whose image $\mathrm{e}(A)$ is the sub-groupoid $\mathbf{lift}_n^{(n)}(A)$ of $\mathbf{lift}_n(A)$ of strict isomorphisms of objects of co-height n in $\mathbf{lift}_n(A)$, i.e. objects F with p-series of form

$$[p]_F(x) = \sum_{0 \leq i \leq n}{}^F (a_i x^{p^i}).$$

Now the functors $A \mapsto ob(\mathbf{lift}_n(A))$ resp. $A \mapsto ob(im(\mathrm{e}(A)))$ are pro-represented by $\widehat{v_n^{-1}BP}_*$ and $\widehat{E(n)}_*$ respectively. One then gets the following

Theorem 3.3 *There is an idempotent continuous homomorphism $e_0 : \widehat{v_n^{-1}BP}_* \to \widehat{v_n^{-1}BP}_*$ which factors as $\widehat{v_n^{-1}BP}_* \xrightarrow{\pi} \widehat{E(n)}_* \xrightarrow{\gamma} \widehat{v_n^{-1}BP}_*$ where π denotes the canonical projection and γ is injective. Moreover, there is a unique $*$- isomorphism*

$$\Phi_n : \gamma_*(H_n) \xrightarrow{\simeq} G_n$$

over $\widehat{v_n^{-1}BP}_*$.

Using 3.3.one then obtains (see [7]) :

Theorem 3.4 *There is a unique idempotent multiplicative natural transformation*

$$\mathbf{E}_n : \widehat{v_n^{-1}BP}^*(-) \to \widehat{v_n^{-1}BP}^*(-)$$

such that on $\mathbf{C}P_\infty$ *we have*

$$\mathbf{E}_n(y) = \Phi_n^{-1}(y).$$

Moreover, there is a canonical natural isomorphism

$$\widehat{E(n)}^*(-) \cong im\left[\mathbf{E}_n : \widehat{v_n^{-1}BP}^*(-) \to \widehat{v_n^{-1}BP}^*(-)\right].$$

Now from 3.4. one deduces easily the

Corollary 3.1 *There is a splitting of* $\widehat{E(n)}$*-module spectra*

$$\widehat{v_n^{-1}BP} \simeq \prod_\alpha \Sigma^{\sigma(\alpha)}\widehat{E(n)}$$

and the natural morphism of ring spectra $\widehat{v_n^{-1}BP} \to \widehat{E(n)}$ *splits as a morphism of* $\widehat{E(n)}$*-module spectra.*

We remark that the same methods also produce splittings

$$\widehat{v_n^{-1}P(m)} \simeq \prod_\gamma \Sigma^{\sigma(\gamma)}\widehat{E(m,n)}$$

of $\widehat{E(m,n)}$-module spectra. In particular, if $n = m$ one obtains

$$\widehat{v_n^{-1}P(n)} = \widehat{B(n)} \simeq \prod_\gamma \Sigma^{\sigma(\gamma)}K(n)$$

and if $n = 1$ one sees that the p-adic completion of the Adams summand G of $K\mathbf{Z}_{(p)}$ completely determines $\widehat{v_1^{-1}BP}$, a partial converse to the classical Conner-Floyd theorem.

The results cited above may be used to give conceptual proofs of some change of rings isomorphisms of [32] which are the starting point for the important work [33]. Let $\widehat{\Gamma}(n)_*$ denote the Hopf algebroid $\widehat{v_n^{-1}BP}_*(\widehat{v_n^{-1}BP})$ and write $\widehat{\Sigma}(n)_*$ for $\widehat{E(n)}_*(\widehat{E(n)}_*)$. Then the Hopf algebroids $\widehat{\Gamma}(n)_*$ and $\widehat{\Sigma}(n)_*$ are seen to be equivalent in the following sense [44]: Let (A_1,Γ_1) and (A_2,Γ_2) be Hopf algebroids and $f,g : (A_1,\Gamma_1) \to (A_2,\Gamma_2)$ be two morphisms. A natural equivalence from f to g is a ring homomorphism $H : \Gamma_1 \to A_2$ such that $H \circ \eta_L = \epsilon \circ f \circ \eta_L$, $H \circ \eta_R = \epsilon \circ g \circ \eta_R$ and $(f, \eta_R \circ H) \circ \Delta = (\eta_L \circ H, g) \circ \Delta$ where $\Delta : \Gamma_1 \to \Gamma_1 \otimes_{A_1} \Gamma_1$ denotes the diagonal. Then $f : (A_1,\Gamma_1) \to (A_2,\Gamma_2)$ is an equivalence if there is a morphism $h : (A_2,\Gamma_2) \to (A_1,\Gamma_1)$ such that $f \circ h$ and $h \circ f$ are equivalent to the respective identity morphisms. Given a left Γ_1-comodule N one can define a Γ_2-comodule $f^*(N)$ by $f^*(N) = A_2 \otimes_{A_1} N$. If $f : (A_1,\Gamma_1) \to (A_2,\Gamma_2)$ is an equivalence it follows that there is an induced natural isomorphism

$$Ext^*_{\Gamma_1}(A_1, N) \cong Ext^*_{\Gamma_2}(A_2, f^*(N)).$$

In particular this implies (see [7]):

Theorem 3.5 *For any $\widehat{\Gamma}(n)_*$-comodule N, there is a natural isomorphism*

$$Ext^*_{\widehat{\Gamma}(n)_*}(\widehat{v_n^{-1}BP_*}, N) \cong Ext^*_{\widehat{\Sigma}(n)_*}(\widehat{E(n)}_*, \widehat{E(n)}_* \otimes_{\widehat{v_n^{-1}BP_*}} N).$$

Now this theorem has two important corollaries which form the main results of [32] (see also [37],[38]):

Corollary 3.2 *Let N be a BP_*BP-comodule in which every element is I_n-torsion and v_n acts bijectively. Then there is a natural isomorphism*

$$Ext^*_{BP_*BP}(BP_*, N) \cong Ext^*_{\Sigma(n)_*}(E(n)_*, E(n)_* \otimes_{BP_*} N).$$

Corollary 3.3 *The natural projection $BP_* \to K(n)_*$ induces an isomorphism*

$$Ext^*_{BP_*BP}(BP_*, v_n^{-1}BP_*/I_n) \cong Ext^*_{K(n)_*K(n)}(K(n)_*, K(n)_*).$$

The proof of the first corollary from theorem 3.6. uses in an essential manner the fact that $\widehat{v_n^{-1}BP_*}$ is faithfully flat on the category of finitely generated $\widehat{\Gamma}(n)_*$-comodules. In corollary 3.2., $K(n)_*K(n)$ denotes the Hopf algebroid $K(n)_*(K(n))$ modulo the generators a_i.

It turns out (see [44]) that $Ext^*_{K(n)_*K(n)}(K(n)_*, K(n)_*)$ admits an interpretation in terms of group cohomology: There is an isomorphism

$$\mathbf{F}_{p^n} \otimes Ext^*_{K(n)_*K(n)}(K(n)_*, K(n)_*) \cong H^*_c(S_n; \mathbf{F}_{p^n})$$

where H^*_c means continuous cohomology (with trivial action on $\mathbf{F}_{p^n}$) and S_n is the group considered in section 2. Using the second corollary above these cohomology groups form the input of the chromatic spectral sequence of [33] which converges to $Ext^*_{BP_*(BP)}(BP_*, BP_*)$, the E_2-term of the Adams-Novikov spectral sequence converging to $\pi_*(S^0)_{(p)}$.

There is an interesting connection between $K(n)$ and $\widehat{E(n)}$ which makes essential use of the Bockstein operations considered at the end of section 2. Let $L_1 = (B\mathbf{Z}_p)^{[2p^n-1]}$ be the $2p^n - 1$ skeleton of $B\mathbf{Z}_p$. Then

$$K(n)^*(L_1^+) \cong K(n)^*[y]/(y^{p^n}) \otimes \Lambda(z_1)$$

where $|y| = 2$ and $|z_1| = 1$. We denote the class of y by y_1. The Q_i are characterized by the following properties:

1. For all $0 \leq k \leq n-1$, Q_k is a $K(n)^*$-derivation

$$Q_k : K(n)^*(L_1^+) \to K(n)^*(L_1^+).$$

2. $Q_k(z_1) = y_1^{p^k}$

3. $Q_k(y_1) = 0$.

In [52],[53],[54],[55] A. Robinson has described a theory of A_∞-ring spectra and their module spectra . In particular he showed that at an odd prime p, $K(n)$ admits uncountably many distinct A_∞-structures compatible with its canonical ring structure. Using Robinson's theory, Baker was able to prove (see [6]) that $\widehat{E(n)}$ admits a unique topological A_∞-structure compatible with its canonical product and that the canonical morphism of ring spectra $\widehat{E(n)} \to K(n)$ is a A_∞-morphism (whichever A_∞-structure on $K(n)$ we take). Moreover, he shows that there is an inverse system of A_∞- module spectra over $\widehat{E(n)}$

$$* \leftarrow K(n) = E(n)/I_n \leftarrow \cdots \leftarrow E(n)/I_n^k \leftarrow E(n)/I_n^{k+1} \leftarrow \cdots \tag{1}$$

whose homotopy inverse limit is $\widehat{E(n)}$. Now there is a cofibre sequence of A_∞-module spectra over $\widehat{E(n)}$ (see [8])

$$\bigvee_{0\leq k\leq n-1} \Sigma^{2p^k-2}K(n) \to E(n)/I_n^2 \to K(n) \tag{2}$$

which realises the exact sequence of $\widehat{E(n)}_*$-modules

$$\bigoplus_{0\leq k\leq n-1} \Sigma^{2p^k-2}K(n)_* \cong I_n/I_n^2 \to E(n)_*/I_n^2 \to K(n)_*,$$

where for the first arrow we use the n homomorphisms defined by $1 \mapsto \overline{v_k} \in E(n)_*/I_n^2$. We denote the cofibre map of (3.2.) by Q. Then $Q = \bigvee_k Q_k$ and the Q_k are just the Bocksteins considered above [8]. More generally, for each $k \geq 1$ one can define higher order Bocksteins $Q_v^k : E(n)/I_n^k \to \Sigma^v K(n)$ where $v = v_0^{r_0}v_1^{r_1}\cdots v_{n-1}^{r_{n-1}}$ with $r_0 + r_1 + \cdots r_{n-1} = k-1$ whose wedge $Q^k = \bigvee_v Q_v^k$ is the coboundary of a cofibre sequence

$$\bigvee_v \Sigma^v K(n) \to E(n)/I_n^{k+1} \to E(n)/I_n^k$$

and which admit a similar characterisation as the ordinary Bocksteins. These exact triangles fit into the tower (3.1.) and by applying the functor $[X,-]$ to this tower one gets a spectral sequence

$$E_1^{s,*} = I_n^s/I_n^{s+1} \otimes_{K(n)_*} K(n)^*(X) \Rightarrow \widehat{E(n)}^*(X)$$

with differential

$$d_1 = \widehat{Q}^{s+1} : E_1^{s,*}(X) \to E_1^{s+1.*}(X)$$

and converging to $\widehat{E(n)}^*(X)$, see [8]. It generalises the classical Bockstein spectral sequence with mod p coefficients. This spectral sequence has recently found an application in the work of J. Hunton [17].

4 Some examples

In this section we will look at a few examples of spaces or spectra X for which $K(n)^*(X)$ is known, with the aim to persuade a sceptical reader of the computability of Morava

K-theories. Whereas most of these calculations are difficult to carry out in detail, almost all of them use in a crucial way the Künneth isomorphism

$$K(n)_*(X \times Y) \cong K(n)_*(X) \otimes_{K(n)_*} K(n)_*(Y).$$

The main example we will mention here is the computation of the Morava K-theory of Eilenberg-Mac Lane spaces due to Ravenel and Wilson (see [50]). First, we consider $K_1 = K(\mathbf{Z}/(p^j), 1)$. As for any complex-oriented theory, $K(n)_*(\mathbf{CP}_\infty)$ is a free $K(n)_*$-module on generators $\beta_i \in K(n)_{2i}(\mathbf{CP}_\infty)$ where the β_i are dual to $y^i \in K(n)^*(\mathbf{CP}_\infty) \cong K(n)_*[[y]]$. Let $*$ denote the product induced on $K(n)_*(-)$ by the H-space structure of $\mathbf{CP}_\infty$ or K_1. There is a fibration

$$S^1 \to K_1 \xrightarrow{\delta} \mathbf{CP}_\infty$$

and by looking at the associated Gysin sequence it is not difficult to prove [50]:

Theorem 4.1 *Let $K_1 = K(\mathbf{Z}/(p^j), 1)$. Then*

1. *The map δ induces a monomorphism of Hopf algebras*

$$\delta_* : K(n)_*(K_1) \to K(n)_*(\mathbf{CP}_\infty).$$

2. *As a $K(n)_*$-module, $K(n)_*(K_1)$ is free on $a_m \in K(n)_{2m}(K_1)$, where $0 \leq m < p^{nj}$ and $\delta_*(a_m) = \beta_m$.*
3. *The coproduct ψ is given by*

$$\psi(a_m) = \sum_{i=0}^{m} a_i \otimes a_{m-i}.$$

4. *As an algebra, $K(n)_*(K_1)$ is generated by the elements $a_{(i)} = a_{p^i}$ for $0 \leq i < nj$ subject to the relations $a_{(n+i-1)}^{*p} = v_n^{p^i} a_{(i)}$ where $a_{(i)} = 0$ for $i < 0$.*

Now let us write K_q for the spaces $K(\mathbf{Z}/(p^j), q)$. Clearly, K_* is the representing Ω-spectrum of ordinary $\mathbf{Z}/(p^j)$- cohomology. The cup-product in $H^*(-; \mathbf{Z}/(p^j))$ produces a pairing $K_i \wedge K_j \to K_{i+j}$ which in turn induces maps

$$\circ : K(n)_*(K_i) \otimes_{K(n)_*} K(n)_*(K_j) \to K(n)_*(K_{i+j})$$

which satisfy a lot of compatibility conditions . The Hopf algebras $K(n)_*(K_q)$ together with this circle-product form a Hopf ring $K(n)_*(K_*) = \{K(n)_*(K_q)\}_{q \geq 0}$ in the sense of [50], [51]. Now using this Hopf-ring structure in connection with a highly non-collapsing bar spectral sequence and the theorem above Ravenel and Wilson were able to compute $K(n)_*(K_q) = K(n)_*(K(\mathbf{Z}/(p^j), q))$ for all odd p and all j. Here we will describe their results only for the case $j = 1$. For any sequence $I = (i_1, i_2, \cdots, i_q)$ where $0 \leq i_k < n$ we define $a_I \in K(n)_*(K_q)$ by the iterated circle product

$$a_I = a_{(i_1)} \circ a_{(i_2)} \circ \cdots \circ a_{(i_q)}.$$

Then one has

Theorem 4.2 *Let p be an odd prime and $K_* = K(\mathbf{Z}/(p), *)$. Then, as $K(n)_*$- algebras, $K(n)_*(K_*)$ may be described as follows:*

1. *$K(n)_*(K_0) \cong K(n)_*[\mathbf{Z}/(p)]$, the group ring of $\mathbf{Z}/(p)$ over $K(n)_*$.*
2. *For $0 < q < n$ there are isomorphisms*

$$K(n)_*(K_q) \cong \bigotimes_I K(n)_*[a_I]/(a_I^{p^{\rho(I)}})$$

   where $\rho(I) = 1 + max\{\{0\} \cup \{s+1 | i_{q-s} = n-1-s\}\}$ and $0 < i_1 < ... < i_q < n$.
3. *If $q = n$ set $I = (0, 1, \cdots, n-1)$. Then*

$$K(n)_*(K_n) \cong K(n)_*[a_I]/(a_I^{*p} + (-1)^q v_n a_I).$$

4. *$K(n)_*(K_q) \cong K(n)_*$ if $q > n$.*

In fact, there is a much more conceptual and elegant way to formulate the theorem above (see [50]): The Hopf ring $K(n)_*(K_*)$ is the free $K(n)_*[\mathbf{Z}/(p)]$- Hopf ring on the Hopf algebra $K(n)_*(K_1)$. Let us also mention that in [50], these results are used to compute $v_n^{-1}BP_*(K(\mathbf{Z}/(p), n))$ which, applying the methods briefly mentioned at the beginning of section 3, allows them to prove the Conner-Floyd conjecture.

Observe that there is an isomorphism

$$lim_j K(n)_*(K(\mathbf{Z}/(p^j), q)) \cong K(n)_*(K(\mathbf{Z}, q+1)),$$

so, by the Künneth isomorphism, $K(n)_*(BG)$ is known for all finitely generated abelian groups G. It is interesting to observe that through the eyes of Morava K-theories, the Eilenberg-MacLane spaces for finite abelian groups appear as finite complexes.

If G is an arbitrary finite group one has the following general result of Ravenel [48]:

Theorem 4.3 *For any finite group G, $K(n)^*(BG)$ is finitely generated as a module over $K(n)^*$.*

If $n = 1$, $K(1)$ is a summand of mod p complex K-theory and Atiyah's description of $K^*(BG)$ in terms of the complex representation ring may be used to show that the rank of $K(1)^*(BG)$ is the number of conjugacy classes of p-elements in G (see [23],[48]). In [23], N. Kuhn has proved the following generalisation of this:

Theorem 4.4 *Let G be a finite group with an abelian p-Sylow subgroup P, and let $W = N_G(P)/C_G(P)$. Then*

$$rank_{K(n)^*} K(n)^*(BG) = |P^n/W|,$$

the number of W-orbits in P^n.

The question of finding the group-theoretic significance of the rank of $K(n)^*(BG)$ is clearly a very interesting one and actually several people are working on this problem. Among other things, the interest in this question is stimulated by the fact that although the Morava K-theories are fairly well understood today, one does not know any good model for the spaces representing them. One then hopes that a better understanding of $K(n)_*(BG)$ in terms of G might furnish some ideas in this direction.Let us also mention in this context the following result of Hopkins, Kuhn and Ravenel (see [24]): For topological groups Γ, G let $Hom(\Gamma, G)$ denote the space of continuous homomorphisms. Letting act G on itself by conjugation this becomes a left G-space. Let G be a finite group. Then $Hom(\hat{\mathbf{Z}}_p^n, G)$ is the set of n-tuples of G generating an abelian p-group. One now has the

Theorem 4.5 *Let G be a finite group. Then*

$$dim_{K(n)^*}K(n)^{even}(BG) - dim_{K(n)^*}K(n)^{odd}(BG) = |Hom(\hat{\mathbf{Z}}_p^n, G)/G|.$$

There are a lot of other spaces X where $K(n)_*(X)$ is known. As examples, let us only mention the computation of $K(n)_*(\Omega^2 S^{2r+1})$ by Yamaguchi [71] , the recent description of $K(m)_*(\Omega^2 SU(n+1)$ by Ravenel in [49] and the work [16], [17] of J. Hunton where (among a lot of other things), he develops a method for computing the Morava K-theories of classifying spaces of wreath products $G \wr C_p$, C_p a cyclic group on p elements.

5 The connected cover of $K(n)$

In this section we will review some properties of $k(n)$, the connected cover of $K(n)$. $k(n)$ is a ring spectrum (non-commutative if $p = 2$) with coefficient ring $k(n)_* = \mathbf{F}_p[v_n]$ and $v_n^{-1}k(n) = K(n)$ and there are cofibrations

$$\cdots \longrightarrow \Sigma^{2p^n-2}k(n) \xrightarrow{v_n} k(n) \xrightarrow{\pi_n} H\mathbf{F}_p \xrightarrow{\overline{Q}_n} \Sigma^{2p^n-1}k(n) \longrightarrow \cdots \qquad (3)$$

where $\pi_n : k(n) \to H\mathbf{F}_p$ denotes the Thom map.

Let $\mathcal{A}^*(p)$ denote the mod p Steenrod algebra. Then (see [4]) π_n induces an isomorphism

$$H^*(k(n); \mathbf{F}_p) \cong \mathcal{A}^*(p)/\mathcal{A}^*(p)Q_n$$

and so $Q_n = \pi_n \overline{Q}_n$, where $Q_n \in \mathcal{A}^*(p)$.

Because $k(n)^*$ is a principal ideal domain, $k(n)^*(X)$ decomposes as a $k(n)^*$-module into copies of $\mathbf{F}_p[v_n]$ and of the quotients $\mathbf{F}_p[v_n]/(v_n^s)$, where $s \geq 1$. The free part of $k(n)^*(X)$ is detected by $K(n)^*(X)$ while the torsion part is analyzed by the Bockstein spectral sequence $\{E_r, d_r\}$ associated to the exact triangle 6.1.. One has $E_1 = H^*(X; \mathbf{F}_p)$ and $d_1 = Q_n$ (resp. $d_1 = Sq^{\Delta_{n+1}}$ if $p = 2$) where Q_n denotes the Milnor operation which is inductively defined by $Q_0 = \beta$ and $Q_n = \mathcal{P}^{p^{n-1}}Q_{n-1} - Q_{n-1}\mathcal{P}^{p^{n-1}}$. Let

$$T_r^*(X) = ker\{v_n^r : k(n)^*(X) \to k(n)^*(X)\}$$

and set

$$T^*(X) = \bigcup_{r \geq 1} T_r^*(X).$$

$T^*(X)$ is the torsion part of $k(n)^*(X)$. Then

$$E_\infty \cong k(n)^*(X)/\left(T^*(X) + v_n k(n)^*(X)\right)$$

and there is a short exact sequence

$$0 \to v_n^{r-1}k(n)^*(X)/v_n^r k(n)^*(X) \to E_r^* \to T_r^*/T_{r-1}^* \to 0.$$

The spectral sequence $\{E_r, d_r\}$ is a spectral sequence of algebras and it can be identified with the Atiyah-Hirzebruch spectral sequence for $k(n)^*$. A detailed study of $k(n)^*(X)$ and the associated spectral sequence appears as the main tool in the paper [20] of R.M. Kane where he proves that for a connected, simply connected mod 2 finite H-space X, $Q^{even}H^*(X;\mathbf{F}_2) = 0$ where $QH^*(X;\mathbf{F}_2)$ denotes the module of indecomposables . Another application of this Bockstein spectral sequence appears in [71] where $k(n)_*(\Omega^2 S^{2r+1})$ is calculated.

The algebra $k(n)^*(k(n))$ of stable $k(n)$-operations has been studied by Yagita in [67] and by Lellmann in [31]. To describe it, one needs to define some algebras associated to it. For any spectrum X, define

$$\mathcal{Z}^*(X) = ker\{Q_n : H^*(X;\mathbf{F}_p) \to H^*(X;\mathbf{F}_p)\}$$

$$\mathcal{B}^*(X) = im\{Q_n : H^*(X;\mathbf{F}_p) \to H^*(X;\mathbf{F}_p)\}$$

and $\mathcal{H}^*(X) = \mathcal{Z}^*(X)/\mathcal{B}^*(X)$. $\mathcal{Z}^*(k(n))$ inherits an algebra structure from $\mathcal{A}^*(p)$ with respect to which $(\pi_n)_*$ is a homomorphism of algebras. Let

$$kP(n)_*P(n)k = k(n)_* \otimes_{P(n)_*} P(n)_*(P(n)) \otimes_{P(n)_*} k(n)_*.$$

$kP(n)_*P(n)k$ inherits a Hopf algebra structure from $P(n)_*(P(n))$ and we define $L^*(n)$ as the dual $k(n)_*$-Hopf algebra. The canonical map $P(n)_*(P(n)) \to k(n)_*(k(n))$ factors to give a map $\eta : kP(n)_*P(n)k \to k(n)_*(k(n))$ and we write κ for the composition

$$\kappa : k(n)^*(k(n)) \to Hom_{k(n)_*}(k(n)_*(k(n)), k(n)_*) \xrightarrow{\eta^*} L^*(n).$$

Using these notations one then has (see [31]):

Theorem 5.1 *There is a surjective algebra homomorphism $\pi_* : L^*(n) \to \mathcal{H}^*(k(n))$ whose kernel is the ideal of v_n-divisible elements and the diagram*

$$\begin{array}{ccc} k(n)^*(k(n)) & \xrightarrow{\kappa} & L^*(n) \\ \downarrow{\scriptstyle (\pi_n)_*} & & \downarrow{\scriptstyle \pi_*} \\ \mathcal{Z}^*(k(n)) & \xrightarrow{pr} & \mathcal{H}^*(k(n)) \end{array}$$

is a pullback diagram of algebras.

This has been proved in [31] for p odd but it also holds for $p = 2$, see [26]. In [67], Yagita described $k(n)^*(k(n))$ by generators and relations. These may also be deduced from the theorem above.

Notice that the structure of the algebra $k(n)_*(k(n))$ is also known (see [67] for the case p odd and [27] for $p = 2$).

We say that a spectrum X has $k(n)^*$-exponent $\leq e$, $exp_{k(n)^*}(X) \leq e$, if

$$T^*(X) = ker\{v_n^e : k(n)^*(X) \to k(n)^*(X)\} = T_e^*(X)$$

and we define $exp_{k(n)_*}(X) \leq e$ similarly. Let $k(n)^{[rq]}$ denote the rq^{th} Postnikov factor of the spectrum $k(n)$ where $r \geq 0$ and $q = 2(p^n - 1)$. Thus $k(n)^{[rq]}$ is again a (commutative) ring spectrum and $\pi_i(k(n)^{[rq]}) = \pi_i(k(n))$ if $i \leq rq$ and $\pi_i(k(n)^{[rq]}) = 0$ if $i > rq$. In particular , $k(n)^{[0]} = H\mathbf{F}_p$. Using the fact that the Postnikov factors of $k(n)$ are related to the Bockstein spectral sequence the following splitting theorem for $k(n) \wedge X$ may be proved:

Theorem 5.2 *Let X be a locally finite connective spectrum and suppose $e \geq 1$. Then the following are equivalent:*

1. $exp_{k(n)_*}(X) \leq e$
2. $exp_{k(n)^*}(X) \leq e$
3. *There is an equivalence of $k(n)$-module spectra*

$$k(n) \wedge X \sim \bigvee_{r=0}^{e-1} \bigvee_{i_r} \Sigma^{n(i_r)}(k(n)^{[rq]}) \vee \bigvee_{i_e} \Sigma^{n(i_e)} k(n)$$

This reflects nicely the structure of $k(n)_*(X)$ for spectra of exponent $\leq e$. For $e = 1$ it appears in [31] and the general case has been proved in [25] where one also finds a similar splitting result for the spectra $k(n)^{[rq]} \wedge X$. An example for a spectrum of exponent ≤ 1 is $k(n)$ itself or the spectrum $B(\mathbf{Z}_p)^r$.

Observe that because $k(n)_*$ is not a (graded) field, there is in general no Künneth formula for $k(n)_*(-)$. However, as one would expect, the following useful theorem holds [31]:

Theorem 5.3 *Let X and Y be locally finite CW-spectra. Then there exists a short exact sequence of $\Lambda = k(n)_*$-modules*

$$0 \to k(n)_*(X) \otimes_\Lambda k(n)_*(Y) \to k(n)_*(X \wedge Y) \xrightarrow{\delta} Tor_\Lambda(k(n)_*(X), k(n)_*(Y)) \to 0,$$

where δ is a map of degree (-1).

6 Uniqueness properties

As we have seen in the previous sections, the Morava K-theories are strongly related to the formal group law F_n and so one may ask if $K(n)^*(-)$ is determined by F_n. Here, we will consider this question in a more general setting. First, for any field k of positive characteristic, we define $K(n)^*(-;k)$, the nth Morava K-theory with coefficients k, by

$$K(n)^*(-;k) = Hom_{\mathbf{F}_p}(K(n)_*(-), k).$$

By a $\mathbf{Z}_2$ -graded ring theory with coefficients in a commutative (ungraded) ring A we mean a $\mathbf{Z}_2$ -graded cohomology theory $T^\bullet(-)$ endowed with an associative product with two-sided unit such that

$$T^\bullet(S^0) = \begin{cases} A & \text{if } \bullet = 0 \\ 0 & \text{if } \bullet = 1. \end{cases}$$

Observe that we do not assume full commutativity of the product on $T^\bullet(-)$, however we always assume that the ring $T^\bullet(PC_\infty \times PC_\infty)$ is commutative. This implies that in the case $T^\bullet(-)$ is C-orientable, the whole theory of general Chern-classes applies, see [12]. Clearly, for all primes p, the $\mathbf{Z}_2$-graded versions of the Morava K-theories are typical examples of such theories.

Because the coefficient ring A of $T^\bullet(-)$ is concentrated in dimension 0, $T^\bullet(-)$ is C-orientable and thus determines an isomorphism class of formal group laws $[F_T(x,y)]$ over the ring A. In particular, if $A = k$ is a field of positive characteristic p, this formal group law is of positive height n . The following classification theorem shows that this correspondence is a bijection and that, moreover, any $\mathbf{Z}_2$ -graded ring theory with coefficients k is essentially a Morava K-theory with a possibly exotic product.

Theorem 6.1 *Let k be a field of positive characteristic p and let $T^\bullet(-)$ be a $\mathbf{Z}_2$ -graded ring theory with coefficient ring k and formal group law $F_T(x,y)$ of height n. Then there exists a product μ on $K(n)^\bullet(X;k)$, unique up to isomorphism, and a natural equivalence of $\mathbf{Z}_2$ -graded ring theories*

$$T^\bullet(X) \xrightarrow{\sim} K(n)^\bullet_\mu(X;k)$$

where $K(n)^\bullet_\mu(X;k)$ denotes Morava K-theory endowed with the product μ. If p is odd, there is a bijection between the set of isomorphism classes of products on $K(n)^\bullet(X;k)$ and the set $FG(k)^n$ of isomorphism classes of formal group laws of height n over k.Moreover, all these products are commutative. If $p = 2$, this correspondence is onto but not injective: To any element of $FG(k)^n$ there exist exactly two isomorphism classes of products on $K(n)^\bullet(X;k)$ which are generated by some non-commutative product μ and its opposite μ^{opp}.

This theorem is just a reformulation of theorem (3.2) of [65] with the exception of the last sentence concerning the case $p = 2$, which was only conjectured there. However, using the results of [26], the same methods used to prove theorem (3.2) of [65] for the case p odd are easily seen to carry over to the case $p = 2$. Notice that if we set $K(\infty) = H\mathbf{F}_p$ the theorem holds also for $n = \infty$: In this case, $FG(k)^\infty$ consists of only one element, the class of the additive formal group law.

Corollary 6.1 *Two $\mathbf{Z}_2$ -graded ring theories with coefficient ring k a field of positive characteristic are isomorphic as cohomology theories with values in the category of k-vector spaces if and only if their formal group laws are of the same height.*

For $FG(k)^n$, there are several more or less explicit descriptions available, see [14]. Let us briefly recall one of them. Let $\overline{k}_{sep}$ denote a separable closure of k and set $\Gamma = Gal(\overline{k}_{sep} : k)$. Then there is an isomorphism

$$FG(k)^n \xrightarrow{\sim} H^1(\Gamma; S_n)$$

where $S_n \cong Aut_{\overline{k}_{sep}}(F_n)$, i.e. S_n is isomorphic to the group of strict units of the maximal order in the central division algebra of invariant $1/n$ and rank n^2 over $\widehat{\mathbf{Q}}_p$. For example, if $n = 1$, S_1 is isomorphic to the group of strict units of $\widehat{\mathbf{Z}}_p^*$. If, moreover,

$k = \mathbf{F}_q$, $q = p^n$, then Γ is topologically generated by the Frobenius homomorphism $\sigma : \alpha \mapsto \alpha^q$ and so in this case one gets bijections

$$FG(\mathbf{F}_q)^1 \approx H^1(\Gamma; \widehat{\mathbf{Z}}_p^*) \approx Hom_{cont}(\Gamma, \widehat{\mathbf{Z}}_p^*) \approx \widehat{\mathbf{Z}}_p^*.$$

Hence, the set of isomorphism classes of $\mathbf{Z}_2$ -graded ring theories with coefficient ring $\mathbf{F}_q$ and formal group of height 1 (resp. the set of isomorphism classes of products on $K(1)^\bullet(-;\mathbf{F}_p)$) is in $1-1$ correspondence with $\widehat{\mathbf{Z}}_p^*$ for p odd and with $\mathbf{Z}_2 \times \widehat{\mathbf{Z}}_2^*$ if $p=2$.

Consider a field extension $k \subset K$ and let $T_1^\bullet(-)$ and $T_2^\bullet(-)$ be $\mathbf{Z}_2$-graded ring theories with coefficients k.We will say that $T_1^\bullet(-)$ is a (twisted) (K/k)- *form* of $T_2^\bullet(-)$, if there is an isomorphism of $\mathbf{Z}_2$-graded ring theories

$$T_1^\bullet(-) \otimes_k K \xrightarrow{\sim} T_2^\bullet(-) \otimes_k K$$

over the category $\mathcal{CW}_f$ of spaces of the homotopy type of a finite CW-complex. Because all coefficients in sight are fields, this isomorphism clearly extends to the category of all complexes.

Now let $\overline{k}_{sep}$ be a separable closure of the field k of characteristic p. Then it is well known that over $\overline{k}_{sep}$, formal group laws are isomorphic if and only if they are of the same height. Combined with theorem 6.1. this implies

Corollary 6.2 *Let k be a field of characteristic $p > 0$. Then all $\mathbf{Z}_2$ -graded ring theories with coefficients k and formal group of height n are $(\overline{k}_{sep}/k)$-forms of the nth Morava K-theory $K(n)^\bullet(-;k)$.*

Notice that in the case $p=2$ of the above corollary, both products on $K(n)^\bullet(-;k)$ have to be considered. Part of this has also been proved in [39].

Corollary 6.2. suggests that it should be possible to recover the theories $K(n)_\mu^\bullet(-;k)$, μ some possibly exotic product on $K(n)^\bullet(-;k)$, in some sense from $K(n)^\bullet(-;\overline{k}_{sep})$. This is in fact possible using the theory of Galois descent (see e.g. [21]).

Let k be a field of positive characteristic p and let K/k be a Galois extension of k with Galois group $\Gamma = Gal(K/k)$. Let $Iso_{K/k}(F)$ denote the set of isomorphism classes of (K/k)–forms of the formal group law F. Then there is a bijection (see [14])

$$\Phi : Iso_{K/k}(F) \xrightarrow{\sim} H^1(\Gamma; Aut_K(F))$$

where $H^1(\Gamma; Aut_K(F))$ denotes the first Galois cohomology group and Γ acts on $Aut_K(F))$ by acting on the coefficients of power series, i.e., $\sigma(\alpha(x)) = \sigma_*\alpha(x)$, where $\sigma \in \Gamma$. As was shown in [65], theorem (3.13), for any extension field K of k there is an isomorphism of groups

$$Aut_K(F) \cong Aut(K(n)_\mu^\bullet(-;K))$$

where $Aut(K(n)_\mu^\bullet(-;K))$ denotes the group of multiplicative automorphisms of $K(n)_\mu^\bullet(-;K)$. This fact together with the elements of the theory of Galois descent allows us to prove

Theorem 6.2 *Let $\overline{k}_{sep}$ be a separable closure of the field k of characteristic $p > 0$ and let μ be some product on $K(n)^\bullet(-;k)$. Then there is an action of $\Gamma = Gal(\overline{k}_{sep}/k)$ on the Morava K-theory $K(n)^\bullet(-;\overline{k}_{sep})$ by k-linear automorphisms such that*

$$K(n)_\mu^\bullet(-;k) \cong K(n)^\bullet(-;\overline{k}_{sep})^\Gamma$$

as $\mathbf{Z}_2$-graded ring theories.

Let us remark that there are also graded versions of the above results and, by passing to connective covers, uniqueness theorems for the connected version of Morava K-theory. In this context we observe that in [42], Pazhitnov proves the following

Theorem 6.3 *The homotopy type of a commutative ring spectrum E with coefficient ring $\pi_*(E) = \mathbf{F}_p[t]/(t^s)$, where $2 < s \leq \infty$, is determined by the integer $dim(t) = 2k$ and the first nontrivial k-invariant. There is a positive integer n such that E is homotopy equivalent to a sum of suspensions of Postnikov stages of $k(n)$.*

7 Morava K-theories and stable homotopy

In this section we will very briefly discuss some results concerning self maps for finite spectra taken from recent work of Devinatz, Hopkins and Smith (see [11], [15]), which demonstrate the importance of the spectra $K(n)$ in the scope of stable homotopy theory. Their investigations have strongly been motivated by a series of conjectures of Ravenel (see [47]). Non-nilpotent self maps of finite spectra are of great importance in the light of the chromatic spectral sequence which suggests possibilities to organize the stable homotopy groups of the spheres into periodic families associated with the indecomposables of the ring BP_* (see [44]).

Let us write $K(0)$ for $H\mathbf{Q}$ and $K(\infty)$ for $H\mathbf{F}_p$. As a first result from [15] let us mention the

Theorem 7.1 *If $f : \Sigma^k X \to X$ is an endomorphism of the finite spectrum X which induces the trivial map in $K(n)^*(-)$ for all $n < \infty$ and all p, then f is nilpotent.*

When X is the sphere spectrum, this is Nishida's theorem which says that each element of positive dimension of the stable homotopy ring $\pi_*(S^0)$ is nilpotent.

Now we fix the prime p and work in the category $\mathcal{C}_0$ of p-local finite spectra. Let X be such a spectrum and let $n \geq 1$. Then a map $f : \Sigma^k X \to X$ is called a *v_n-self map* if $K(n)_*(f)$ is an isomorphism and $K(m)_*(f)$ is nilpotent for $m \neq n$. If $n = 0$, a v_0-self map is a map inducing multiplication by p^j in rational cohomology, for some j. Clearly, v_n-self maps represent (if they exist) a simple class of non-nilpotent endomorphisms of X.

To be able to answer the question about existence of such v_n-self maps one has first to consider certain subcategories of the stable homotopy category $\mathcal{S}$. If X is a finite spectrum we know by a result of Ravenel [47] that

$$rank_{K(n)^*} K(n)^*(X) \leq rank_{K(n+1)^*} K(n+1)^*(X).$$

This allows us to define the *type* of a p-local finite spectrum to be the smallest integer n such that $K(n)^*(X) \neq 0$. Let $\mathcal{C}_n$ denote the full subcategory of $\mathcal{C}_0$ of $K(n-1)-$ acyclic spectra. It is a non-trivial fact proved first by S. Mitchell [35] that there are strict inclusions $\mathcal{C}_{n+1} \subset \mathcal{C}_n$. Now it is one of the main consequences of [11] (see [15]) that these categories $\mathcal{C}_n$ play a very interesting rôle inside $\mathcal{S}$.

Theorem 7.2 *Let C be a full subcategory of C_0 which is closed under cofibrations (i.e. if two of three terms in a cofibre sequence lie in C then so does the third) and under retracts (i.e. if X is an object of C then any retract of X is an object of C). Then there exists an integer $n \geq 0$ such that $C = C_n$.*

As a rather immediate application of this theorem one obtains [15]

Theorem 7.3 *A p-local finite spectrum X admits a v_n-self map if and only if X is an object of C_n.*

Now in fact, Hopkins and Smith also show that such self maps are unique in the sense that if f and g are two v_n-self maps of X, then some iterate of f is homotopic to some iterate of g. Moreover, they prove that the v_n-self maps generate the centers of the homotopy endomorphism rings of finite spectra modulo nilpotents and that these endomorphism rings have Krull dimension 1.

Let us finally cite another consequence of the work [11],[15] which again underlines the special rôle of the Morava K-theories:

Theorem 7.4 *Let E be a ring spectrum with the property that for all X, $E \wedge X$ is equivalent to a wedge of suspensions of E. Then there exists an n such that E is (non-multiplicatively) homotopy equivalent to a wedge of suspensions of $K(n)$.*

In fact this means that the Morava K-theories (with all possible products, see the last section) and ordinary cohomology with field coefficients are essentially the only homology theories where a Künneth isomorphism holds without restrictions.

References

[1] Adams, J.F.:*Stable homotopy and generalised homology,* Univ. of Chicagao press, Chicago, Illinois and London (1974).

[2] Araki, S.:*Typical formal groups in complex cobordism and K-theory,* Lecture Notes in Math., Kyoto Univ. 6, Kinokuniya Book Store, 1973.

[3] Baas, N.A.:*On bordism theory of manifolds with singularities* , Math. Scand. **33** (1973), 279-302.

[4] Baas, N.A. and Madsen, I.:*On the realization of certain modules over the Steenrod algebra,* Math. Scand **31** (1972), 220-224.

[5] Baker,A.:*Some families of operations in Morava K-theory,* Amer. J. Math. **111**(1989),95-109.

[6] —:*A_∞-structures on some spectra related to Morava K-theories,* preprint Manchester Univ., (1988).

[7] Baker,A. and Würgler,U.:*Liftings of formal groups and the Artinian completion of $v_n^{-1}BP$,* Math. Proc. Camb. Phil. Soc. **106** (1989),511-530.

[8] —:*Bockstein operations in Morava K-theory,* preprint 1989.

[9] Brown, E.H. and Peterson, F.P.: *A spectrum whose $\mathbf{Z}_p$-cohomology is the algebra of reduced p-th powers,* Topology **5** (1966), 149-154.

[10] Cartier,P.: *Modules associés à un groupe formel commutatif, courbes typiques,* C. R. Acad. Sci. Paris Série A **265**(1965), 129-132.

[11] Devinatz,E.S., Hopkins,M.J. and Smith, J.H.: *Nilpotence and stable homotopy I,* Ann. Math. **128**(1988), 207-241.

[12] Dold,A.:*Chern classes in general cohomology.* Symp. Math. **V**(1970),385-410 .

[13] Fröhlich,A.: *Formal groups,* Lecture Notes in Math. **74**(1968).

[14] Hazewinkel M.: *Formal groups and applications.* Academic press, 1978.

[15] Hopkins, J.R.:*Global methods in homotopy theory,* Proc. Durham Symp. 1985, Cambridge Univ. Press (1987), 73-96.

[16] Hunton, J.:*The Morava K-theories of wreath products,* Preprint Cambridge Univ. (1989).

[17] —: Ph.D. Thesis, Cambridge Univ. (1989).

[18] Johnson,D.C. and Wilson, W.S.:*BP-operations and Morava's extraordinary K-theories,* Math.Z. **144**(1975),55-75.

[19] —: *The Brown-Peterson homology of elementary p-groups,* Amer. J. Math. **107**(1984), 427-453.

[20] Kane, R.M.:*Implications in Morava K-theory,* Mem. Amer. Math. Soc. **59** (1986), No.340 .

[21] Knus M.,and Ojanguren, M.: *Théorie de la descente et algébres d'Azumaya.* Lecture Notes in Mathematics **389**, 1974.

[22] Kuhn, N.J.:*Morava K-theories and infinite loop spaces,* Springer Lect. Notes in Math. **1370**(1989),243-257.

[23] —:*The Morava K-theories of some classifying spaces,* TAMS **304**(1987),193-205.

[24] —:*Character rings in algebraic topology,* London Math. Soc. Lecture Notes **139** (1989), 111-126.

[25] Kultze,R.:*Die Postnikov-Faktoren von k(n),* Manuskript, Universität Frankfurt (1989).

[26] Kultze,R. and Würgler,U.:*A Note on the algebra $P(n)_*(P(n))$ for the prime* 2, Manuscripta Math. **57**(1987),195-203.

[27] —:*The algebra $k(n)_*(k(n))$ for the prime* 2, Arch. Math. **51**(1988),141-146.

[28] Landweber, P.S.: *$BP_*(BP)$ and typical formal groups,* Osaka J. Math. **12**(1975),357-363.

[29] —:*Homological properties of comodules over* $MU_*(MU)$ *and* $BP_*(BP)$, Amer. J. Math. **98**(1976),591-610.

[30] Lazard,M.: *Sur les groupes de Lie formels á un paramètre,* Bull. Soc. Math. France **83**, 251-274.

[31] Lellmann,W.:*Connected Morava K-theories,* Math. Z. **179** (1982), 387-399.

[32] Miller, H.R. and Ravenel, D.C.:*Morava stabilizer algebras and the localization of Novikov's* E_2*-term,*Duke Math. J. **44**(1977), 433-447.

[33] Miller, H.R., Ravenel, D.C. and Wilson, W.S.: *Periodic phenomena in the Adams-Novikov spectral sequence,* Ann. of Math. (2)**106** (1977), 459-516.

[34] Mischenko : Appendix 1 in Novikov [41].

[35] Mitchell, S.A.:*Finite complexes with A(n)-free cohomology,* Topology **24**(1985), 227-248.

[36] Morava, J.:*A product for odd-primary bordism of manifolds with singularities,* Topology **18**(1979), 177-186.

[37] —,*Completions of complex cobordism,* Lecture Notes in Math. **658**(1978),349-361.

[38] —,*Noetherian localisations of categories of cobordism comodules,* Ann. of Math. **121**(1985), 1-39.

[39] —,*Forms of K-theory,* Math. Z. **201**(1989),401-428.

[40] Mironov, O.K.: *Existence of multiplicative structures in the theory of cobordism with singularities,* Izv. Akad. Nauk SSSR Ser. Mat. **39**(1975),No.5, 1065-1092.

[41] Novikov,S.P.:*The methods of algebraic topology from the viewpoint of complex cobordism theories,* Math. USSR Izv. (1967), 827-913.

[42] Pazhitmov,A.V.: *Uniqueness theorems for generalized cohomology theories,* Math. USSR Izvestiyah **22**(1984),483-506.

[43] Quillen, D.G.: *On the formal group laws of unoriented and complex cobordism theory,* Bull. Amer. Math. Soc. **75**(1969),1293-1298.

[44] Ravenel,D.C.: *Complex cobordism and stable homotopy groups of spheres,*Academic Press (1986).

[45] —,*The structure of* $BP_*(BP)$ *modulo an invariant prime ideal,* Topology **15**(1976),149-153.

[46] —,*The structure of Morava stabilizer algebras,* Invent. Math. **37**(1976),109-120.

[47] —,*Localization with respect to certain periodic homology theories,* Amer. J. Math. **106**(1984),351-414.

[48] —,*Morava K-theories and finite groups*, Contemp. Math. AMS **12** (1982), 289-292.

[49] —,*The homology and Morava K-theory of $\Omega^2 SU(n)$*, preprint Univ. of Rochester (1989).

[50] Ravenel,D.C. and Wilson,S.W.:*The Morava K-theories of Eilenberg-MacLane spaces and the Conner-Floyd conjecture*, Amer. J. Math. **102**(1980),691-748.

[51] —,*The Hopf ring for complex cobordism*, J. Pure Appl. Algebra **9**(1977),241-280.

[52] Robinson, A.: *Obstruction theory and the strict associativity of Morava K-theories*, London Math. Soc. Lecture Notes **139** (1989), 143-152.

[53] —: *Derived tensor products in stable homotopy theory*, Topology **22**(1983),1-18.

[54] —: *Spectra of derived module homomorphisms*, Math. Proc. Camb. Philos. Soc. **101**(1987), 249-257.

[55] —:*Composition products in RHom and ring spectra of derived homomorphisms*, Springer Lecture Notes in Math. **1370**(1989), 374-386.

[56] Sanders, J.P.: *The category of H-modules over a spectrum*, Mem. Am. Math. Soc. **141**(1974).

[57] Shimada,N and Yagita,N.: *Multiplication in the complex bordism theory with singularities*, Publ. Res. Inst. Math. Sci. **12** (1976/1977),No.1, 259-293.

[58] Wilson,S.W.:*Brown-Peterson homology, an introduction and sampler*, Regional Conference series in Math. No. 48, AMS, Providence, Rhode Island (1980).

[59] —:*The Hopf ring for Morava K-theory*, Pub. RIMS Kyoto Univ. **20**(1984), 1025-1036.

[60] —:*The complex cobordism of BO_n*, J. London Math. Soc. **29**(1984), 352-366.

[61] Würgler,U.: *Cobordism theories of unitary manifolds with singularities and formal group laws*, Math. Z. **150**(1976),239-260.

[62] —: *On products in a family of cohomology theories associated to the invariant prime ideals of $\pi_*(BP)$*, Comment. Math. Helv. **52** (1977),457-481.

[63] —:*On the relation of Morava K-theories to Brown-Peterson homology*, Monographie no. 26 de L'Enseignement Math.(1978),269-280.

[64] —:*A splitting theorem for certain cohomology theories associated to $BP^*(-)$*, Manuscripta Math. **29**(1979), 93-111.

[65] —:*On a class of 2-periodic cohomology theories*, Math. Ann. **267**(1984), 251-269.

[66] —:*Commutative ring-spectra of characteristic 2*, Comment. Math. Helv. **61**(1986), 33-45.

[67] Yagita, N.:*On the Steenrod algebra of Morava K-theory,* J. London Math. Soc. **22**(1980), 423-438.

[68] —,*The exact functor theorem for BP_*/I_n-theory,* Proc. Japan Acad. **52**(1976),1-3.

[69] —,*On the algebraic structure of cobordism operations with singularities,* J. London Math. Soc. **16**(1977),131-141.

[70] —,*A topological note on the Adams spectral sequence based on Morava's K-theory,* Proc. Am. Math. Soc. **72**(1978),613-617.

[71] Yamaguchi, A.: *Morava K-theory of double loop spaces of spheres,* Math. Z. **199** (1988),511-523.

Examples of Lack of Rigidity in Crystallographic Groups

Frank Connolly and Tadeusz Koźniewski

§1. Introduction.

Let Γ be a crystallographic group of rank n, i.e. a discrete cocompact group of isometries of $\mathbf{R}^n$. The rigidity conjecture concerning Γ asserts, roughly, that any "suitable" subgroup of $Homeo(R^n)$ which is isomorphic to Γ is in fact conjugate to Γ. More precisely, it asserts that the structure set, $S(\Gamma)$, consists of one element. Here $S(\Gamma)$ is the set of homeomorphism classes of crystallographic Γ manifolds whose topological torsion vanishes. (Definitions are given in §4 below).

Let G be the holonomy group of Γ. This conjecture is proved in [CK3] when G has odd order, subject to mild dimensional hypotheses. Earlier it had been proved when Γ was free abelian [HS] and when Γ was torsion free [FH2]. The topological torsion mentioned above is in the equivariant topological Whitehead group $Wh_G^{top,\rho}(M_\Gamma)$ introduced by Steinberger and West [StW],[St]. Here M_Γ denotes $\mathbf{R}^n/A$ where A denotes the translation subgroup of Γ.

The Whitehead group above has an involution on it (obtained by reversing the direction of h-cobordisms) and a crucial step in the above mentioned proof is the following vanishing result:

$$(*)_\Gamma \qquad \hat{H}^*(\mathbf{Z}/2\mathbf{Z}; Wh_G^{top,\rho}(M_\Gamma)) = 0$$

if $|G|$ is odd.

The purpose of this note is to show that $(*)_\Gamma$ is false when $|G|$ is even and that, even when $|G|$ is odd, the piecewise-linear analogue:

$$(*)_\Gamma^{pl} \qquad \hat{H}^*(\mathbf{Z}/2\mathbf{Z}; Wh_G^{pl,\rho}(M_\Gamma)) = 0$$

is similarly false. This answers questions raised orally with us by S.Weinberger and F.T.Farrell.

We then show that the failures of $(*)_\Gamma^{pl}$ and $(*)_\Gamma$ yield immediate failures in the corresponding rigidity conjectures. That is to say, we give an example of a crystallographic group Γ for which the PL analogue, $S_{pl}(\Gamma)$, is not trivial, even though $|G|$ is odd. We further provide an example where $|G|$ is even and $S(\Gamma) \neq 0$.

§2. An example where $(*)^{pl}_{\Gamma}$ fails.

Let G be cyclic of order five acting on the cyclotomic integers $\mathbf{Z}[\zeta_5]$ via multiplication by powers of ζ_5. For any $n > 1$, let A be the sum of n copies of $\mathbf{Z}[\zeta_5]$. We form the semi direct product group:

$$\Gamma = A \rtimes G\,. \tag{2.1}$$

Γ is a crystallographic group of rank 4n with holonomy group G. We are going to prove that $\hat{H}^*(\mathbf{Z}/2\mathbf{Z}; Wh_G^{pl,\rho}(M_\Gamma)) \neq 0$.

First of all, for any crystallographic group Γ, we have the following calculation from [CK2]:

$$Wh_G^{pl,\rho}(M_\Gamma) = \sum_H Wh(N_\Gamma(H)/H) \tag{2.2}$$

where H runs over a set of conjugacy classes of those finite subgroups of Γ for which $rk\ Z_A(H) > rk\ Z_A(K)$ if $H \subsetneqq K$. This last inequality is the algebraic way of specifying the isotropy groups of Γ.

In the present case, $|H| = 1$ or 5, and if $H \neq 1$ then $N_\Gamma(H)/H \cong\ Z_A(H)\ = 1$ so that (2.1) here reduces to

$$Wh_G^{pl,\rho}(M_\Gamma)\ =\ Wh(\Gamma)\,. \tag{2.3}$$

The "forget control map" $Wh(\Gamma)_c \to Wh(\Gamma)$ (where M_Γ/G is the control space) is an isomorphism in this case by results of the Ph.D. thesis of G. Tsapogas [T]. This follows from the fact that A contains no one-dimensional G-submodules. So now we apply the spectral sequence of F. Quinn ([Q1],[Q2]) which computes $Wh(\Gamma)_c$. We have: $E^2_{pq} = H^G_p(M_\Gamma; \tilde{K}_q(\mathbf{Z}G_x))$. Since $\tilde{K}_q(\mathbf{Z}G_x) = 0$ if $G_x = 1$, we get: $E^2_{pq} = H_p(M^G_\Gamma; \tilde{K}_q(\mathbf{Z}G))$. By [CK1], Lemma 2.2, each component of M^G_Γ is a torus of dimension equal to $rk\ Z_A(G)$. Since $Z_A(G) = 0$, M^G_Γ is discrete and we get: $E^2_{pq} = 0$ if $p \neq 0$. Since $Wh(G) = \mathbf{Z}$, and the algebraic involution on $Wh(G)$ is trivial (by Milnor [M])we get:

$$Wh(\Gamma) \cong Wh(\Gamma)_c \cong H_0(M^G_\Gamma; Wh(G)) \cong H_0(M^G_\Gamma; \mathbf{Z}), \tag{2.4}$$

and the involution is trivial on $Wh(\Gamma)$. According to [CK1], the number of components of M^G_Γ is equal to $|H^1(G; A)| = |\mathbf{Z}(\zeta_5)/(1-\zeta_5)|^n = 5^n$. So from (2.3) and (2.4) we get isomorphisms:

$$\sum_{i=1}^{5^n} (Wh(G))_i \to Wh(\Gamma) \to Wh_G^{pl,\rho}(M_\Gamma)\,. \tag{2.5}$$

By [CL] section 2, these maps preserve the involutions if the left hand group has trivial involution. This uses the fact that $dim\ M_\Gamma$ is even and the fact that the algebraic involution on $Wh(G)$ is trivial. Hence we obtain an isomorphism:

$$\hat{H}^i(\mathbf{Z}/2\mathbf{Z}; Wh_G^{pl,\rho}(M_\Gamma)) = \sum_{i=1}^{5^n} \hat{H}^i(\mathbf{Z}/2\mathbf{Z}; Wh(G)) = \begin{cases} 0 & \text{if} \quad i \text{ is odd} \\ \text{an } \mathbf{F}_2 \text{ vector space} & \\ \text{of dimension } 5^n & \text{if} \quad i \text{ is even} \end{cases}$$

In particular:

(2.6) $$\hat{H}^0(\mathbf{Z}/2\mathbf{Z}; Wh_G^{pl,p}(M_\Gamma)) \neq 0.$$

This is the non-vanishing result we sought.

§3. An example where $(*)_\Gamma$ fails.

Let G be the cyclic group of order four acting on the Gaussian integers, $\mathbf{Z}[i]$, via multiplication by powers of i. Let n be a postive integer. Let A be the direct sum of n copies of $\mathbf{Z}[i]$ and two copies of $\mathbf{Z}$, the trivial G-module. Set $\Gamma = (A \rtimes G) \times K$ where K is the fundamental group of the Klein bottle. That is to say, $K = T \rtimes_\kappa T$ where $\kappa : T \to T$ is the non trivial automorphism. The holonomy group of Γ is $G_\Gamma = G \times G'$ where G' is the cyclic group of order two. We are going to prove that

$$\hat{H}^*(\mathbf{Z}/2\mathbf{Z}; Wh_{G_\Gamma}^{top,p}(M_\Gamma)) \neq 0 \ .$$

Let $\mathbf{N}$ be the monoid of positive integers: $\{1, 2, 3, ...\}$. According to [CdaS], the nil-K theory, $NK_*(R)$ is a $\mathbf{Z}[\mathbf{N}]$ module in a natural way, for any ring R.

In the present case, we claim there is an isomorphism of $\mathbf{Z}[\mathbf{N}]$ modules:

(3.1) $$\delta : \mathbf{F}_2[\mathbf{N}] \cong NK_0(\mathbf{Z}G) \ .$$

To see this observe that the boundary map in Nil-K-theory of the Meyer-Vietoris sequence of the cartesian square:

$$\begin{array}{ccc} \mathbf{Z}G & \to & \mathbf{Z}[i] \\ \downarrow & & \downarrow \\ \mathbf{Z}[\mathbf{Z}/2\mathbf{Z}] & \to & \mathbf{F}_2[\mathbf{Z}/2\mathbf{Z}] \end{array}$$

provides an isomorphism:

(3.2) $$d : NK_1(\mathbf{F}_2[\mathbf{Z}/2\mathbf{Z}]) \cong NK_0(\mathbf{Z}G) \ .$$

because the groups $NK_j(\mathbf{Z}[\mathbf{Z}/2\mathbf{Z}], NK_j(\mathbf{Z}[i]), j = 0, 1$ vanish. d preserves the algebraic involutions so the involution on $\mathbf{Z}G$ acts trivially on $NK_0(\mathbf{Z}G)$.

By a result of Bass-Murthy ([BM], 7.6):

$$NK_1(\mathbf{F}_2[\mathbf{Z}/2\mathbf{Z}]) \cong NU(\mathbf{F}_2[\mathbf{Z}/2\mathbf{Z}]) \ .$$

But $NU(\mathbf{F}_2[\mathbf{Z}/2\mathbf{Z}]) = \{1 + \epsilon_1 \nu x + \epsilon_2 \nu x^2 + ... \mid \nu = 1 - t, \epsilon_i = 0 \text{ or } 1\}$ where $\mathbf{Z}/2\mathbf{Z} = \{1, t\}$. The action of an element $s \in \mathbf{N}$ on a unit $p(x) \in NU(R)$ sends $p(x)$ to $p(x^s)$. The map of $\mathbf{Z}[\mathbf{N}]$ modules: $\mathbf{F}_2[\mathbf{N}] \to NU(\mathbf{F}_2[\mathbf{Z}/2\mathbf{Z}])$ which sends 1 to $1 + \nu x$ is easily seen to be an isomorphism, and this provides the isomorphism of (3.1).

By [Sw], [C] and [CdaS] $\tilde{K}_i(\mathbf{Z}G) = 0, \ i \leq 0$ and $NK_{-1}(\mathbf{Z}G) = 0$. This implies that

(3.3) $$\tilde{K}_0(\mathbf{Z}G \times T) = NK_0(\mathbf{Z}G) \oplus NK_0(\mathbf{Z}G) \ ,$$

$$(3.4) \qquad \tilde{K}_{-1}(\mathbf{Z}G \times T) = 0$$

and the automorphism $\alpha = (1 \times \kappa)_* : \tilde{K}_0(\mathbf{Z}G \times T) \to \tilde{K}_0(\mathbf{Z}G \times T)$ interchanges the two summands of (3.3). Therefore

$$(3.5) \qquad coker(1-\alpha) \cong NK_0(\mathbf{Z}G) .$$

The exact sequence of [FH1]:

$$\tilde{K}_0(\mathbf{Z}G \times T) \to \tilde{K}_0(\mathbf{Z}G \times T) \to \frac{\tilde{K}_0(\mathbf{Z}G \times T \times_\kappa T)}{NK_0(\mathbf{Z}G \times T, \alpha) \oplus NK_0(\mathbf{Z}G \times T, \alpha^{-1})} \to K_{-1}(\mathbf{Z}G \times T)$$

together with (3.4) implies that the inclusion map

$$(3.6) \qquad 0 \to coker(1-\alpha) \to \tilde{K}_0(\mathbf{Z}G \times T \times_\kappa T)$$

induces an isomorphism of Tate cohomology groups. Therefore by (3.1), (3.5), (3.6) we get, for any i

$$(3.7) \qquad \hat{H}^i(\mathbf{Z}/2\mathbf{Z}; \tilde{K}_0(\mathbf{Z}G \times K)) \cong \mathbf{F}_2[\mathrm{N}] .$$

Now we turn to $Wh^{pl,\rho}_{G_\Gamma}(M_\Gamma)$. Since $N_\Gamma(H)/H$ is free abelian if H is a finite subgroup of order 4, or if H is an isotropy group of order 2, the formula (2.2) yields:

$$Wh^{pl,\rho}_{G_\Gamma}(M_\Gamma) = Wh(\Gamma) .$$

But this time $Wh(\Gamma)_c = 0$. To see this, note that E^2 term of Quinn's spectral sequence vanishes because $\tilde{K}_q(\mathbf{Z}H) = 0$ if $q \leq 1$ and $H = \{1\}$ or $\mathbf{Z}/4\mathbf{Z}$ (Carter [C], Swan [Sw], Milnor[M]). Now, according to [CK2], in this case we have:

$$(3.8) \qquad \epsilon : Wh(\Gamma) \cong Wh(\Gamma)/Wh(\Gamma)_c \cong Wh^{top,\rho}_{G}(M_\Gamma) .$$

The isomorphism in (3.8) preserves the involutions because Γ has even rank (see[CL], section 2). The split monomorphism

$$j : T \times G \times K \to A \rtimes G \times K$$

where T goes to a trivial summand of A, yields a split monomorphism of Whitehead groups

$$(3.9) \qquad j_* : \hat{H}^i(\mathbf{Z}/2\mathbf{Z};\ Wh(T \times G \times K)) \to \hat{H}^i(\mathbf{Z}/2\mathbf{Z};\ Wh(\Gamma)) .$$

The fundamental theorem of algebraic K-theory [B] then yields a split monomorphism:

$$(3.10) \qquad i : \tilde{K}_0(G \times K) \to Wh(T \times G \times K) .$$

By combining (3.8), (3.9) and (3.10) we obtain a split monomorphism:

$$(3.11) \qquad \epsilon_* j_* i_* : \hat{H}^i(\mathbf{Z}/2\mathbf{Z};\ Wh(T \times G \times K)) \to \hat{H}^i(\mathbf{Z}/2\mathbf{Z};\ Wh^{top,\rho}_{G_\Gamma}(M_\Gamma)) \quad \text{for all i.}$$

By 3.3,

$$\epsilon_* j_* i_* \delta_* : \mathbf{F}_2[\mathbf{N}] \to \hat{H}^i(\mathbf{Z}/2\mathbf{Z};\ Wh^{top,\rho}_{G_\Gamma}(M_\Gamma))$$

is a split monomorphism. In particular:

$$\hat{H}^i(\mathbf{Z}/2\mathbf{Z};\ Wh^{top,\rho}_{G_\Gamma}(M_\Gamma)) \neq 0. \tag{3.12}$$

This is the non vanishing result we were seeking.

§4. Geometric Consequences of the Calculations.

Here we show that $S(\Gamma) \neq 0$ and $S^{pl}(\Gamma) \neq 0$ for certain Γ, as explained in §1. These examples are all coming from the nonvanishing of relevant Whitehead torsions. Examples of a rather different flavor, due to the nonvanishing of relevant UNil groups, are also possible, as has been pointed out by S. Weinberger [W].

To begin, we give a careful definition of the structure sets we are using. Let $(\tilde{M}, \Gamma)$ be a topological manifold with a properly discontinuous Γ-action for which $\tilde{M}/\Gamma$ is compact and for which each fixed set $\tilde{M}^H$ is a contractible, locally flat submanifold in any bigger fixed set $\tilde{M}^K$, $K \subset H$. The standard example is $(\tilde{M}_\Gamma, \Gamma) = \mathbf{R}^n$ with the isometric action.

According to [CK1] there is a Γ-map, unique up to equivariant homotopy:

$$\tilde{J} : \tilde{M} \to \tilde{M}_\Gamma\ .$$

We write M for $\tilde{M}/A$, M_Γ for$\tilde{M}_\Gamma/A$; $\tilde{J}$ induces a G-homotopy equivalence $J : M \to M_\Gamma$ whose torsion can be measured in $Wh^{top,\rho}_G(M_\Gamma)$. If $\tilde{J}$ can be chosen isovariant, we say $(\tilde{M}, \Gamma)$ is a crystallographic manifold.

$S(\Gamma)$ is the set of equivariant homeomorphism classes of crystallographic manifolds whose torsion, in $Wh^{top,\rho}_G(M_\Gamma)$ is zero. If we wish to drop the torsion condition, we write $S^h(\Gamma)$ for the set of equivariant h-cobordism classes of such manifolds. If we wish to consider PL-manifolds and PL-actions, up to PL homeomorphism we write $S_{pl}(\Gamma)$, this time requiring the torsion to vanish in $Wh^{pl,\rho}_G(M_\Gamma)$.

We will be using the following two exact Rothenberg sequences of structure sets:

$$\begin{aligned} S(M_\Gamma \times I) &\to S^h(M_\Gamma \times I) \to \hat{H}^0(\mathbf{Z}/2\mathbf{Z};\ Wh^{top,\rho}_G(M_\Gamma)) \to \\ &\to S(\Gamma) \to S^h(\Gamma) \to \hat{H}^1(\mathbf{Z}/2\mathbf{Z};\ Wh^{top,\rho}_G(M_\Gamma)) \end{aligned} \tag{4.1}$$

$$\begin{aligned} S_{pl}(M_\Gamma \times I) &\to S^h_{pl}(M_\Gamma \times I) \to \hat{H}^0(\mathbf{Z}/2\mathbf{Z};\ Wh^{pl,\rho}_G(M_\Gamma)) \to \\ &\to S_{pl}(\Gamma) \to S^h_{pl}(\Gamma) \to \hat{H}^1(\mathbf{Z}/2\mathbf{Z};\ Wh^{pl,\rho}_G(M_\Gamma))\ . \end{aligned} \tag{4.2}$$

For a proof of exactness of (4.1) see [CK2]; the proof of the exactness of (4.2) follows in a formally identical manner. Here $S(M_\Gamma \times I)$ means the G-structures on $M_\Gamma \times I$ which are homeomorphisms over $M_\Gamma \times \partial I$;other structure sets are defined similarly.

First suppose that Γ is the group defined in §2. By (2.5) and (4.2) either $S_{pl}(\Gamma)$ or $S^h_{pl}(M_\Gamma \times I)$ is non trivial. If $S^h_{pl}(M_\Gamma \times I) \neq 0$, then an easy application of Farrell's thesis [F] implies that $S^h_{pl}(\Gamma \times T) \neq 0$ and it also yields an exact sequence:

$$0 \to S^h_{pl}(\Gamma \times T) \to S_{pl}(\Gamma \times T \times T)$$

So either $S_{pl}(\Gamma)$ or $S_{pl}(\Gamma \times T \times T)$ is $\neq 0$.

Next suppose Γ is the group discussed in §3. By (3.8), (4.1) and the argument in the previous paragraph, either $S(\Gamma)$ or $S(\Gamma \times T^2)$ is $\neq 0$. These are the failures to the rigidity conjectures mentioned in §1.

§6 References

[B] Bass, H.:Algebraic K-Theory. New York: W.A.Benjamin Inc., 1968

[BM] Bass, H., Murthy, P.: Grothendieck groups and Picard groups of Abelian group rings. Annals of Math.(2)**86**,16-73 (1967)

[C] Carter, D.: Lower K-theory of finite groups. Comm. Algebra **8** 1927-1937 (1980)

[CdaS] Connolly, F., daSilva, M.:$N^iK_0(\mathbf{Z}\pi)$ is a finitely generated $\mathbf{ZN^i}$ module for any finite group π. (to appear)

[CK1] Connolly, F., Koźniewski, T.: Finiteness properties of classifying spaces of proper Γ actions. Journal of Pure and Applied Algebra **41**, 17-36 (1986)

[CK2] Connolly, F., Koźniewski, T.:Rigidity and Crystallographic Groups, I. Inventiones Math.**99** 25-49 (1990)

[CK3] Connolly, F., Koźniewski, T.:Rigidity and Crystallographic Groups, II. (in preparation)

[CL] Connolly, F., Lück, W.: The involution on the Equivariant Whitehead Group. Journal of K-Theory,(to appear, 1990)

[F] Farrell, F.T. :The obstruction to fibering a manifold over a circle. Indiana Univ. Math. J. **21**,3125-346 (1971)

[FH1] Farrell, F.T., Hsiang, W.C.: A formula for $K_1(R_\alpha[T])$. Proc. Symp. Pure Math. vol. 17 (1970)

[FH2] Farrell, F.T., Hsiang, W.C.: Topological Characterization of flat and almost flat manifolds, $M^n, n \neq 3, 4$. Amer. Jour. Math.**105**,641-672 (1983)

[HS] Hsiang, W.C., Shaneson, J.: Fake Tori. In: Topology of Manifolds. Chicago, Markham 1970 pp. 18-51

[M] Milnor, J.W.: Whitehead Torsion. Bulletin of the Amer. Math. Soc. **72**, 358-426 (1966)

[Q1] Quinn, F.: Ends of maps II. Inventiones Math.**68**,353-424 (1982)

[Q2] Quinn, F.: Algebraic K-theory of poly-(finite or cyclic) groups, Bulletin of the Amer. Math. Soc.**12**, 221-226 (1985).

[St] Steinberger, M. : The equivariant topological s-cobordism theorem. Inventiones Math. **91**, 61-104 (1988)

[StW] Steinberger, M., West, J.:Equivariant h-cobordisms and finiteness obstructions. Bulletin of the Amer. Math. Soc.**12**, 217-220 (1985)

[Sw] Swan, R.: The Grothendieck ring of a finite group. Topology **2**, 85-110 (1963)

[Ts] Tsapogas, G. : On the K-theory of crystallographic groups, Ph.D. dissertation, University of Notre Dame, 1990.
[W] Weinberger, S. : Private communication

Frank Connolly *
Department of Mathematics
University of Notre Dame
Notre Dame, Indiana 46556, USA

Tadeusz Koźniewski **
Instytut Matematyki
Warsaw University
PKiN IXp, 00-901 Warszawa, Poland

* Partially supported by NSF Grant DMS-90-01729
** Partially supported by Polish Scientific Grant RP I.10

SUR LA TOPOLOGIE DES BRAS ARTICULES

Jean-Claude HAUSMANN

(0.1) Considérons l'application $\beta_a : (S^{k-1})^n \longrightarrow \mathbf{R}^k$ définie par

$$\beta_a(z_1,\dots z_n) = \sum a_i \cdot z_i \qquad (a_i > 0).$$

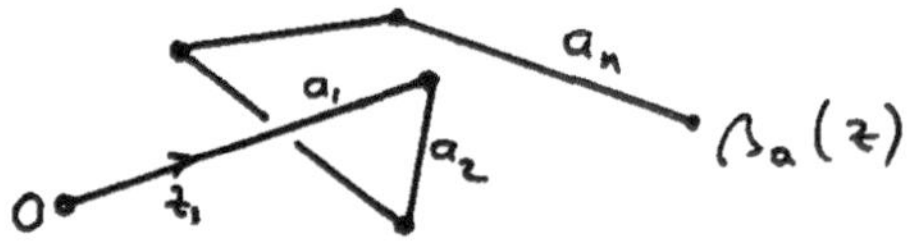

Nous appellerons β_a le **bras articulé** dans $\mathbf{R}^k$, de longueur n et de type a = = $(a_1,\dots,a_n)$. Dans cet article nous démontrons quelques résultats concernant les points critiques de β_a ainsi que sur les pré-images $\beta_a^{-1}(\{q\})$, $q \in \mathbf{R}^k$. Il est possible que ce genre d'information soit utile en robotique (voir [Go1 et 2]). En tout cas, on verra que c'est la source d'exemples et de problèmes intéressants de topologie différentielle.

L'application β_a est transverse à $\mathbf{R}_{>0}\times\{0\}$ (voir (1.3) ci-dessous). Nous appelerons V_a la préimage de ce rayon : $V_a = \beta_a^{-1}(\mathbf{R}_{>0}\times\{0\})$, qui est donc une sous-variété de codimension k-1 de $(S^{k-1})^n$. On dénote par $\gamma_a : V_a \longrightarrow \mathbf{R}$ la première composante de $\beta_a | V_a$. L'application γ_a est le **bras articulé à extrémité coullissante** de longueur n, de type a, dans $\mathbf{R}^k$.

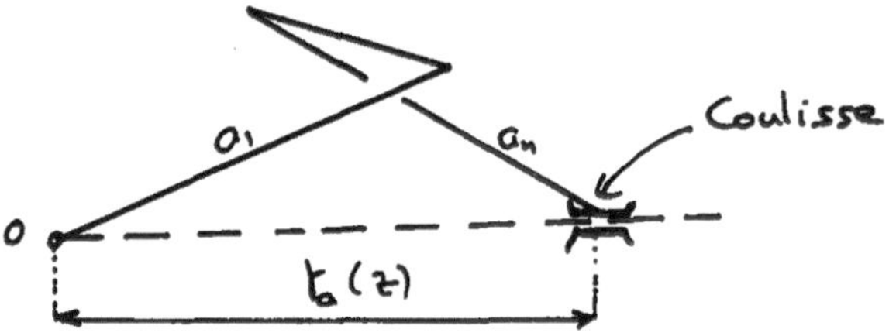

(0.2) Remarques

a) Comme β_a est la restriction à $(S^{k-1})^n$ d'une application linéaire de $(\mathbf{R}^k)^n \longrightarrow \mathbf{R}^k$, les espaces $\beta_a^{-1}(\{q\})$ sont l'intersection dans $(\mathbf{R}^k)^n$ d'un sous-espace affine de codimension k avec le produit de sphères $(S^{k-1})^n$. En particulier, ce sont des ensemble algébriques réels de $(\mathbf{R}^k)^n$. Nous nous contenterons cependant de considérer

$\beta_a^{-1}(\{q\})$ comme espace topologique ou, dans le cas où q est une valeur régulière, comme variété différentiable.

b) L'espace $\beta_a^{-1}(\{q\})$ est l'espace de configurations du système articulé suivant dans $\mathbf{R}^k$:

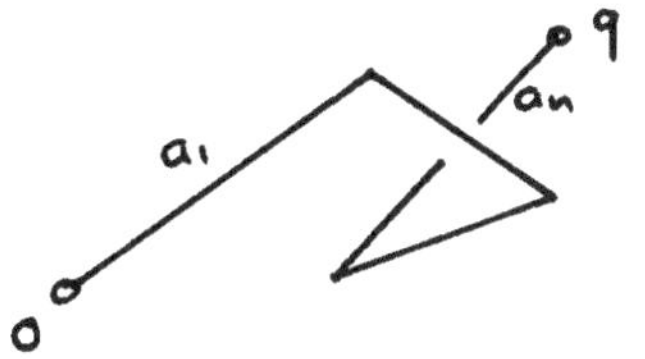

Ces espaces ont été étudiés, en tout cas pour $k = 2$ et $n \leq 5$, par W. Thurston, [TW], K. Walker [Wa], A Wenger [We]. Pour une étude du point de vue de la géométrie algébrique voir [GN].

1 SYMETRIES ET ACTIONS

(1.1) Pour $\mu > 0$, on a $\beta_{\mu a}(z) = \mu \cdot \beta_a(z)$. Il en résulte que les propriétés de β_a qui nous intéressent seront les mêmes que celles de $\beta_{\mu a}$. On ne restreint donc pas la généralité en supposant, par exemple, que $\sum a_i = 1$. Le vecteur a est alors un élément du simplexe standard Δ^{n-1}. (Nous noterons les coordonnées de Δ^{n-1} de 1 à n au lieu de la convention habituelle qui est 0,...,n-1.)

(1.2) Le groupe symétrique $\mathfrak{S}_n$ agit sur $(S^{k-1})^n$ et sur Δ^{n-1}, par permutation des coordonnées. Si $\sigma \in \mathfrak{S}_n$, on a

$$\beta_{\sigma a}(\sigma z) = \beta_a(z)$$

Par exemple, si $a_1 = a_2 = \dots = a_n$, l'espace $\beta_a^{-1}(\{q\})$ est un sous-espace de $(S^{k-1})^n$ invariant par permutation de coordonnées. Cela donne d'intéressants exemples d'actions algébriques du groupe symétrique.

(1.3) Considérons d'une part l'action standard de SO_k sur $\mathbf{R}^k$ et d'autre part celle de ce même groupe sur $(S^{k-1})^n$, diagonalement, par l'action standard sur chaque facteur. Pour $\alpha \in SO_k$, on a

$$\beta_a(\alpha \cdot z) = \alpha \cdot \beta_a(z)$$

Soit $0 \neq q \in \mathbf{R}^k$ et soit $z_0 \in \beta_a^{-1}(\{q\})$. On considère l'application $SO_k \longrightarrow (S^{k-1})^n$ envoyant α sur αz_0. Sa composition avec β_a envoie α sur αq. Cette dernière application est une submersion sur la sphère de rayon $\|q\|$. On en déduit que β_a est transverse aux rayons de $\mathbf{R}^k$ en particulier au rayon $\mathbf{R}_{>0} \times \{0\}$, comme annoncé dans la définition de V_a au paragraphe 0.

(1.4) Si $0 \neq q \in \mathbf{R}^k$. Identifions le stabilisateur de q dans SO_k avec SO_{k-1}. Cela donne une action de SO_{k-1} sur $\beta_a^{-1}(\{q\})$:

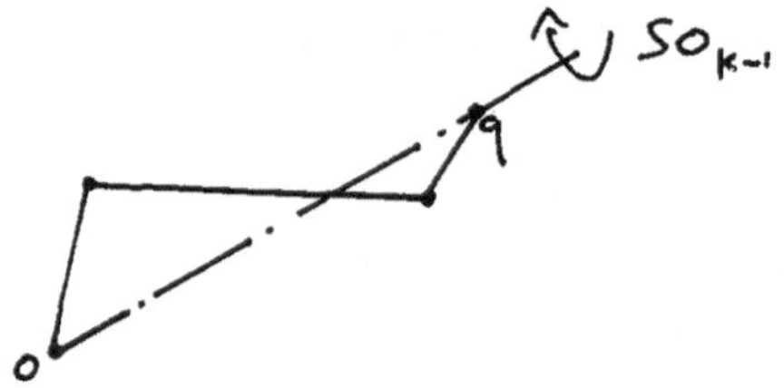

En particulier, on a une action de SO_{k-1} sur V_a telle que $\gamma_a(\alpha z) = \gamma_a(z)$.

3. LES POINTS CRITIQUES DE β_a

Dans ce paragraphe, nous déterminons l'ensemble $\mathrm{Crit}(\beta_a)$ des points critiques d'un bras articulé β_a dans $\mathbf{R}^k$ ainsi que l'ensemble $\mathrm{Crit}(\gamma_a)$ de ceux du bras articulé à extrémité coulissante associé γ_a. On démontre que γ_a est une fonction de Morse.

(3.1) <u>Théorème</u> Supposons que $a_i \neq 0$ pour tout i. Le point $z = (z_1,\ldots,z_n)$ est un point critique de $\beta_a : (S^{k-1})^n \longrightarrow \mathbf{R}^k$ si et seulement si $z_i = \pm z_j$, pour tout i,j.

Le théorème (3.1) implique que z est un tel point critique si et seulement si $\beta_a(z)$ est une **configuration alignée** :

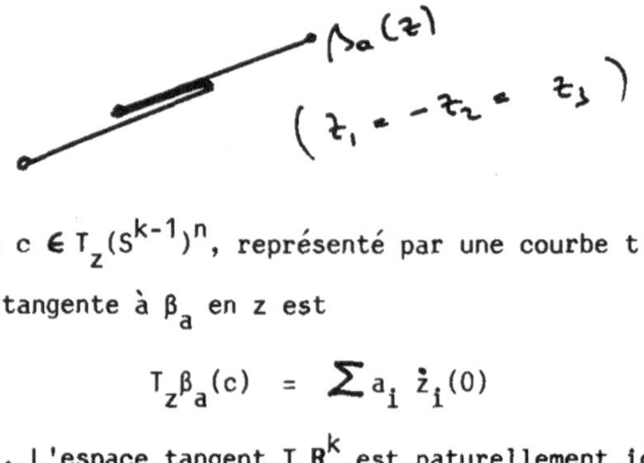

<u>Preuve</u> : Soit $c \in T_z(S^{k-1})^n$, représenté par une courbe $t \longmapsto z(t)$. L'image de c par l'application tangente à β_a en z est

$$T_z\beta_a(c) = \sum a_i \dot{z}_i(0)$$

Soit $q = \beta_a(z)$. L'espace tangent $T_q\mathbf{R}^k$ est naturellement identifié à $\mathbf{R}^k$. Via cette identification, l'image de $T_z\beta_a$ est, vu la formule ci-dessus, le sous-espace vectoriel de $\mathbf{R}^k$ engendré par la réunion des suppléments orthogonaux au vecteurs z_i. Le théorème (3.1) en découle immédiatement.

Le résultat pour les points critiques de $\gamma_a : V_a \longrightarrow]0,1]$ est analogue. Considérons la sphère $S^0 = \{\pm 1\}$ comme incluse dans S^{k-1}, en identifiant ± 1 à $(\pm 1,0,\ldots,0)$. Si $p = (\pm 1,\pm 1,..,\pm 1) \in (S^0)^n$, on dénote par ind(p) le nombre de composantes égales à +1.

(3.2) **Théorème** $\gamma_a : V_a \longrightarrow]0,1]$ est une fonction de Morse avec $\mathrm{Crit}(\gamma_a) = V_a \cap (S^0)^n$. L'indice du point critique $p = (\pm1,\pm1,..,\pm1)$ est égal à $(k-1)(\mathrm{ind}(p) - 1)$.

Ce théorème a été obtenu par K. Walker pour dans le cas $k = 2$ [Wa]. En fait, la démonstration de Walker n'est pas vraiment complète; elle ne tient pas compte par exemple du cas 2 ci-dessous. Notre preuve repose sur un principe différent.

Preuve : Il résulte de la définition de V_a que, pour $z \in V_a$, l'application tangente à β_a se factorise ($y = f(z)$) :

$$T_z\beta_a = T_z\gamma_a \oplus T_z\beta_a|\nu_zV_a \quad : \quad T_zV_a \oplus \nu_zV_a \longrightarrow T_y(\mathbf{R}\times 0) \oplus T_y(0\times\mathbf{R}^{k-1}))$$

Comme β_a est transverse à $]0,1]\times 0$, l'application

$$T_z\beta_a|\nu_zV_a \quad : \quad \nu_zV_a \longrightarrow T_y(0\times\mathbf{R}^{k-1}))$$

est surjective. On a donc $\mathrm{Crit}(\gamma_a) = \mathrm{Crit}(\beta_a) \cap V_a$. Par le théorème 3.1, on a $\mathrm{Crit}(\beta_a) \cap V_a = V_a \cap (S^0)^n$.

Soit $p \in \mathrm{Crit}(\gamma_a)$. La démonstration que p est un point critique non-dégénéré se fait par récurrence sur la longueur n du bras.

Cas 1 $n = 2$: Posons $a = (A,B)$. Comme $\gamma_a(p)$ est une position alignée, un système de coordonnées (carte de V_a) au voisinage de p est donné par

$$(z_1,z_2) \mapsto x = (x_1,\dots,x_{k-1}) \in \mathbf{R}^{k-1},$$

où x est la projection sur $\{0\}\times\mathbf{R}^{k-1}$ du point Az_1, l'unique articulation de γ_a. Les trois cas de figure possibles sont :

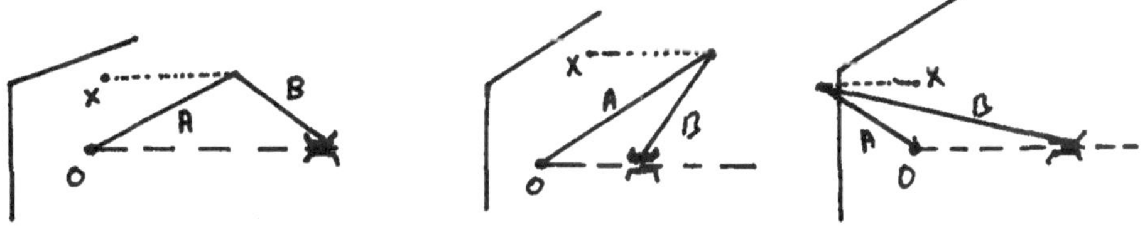

On a donc, pour ces coordonnées :

$$\gamma_a(x) = \pm\sqrt{A^2 - \|x\|^2} \pm \sqrt{B^2 - \|x\|^2}$$

Les dérivées premières :

$$\frac{\partial\gamma_a}{\partial x_i} = \frac{\mp\, x_i}{\sqrt{A^2 - \|x\|^2}} \mp \frac{x_i}{\sqrt{B^2 - \|x\|^2}}$$

s'annulent pour $x_i = 0$, qui sont les coordonnées de p. Les dérivées secondes en ce point valent :

$$\frac{\partial^2\gamma_a}{\partial x_j \partial x_i}(0) = 0 \quad \text{si } i \neq j$$

$$\frac{\partial^2\gamma_a}{\partial x_i^2}(0) = \mp \frac{1}{A} \mp \frac{1}{B}$$

La matrice hessienne en p :

$$\mathcal{H}(\gamma_a,p) = \left(\frac{\partial^2\gamma_a}{\partial x_j \partial x_i}(0)\right)$$

est donc diagonale. Son déterminant ne pourrait s'annuler que si les signes sont opposés dans la formule donnant γ_a et si $A = B$. Mais alors $\gamma_a(p) = 0$ ce qui contredit le fait que $p \in V_a$. Cela prouve que p est non-dégénéré dans le cas $n = 2$.

Cas 2 : $n \geq 3$ et $a_i = a_j$ pour tout i,j et $z = (1,-1,1,-1,\ldots,-1,1)$: On est donc dans la situation :

Comme tous les a_i sont égaux, on a $\gamma_a(z) = \gamma_a(\sigma z)$ pour tout $\sigma \in \mathfrak{S}_n$, où l'action du groupe symétrique sur V_a est celle donnée en (1.2). Pour démontrer que le point critique p est non-dégénéré, il suffit donc de démontrer que $\sigma{\cdot}p$ l'est, où σ est la transposition (2,3). La configuration pour $\gamma_a(\sigma p)$ est :

On est donc placé dans le cas 3 ci-dessous.

Cas 3 : Ce sera le cas où $n \geq 3$ et où les conditions de cas 2 ne sont pas vérifiées. Les articulations de la configuration $\gamma_a(p)$ occupent au moins 3 points distincts. On peut donc décomposer le bras articulé γ_a en deux sous-bras de la manière suivante :

$$a = (a_1,\ldots,a_n), \quad a' = (a_1,\ldots a_s), \quad a'' = (a_{s+1},\ldots,a_n)$$

$$z = (z_1,\ldots,z_n), \quad z' = (z_1,\ldots z_s), \quad z'' = (z_{s+1},\ldots,z_n)$$

et supposer que $\gamma_{a'}(p') \neq 0 \neq \gamma_a(p)$. Trois cas sont à distinguer :

1) $0 < \gamma_{a'}(p') < \gamma_a(p)$

2) $\gamma_a(p) < \gamma_{a'}(p')$

3) $\gamma_{a'}(p') < 0$

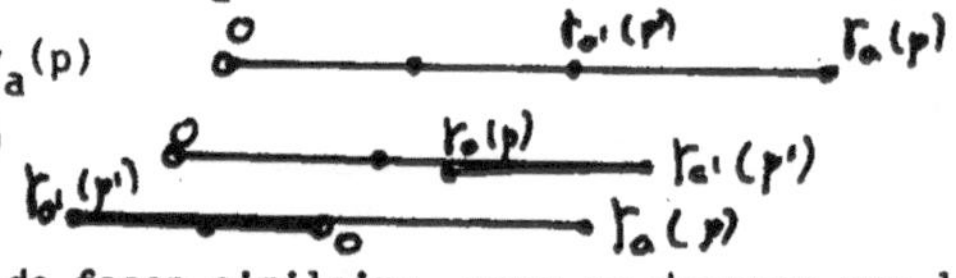

Comme ils se traitent de façon similaire, nous ne donnons que les détails du cas 1. En utilisant les sections locales usuelles de $SO_k \longrightarrow S^{k-1}$, on peut trouver pour tout point z dans un voisinage de p :

- Une unique rotation $\alpha' \in SO_k$ telle que $\alpha'(\beta_{a'}(z')) \in]0,1]\times\{0\}$
- " " " $\alpha'' \in SO_k$ telle que $\alpha''(\gamma_a(z)-\beta_{a'}(z')) \in]0,1]\times\{0\}$

L'application $z \longmapsto (\alpha'(z'),\alpha''(z''),x(\beta_{a'}(z'))$, où α' et α'' agissent diagonalement sur les composantes de z' et z'' et x désigne la projection sur $0\times\mathbf{R}^{k-1}$, donne un difféomorphisme d'un voisinage de p dans V_a sur un voisinage de $(p',p'',0)$ dans $V_{a'} \times V_{a''} \times \mathbf{R}^{k-1}$.

Choissisons des coordonnées u et v au voisinage de p' et p'', dans $V_{a'}$ et $V_{a''}$. On vérifie que dans les coordonnées (u,v,x), la matrice hessienne de γ_a en p est de la forme :

$$\mathcal{H}(\gamma_a,p) = \left(\begin{array}{c|c|c} \mathcal{H}(\gamma_{a'},p') & 0 & 0 \\ \hline 0 & \mathcal{H}(\gamma_{a''},p'') & 0 \\ \hline 0 & 0 & \mathrm{diag}(-\frac{1}{A}-\frac{1}{B}) \end{array}\right)$$

où $A = \gamma_{a'}(p')$ et $B = \gamma_{a''}(p'')$. On peut supposer, par hypothèse de récurrence, $\det\mathcal{H}(\gamma_{a'},p') \neq 0 \neq \det\mathcal{H}(\gamma_{a''},p'')$. On en déduit que $\det\mathcal{H}(\gamma_a,p) \neq 0$ et donc p est un point critique non-dégénéré. La formule pour l'indice de p se démontre aussi par récurrence, utilisant l'expression de $\mathcal{H}(\gamma_a,p)$ ci-dessus.

4. PREIMAGES SPHERIQUES ET ACTIONS

Soit β_a un bras articulé de longueur n, dans $\mathbf{R}^k$.

(4.1) **Proposition** Si $1 - 2\min\{a_i\} < \|u\| < 1$ alors $\beta_a^{-1}(\{u\})$ est difféomorphe à la sphère standard $S^{(n-1)(k-1)-1}$. L'action de SO_{k-1} sur $\beta_a^{-1}(\{u\})$ est différentiablement conjuguée à la restriction sur l'action $S^{(n-1)(k-1)-1}$ de l'action diagonale de SO_{k-1} sur le produit de $(n-1)$ copies de $\mathbf{R}^{k-1}$.

Preuve : On a $\beta_a^{-1}(\{u\}) = \gamma_a^{-1}(\{q\})$, où $q = \|u\|$. Il suffit de démontrer le résultat pour les préimages de γ. Soit $b = 1 - 2\min\{a_i\}$. Observons que si $\gamma_a(z) \in]b,1[$, alors $z \neq (1,1,\dots,1)$ et de plus aucun des z_i ne peut être égal à -1. Par la proposition (3.2) l'intervalle $]b,1[$ ne contient alors que des valeurs régulières. Si $q < q' \in]b,1[$ on peut construire un difféomorphisme SO_{k-1}-équivariant de $\gamma_a^{-1}(\{q\})$ sur $\gamma_a^{-1}(\{q'\})$ en suivant les trajectoires du champ de vecteurs $\mathrm{grad}\gamma_a$ convenablement normalisé. (voir [Mi2, Théorème 3.4]). On peut donc supposer que $\gamma_a^{-1}(\{q\})$ est dans un voisinage convenable de $p = (1,1,\dots,1)$.

Comme p est un maximum non-dégénéré, par la Proposition (3.2), le lemme de Morse [Mi1, Lemme 2.2] assure l'existance d'un système de coordonnées (x) au voisinage de p tel que

$$\gamma_a(x) = 1 - \|x\|^2$$

Les surfaces de niveau de cette fonction sont des sphères standard. On en déduit donc que $\beta_a^{-1}(\{u\})$ est difféomorphe à $S^{(n-1)(k-1)-1}$.

Pour trouver un difféomorphisme SO_{k-1}-équivariant, on procède de la manière suivante : on suppose que $\gamma_a^{-1}(\{q\}) \in U$, où U est le voisinage de p constitué des points $z \in V_a$ tels que chaque composante de z à sa première coordonnée strictement positive. Dans une telle configuration, chaque arête est dirigée vers la droite. L'ouvert U est le domaine d'une carte de V_a

$$X : U \longrightarrow \widehat{U} \subset \mathbf{R}^{(k-1)(n-1)}$$

$$z \longmapsto x(z)$$

où

$x_i(z)$ = projection sur $0 \times \mathbf{R}^{k-1}$ de $\gamma_{(a_1,a_2,\ldots,a_i)}(z_1,z_2,\ldots z_i)$.

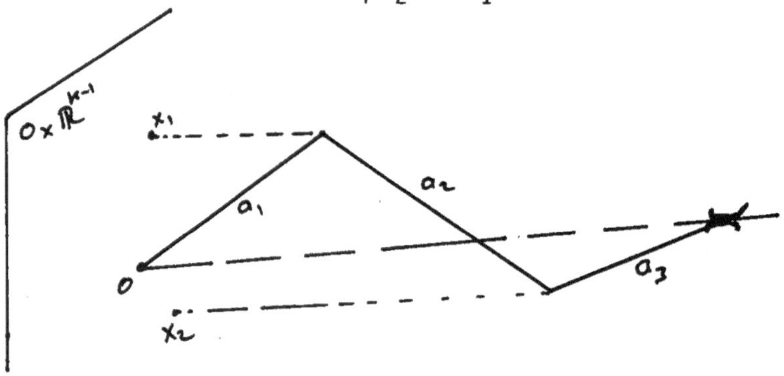

La carte X est SO_{k-1} équivariante pour, sur $\mathbf{R}^{(k-1)(n-1)}$, l'action diagonale de SO_{k-1}. L'image par X de $\gamma_a^{-1}(\{q\})$ est une sous-variété de $\mathbf{R}^{(k-1)(n-1)}$ qui est SO_{k-1}-invariante. (Un tel plongement d'une sphère produit sur celle-ci une action que l'on peut qualifier de quasi-linéaire; on ne peut pas, en général déduire qu'une action quasi-linéaire est différentiablement (ou même topologiquement) conjuguée à une action linéaire (Voir [Ha]).

Soit $x \in \widehat{U}$. L'application $t \longmapsto \gamma_a(tx)$ est à dérivée strictement positive. Donc $X(\gamma_a^{-1}(\{q\}))$ est transverse à chaque rayon de $\mathbf{R}^{(k-1)(n-1)}$. On en déduit que la projection radiale de $\gamma_a^{-1}(\{q\})$ sur une sphère standard dans $\widehat{U}$ est un difféomorphisme SO_{k-1}équivariant. Cela achève la démonstration de la proposition (4.1).

(4.2) **Remarque** : Dans le cas k = 3 (bras articulés dans l'espace) on a $\beta_a^{-1}(\{q\})$ difféomorphe à S^{2n-3} et l'action de $SO_2 = S^1$ conjuguée à l'action standard. On a donc $S^1 \backslash \beta_a^{-1}(\{u\}) = CP^{n-2}$. Observons que chaque orbite de $\gamma_a^{-1}(\{q\})$ a un unique représentant $z = (z_1,\dots,z_n) \in V_a$ tel que $z_n = (x_1,0,x_3)$, avec $x_3 \geqslant 0$. Les points du quotient correspondant à $x_1 > -1$ constituent une cellule ouverte attachée sur $CP^{n-3} = S^1 \backslash \gamma^{-1}_{(a_1,\dots,a_{n-1})}(\{q-a_n\})$. On voit ici apparaître la décomposition cellulaire classique des espace projectif complexes.

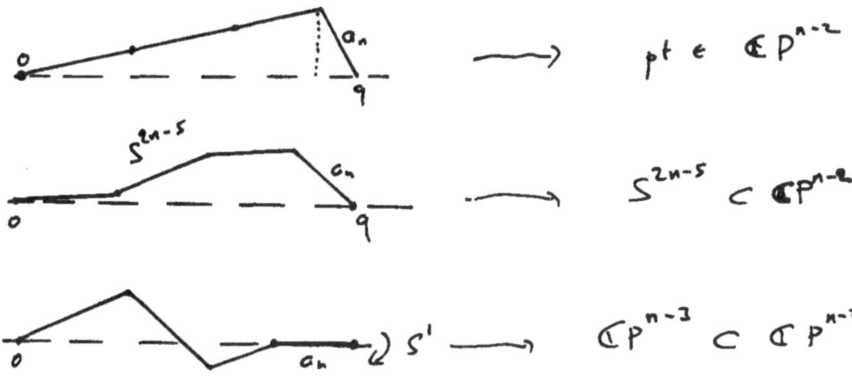

On peut de même considérer le cas k = 2. On a alors $\gamma_a^{-1}(\{q\})$ est difféomorphe à S^{n-2}. On démontre de même que la réflexion par rapport à l'axe horizontal est conjuguée par ce difféomorphisme à l'application antipodale. Le même argument que ci dessus fait apparaître la décomposition cellulaire bien connue de $\mathbb{R}P^{n-2}$.

5. PREIMAGES OF ZERO

Les préimages de 0 pour un bras articulé sont déterminées par les préimages d'un bras articulé de longueur n-1. Pour voir cela, soit $\tau : (\mathbb{R}_{>0})^n \longrightarrow (\mathbb{R}_{>0})^{n-1}$ l'application définie par

$$\tau(a_1,\dots,a_n) = \frac{1}{1-a_n}(a_1,\dots,a_{n-1})$$

La projection de $(S^{k-1})^n \longrightarrow S^{k-1}$ sur le n^e facteur donne, par restriction à $\beta_a^{-1}(\{0\})$ un fibré de fibre $\gamma_{\tau(a)}^{-1}(\{a_n\})$ et de groupe structural SO_{k-1} (avec l'action de SO_{k-1} sur les préimages de $\gamma_{\tau(a)}$ donnée en (1.4)).

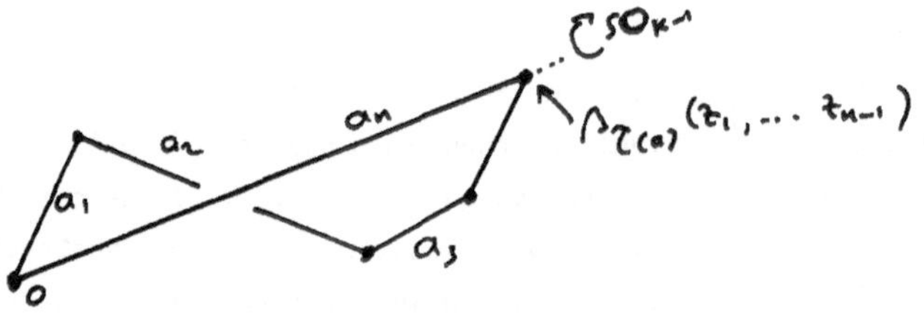

On peut caractériser ce fibré en disant qu'il a même SO_{k-1}-fibré principal associé que le fibré tangent unitaire T^1S^{k-1} à S^{k-1}. Cette assertion est équivalente au résultat suivant :

(5.1) **Proposition** Pour $a = (a_1,\dots,a_n)$, on a

$$\beta_a^{-1}(\{0\}) \;=\; SO_k \times_{SO_{k-1}} \gamma_{\tau(a)}^{-1}(\{a_n\}) .$$

On voit qu'il est de première importance de connaître l'action de SO_{k-1} sur les préimages de $\gamma_{\tau(a)}$, comme cela a été fait dans un cas particulier au paragraphe 4.

Preuve : On définit une application $f : SO_k \times \gamma_{\tau(a)}^{-1}(\{a_n\}) \longrightarrow \beta_a^{-1}(\{0\})$ par

$$f(\alpha,z_1,\dots,z_{n-1}) \;=\; (\alpha z_1,\dots,\alpha z_{n-1},-u)$$

où

$$u \;=\; \frac{1}{\|\beta_{\tau(a)}(\alpha z_1,\dots,\alpha z_{n-1})\|}\;\beta_{\tau(a)}(\alpha z_1,\dots,\alpha z_{n-1})$$

Il est clair que $f(\alpha\,\alpha',z) = f(\alpha,\alpha' z)$ pour $\alpha' \in SO_{k-1}$, d'où une application

$$f : SO_k \times_{SO_{k-1}} \gamma_{\tau(a)}^{-1}(\{a_n\}) \longrightarrow \beta_a^{-1}(\{0\}).$$

En utilisant l'action de SO_k sur $\beta_a^{-1}(\{0\})$, on vérifie facilement que f est un homéomorphisme, et un difféomorphisme si a_n est une valeur régulière de $\gamma_{\tau(a)}$.

6. CLASSIFICATION DES PRÉIMAGES DE γ_a

Soit $\gamma_a : (S^{k-1})^n \longrightarrow]0,1]$ $(a \in \Delta^{n-1})$ un bras à extrémité coulissante de longueur n. Pour $d \in [0,1]$, considérons l'ensemble $H(n,d) \subset \Delta^{n-1}$ qui est l'union des hyperplans de Δ^{n-1} d'équation $\sum \varepsilon_i \cdot a_i = d$, avec $\varepsilon_i \in \{\pm 1\}$. Appelons **chambre** une composante connexe de $\Delta^{n-1} - H(n,d)$.

(6.1) **Proposition** Soit $d > 0$ et $a \in \Delta^{n-1}$. Alors

a) d est un niveau critique de γ_a si et seulement si $a \in \mathbf{H}(n,d)$.

b) Si a, $a' \in \mathrm{int}\Delta^{n-1}$ sont dans une même chambre, les variétés différentiables $\gamma_a^{-1}(\{d\})$ et $\gamma_{a'}^{-1}(\{d\})$ sont canoniquement difféomorphes.

Preuve : Nous avons vu dans le paragraphe 3 que l'ensemble des points critiques de γ_a est constitué par les positions alignées $\{z \in (S^{k-1})^n \mid z_i = \pm 1\}$. Un élément d [0,1] est donc un niveau critique pour γ_a si et seulement si il existe $\varepsilon_i \in \{\pm 1\}$ (i = 1,...,n) tels que $\sum \varepsilon_i \cdot a_i = d$. Cela démontre a).

Si a et a' sont dans la même chambre C, il existe $v > 0$ tel que $u(t) = ta + (1-t)a' \in C$ pour $t \in]-v,1+v[$. L'application différentiable

$$\begin{array}{ccc} (S^{k-1})^n \times]-v,1+v[& \longrightarrow & \mathbf{R}^k \times]-v,1+v[\\ (z,t) & \longmapsto & (\beta_{u(t)}(z),t) \end{array}$$

est alors transverse à $\{(d,0)\}\times]-v,1+v[$. La préimage de cet intervalle donne un cobordisme W entre $\gamma_a^{-1}(\{d\})$ et $\gamma_{a'}^{-1}(\{d\})$. La projection sur $]-v,1+v[$ est une fonction $g : W \to]-v,1+v[$ sans point critique telle que $\gamma_a^{-1}(\{d\}) = g^{-1}(1)$ et $\gamma_{a'}^{-1}(\{d\}) = g^{-1}(0)$. Les trajectoires du champ de gradient de g convenablement normalisé donnent un difféomorphisme de $\gamma_a^{-1}(\{d\})$ sur $\gamma_{a'}^{-1}(\{d\})$ (voir [Mi2,Théorème 3.4]). Remarquons que ce difféomorphisme commute avec l'action de SO_{k-1}.

On a le même résultat pour classifier les espaces $\beta_a^{-1}(\{0\})$. On utilise pour le démontrer la proposition (6.1) ci-dessus et les résultats du paragraphe 5.

L'énoncé précis est :

(6.2) **Proposition**

a) 0 est une valeur critique de β_a si et seulement si $a \in \mathbf{H}(n,0)$.

b) Si a, $a' \in \mathrm{int}\Delta^{n-1}$ sont dans une même chambre, les variétés différentiables $\beta_a^{-1}(\{0\})$ et $\beta_{a'}^{-1}(\{0\})$ sont canoniquement difféomorphes.

Remarquons que $\mathbf{H}(n,d)$ ne dépend pas de k. Le nombre de préimages différentes des bras articulés de longueur n dans $\mathbf{R}^k$ ne dépend donc que de n.

Il est possible de construire le système d'hyperplan $\mathbf{H}(n,d)$ par un procèdé de récurrence sur n que nous allons décrire maintenant. Soit $\mathbf{H}(n)$ la famille d'hyperplans du prisme $\Delta^{n-1} \times [0,1]$ d'équation $\sum \varepsilon_i \cdot a_i = d$, avec $\varepsilon_i \in \{\pm 1\}$. On a $\mathbf{H}(n,d) = \mathbf{H}(n) \cap \Delta^{n-1} \times \{d\}$. Le procèdé de récurrence consiste en deux opérations :

I) <u>$\mathbf{H}(n,0)$ détermine $\mathbf{H}(n)$</u> : Soit

$$A = \{(a_1,\dots,a_{n+1}) \in \Delta^{n-1} \mid \sum \varepsilon_i \cdot a_i = d\},$$

avec $\varepsilon_i \in \{\pm 1\}$, un hyperplan de $\mathbf{H}(n,0)$. A est l'intersection de deux hyperplans $A_\pm$

de **H**(n), d'équation $\sum \varepsilon_i \cdot a_i = \pm d$. L'hyperplan A_+ intersecte la face de $\Delta^{n-1} \times \{1\}$ d'équation $a_i = 0$ si $\varepsilon_i = -1$ et A_- intersecte la face de $\Delta^{n-1} \times \{1\}$ d'équation $a_i = 0$ si $\varepsilon_i = 1$. On obtient ainsi la famille **H**(n) à partir de **H**(n,0).

La détermination des préimages de γ_a à partir de celles de $\beta_a^{-1}(\{0\})$ n'est pas complète. Toutefois, le fait que γ_a est une fonction de Morse donne quelques indications sur ces préimages.

II) <u>**H**(n) détermine **H**(n+1,0)</u> : considérons l'isomorphisme linéaire par morceau

$$h \; : \; \Delta^{n-1} \times [0,1] \longrightarrow \{(a_1,\ldots,a_{n+1}) \in \Delta^n \mid a_{n+1} \leq 1/2\}$$

tel que

$$h((a_1,\ldots,a_n),0) = (a_1,\ldots,a_n,0)$$

et

$$h((a_1,\ldots,a_n),1) = (a_1/2,\ldots,a_n/2,1/2)$$

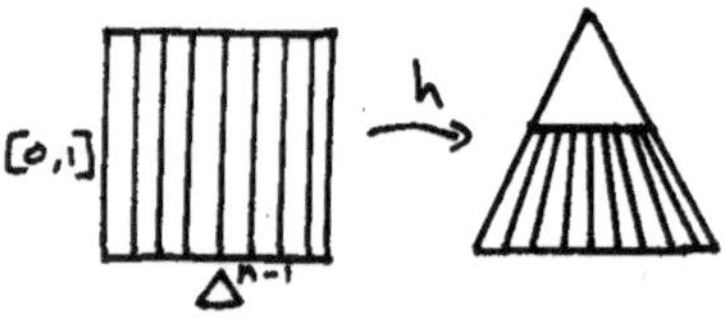

On vérifie aisément que **H**(n+1,0) = h(**H**(n)) $\cup$ $\{a_{n+1} = 1/2\}$. La préimage d'un point a de la chambre $\{a_{n+1} > 1/2\}$ est vide. Celle d'un point h(a,d) où a est un point d'une chambre C de Δ^{n-1} - **H**(n,d) est difféomorphe à $SO_k \times_{SO_{k-1}} \gamma_a^{-1}(\{d\})$, par la proposition (5.1).

Nous allons illustrer les pas I et II ci-dessus. Il est clair que **H**(2,0) consiste en l'hyperplan $\{a_1 - a_2 = 0\}$ de Δ^1, c'est-à-dire le point $\{a_1 = a_2 = 1/2\}$. On a $\beta_a^{-1}(\{0\}) = \emptyset$ si $a \notin$ **H**(2,0) et $\beta_a^{-1}(\{0\}) = S^{k-1}$ si a = (1/2,1/2).

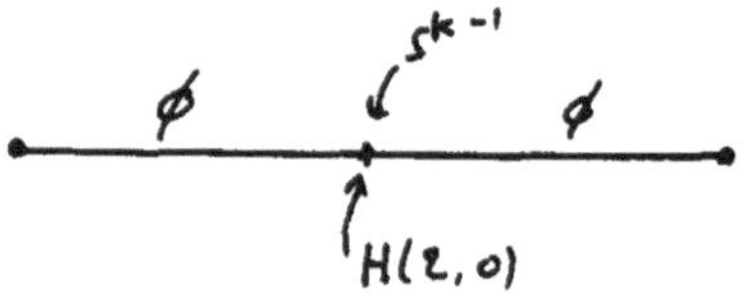

En appliquant le pas I), on trouve, pour **H**(2) :

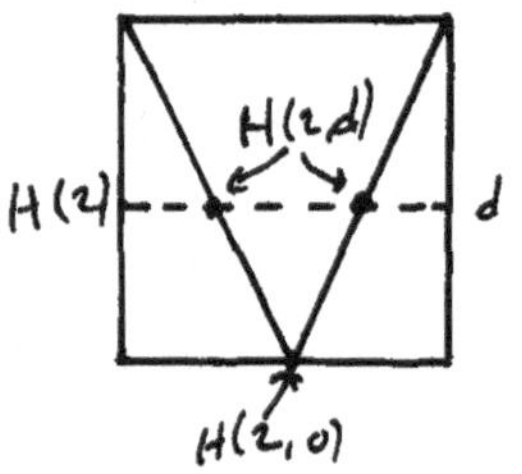

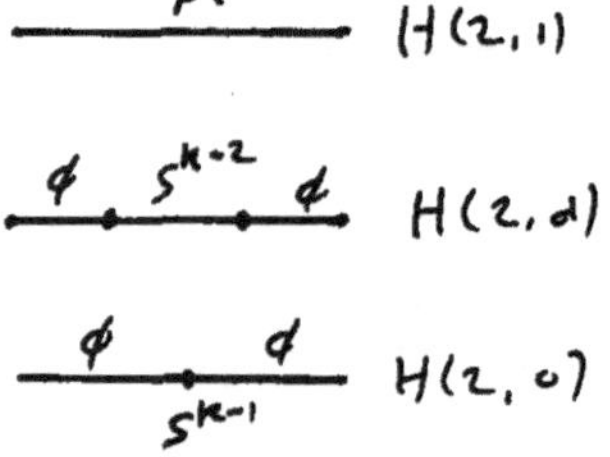

On applique maintenant le pas II) pour obtenir H(3,0). Comme l'action de SO_{k-1} sur S^{k-2} est l'action standard (proposition (4.1)), la préimage non-vide générique de γ_a sera le fibré tangent unitaire T^1S^{k-1} à la sphère S^{k-1}.

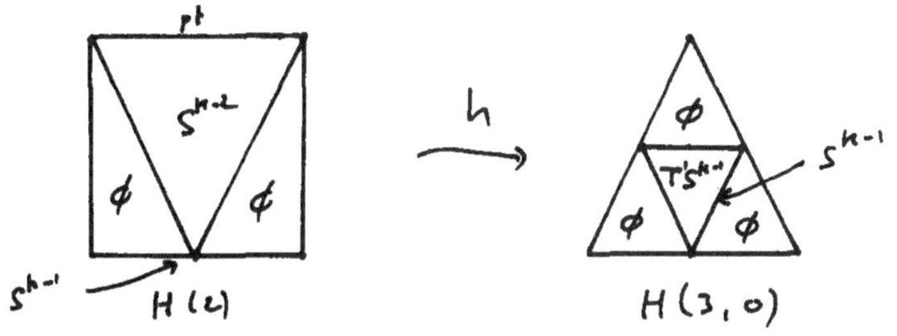

Le pas I) appliqué à cette figure donne, pour H(3), une famille de 6 hyperplans de $\Delta^2 \times [0,1]$. Nous avons dessiné ci-dessous quelques familles H(3,d) correspondantes. On observe un point triple en ((1/3,1/3,1/3),1/3), c'est-à-dire que pour a = (1/3,1/3,1/3), $\gamma_a^{-1}(\{1/3\})$ contient 3 points critiques. Pour les autres points doubles (a,d), $\gamma_a^{-1}(\{d\})$ contient deux points critiques. Les dessins des espaces $\gamma_a^{-1}(\{d\})$ pour les niveaux critiques d sont ceux pour le cas k = 2 où l'on obtient les graphes :

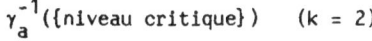

$\gamma_a^{-1}(\{\text{niveau critique}\})$ (k = 2)

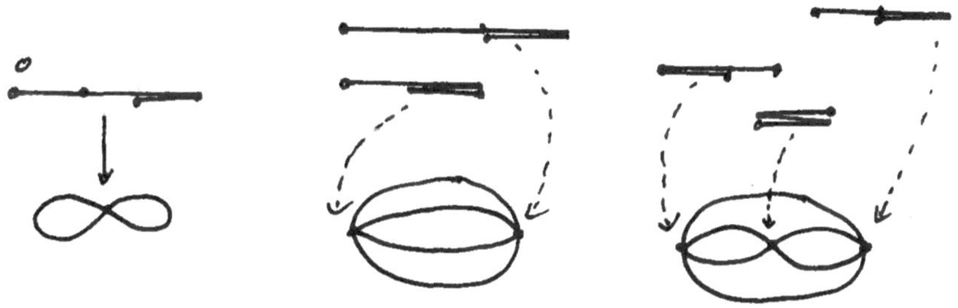

1 point critique 2 points critiques 3 points critiques

QUELQUES FAMILLES H(3,d) :

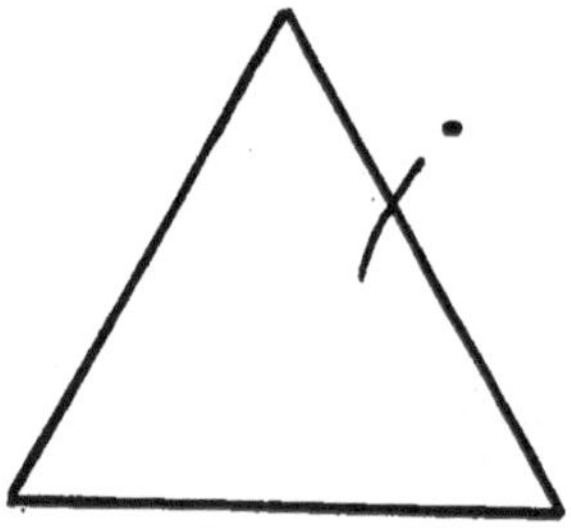

$d = 1$

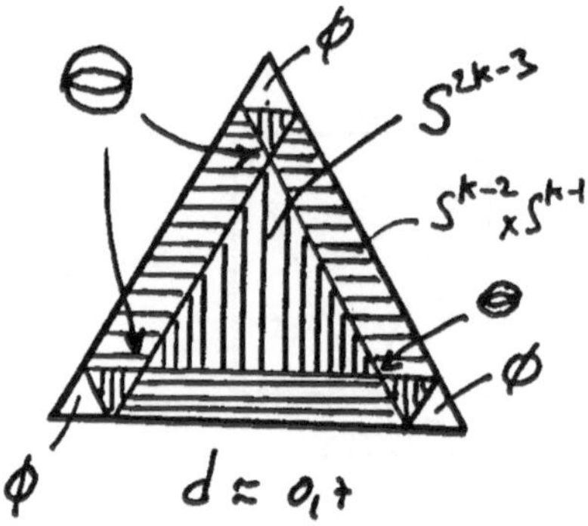

$d \simeq 0,7$

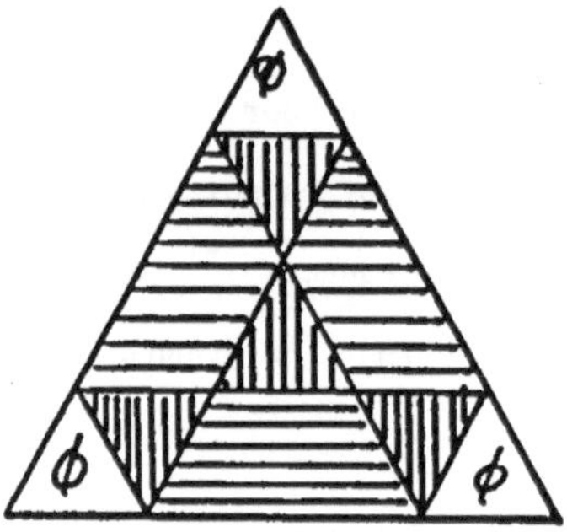

$d = 0.5$

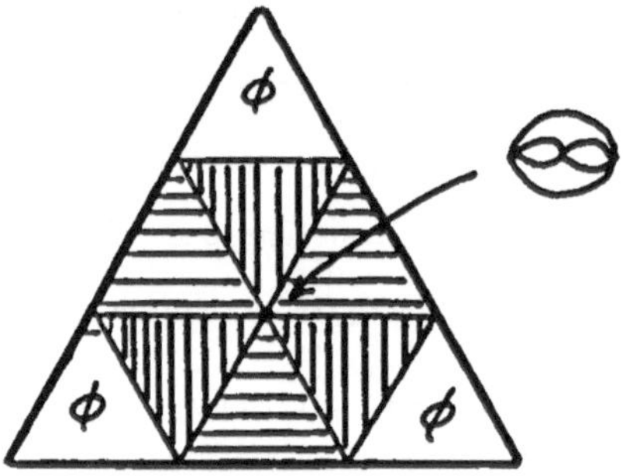

$d = \frac{1}{3}$

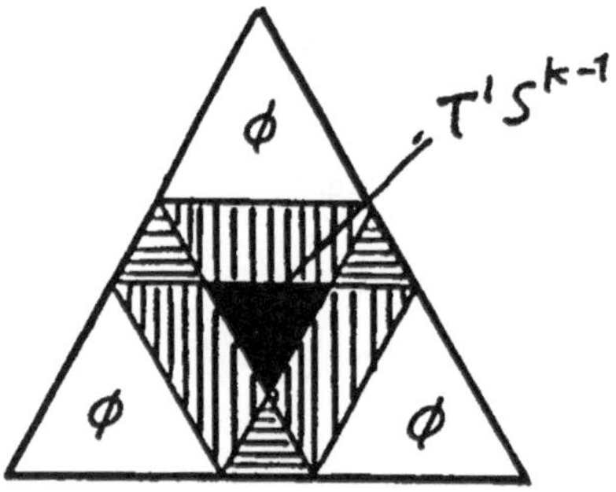

$0 < d < \frac{1}{3}$

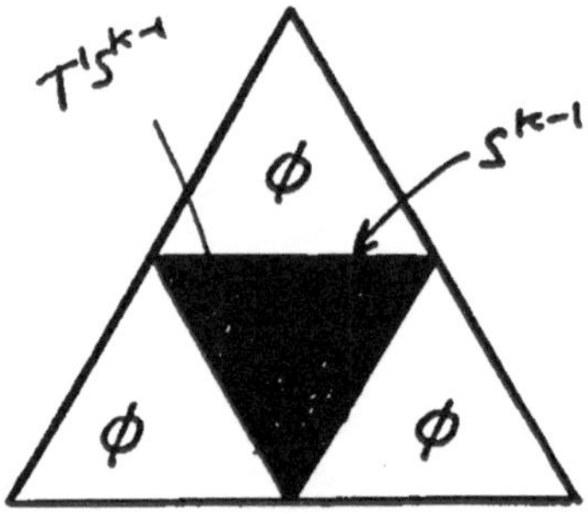

$d = 0$

BIBLIOGRAPHIE

[GN] GIBSON C.G.-NEWSTEAD P.E. On the geometry of the 4-bar mechanism. Acta Applic. Math. 7 (1986) 113-135

[Go1] GOTTLIEB D. Robots and fibre bundles Bull. Soc. Math de Belgique 37 (1987) 219-223

[Go2] GOTTLIEB D. Topology and the robot arm Acta Applic. Math. 11 (1988) 117-121

[Ha] HAUSMANN J-Cl. Action quasi-linéaires sur les sphères. A paraître.

[Mi1] MILNOR J. Morse Theory. Ann. of Math. Study 51, Princeton Univ. Press 1969

[Mi2] MILNOR J. Lectures on the h-cobordism theorem Princeton Univ. Press, 1965

[TW] THURSTON W.-WEEKS J. The mathematics of the three-dimensional manifolds Scientific American, July 1986, 94-106

[Wa] WALKER K. Configuration spaces of linkages Bachelor Thesis, Princeton 1985

[We] WENGER A. Etudes des espaces de configurations de certains systèmes articulés. Trav. Diplôme, Univ. de Genève, 1988.

POST SCRIPTUM : La construction et l'usage de fonctions de Morse sur l'espace de configurations de bras articulés (dans S^3) ont été récemment développés pour étudier l'espace des SU(2)-représentations de $\pi_1(V)$, où V est une 3-sphère d'homologie seifertique. On peut utiliser ces résultats pour calculer l'homologie de Floer de V via les techniques de Fintushel-Stern. Voir : A. KIRK et E. KLASSEN, Representation spaces of Seifert fibered homology spheres, preprint.

Section de Mathématiques
Université de Genève

SEMICONTRACTIBLE LINK MAPS AND THEIR SUSPENSIONS

Ulrich Koschorke*
Mathematical Sciences Research Institute, Berkeley, and
Mathematik V, Universität GH, D5900 Siegen

Introduction.

Given dimensions $p_1, \ldots, p_r$ and m, a map

$$f = f_1 \amalg \ldots \amalg f_r \quad : \quad S^{p_1} \amalg \ldots \amalg S^{p_r} \longrightarrow S^m$$

whose components have pairwise disjoint images (i.e. $f_i(S^{p_i}) \cap f_j(S^{p_j}) = \emptyset$ for $1 \leq i \neq j \leq r$) is called a *link map*, and a deformation through such link maps is called a *link* homotopy. This equivalence relation was introduced by J. Milnor in 1954 in his celebrated paper **[M]** which gave enormous impulses to classical link theory.

In higher dimensions, however, there was hardly any progress on link homotopy questions for thirty years. This was mainly due to

(i) the lack of examples of interesting link maps which do not embed at least one component sphere; and

(ii) the lack of interesting link homotopy invariants (other than the rather weak α-invariant which generalizes the classical linking number and takes values in a - usually finite - stable homotopy group of spheres).

But since 1984 there have been dramatic new developments. They started with work of R. Fenn and D. Rolfsen **[FR]** which produced

(i) a new link map

$$(I.1) \qquad f = f_+ \amalg f_- : S^2 \amalg S^2 \longrightarrow S^4$$

whose construction is based on the following property of the classical Whitehead link w:

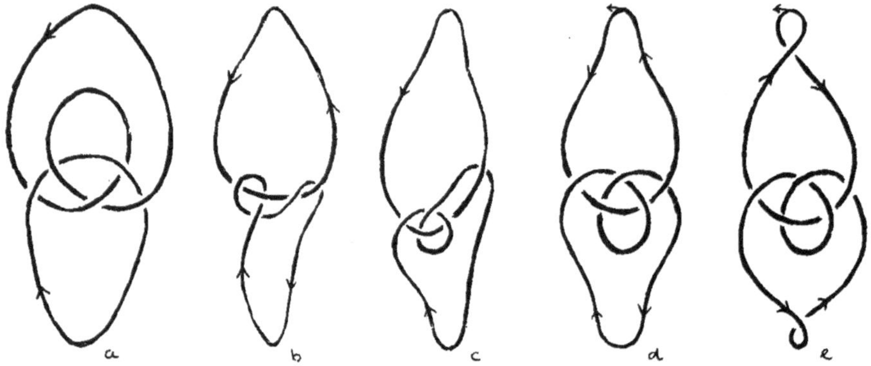

Figure I.2. Various isotopic versions of the Whitehead link w

*Research supported in part by the Deutsche Forschungsgemeinschaft.

(*I*.3) each component allows a nullhomotopy in the complement of the image of the other component.

Thus we can deform a trivial (i.e. locally constant) link map $*$ to w by one of these nullhomotopies and then back to $*$ by the other one. The (level preserving) track

$$(S^1 \amalg S^1) \times I \longrightarrow \mathbb{R}^3 \times I \subset S^4$$

of this link homotopy induces the Fenn–Rolfsen link map f_{FR}.

(ii) a new link homotopy invariant

$$(I.4) \qquad \beta : LM^4_{2,2} \longrightarrow \mathbb{Z}_2$$

where $LM^4_{2,2}$ denotes the set of all link homotopy classes of link maps as in (*I*.1). Any such class can be represented by a self-transverse immersion, and each double point

$$v = f_+(u) = f_+(u'), \quad u \neq u' \in S^2,$$

has a multiplicity n_v defined as follows: pick any path in S^2 from u to u', compose it with f_+ and let n_v be the linking number of the resulting loop in S^4 with the other component map of f. Unfortunately, this integer changes its sign if we interchange the rôles of u and u', and there is no natural way to order the two branches of f_+ through v. Thus Fenn and Rolfsen reduced n_v mod 2 and defined $\beta(f)$ to be the resulting sum over all double points of f_+. Clearly f_{FR} has β-invariant 1 and hence its link homotopy class differs from the trivial one.

As late as 1985 it was conjectured that these might be the only elements of $LM^4_{2,2}$. Thus it came as a surprise when a year later P. Kirk discovered the link homotopy invariant

$$(I.5) \qquad \sigma = (\sigma_+, \sigma_-) : LM^4_{2,2} \longrightarrow \mathbb{Z}[X]/\mathbb{Z}X^0 \oplus \mathbb{Z}[X]/\mathbb{Z}X^0$$

and determined its infinitely generated image [**Ki1**]. $\sigma_\pm(f)$ is simply the sum of $\text{sign}(v)X^{|n_v|}$ over all double points v of $f_\pm$, where $\text{sign}(v)$ equals +1 or -1 according to whether the orientation induced by the two branches of f through v agrees with the standard orientation of S^4 or not.

Finally, in 1987 link homotopy classification was accomplished in a large metastable dimension range (cf. [**Ko3**] and also [**Ko4**]). This involved

(i) constructing many examples of link homotopies and, by taking tracks, of link maps; and

(ii) showing that the $\tilde{\beta}$-invariant (which generalizes and refines $\sigma_\pm$) is the only obstruction to deforming a given component of a link map to an embedding.

The proofs are very geometric and use embedding and isotopy arguments which do not work in the case of $LM^4_{2,2}$, and for good reason: link maps provide a setting for an elementary proof that the Whitney trick does not work in dimension 4 (see [**Ki2**]). However, we can exploit specifically low-dimensional phenomena such as the asphericity theorem of Papakyriakopoulos [**P**]. This leads to a well-defined suspension homomorphism

$$(I.6) \qquad J : \pi_0(\mathcal{L}_+ \cap \mathcal{L}_-) \longrightarrow LM^4_{2,2}$$

which generalizes the construction of Fenn and Rolfsen to all *semicontractible* link maps (i.e. link maps that satisfy condition (*I*.3) above). A version of J was already studied by G. T. Jin [**J**] who established e.g. a very close relationship between $\sigma \circ J$ and T. Cochran's derived invariants for semiboundary links (see also [**Ki1**], p. 41f).

In this paper we study link maps in dimensions 3 and 4 in more detail. In section 1 we introduce an operation called *shuffled sums*. In many settings it allows us to produce a large number of link maps or links even if we start from just a few prototypes. This operation combines the following three ingredients: connected sums, running through the same components repeatedly, and permuting the order of these runs.

Our notion of shuffled sums applies in particular to classical knots and links and hopefully presents an interesting new point of view. E.g. a classical link is much more likely to be composite with respect to shuffled sums than with respect to traditional connected sums; thus prime knots or links in the shuffled sense are fewer and hence more likely to fit into some enumeration scheme. On the other hand, given a fixed link, it is an interesting problem to determine which links can be obtained as shuffled sums of it.

In any case our operation produces an enormous number of semicontractible link maps. Consider e.g. the (very partial) sequence

$$(I.7) \qquad \ldots\ \ell(1,3;w),\ \ell(1,2;w),\ \ell(1,1;w),\ \ell(2,1;w),\ \ell(3,1;w),\ \ldots$$

of shuffled sums having the Whitehead link as their only summand. The corresponding suspensions $J(\ell(c_+, c_-; w))$ form the basis (over $\mathbb{Z}$) of a subgroup of the abelian semigroup $LM^4_{2,2}$, and the σ-invariant restricts to an isomorphism from this subgroup onto $\sigma(LM^4_{2,2})$. This is an easy consequence of our detailed analysis (given in section 2) of the suspension homomorphism J and its compatibilities e.g. with shuffled sums (see also **[Ki1]**). The images of J and of a natural track homomorphism are further canonical subgroups of $LM^4_{2,2}$. However, it is not known whether $LM^4_{2,2}$ itself is a full group: so far the existence of additive inverses has been established only if we classify link maps as in $(I.1)$ up to *link concordance*; incidentally, σ is still a well-defined *concordance* invariant (see **[Ko2]**, 1.11 and 3.9).

§1. Shuffled sums of classical links and link maps.

Let D^3 denote the closed unit ball in euclidean (x,y,z)-space $\mathbb{R}^3$, and fix base points $*_1, \ldots, *_r$ which are lined up consecutively along the equator, i.e. along the intersection of the boundary sphere $S^2 = \partial D^3$ with the (x,y)-plane. Also, we fix a base point $*$ in the standard unit circle S^1.

DEFINITION 1.1: A continuous map

$$\ell = \ell_1 \amalg \ldots \amalg \ell_r : S^1 \amalg \ldots \amalg S^1 \longrightarrow D^3$$

is called a *base point preserving link map* (with r components and in classical dimensions) if

$$\ell_i\{*\} = \{*_i\} = \ell_i(S^1) \cap \partial D^3 \quad, \ i = 1, \ldots, r,$$

and if the component maps ℓ_i have pairwise disjoint images.

If in addition ℓ is a smooth embedding which takes all positively oriented tangent vectors at $*$ into the negative z-direction, then ℓ is called a *base point preserving link*.

Usually we will consider base point preserving link maps (or links) only up to *base point preserving link homotopy* (or *isotopy*), i.e. up to deformations which yield a base point preserving link map (or link) at every deformation parameter t. E.g. rotation by 180° defines a base point preserving link homotopy (but not a *base point preserving isotopy*) which reverses the orientation of both components of the Whitehead link (see Figure I.2, d and e).

We are going to assign a new base point preserving link map (or link) $\ell(T)$ as above to any set T of data consisting of

(i) an r-tuple $c_1, \ldots, c_r$ of elements in the fundamental group $\pi_1(\bigvee^k S^1)$ or, equivalently, in the nonabelian free group with the generators $\iota_1, \ldots, \iota_k$ which are represented by the obvious inclusions of S^1, and

(ii) a k-tuple $\ell^1, \ldots, \ell^k$ of base point preserving r-component link maps (or links) in classical dimensions.

Here the number k of wedge factors in $S^1 \vee \ldots \vee S^1$ and of summands ℓ^j can be any positive integer.

Pick any sequence of numbers

$$-1 \le z_-^1 < z_+^1 < \cdots < z_-^j < z_+^j < z_-^{j+1} < \cdots < z_+^k \le 1.$$

For $j = 1, \ldots, k$ the obvious affine transformation $[-1,1] \approx [z_-^j, z_+^j]$ induces a homeomorphism h^j from D^3 onto

$$D^{(j)} := \{(x,y,z) \in D^3 \mid x^2 + y^2 \le 1, \quad z_-^j \cdot g(x,y) \le z \le z_+^j \cdot g(x,y)\}$$

where $g(x,y) = \sqrt{1 - x^2 - y^2}$. Note that h^j fixes the equator which is also the intersection locus $D^{(j)} \cap D^{(j')}$ for $j \neq j'$.

We define the i-th component of $\ell(T)$ by

$$(1.2) \qquad \ell_i(T) : S^1 \xrightarrow{c_i} \bigvee_{j=1}^{k} S^1 \xrightarrow{\vee h^j \circ \ell_i^j} D^3,$$

$i = 1, \ldots, r$. In other words, we identify each "summand" $\ell^j = \ell_1^j \amalg \ldots \amalg \ell_r^j$, $j = 1, \ldots, k$, with its image link map in $D^{(j)}$ and then run through the i-th component maps $\ell_i^1, \ldots, \ell_i^k$ in the order and with the winding numbers given by the order of appearance of the generators $\iota_1, \ldots, \iota_k$ and by their exponents in the word $c_i = \iota_{j_1}^{n_1} \ldots \iota_{j_s}^{n_s}$.

Clearly, the base point preserving link homotopy class of $\ell(T)$ is independent of all choices made above and depends only on $c_1, \ldots, c_r$ and on the base point preserving link homotopy classes of $\ell^1, \ldots, \ell^k$. It even remains unchanged by simultaneous permutations of the "summands" $\ell^1, \ldots, \ell^k$ and of the wedge factors in $\bigvee^k S^1$ provided $r \le 2$; indeed, in this case we

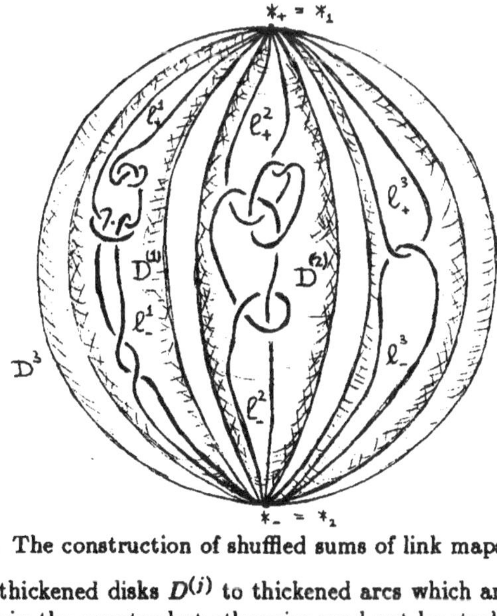

Figure 1.3. The construction of shuffled sums of link maps with two components

may retract the thickened disks $D^{(j)}$ to thickened arcs which are attached to the equator at the base point(s) in the equator but otherwise need not be stacked on top of each other with respect to the z-direction.

Whatever the number of components, the corresponding construction for base point preserving *links* requires a little extra care in order to avoid selfintersections near the base points or when we run through the same component ℓ_i^j several times. After a small isotopy we may assume that the oriented knot $\ell_i^j(S^1)$ consists of an arc A_i^j (which lies entirely in the ball $D(1-\epsilon)$ of radius $1-\epsilon$ around 0) and of a half-circle (which lies both in $D^3 - \overset{\circ}{D}(1-\epsilon)$ and in the plane through $*_i$ which is perpendicular to the equator). Moreover, there is a Seifert surface of $\ell_i^j(S^1)$ which, near $*_i$, lies in this plane; the resulting normal vector field along $\ell_i^j(S^1)$ points in the positive equatorial direction near $*_i$ and allows us to push the arc A_i^j out to disjoint copies of itself. Now, given the word c_i, we replace each occurence of ι_j (or ι_j^{-1}) by a run (in the positive or negative sense, resp.) through one of these pushed out copies of A_i^j; as we proceed from one such letter to the next one, we increase the level of the beginning and end points of our arcs with respect to the positive direction along the equator. Thus the straight connections between the end point of one arc and the starting point of the next one, as well as the initial and final connections from and to the base point $*_i$ create no selfintersections and can be smoothed in a canonical fashion. Also, running forward through one pushed copy of A_i^j and then immediately backward through another one is isotopic to the straight connection of end points. Thus the base point preserving isotopy class of $\ell(T)$ is independent of the choices involved in our construction and depends only on the base point preserving isotopy classes of $\ell^1, \ldots, \ell^k$ and on the elements $c_1, \ldots, c_r$ in the free group $\pi_1(\bigvee^k S^1)$. Moreover, if we drop any trivial summands ℓ^j and simultaneously collapse the corresponding j^{th} wedge factors S^1, the class of $\ell(T)$ does not change.

EXAMPLE 1.4: CONNECTED SUMS: In the special case $c_1 = \cdots = c_r = \iota_1 \bullet \cdots \bullet \iota_k$ our construction yields the *connected sum*

$$\ell^1 + \cdots + \ell^k := \ell(c_1, \ldots, c_r; \ell^1, \ldots, \ell^k).$$

Clearly, up to base point preserving link homotopy (or isotopy, resp.), this addition is associative and has the trivial link (map) as a unit.

EXAMPLE 1.5: In the case of *knots* (i.e. links with one component) the base point preserving and the base point free isotopy classifications in D^3, $\mathbb{R}^3$ and S^3 all coincide and the previous example just describes the classical connected sum operation. However, shuffled sums can do much more; e.g. if $k = 1$, they produce already many cable knots based on a given companion ℓ.

EXAMPLE 1.6: Let h be the Hopf link, i.e. h consists of a trivial knot and its meridian. Then it is easy to see that

$$\ell(\iota, \iota^k; h) = h + \cdots + h \quad (k \text{ summands}).$$

However, no such relation can hold for the Whitehead link when $k > 1$ (see example 2.5).

REMARK 1.7: Many interesting families of links or link maps are closed under shuffled sums. E.g. if each summand $\ell^1, \ldots, \ell^k$ is a link with unknotted components (or a link map all of whose sublink maps with s components allow a base point preserving nullhomotopy, $s \leq r$), then so is every shuffled sum of $\ell^1, \ldots, \ell^k$.

§2. The suspension homomorphism.

In this section we focus our attention on a family of classical two-component link maps which is closed under shuffled sums and which provides the natural setting for suspensions to $LM^4_{2,2}$. It is convenient to label components, base points, etc. by the subscripts $+$ and $-$ instead of 1 and 2.

DEFINITION 2.1: A base point preserving link map

$$\ell = \ell_+ \amalg \ell_- : S^1 \amalg S^1 \longrightarrow D^3$$

is called *+-contractible* if ℓ_+ is null homotopic in $D^3 - \ell_-(S^1)$ or, in other words, if ℓ can be extended to a *+-contraction* $L_+ \amalg \ell_- : D^2 \amalg S^1 \longrightarrow D^3$, i.e. $L_+(D^2) \cap \ell_-(S^1) = \emptyset$. If ℓ_- is null homotopic in $D^3 - \ell_+(S^1)$, then ℓ is called *--contractible*. If both conditions hold, then we call ℓ *semicontractible*.

Let $\mathcal{L}$ denote the space (endowed with the natural metric topology) of all base point preserving link maps $\ell : S^1 \amalg S^1 \longrightarrow D^3$, and let $\mathcal{L}_+$ and $\mathcal{L}_-$ be the subspaces of $\mathcal{L}$ formed by the link maps which are +-contractible and --contractible, resp. The following central fact is an easy consequence of the asphericity theorem of Papakyriakopoulos [P].

PROPOSITION 2.2. *The spaces $\mathcal{L}_+$ and $\mathcal{L}_-$ of ±-contractible link maps are open in $\mathcal{L}$ and simply connected.*

More precisely, given any path in $\mathcal{L}_+$ and any +-contractions of its starting and end points, there is a path of +-contractions which extends all these data. The corresponding statement holds also for $\mathcal{L}_-$.

PROOF: Every +-contraction L extending a link map ℓ induces +-contractions of all maps ℓ' which are ϵ-close to ℓ, provided 2ϵ is smaller than the distance between $L_+(D^2)$ and $\ell_-(S^1)$. Indeed, the composition of the nullhomotopy L_+ with the straight linear homotopy from ℓ_+ to ℓ'_+ avoids the ϵ-neighborhood of $\ell_-(S^1)$ and, in particular, the image of ℓ'_-.

This shows that $\mathcal{L}_+$ is open in $\mathcal{L}$. Moreover, any given path in $\mathcal{L}_+$ extends to a path of +-contractions, at least when restricted to each subinterval $[t_i, t_{i+1}]$ of a suitable finite subdivision

$$0 = t_0 < t_1 < \cdots < t_i < t_{i+1} < \cdots < t_n = 1$$

of the parameter interval. Still, at each of the transition points t_i we have to compare two possibly different +-contractions of the same link map ℓ_{t_i}. But any two contractions

$$L_+, L'_+ : D^2 \longrightarrow D^3 - \ell_-(S^1)$$

of a +-contractible link map ℓ can be deformed into one another through such contractions. Indeed, they define a map on $S^2 = D^2 \cup D^2$ which must be null homotopic in $S^3 - \ell_-(S^1)$ (see [P], Corollary 26.2) and, after a suitable retraction, also in $D^3 - \ell_-(S^1)$.

In particular, every loop in $\mathcal{L}_+$ extends to a loop in the space of all +-contractions

$$L : D^2 \amalg S^1 \longrightarrow D^3.$$

But this space itself is canonically contractible: first compose L_+ with the linear retraction of D^2 to its base point $* \in S^1$, and then contract ℓ_- in $D^3 - \{*_+\}$. Therefore, $\mathcal{L}_+$ is at least 1-connected. ∎

Our interest in semicontractible link maps stems from the following construction. Each ℓ in the intersection $\mathcal{L}_+ \cap \mathcal{L}_-$ extends to a +-contraction and a --contraction which, as above, determine paths p_+ in $\mathcal{L}_+$ and p_- in $\mathcal{L}_-$ joining ℓ to the constant map $*$. These paths fit together to yield a loop $p = p_+ \circ p_-^{-1}$ in $\mathcal{L}_+ \cup \mathcal{L}_-$. It follows from Proposition 2.2 that the resulting class $[p]$ in the fundamental group of $\mathcal{L}_+ \cup \mathcal{L}_-$ is independent of the choices made above and depends only on the path component of ℓ in $\mathcal{L}_+ \cap \mathcal{L}_-$. Thus we get

COROLLARY 2.3. *The suspension construction above leads to the well-defined homomorphisms*

$$J : \pi_0(\mathcal{L}_+ \cap \mathcal{L}_-) \xrightarrow{\ j\ } \pi_1(\mathcal{L}) \xrightarrow{\text{track}} LM^4_{2,2}$$

of semigroups.

Here the additive structures on the domain and the target of the composite homomorphism J are given by connected sums (see 1.4 and [Ko1], 2.3) which, incidentally, make $\mathcal{L}$, $\mathcal{L}_+$, $\mathcal{L}_-$ and $\mathcal{L}_+ \cap \mathcal{L}_-$ into H-spaces (hence $\pi_1(\mathcal{L})$ is abelian, and so is $LM^4_{2,2}$). More generally, all these spaces are closed under shuffled sums (see also Remark 1.7).

This is particularly interesting in the case of $\mathcal{L}_+ \cap \mathcal{L}_-$ since it allows us to produce large numbers of semicontractible link maps which lie in different path components of $\mathcal{L}_+ \cap \mathcal{L}_-$. All we need is to start from a few semicontractible summands, and these are not hard to find: take e.g. the links $5^2_1 (= w), 7^2_3, 8^2_{13}, 9^2_4, \dots$ in Rolfsen's tables [R], p. 416ff, or any other link with zero linking number and with unknotted components. Note on the other hand that smoothly embedded links are dense in $\mathcal{L}_+ \cap \mathcal{L}_-$, and that selfintersections of semicontractible link maps are subject to severe restrictions. In general, a double point of ℓ_-, say, complicates the fundamental group of $D^3 - \ell_-(S^1)$ considerably and hence reduces the chances of ℓ_+ to be null homotopic; e.g. if $\ell_-(S^1)$ is a figure 8 in the plane (as opposed to a trivial knot), then ℓ_+ must vanish in a nonabelian free group with two generators (and not just in $\mathbb{Z}$). This makes it often hard to decide whether two nonisotopic links lie in the same path component of $\mathcal{L}_+ \cap \mathcal{L}_-$. One of the best available approaches to this question involves the suspension homomorphism J, composed with Kirk's σ-invariant (for a refined version $\tilde{\sigma} \circ \tilde{J}$ see 2.9). Unfortunately, most of the presumably very interesting noncommutative aspects of $\pi_0(\mathcal{L}_+ \cap \mathcal{L}_-)$ get lost in this process.

PROPOSITION 2.4. *Consider the data* $T = (c_+, c_-; \ell^1, \dots, \ell^k)$ *where* $c_\pm \in \pi_1(\bigvee^k S^1)$ *and* $\ell^1, \dots, \ell^k \in \pi_0(\mathcal{L}_+ \cap \mathcal{L}_-)$.

Then the image $\tilde{J}(\ell(T)) \in \pi_1(\mathcal{L})$ *of the corresponding shuffled sum under the refined suspension homomorphism* $\tilde{J}$ *depends only on* $\ell^1, \dots, \ell^k$ *and on the k-tuples*

$$(c^1_\pm, \dots, c^k_\pm) \in H_1(\bigvee^k S^1) = \mathbb{Z}^k$$

consisting of the winding numbers of $c_\pm$ *around the factor circles in the wedge* $\bigvee^k S^1$.

Moreover, we have

$$\sigma_\pm(J(\ell(T)))(X) = \sum_{j=1}^k (c^j_\pm)^2 \sigma_\pm(J(\ell^j))(X^{|c^j_\mp|}).$$

PROOF: Partition the domain of $c_\pm$ by the inverse images $a^j_\pm$, $j = 1, \dots, k$, of the factor circles of the wedge under $c_\pm$. Then $\ell(T)_\pm$ maps $a^j_\pm$ into $D^{(j)}$, and the same holds at every t-level, $t \in I$, of the link homotopy

$$p : (S^1 \amalg S^1) \times I \longrightarrow D^3$$

which defines $\tilde{J}(\ell(T))$. Since the balls $D^{(j)}$ intersect only in the base points $*_\pm$, we can reparametrize the t-levels in each $a^j_\pm \times I$ until p is just the composite of the link homotopies p_j which agree with p on $a^j_\pm \times I$ and are constant everywhere else. But $[p_j]$ depends only on $c^j_\pm$ and ℓ^j. Note that this argument does not work in $\pi_1(\mathcal{L}_+ \cup \mathcal{L}_-)$.

The formula for the σ-invariant of $J(\ell(T))$ follows as in Proposition 3.2 of [Ki1]: every double point with multiplicity n of a selftransverse immersion representing $J(\ell^j)_+$ leads in the shuffled sum to $(c^j_+)^2$ double points with multiplicity $|c^j_-| \cdot n$ and with the same sign. ∎

EXAMPLE 2.5: Let $k = 1$ (thus $(c_+, c_-) \in \mathbb{Z} \oplus \mathbb{Z}$) and let $\ell = w$ be the Whitehead link. Then

$$\sigma(J(\ell(c_+, c_-; w))) = ((c_+)^2 \cdot X^{|c_-|}; -(c_-)^2 \cdot X^{|c_+|})$$

(we may neglect constant polynomials in X). In particular, if $c_+ \geq 1$, $c_- \geq 1$, and c_+ or c_- equals 1 we obtain a basis over $\mathbb{Z}$ of the image $\sigma(LM^4_{2,2})$ (which was determined in **[Ki1]**, Theorem 3.5).

We conclude this paper with a few more details on the partial homomorphisms in 2.3.

FACT 2.6: We may interpret $\pi_1(\mathcal{L})$ and $LM^4_{2,2}$ as the semigroups of all base point preserving link homotopies and link concordances, resp., from the trivial link map $*$ to itself, taken up to *homotopies* through such maps (use the techniques of the proof of Proposition 2.3 in **[Ko1]**). Thus the level preserving features of the track of an element $p \in \pi_1(\mathcal{L})$ are forgotten in $LM^4_{2,2}$. E.g. a rotation argument which is not possible in $\pi_1(\mathcal{L})$ shows that $p + \bar{p}$ becomes trivial in $LM^4_{2,2}$; here the involution $^-$ on $\mathcal{L}$ is defined by reflecting both the domains S^1 and the range D^3 in a base point preserving fashion.

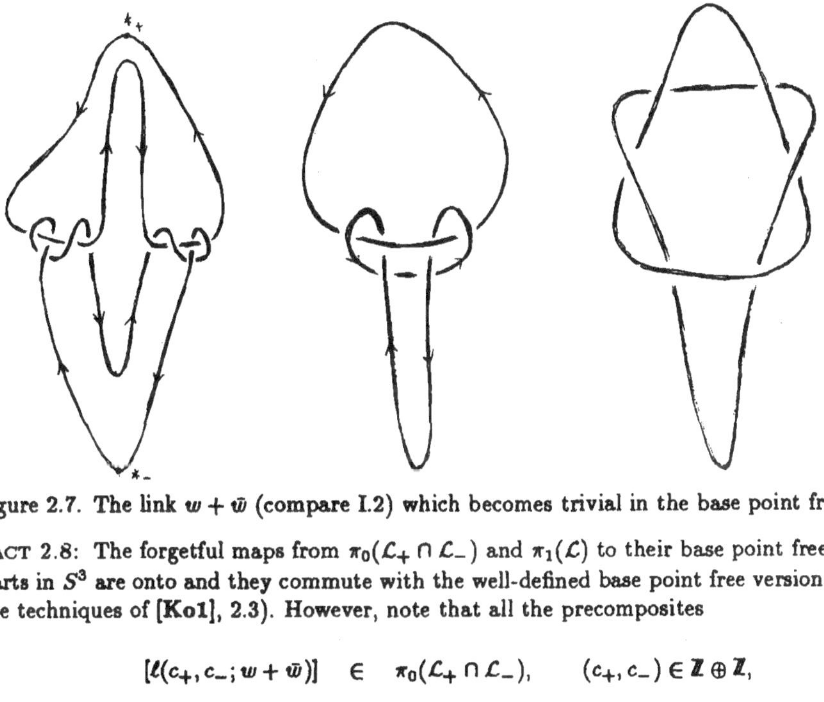

Figure 2.7. The link $w + \bar{w}$ (compare I.2) which becomes trivial in the base point free setting.

FACT 2.8: The forgetful maps from $\pi_0(\mathcal{L}_+ \cap \mathcal{L}_-)$ and $\pi_1(\mathcal{L})$ to their base point free counterparts in S^3 are onto and they commute with the well-defined base point free version of $\tilde{J}$ (use the techniques of **[Ko1]**, 2.3). However, note that all the precomposites

$$[\ell(c_+, c_-; w + \bar{w})] \quad \in \quad \pi_0(\mathcal{L}_+ \cap \mathcal{L}_-), \qquad (c_+, c_-) \in \mathbb{Z} \oplus \mathbb{Z},$$

of the link in the figure above become trivial in the base point free setting.

On the other hand, $LM^4_{2,2}$ is even isomorphic to its base point preserving analogon $BLM^4_{2,2}$ (see **[Ko4]**, 1.7); equivalently, we may interpret it both in terms of base point free and base point preserving selfconcordances of $*_\pm$.

In particular, the subgroups

$$J(\pi_0(\mathcal{L}_+ \cap \mathcal{L}_-)) \quad \subset \quad \text{track}(\pi_1(\mathcal{L}))$$

of the abelian semigroup $LM^4_{2,2}$ are independent of whether they are defined in the base point preserving or in the base point free setting.

FACT 2.9: As P. Kirk has shown, there exists a well-defined invariant $\tilde{\sigma} = (\tilde{\sigma}_+, \tilde{\sigma}_-)$ which fits into the commuting diagram of homomorphisms

$$\begin{array}{ccc} \pi_1(\mathcal{L}) & \xrightarrow{\tilde{\sigma}} & \mathbb{Z}[X, X^{-1}]/\mathbb{Z}X^0 \oplus \mathbb{Z}[X, X^{-1}]/\mathbb{Z}X^0 \\ \big\downarrow \text{track} & & \big\downarrow \text{fold} \oplus \text{fold} \\ LM^4_{2,2} & \xrightarrow{\sigma} & \mathbb{Z}[X]/\mathbb{Z}X^0 \oplus \mathbb{Z}[X]/\mathbb{Z}X^0 \end{array}$$

where $\text{fold}(X^n) := X^{|n|}$. Indeed, the track of any loop p in $\mathcal{L}$ can be approximated by a selftransverse immersion f with good embedding properties e.g. near the base point $*$ of S^1. Thus the two preimages $u \neq u'$ of every double point v of f lie already in a copy of $(S^1 - \{*\}) \times I = \mathbb{R} \times I$. If they have equal $\mathbb{R}$-coordinates then the linking number n_v used in I.4 and I.5 vanishes, since f nearly preserves also the I-levels and therefore u is very close to u'. If the $\mathbb{R}$-coordinates of u and u' differ then they determine a canonical ordering of the two branches of f through v and hence a well-defined *integer* n_v. Thus we can define $\tilde{\sigma}_\pm[p]$ as the sum of $\text{sign}(v) \cdot X^{n_v}$ over all double points of $f_\pm$.

Given any integer degrees $c_\pm$, it is not hard to prove for p (and f) as above that $\tilde{\sigma}_\pm(\ell(c_+, c_-; p))$ equals the sum of

$$\frac{1}{2}\,\text{sign}(v)(|c_\pm|(|c_\pm| + 1)X^{c_+c_-n_v/|c_\pm|} + |c_\pm|(|c_\pm| - 1)X^{-c_+c_-n_v/|c_\pm|})$$

over all the double points v of $f_\pm$. This precomposition formula implies in particular that the elements

$$\tilde{\sigma} \circ \tilde{J}(\ell(c_+, c_-; w + \bar{w}) \mid c_\pm \geq 1;\ c_+ \text{ or } c_- \text{ equals } 1\},$$

together with $(X - X^{-1}, 0)$, from a $\mathbb{Z}$-basis of the kernel of fold $\oplus$ fold (see also 2.7 and 2.8).

This has the following two interesting corollaries. First, there is an infinite number of additively independent elements both in $\pi_0(\mathcal{L}_+ \cap \mathcal{L}_-)$ and in $\pi_1(\mathcal{L})$ which all become trivial in the corresponding base point free setting as well as in $LM^4_{2,2}$. In particular, the kernel of the track homomorphism in 2.3 contains an infinitely generated free abelian subgroup.

Secondly, the full $\tilde{\sigma}$-invariant depends crucially on the base point and level preserving aspects of the track of $[p] \in \pi_1(\mathcal{L})$. The only possibly new additive invariant (besides Kirk's original σ-invariant) induced by $\tilde{\sigma}$ in the base point free setting or on $LM^4_{2,2}$, takes values in a cyclic group. At present, there is no other invariant known which might detect nontrivial elements in the kernel of σ. On the other hand, it is hard to believe that σ should be injective on $LM^4_{2,2}$, since this would imply e.g. that every link homotopy class remains unchanged if we precompose one or both of the component maps with a reflection.

References.

[D] U. Dahlmeier, "Gewisse Verschlingungen und ihre Jin-Suspensionen," Diplomarbeit, Universitaet Siegen, 1989.

[FR] R. Fenn and D. Rolfsen, *Spheres may link homotopically in 4-space*, J. London Math. Soc. (2) **34** (1986), 177–184.

[J] G. T. Jin, *Invaraints of two-component links*, Thesis, Brandeis University (1988).

[Ki1] P. Kirk, *Link maps in the 4-sphere*, Proc. Siegen Topology Symp. LNiM 1350, Springer Verlag (1988).

[Ki2] ———, *Link homotopy with one codimension 2 component*, Trans. AMS, to appear.

[Ko1] U. Koschorke, *Link maps and the geometry of their invariants*, Manuscr. Math. **61** (1988), 383–415.

[Ko2] ———, *Multiple point invariants of link maps*, Proc. Second Siegen Topology Symposium 1987, Springer LNiM 1350 (1988), 44–86.

[Ko3] ———, *On link maps and their homotopy classification*, Math. Annalen, to appear.

[Ko4] ———, *Link homotopy with many components*, Topology, to appear.

[M] J. Milnor, *Link groups*, Ann. of Math. **59** (1954), 177–195.

[P] C. D. Papakyriakopoulos, *Dehn's lemma and asphericity of knots*, Ann. of Math. **66** (1957), 1–26.

[R] D. Rolfsen, "Knots and Links," Math. Lect. Series 7, Publish or Perish, 1976.

[S] G. P. Scott, *Homotopy links*, Abh. Math. Sem. Hamburg **32** (1968), 186–190.

The KO-Assembly Map and Positive Scalar Curvature

JONATHAN ROSENBERG

Department of Mathematics, University of Maryland
College Park, Maryland 20742, U.S.A.

Abstract. We state a geometrically appealing conjecture about when a closed manifold with finite fundamental group π admits a Riemannian metric with positive scalar curvature: this should happen exactly when there are no KO_*-valued obstructions coming from Dirac operators. When the universal cover does **not** have a spin structure, the conjecture says there should always be a metric of positive scalar curvature, and we prove this if the dimension is ≥ 5 and if all Sylow subgroups of π are cyclic. In the spin case, the conjecture is closely tied to the structure of the assembly map $KO_*(B\pi) \to KO_*(\mathbf{R}\pi)$, and we compute this map explicitly for all finite groups π. Finally, we give some evidence for the conjecture in the case of spin manifolds with $\pi = \mathbf{Z}/2$.

§0. INTRODUCTION

This paper is a continuation of my previous papers [**R1**], [**R2**], and [**R3**], but with an emphasis on manifolds with *finite* fundamental group. In other words, I shall try to answer the following question: given a smooth closed connected manifold M^n with finite fundamental group π, when does it admit a metric of positive scalar curvature? A few very partial results on this problem were given in [**R2**] and [**R3**], and some further cases were studied in [**KS1**] and [**KS2**]. Extrapolating from these and other cases, I would like to make here a somewhat audacious but intuitively appealing conjecture:

CONJECTURE 0.1. *A closed manifold M^n with finite fundamental group admits a metric of positive scalar curvature if and only if all (KO_*-valued) index obstructions associated to Dirac operators with coefficients in flat bundles (on M and it covers) vanish, at least if $n \geq 5$.*

The rest of this paper will be devoted to explaining exactly what are the obstructions described in the Conjecture, and to proving that the Conjecture is valid in many cases. As explained in [**GL2**] and in [**R2**], the problem naturally splits into two cases, depending on whether or not $w_2(\tilde{M})$, where $\tilde{M}$ is the universal cover of M, vanishes. If $w_2(\tilde{M}) \neq 0$, so that $\tilde{M}$ (and *a fortiori* M) doesn't admit a spin structure, then there are no Dirac operators with coefficients in flat bundles defined on M or on any of its covers. Thus the Conjecture reduces to:

CONJECTURE 0.2. *If M^n is a closed connected manifold with finite fundamental group π, and if $w_2(\tilde{M}) \neq 0$ and $n \geq 5$, then M admits a metric of positive scalar curvature.*

Section 1 will be devoted to the proof of an interesting case of Conjecture 0.2. I would like to thank the referee for some corrections to the proofs and improvements in the exposition. By the way, the condition in Conjecture 0.2 that π be finite cannot be

Partially supported by NSF Grants DMS–8400900 and DMS–8700551. This paper is in final form and is not merely an announcement of work to appear elsewhere.

omitted in general, as shown by the example in [GL3, p. 186] of $\mathbf{CP}^2\#T^4$. (The reader concerned about the fact that this example has the exceptional dimension 4 can easly replace it by $(\mathbf{CP}^2 \times S^2)\#T^6$.)

The rest of the paper, §§2 and 3, will deal with the spin case, that is, the case where $w_2(M) = 0$. Section 2 actually involves no geometry, only pure algebraic topology and algebra, and may have some independent interest because of its parallels to known results about assembly maps in L-theory and algebraic K-theory. Theorem 2.5 was proved during a visit to Århus in 1985, and I would like to thank Ib Madsen and Gunnar Carlsson for helpful suggestions at that time.

The concluding section, §3, returns to the problem of positive scalar curvature. Here Conjecture 0.1 is restated in the spin case, using the language of §2, and we give some evidence for the Conjecture in the "hard case" of spin manifolds with fundamental groups of *even* order. We also briefly indicate how to interpret the Conjecture when $w_2(\tilde{M}) = 0$ but $w_2(M) \neq 0$, though there are substantial technical difficulties in getting any good results for this case.

§1. Positive scalar curvature when the universal cover is non-spin

The object of this section is to give some evidence for Conjecture 0.2 above. In fact, this conjecture was proved in [**R2**, Theorem 2.14] in the case where π is cyclic of odd order, and this result was strengthened in [**KS1**] to cover the case of any group of odd order with periodic cohomology (or equivalently, with all Sylow subgroups cyclic). One of the technical advances in [**KS1**] was Corollary 1.6 of that paper, which showed that the conjecture holds for a finite group π if and only if it holds for all its Sylow subgroups. However, as is clear from [**R2**], [**R3**], [**KS1**], and [**KS2**], it is much harder to prove results for even-order groups than for the odd order case. Thus the following theorem is in a way much more convincing evidence for Conjecture 0.2.

THEOREM 1.1. *If M^n is a closed orientable connected manifold with cyclic finite fundamental group π, and if $w_2(\tilde{M}) \neq 0$ and $n \geq 5$, then M admits a metric of positive scalar curvature.*

COROLLARY 1.2. *If M^n is a closed orientable connected manifold with a finite fundamental group π, all of whose Sylow subgroups are cyclic, and if $w_2(\tilde{M}) \neq 0$ and $n \geq 5$, then M admits a metric of positive scalar curvature.*

PROOF OF COROLLARY: This follows immediately from the Theorem and from [**KS1**, Proposition 1.5]. ∎

PROOF OF THEOREM: Because of the results of [**R2**] and [**KS1**] just quoted, it's enough to consider the case where our cyclic group has order a power of two. We begin with the key case where π is of order 2. By [**R2**, Theorem 2.13], it is enough to exhibit an oriented Riemannian manifold X^n of positive scalar curvature, together with a map $X^n \to \mathbf{RP}^\infty$, in every class in $\Omega_n(\mathbf{RP}^\infty)$, for all $n \geq 5$. For this we use the well-known isomorphism of [S, pp. 216–217]:

$$\Omega_n(\mathbf{RP}^\infty) \cong \Omega_n \oplus \mathfrak{N}_{n-1}.$$

The summand of Ω_n corresponds to the case where X is simply connected (or at least the map $X^n \to \mathbf{RP}^\infty$ is null-homotopic), so this case is handled by [**GL2**, Proof of Theorem C]. So it remains to deal with the summand $\mathfrak{N}_{n-1}$. Suppose Y^{n-1} represents a class in $\mathfrak{N}_{n-1}$. By the analysis in [S, pp. 216–217], the corresponding element of $\Omega_n(\mathbf{RP}^\infty)$ is represented by $f : X^n \to \mathbf{RP}^\infty$, where Y is the submanifold of X of codimension 1 which is dual to the line bundle defined by f. Note that Y doesn't determine (X, f) uniquely; however, the class of (X, f) in $\Omega_n(\mathbf{RP}^\infty)$ is determined up to an element of Ω_n (which we can "subtract off" by what we already know). Now given the manifold Y^{n-1}, if Y is orientable, we can simply orient Y and take $X = Y \times S^1$, with f factoring through S^1 and inducing a surjection on π_1. If Y has a metric of positive scalar curvature, we can give X a product metric, and then X will have positive scalar curvature as well. So suppose Y is not orientable, and let $\tilde{Y}$ be its orientable double cover, which carries a canonical orientation-reversing involution τ. Let σ be the orientation-reversing involution on S^1 defined by complex conjugation on the unit circle in $\mathbf{C}$. Then $\tau \times \sigma$ is an orientation-**preserving** involution on $\tilde{Y} \times S^1$, so $X = (\tilde{Y} \times S^1)/(\tau \times \sigma)$ can be oriented. Furthermore, there is a map $\pi_1(X) \to \mathbf{Z}/2$, and thus a map $f : X \to \mathbf{RP}^\infty$, associated to this construction of X, for which Y is the dual submanifold. Finally, if Y has a metric of positive scalar curvature, we lift the metric to $\tilde{Y}$ and give $\tilde{Y} \times S^1$ the product metric, and this descends to a metric of positive scalar curvature on X.

Hence to complete the proof for the case where π has order 2, it will suffice to construct additive generators with positive scalar curvature for $\mathfrak{N}_n$, for all $n \geq 4$. In fact, since the property of positive scalar curvature is preserved under taking products, it's in fact enough to find **multiplicative** generators for $\mathfrak{N}_*$ with positive scalar curvature. But by the structure theory for unoriented bordism (see for instance [S, pp. 96–98]), $\mathfrak{N}_*$ is a polynomial algebra over the field $\mathbf{F}_2$ of two elements, with generators represented by even-dimensional real projective spaces and by hypersurfaces of degree $(1, 1)$ in products of pairs of real projective spaces. These manifolds all have natural metrics of positive scalar curvature (cf. [**GL2**, p. 43]), so this completes the first part of the proof.

Now we have to go on to the case where the order of π is any positive power of 2. The key fact we need, which is proved in [S, pp. 209–212 and 233–236], is that the oriented bordism spectrum is Eilenberg-MacLane at 2, and thus that for π a 2-group, the Atiyah-Hirzebruch spectral sequence

$$H_*(\pi, \Omega_*) \Longrightarrow \Omega_*(B\pi)$$

collapses, and

$$\Omega_n(B\pi) \cong \oplus_{p+q=n} H_p(B\pi, \Omega_q). \tag{1.3}$$

Note that the natural map $\Omega_n(B\pi) \to H_n(B\pi, \mathbf{Z})$ corresponds to projection onto the $(p = n,\ q = 0)$ summand.

In order to facilitate future improvements of Theorem 1.1, we first prove the following:

LEMMA 1.4. *Let π be a finite 2-group and let M be a closed connected oriented n-manifold with fundamental group π such that $w_2(\tilde{M}) \neq 0$, $n \geq 5$, and the bordism class of M maps to zero in $H_n(B\pi, \mathbf{Z})$. Then M admits a metric of positive scalar curvature.*

PROOF OF LEMMA: We need to produce enough manifolds of positive scalar curvature to generate the summands in (1.3) other than the ($p = n$, $q = 0$) summand. These are of two types, copies of $H_p(B\pi, \mathbf{Z})$ in bidegrees (p, q) with $q \geq 4$ divisible by 4, and copies of $H_p(B\pi, \mathbf{Z}/2)$ in bidegrees (p, q) for which Ω_q contains a $\mathbf{Z}/2$ summand.

The summands of the first type are no problem, since they correspond to oriented bordism classes (over $B\pi$) of the form

$$N^p \times Y^{4t} \xrightarrow{\phi} B\pi,$$

where ϕ only depends on the first coordinate, where Y^{4t} is a generator for a torsion-free summand in Ω_{4t}, where $N^p \xrightarrow{\phi} B\pi$ generates a cyclic summand in $H_p(B\pi, \mathbf{Z})$, and where $p + 4t = n \geq 5$. Since $t \geq 1$, then by [GL2, Theorem C], we may choose Y^{4t} to have positive scalar curvature, and then so does $N^p \times Y^{4t}$ for suitable product metric.

Consider now the summands of $\Omega_n(B\pi)$ coming from $H_*(\pi, \mathbf{Z}/2)$. If a class in $H_*(\pi, \mathbf{Z}/2)$ is the reduction of an integral class, it can be realized by some $N^p \xrightarrow{\phi} B\pi$ with N^p a closed oriented p-manifold, and as before, the corresponding classes in $\Omega_*(B\pi)$ are represented by $N^p \times Y \xrightarrow{\phi} B\pi$, where ϕ only depends on the first coordinate, where Y is a closed oriented manifold giving a 2-torsion summand in Ω_*. Since all such Y's can be chosen to admit metrics of positive scalar curvature [GL2, Theorem C], so can $N \times Y$. So it remains to deal with classes in $H_*(\pi, \mathbf{Z}/2)$ which are **not** reductions of integral classes. Such classes only occur in even degree and **cannot** be represented by oriented manifolds mapping into $B\pi$. They can, however, be represented by **non-orientable** manifolds, since $\mathfrak{N}_*(B\pi)$ surjects onto $H_*(\pi, \mathbf{Z}/2)$. Thus consider a class in $\Omega_*(B\pi)$ corresponding to $\phi_*([N]) \times Y$, where $N^p \xrightarrow{\phi} B\pi$, N is non-orientable, $[N]$ is its $\mathbf{Z}/2$-fundamental class, and Y is an orientable manifold giving a $\mathbf{Z}/2$-torsion class in Ω_*. Fortunately, we can construct an oriented manifold mapping into $B\pi$ and defining the same bordism class.

Namely, observe that the metrics of positive scalar curvature on the standard generators of the torsion classes in Ω_*, the Dold manifolds appearing in the proof of [GL2, Theorem C], admit orientation-reversing (not necessarily free) involutions. If we choose such an involution σ' on Y and let σ be the orientation-reversing free involution on the oriented double cover $\tilde{N}$ of N, then $\sigma \times \sigma'$ is free and orientation-preserving, and we have a fibration

$$Y \to (\tilde{N} \times Y)/(\sigma \times \sigma') \to N.$$

The composite $(\tilde{N} \times Y)/(\sigma \times \sigma') \to N \xrightarrow{\phi} B\pi$ now represents our class in $\Omega_*(B\pi)$ by an oriented manifold of positive scalar curvature. This completes the proof. ∎

PROOF OF THEOREM 1.1, CONTINUED: Suppose now that π is a cyclic 2-group. By the lemma, it's enough to exhibit an oriented manifold of positive scalar curvature corresponding to each cyclic summand in $H_*(\pi, \mathbf{Z})$. But lens spaces obviously do the trick. ∎

In fact we can improve Corollary 1.2 considerably by allowing a much greater variety of Sylow 2-subgroups. The following two theorems give sample results along these lines.

THEOREM 1.5. *If M^n is a closed orientable connected manifold with fundamental group $\pi = Q$, the quaternion group of order 8, and if $w_2(\tilde{M}) \neq 0$ and $n \geq 5$, then M admits a metric of positive scalar curvature.*

PROOF: By Lemma 1.4, it is enough to exhibit an oriented Riemannian manifold X^n of positive scalar curvature, together with a map $X^n \to BQ$, in every class in $H_n(Q, \mathbf{Z})$, for all $n \geq 5$. So we only have to worry about the case of manifolds of the form $N^n \xrightarrow{\phi} BQ$ generating a cyclic summand in $H_n(Q, \mathbf{Z})$. By [CE, pp. 253–254], such summands occur only for n odd. If $n \equiv 3 \pmod 4$, there is only one such summand, generated by a quaternionic lens space, which can be given a metric of constant positive sectional curvature. If $n \equiv 1 \pmod 4$, there are two such summands, each of order 2, and since one can be taken to the other by an automorphism of Q, we only have to worry about one of them. Such a summand is represented by a submanifold of codimension 2 in a quaternionic lens space S^{4n-1}/Q, dual to a flat complex line bundle. Note that $Q \lhd H$, where H is the normalizer of a maximal torus in $SU(2)$, which also acts freely on S^{4n-1}, and that $H/Q \cong S^1$. Thus we have a fibration

$$S^1 \to S^{4n-1}/Q \to S^{4n-1}/H = \mathbf{C}\mathrm{P}^{2n-1}/(\mathbf{Z}/2),$$

and it's easy to see that the appropriate flat line bundle on S^{4n-1}/Q is pulled back from the quotient. Thus N^{4n-3} projects onto the submanifold R of $\mathbf{C}\mathrm{P}^{2n-1}/(\mathbf{Z}/2)$ dual to the non-trivial flat line bundle on this manifold. By [BB, Theorem C], it's enough to show that R admits a metric of positive scalar curvature. But a little calculation shows that R is a homogeneous manifold for a "large" compact Lie group (its double cover is a homogeneous complex quadric hypersurface in $\mathbf{C}\mathrm{P}^{2n-1}$) and so this is easy to check (it even has a metric with non-negative sectional curvature). This completes the proof. ∎

THEOREM 1.6. *If M^n is a closed orientable connected manifold with fundamental group π a product of $k \leq 4$ cyclic groups of order 2, and if $w_2(\tilde{M}) \neq 0$ and $n \geq 5$, then M admits a metric of positive scalar curvature.*

PROOF: The proof proceeds like that of Theorem 1.5. The same reasoning will work, provided we can realize every additive generator in $H_n((\mathbf{Z}/2)^k, \mathbf{Z})$, $n \geq 5$, by an orientable manifold of positive scalar curvature mapping into $B\pi = (\mathbf{R}\mathrm{P}^\infty)^k$. However, these classes are all represented by products of either odd-dimensional real projective spaces (which are orientable) or else manifolds of the form $(S^{2m} \times S^{2j})/(\sigma \times \sigma')$, where σ and σ' are the antipodal involutions on even-dimensional spheres. (The latter represent the Tor terms in the Künneth formula.) In any event, these manifolds all have non-negative sectional curvature, and zero curvature only occurs in the case of a torus $(S^1)^k$, which can't have the requisite dimension if $k \leq 4$. (Compare [**R3**, Theorem 3.6].) ∎

§2. THE *KO*-ASSEMBLY MAP

For applications to the positive scalar curvature problem in §3, we need now to examine in some detail the "assembly map" $\beta : KO_*(B\pi) \to KO_*(C^*_{\mathbf{R}}(\pi))$ introduced in [K] (in the complex case) and in [R3, §2]. Here π is any group with the discrete topology (for the moment not necessarily finite), $B\pi$ is a $K(\pi, 1)$-space, and $C^*_{\mathbf{R}}(\pi)$ is some C*-completion of the real group ring $\mathbf{R}\pi$. We do not need to worry about which C*-completion is to be used, though in practice the usual choice would be the reduced C*-algebra, that is, the completion of the group ring in the operator norm for its action on $\ell^2(\pi)$ by convolution. Since the space $B\pi$ will rarely be a finite complex, the *KO*-homology groups $KO_*(B\pi)$ are to be interpreted as what Kasparov called $RKO_*(B\pi)$, that is, as the inductive limit

$$\varinjlim_{X \hookrightarrow B\pi} KO_*(X),$$

where X runs over the finite subcomplexes of $B\pi$.

The first result of this section, for which we don't claim any originality (in fact the theorem is known to most workers in the subject, though it seems never to have been stated anywhere in print) identifies Kasparov's β with a map with a homotopy-theoretic construction similar to that used in [L]. Recall that from the point of view of a homotopy theorist, we may also identify $KO_*(B\pi)$ with $\pi_*(B\pi^+ \wedge (BO \times \mathbf{Z}))$, since $BO \times \mathbf{Z}$ is the classifying space for real K-theory. Here the $^+$ means that a disjoint basepoint is to be added—this is to avoid getting the reduced homology groups. (More accurately, the homotopy groups here are those of spectra, and we use the usual periodic spectrum whose zeroth space is the infinite loop space $BO \times \mathbf{Z}$.) Similarly, $KO_*(C^*_{\mathbf{R}}(\pi)) \cong \pi_*(BO(C^*_{\mathbf{R}}(\pi)) \times KO_0(C^*_{\mathbf{R}}(\pi)))$. Thus to construct an assembly map between these homotopy groups, it suffices to construct a map of spaces (or of spectra)

$$\gamma : B\pi^+ \wedge (BO \times \mathbf{Z}) \to BO(C^*_{\mathbf{R}}(\pi)) \times KO_0(C^*_{\mathbf{R}}(\pi)). \tag{2.1}$$

The actual assembly map itself will then be the induced map γ_* on homotopy groups.

THEOREM 2.2 (FOLKLORE). *The map $\beta : KO_*(B\pi) \to KO_*(C^*_{\mathbf{R}}(\pi))$ introduced by Kasparov coincides with the assembly map γ_*, where γ as in (2.1) is constructed as the composite*

$$\mu \circ (B\iota \wedge id_{BO\times\mathbf{Z}}).$$

Here ι is the inclusion $\pi \hookrightarrow O(C^*_{\mathbf{R}}(\pi))$ and

$$\mu : (BO(C^*_{\mathbf{R}}(\pi)) \times KO_0(C^*_{\mathbf{R}}(\pi))) \wedge (BO \times \mathbf{Z}) \to BO(C^*_{\mathbf{R}}(\pi)) \times KO_0(C^*_{\mathbf{R}}(\pi))$$

is the multiplication map corresponding to the action of $KO_*(\mathbf{R})$ on $KO_*(C^*_{\mathbf{R}}(\pi))$.

PROOF: Let's go back to the original definition of Kasparov's map. For convenience set $A = C^*_{\mathbf{R}}(\pi)$. There is a canonical flat A-line bundle $\mathcal{V}_{B\pi}$ on $B\pi$ defined as $E\pi \times_\pi A$, and given $X \hookrightarrow B\pi$, this pulls back to an A-line bundle $\mathcal{V}_X$ on X, which has a class $[\mathcal{V}_X]$ in $KO^0(X; A)$. The map β is obtained upon passage to the limit over X from the slant (or Kasparov) product

$$KO_*(X) \xrightarrow{\otimes_X [\mathcal{V}_X]} KO_*(A).$$

On the other hand, products in homology and cohomology theories, such as this slant product pairing, come homotopy-theoretically from pairings of spectra. Thus given a class $x \in KO^0(X; A)$, it corresponds to the homotopy class of the classifying map $f_x : X \to (BO(A) \times KO_0(A))$ (or of the corresponding pointed map on X^+). The associated pairing

$$(2.3) \qquad KO_*(X) \xrightarrow{\otimes x f_x} KO_*(A)$$

is then given by δ_*, where δ is the composite

$$X^+ \wedge (BO \times \mathbf{Z}) \xrightarrow{f_x \wedge id} (BO(A) \times KO_0(A)) \wedge (BO \times \mathbf{Z}) \xrightarrow{\mu} (BO(A) \times KO_0(A)).$$

We apply this with $x = [\mathcal{V}_X]$, for which the classifying map f_x is clearly just the composite

$$X \hookrightarrow B\pi \xrightarrow{B\iota} (BO(A) \times KO_0(A)).$$

The result now follows on taking the limit over X. ∎

We proceed now to compute β explicitly in the case where π is a finite group. In this case, the real group ring $\mathbf{R}\pi$ is finite-dimensional, and although there are many different Banach algebra norms on this algebra, they are all equivalent and give the same K-theory. Furthermore, by Maschke's Theorem, $\mathbf{R}\pi$ is semisimple; hence by Wedderburn theory, this algebra is a finite direct sum of matrix algebras over $\mathbf{R}$, $\mathbf{C}$, and $\mathbf{H}$ (the reals, complexes, and quaternions). There is one summand of given type for each irreducible representation of π of the same type.

Since K-theory is invariant under Morita equivalence, we instantly deduce:

LEMMA 2.4. *For π a finite group,*

$$KO_*(C^*_{\mathbf{R}}(\pi)) = KO_*(\mathbf{R}\pi) \cong (\oplus_r KO_*(pt)) \oplus (\oplus_c K_*(pt)) \oplus (\oplus_h KSp_*(pt)).$$

Here r, c, and h are the numbers of irreducible representations of π of real, complex, and quaternionic type (respectively). ∎

Note that it follows that all torsion-free summands in $KO_*(\mathbf{R}\pi)$ occur in even degree (in fact divisible by 4, except for summands associated to representations of complex type), and that all torsion is of order 2 and occurs in degrees 1 and 2 (mod 8) (if coming from representations of real type) and in degrees 5 and 6 (mod 8) (if coming from representations of quaternionic type). On the other hand, for π finite, $\widetilde{KO}_*(B\pi)$ consists **entirely** of torsion, and thus its image under β can hit **only** the $\mathbf{Z}/2$ summands in degrees 1, 2, 5, or 6 (mod 8). The following theorem now completely describes the map.

THEOREM 2.5. *If π is a finite 2-group, the map $\beta : KO_*(B\pi) \to KO_*(\mathbf{R}\pi)$ gives a split surjection onto each $\mathbf{Z}/2$ summand on the right. If π is a general finite group, the image of β consists exactly of the $KO_*(pt)$ summand corresponding to the trivial representation, plus the image under the map induced by the inclusion of the Sylow 2-subgroup π_2 of all torsion in $KO_*(\pi_2)$.*

PROOF: First note that we can reduce immediately to the case of a 2-group, because of the fact that all torsion in $KO_*(\mathbf{R}\pi)$ is of order 2 and because of commutativity of the diagram

$$\begin{array}{ccc} KO_*(B\pi_2) & \xrightarrow{\beta_{\pi_2}} & KO_*(\mathbf{R}\pi_2) \\ {\scriptstyle i_*}\downarrow & & {\scriptstyle i_*}\downarrow \\ KO_*(B\pi) & \xrightarrow{\beta_\pi} & KO_*(\mathbf{R}\pi) \end{array}$$

(cf. [**R1**, proof of Proposition 2.7]) and the fact that the map on the left is a split epimorphism when localized at the prime 2 (with the transfer as a splitting).

Thus suppose π is a 2-group. The idea is to use the results of [**AS**], which describe $KO^*(B\pi)$ (and in fact the pro-ring $\{KO^*(X) : X \hookrightarrow B\pi\}$), together with the universal coefficient theorem for KO, due to Yosimura [**Y**]. (The latter also works for real C*-algebras such as $\mathbf{R}\pi$—see [**MR**].) Since the Atiyah-Segal results refer to I-adic completions, where I is the augmentation ideal of the representation ring $R(\pi)$, we will also have to use the following well-known fact:

LEMMA 2.6. *If G is a p-group and I is the augmentation ideal in the representation ring $R(G)$, then $I \otimes_{\mathbf{Z}} \mathbf{Z}/p$ is nilpotent.*

PROOF OF LEMMA: This is a special case of [Se, Lemma 3.6], which asserts that if one has a split extension $S \rightarrowtail G \twoheadrightarrow P$ of a p-group P by a cyclic group S of order prime to p, then the restriction map $r : R(G) \otimes_{\mathbf{Z}} \mathbf{Z}/p \to (R(S) \otimes_{\mathbf{Z}} \mathbf{Z}/p)^P$ is surjective with kernel the nilradical of $R(G) \otimes_{\mathbf{Z}} \mathbf{Z}/p$. Take $S = \{1\}$, $P = G$; then $I(G)$ is just the kernel of $r : R(G) \to R(S)$, and so $I(G) \otimes_{\mathbf{Z}} \mathbf{Z}/p \subseteq \text{nilrad } R(G) \otimes_{\mathbf{Z}} \mathbf{Z}/p$. This of course means $I(G)$ is nilpotent mod p, as claimed. ∎

PROOF OF THEOREM (CONTINUED): We return to the case of π a 2-group. By Lemma 2.6, $I(\pi)$ acts nilpotently on $R(\pi) \otimes_{\mathbf{Z}} \mathbf{Z}/2$. Thus some power of $I(\pi)$ acts trivially on each $\mathbf{Z}/2$ summand in $KO_\pi^{-j}(pt)$ (such summands can occur for $j \equiv 1$, 2, 5, or 6 (mod 8)), and so nothing happens to these summands upon I-adic completion.

We refer to the description of $KO^*(B\pi)$ in [**AS**, p. 17] (though there is one misprint to correct—$KO_\pi^{-6}(pt) = R(\pi)/R_{\mathbf{R}}(\pi)$, not $R(\pi)/R_{\mathbf{H}}(\pi)$). Thus for $j = 1$, 2, 5, or 6 (mod 8), respectively, $KO^{-j}(B\pi)$ is gotten from I-adic completion of $KO_\pi^{-1}(pt) = R_{\mathbf{R}}(\pi)/\rho R(\pi)$, $KO_\pi^{-2}(pt) = R(\pi)/R_{\mathbf{H}}(\pi)$, $KO_\pi^{-5}(pt) = R_{\mathbf{H}}(\pi)/\eta R(\pi)$, and $KO_\pi^{-6}(pt) = R(\pi)/R_{\mathbf{R}}(\pi)$. Matching these up with the description of $KO_*(\mathbf{R}\pi)$, we see that the "dual Kasparov map"

$$\alpha : KO^{-j}(\mathbf{R}\pi) \to KO^{-j}(B\pi)$$

as described in [**K**, §6.2] is an injection on the torsion for $j \equiv 1$, 2, 5, and 6 (mod 8), and even an isomorphism of $\mathbf{F}_2$-vector spaces when $j \equiv 1$ or 5 (mod 8). (Note: As pointed out in the Remarks following [**K**, Corollary 2.15], there is a natural identification of

$KO^{-j}(\mathbf{R}\pi)$ with $KO_{\pi}^{-j}(pt)$, whereby one can identify α with a similar map studied by Atiyah and Segal.)

Now applying the universal coefficient theorem of [Y], together with the fact that $\mathbf{Q}/\mathbf{Z}$ is an injective $\mathbf{Z}$-module, gives a commutative diagram

$$\begin{array}{ccc} KSp^*(\mathbf{R}\pi;\, \mathbf{Q}/\mathbf{Z}) & \xrightarrow{\cong} & \mathrm{Hom}(KO_*(\mathbf{R}\pi),\, \mathbf{Q}/\mathbf{Z}) \\ {\scriptstyle\alpha}\downarrow & & {\scriptstyle\beta^*}\downarrow \\ KSp^*(B\pi;\, \mathbf{Q}/\mathbf{Z}) & \xrightarrow{\cong} & \mathrm{Hom}(KO_*(B\pi),\, \mathbf{Q}/\mathbf{Z}). \end{array}$$

Let $x \in KO_j(\mathbf{R}\pi)$ with $j \equiv 1$ or $5 \pmod 8$ and let $z_1, \dots, z_n$ be an F_2-basis for the torsion subgroup of $KO^{j-3}(\mathbf{R}\pi) \cong KSp^{j+1}(\mathbf{R}\pi)$. The exact sequence

$$\begin{array}{ccccccccc} 0 & \longrightarrow & KSp^j(\mathbf{R}\pi)\otimes_{\mathbf{Z}}\mathbf{Q}/\mathbf{Z} & \longrightarrow & KSp^j(\mathbf{R}\pi;\, \mathbf{Q}/\mathbf{Z}) & \xrightarrow{\cong} & \mathrm{Tor}(KSp^{j+1}(\mathbf{R}\pi),\, \mathbf{Q}/\mathbf{Z}) & \longrightarrow & 0 \\ & & \| & & & & & & \\ & & 0 & & & & & & , \end{array}$$

shows these lift to unique elements $\bar{z}_1, \dots, \bar{z}_n$ in $KSp^j(\mathbf{R}\pi;\, \mathbf{Q}/\mathbf{Z})$, each of order 2. Then $\alpha(\bar{z}_1), \dots, \alpha(\bar{z}_n)$ must be linearly independent elements in $KSp^j(B\pi;\, \mathbf{Q}/\mathbf{Z})$ (since $\alpha(z_1)$, $\dots$, $\alpha(z_n)$ are linearly independent in $KSp^{j+1}(B\pi)$ by the application of Atiyah-Segal). However,

$$KSp^j(B\pi;\, \mathbf{Q}/\mathbf{Z}) \cong \mathrm{Hom}(KO_j(B\pi),\, \mathbf{Q}/\mathbf{Z}),$$

so $\alpha(\bar{z}_1), \dots, \alpha(\bar{z}_n)$ may be viewed as homomorphisms $KO_j(B\pi) \to \mathsf{F}_2$. Since these are linearly independent, we may choose an element $y \in KO_j(B\pi)$ with

$$\langle y,\, \alpha(\bar{z}_j)\rangle = \langle x,\, \bar{z}_j\rangle \in \mathbf{Z}/2 \hookrightarrow \mathbf{Q}/\mathbf{Z},$$

and then

$$\langle \beta(y),\, \bar{z}_j\rangle = \langle x,\, \bar{z}_j\rangle \quad \text{for all } j,$$

hence $x = \beta(y)$.

This proves that for π a 2-group, β is surjective in degrees $\equiv 1$ or $5 \pmod 8$. But multiplication by the generator θ of $KO_1(pt)$ gives isomorphisms

$$KO_1(\mathbf{R}\pi) \xrightarrow{\cong} \mathrm{Tors}(KO_2(\mathbf{R}\pi)), \quad KO_5(\mathbf{R}\pi) \xrightarrow{\cong} \mathrm{Tors}(KO_6(\mathbf{R}\pi)),$$

hence since β commutes with multiplication by θ, β is also surjective onto the torsion in degrees $\equiv 2$ or $6 \pmod 8$. Finally, for the statement about splitting, recall that $KO_1(B\pi)$ and $KO_5(B\pi)$ are countable abelian torsion groups to which we can apply Pontryagin duality. For such a group H (which we give the discrete topology), the dual group $\hat{H} = \mathrm{Hom}(H,\, \mathbf{T})$ is compact and coincides with $\mathrm{Hom}(H,\, \mathbf{Q}/\mathbf{Z})$; furthermore, we can recover H as $\mathrm{Hom}_{\mathrm{cont}}(\hat{H},\, \mathbf{T})$. From the analysis above, $\mathrm{Hom}(KO_1(B\pi),\, \mathbf{Q}/\mathbf{Z})$ and $\mathrm{Hom}(KO_5(B\pi),\, \mathbf{Q}/\mathbf{Z})$ contain $\mathbf{Z}/2$ summands which map isomorphically under β^*. But direct sum decompositions of these dual groups dualize to give direct sum decompositions of $KO_1(B\pi)$ and $KO_5(B\pi)$ that give a splitting of β. ∎

REMARK 2.7.

Unfortunately, we do not know of any way to avoid mention of the map $KO_*(\mathbf{R}\pi_2) \to KO_*(\mathbf{R}\pi)$ in Theorem 2.6, since this map fail to be injective or may fail to be surjective (or both) on the torsion. For instance, if $\pi = S_3$, $\pi_2 = \mathbf{Z}/2$, the map is injective but not surjective on torsion since π has 3 irreducible representations of real type and π_2 has only 2. On the other hand, if $\pi = A_4$, then $\pi_2 \cong \mathbf{Z}/2 \times \mathbf{Z}/2$ is normal, π_2 has 4 irreducible representations of real type, but π has only 2; the map induced by the inclusion in this case is surjective on torsion but not injective.

§3. POSITIVE SCALAR CURVATURE WHEN THE UNIVERSAL COVER IS SPIN

In this final section, we use the results of §2 to return to the positive scalar curvature problem for manifolds with finite fundamental group. In a while we shall mention how to understand Conjecture 0.1 when $\tilde{M}$ has a spin structure but M does not, but first we discuss the case where it is easiest to give specific results, namely the case where M itself is a spin manifold. First we recall one of the main results of [**R3**], which explains why Theorem 2.5 above is relevant to our problem.

THEOREM 3.1 [**R3**, THEOREM 3.4]. *Let M^n be a closed Riemannian manifold with positive scalar curvature and with a spin structure s. Let $f : M \to B\pi$ be the classifying map for the universal cover of M, and let $[M, s] \in KO_n(M)$ be the KO-fundamental class defined by the spin structure. Then $\beta \circ f_*([M, s]) = 0$ in $KO_n(C^*_{\mathbf{R}}(\pi))$.*

The precise meaning of Conjecture 0.1 in the spin case is that we conjecture that the necessary condition for positive scalar curvature in Theorem 3.1 is actually sufficient. Note that there are two ways of viewing the obstructions given by Theorem 3.1. For a spin manifold with finite fundamental group π, one may think of there being a single obstruction to positive scalar curvature in $KO_n(\mathbf{R}\pi)$, or (by Lemma 2.4) of there being a whole family of obstructions, corresponding to indices of Dirac operators with coefficients in the various flat bundles parameterized by the irreducible representations of π. The content of Theorem 2.5 is that when π is a 2-group, **every** irreducible representation of real or quaternionic type gives rise to genuine obstructions to positive scalar curvature (in dimensions 1 and 2 (mod 8) in the real case, 5 and 6 (mod 8) in the quaternionic case) for manifolds with the given fundamental group. This is because every class in $KO_*(B\pi)$ can be realized by a spin manifold with the correct fundamental group, provided we jack up the dimension sufficiently using Bott periodicity.

Now we consider the positive evidence for Conjecture 0.1 in the spin case. When the manifold is simply connected, the conjecture just becomes the original conjecture of Gromov and Lawson [**GL2**] about simply connected manifolds of positive scalar curvature, which it seems has now been verified by Stolz [**Sz**]. And by [**R3**] and [**KS1**], the conjecture in the simply connected case implies the conjecture for manifolds with fundamental group of odd order, provided the Sylow subgroups of the fundamental group are all cyclic. Even more than in the non-spin case treated in §1 above, the crucial case to consider is that where the fundamental group π is a 2-group. Even the simplest case of a cyclic group of order 2 is quite hard; however, we do have the following positive result. Note that when $\pi = \mathbf{Z}/2$, M admits exactly one non-trivial flat line bundle, giving us a "twisted Dirac operator" having (by Lemma 2.4 and Theorem 2.5) a ($\mathbf{Z}/2$)-valued index which is an obstruction to positive scalar curvature in dimensions $\equiv 1$ and $2 \pmod 8$.

THEOREM 3.2. *Let M^n be a closed spin manifold with fundamental group $\mathbf{Z}/2$. If $5 \leq n \leq 15$, then Conjecture 0.1 holds for M; that is, M admits a metric of positive scalar curvature if and only if the KO_*-valued index obstructions associated to the Dirac operator on M and the twisted Dirac operator on M vanish.*

Furthermore, for arbitrary $n \geq 5$, if M^n is a spin boundary (forgetting the fundamental group) and if $[M^n] \in \tilde{\Omega}_n^{\mathrm{Spin}}(\mathbf{RP}^\infty)$ has order greater than 2, then M admits a metric of positive scalar curvature.

PROOF: We use the isomorphism

$$\Omega_n^{\mathrm{Spin}}(\mathbf{RP}^\infty) \cong \Omega_n^{\mathrm{Spin}} \oplus \tilde{\Omega}_n^{\mathrm{Spin}}(\mathbf{RP}^\infty) \cong \Omega_n^{\mathrm{Spin}} \oplus \Omega_{n-1}^{\mathrm{Pin}} \tag{3.3}$$

(the analogue of the decomposition of $\Omega_n(\mathbf{RP}^\infty)$ used in the proof of Theorem 1.1) and the results of [**ABP**] and [**G**]. We may restrict attention to the second summand, since the first summand corresponds to the simply connected case of the positive scalar curvature problem.

Let's handle the second statement first, since it will take care of much of the first statement (the "low-dimensional" case) as well. By [**G**, Corollary 3.5], the subgroup of Ω_*^{Pin} generated by elements of order greater than two is generated by products of certain spin manifolds M_J with $\mathbf{RP}^{4k+2}$'s. Under the isomorphism of (3.3), such products correspond in $\Omega_n^{\mathrm{Spin}}(\mathbf{RP}^\infty)$ with $M_J \times \mathbf{RP}^{4k+3}$ (note $\mathbf{RP}^{4k+3}$ is a spin manifold with fundamental group $\mathbf{Z}/2$). Since $\mathbf{RP}^{4k+3}$ has a metric of positive curvature, $M_J \times \mathbf{RP}^{4k+3}$ has a metric of positive scalar curvature.

Now let's go back to the case where $5 \leq n \leq 15$. [**ABP**, Theorem 5.1] gives us a precise calculation of $\tilde{\Omega}_n^{\mathrm{Spin}}(\mathbf{RP}^\infty) \cong \Omega_{n-1}^{\mathrm{Pin}}$. Consider first the summands coming from $\mathbf{BO}\langle 0\rangle$. Aside from cyclic summands of large order, which we've already handled, we have, in our range of values of n, $\mathbf{Z}/2$-summands in dimensions 9 and 10. These correspond to manifolds for which the KO_*-valued index of the twisted Dirac operator is non-zero (e.g., $M_0^8 \times S^1$ with spin surgery to reduce the fundamental group to $\mathbf{Z}/2$; here M_0^8 is a spin 8-manifold with $\hat{A}$-genus $= 1$). We've seen these manifolds do **not** have metrics of positive scalar curvature.

Next consider the remaining summands, which come from $\mathbf{BO}\langle 8\rangle$ and from $\mathbf{BO}\langle 10\rangle$. We know the cyclic summands of large order correspond to manifolds of positive scalar curvature, and the remaining $\mathbf{Z}/2$-summands occur in dimensions 9 and 10 (coming from $\mathbf{BO}\langle 8\rangle$) and in dimensions 12, 13, and 14 (coming from $\mathbf{BO}\langle 10\rangle$). We can find representatives for all of these with positive scalar curvature. For the generators in dimensions 9 and 10, one can take $\mathbf{HP}^2 \times S^1$ and $\mathbf{HP}^2 \times \bar{S}^1 \times S^1$, with suitable spin surgeries to reduce the fundamental group to $\mathbf{Z}/2$. The generators in dimensions 12, 13, and 14 can be built from a spin manifold M^{10} of positive scalar curvature (representing a $\mathbf{Z}/2$ summand in $\Omega_{10}^{\mathrm{Spin}}$) as $M^{10} \times S^1$ with spin surgery to reduce the fundamental group, as $(M^{10} \times S^2)/\sigma$, σ a suitable free involution, and as $M^{10} \times \mathbf{RP}^3$. ∎

Finally, we explain the meaning of Conjecture 0.1 in the non-spin case. If M is a manifold whose universal cover $\tilde{M}$ has a spin structure, then the sections of any flat vector bundle over M may be identified with a suitable space of vector-valued functions on $\tilde{M}$. As such, there is a Dirac operator acting on them (after tensoring with the spinor bundle on $\tilde{M}$). For some vector bundles, the Dirac operator will map this space back into itself, and thus there is an associated KO_*-valued index of the twisted Dirac operator which will be an obstruction to a metric of positive scalar curvature on M. The meaning of the Conjecture is that these should be the only obstructions. So far we have only paltry evidence for the Conjecture, but it should be possible to test it by using the 2-connected bordism class of the clasifying map $M \to B\pi$, as introduced in **[KS1]**, **[KS2]**.

REFERENCES

[ABP] D. W. Anderson, E. H. Brown, Jr., and F. P. Peterson, *Pin cobordism and related topics*, Comment. Math. Helv. **44** (1969), 462–468.

[AS] M. F. Atiyah and G. B. Segal, *Equivariant K-theory and completion*, J. Differential Geometry **3** (1969), 1–18.

[BB] L. Berard Bergery, *Scalar curvature and isometry group*, in "Proc. Franco-Japanese Seminar on Riemannian Geometry, Kyoto, 1981," to appear.

[CE] H. Cartan and S. Eilenberg, "Homological Algebra," Princeton Math. Ser., no. 19, Princeton Univ. Press, Princeton, N. J., 1956.

[G] V. Giambalvo, *Pin and Pin′ cobordism*, Proc. Amer. Math. Soc. **39** (1973), 395–401.

[GL1] M. Gromov and H. B. Lawson, Jr., *Spin and scalar curvature in the presence of a fundamental group, I*, Ann. of Math. **111** (1980), 209–230.

[GL2] ———, *The classification of simply connected manifolds of positive scalar curvature*, Ann. of Math. **111** (1980), 423–434.

[GL3] ———, *Positive scalar curvature and the Dirac operator on complete Riemannian manifolds*, Publ. Math. I.H.E.S. no. 58 (1983), 83–196.

[K] G. G. Kasparov, *Equivariant KK-theory and the Novikov Conjecture*, Invent. Math. 91 (1988), 147–201.

[KS1] S. Kwasik and R. Schultz, *Positive scalar curvature and periodic fundamental groups*, Math. Annalen (to appear).

[KS2] ———, *Positive scalar curvature and spherical spaceforms*, preprint.

[L] J.-L. Loday, *K-théorie algébrique et représentations des groupes*, Ann. Sci. École Norm. Sup. (4) **9** (1976), 309–377.

[MR] I. Madsen and J. Rosenberg, *The universal coefficient theorem for equivariant K-theory of real and complex C*-algebras*, in "Index Theory of Elliptic Operators, Foliations, and Operator Algebras," J. Kaminker, K. Millett, and C. Schochet, eds., Contemp. Math., no. 70, Amer. Math. Soc., Providence, pp. 145–173.

[Ra] A. Ranicki, "Exact Sequences in the Algebraic Theory of Surgery," Mathematical Notes, no. 26, Princeton Univ. Press, Princeton, N. J., 1981.

[R1] J. Rosenberg, *C*-algebras, positive scalar curvature, and the Novikov Conjecture*, Publ. Math. I.H.E.S. no. 58 (1983), 197–212.

[R2] ———, *C*-algebras, positive scalar curvature, and the Novikov Conjecture, II*, in "Geometric Methods in Operator Algebras," H. Araki and E. G. Effros, eds., Pitman Research Notes in Math., no. 123, Longman/Wiley, Harlow, Essex, England and New York, 1986, pp. 341–374.

[R3] ———, *C*-algebras, positive scalar curvature, and the Novikov Conjecture, III*, Topology **25** (1986), 319–336.

[Se] G. Segal, *The representation ring of a compact Lie group*, Publ. Math. I.H.E.S. no. 34 (1968), 113–128.

[Sz] S. Stolz, *Simply connected manifolds of positive scalar curvature*, preprint.

[S] R. E. Stong, "Notes on Cobordism Theory," Mathematical Notes, no. 7, Princeton Univ. Press, Princeton, N. J., 1968.

[Y] Z. I. Yosimura, *Universal coefficient sequences for cohomology theories of CW-spectra*, Osaka J. Math. **12** (1975), 305–323.

Keywords. positive scalar curvature, real K-theory, spin manifold, spin cobordism, Dirac operator, assembly map
1980 *Mathematics subject classifications*: 53C20, 55N15, 58G12, 46L80, 19L64, 19B28, 57R90

Department of Mathematics, University of Maryland, College Park, MD 20742

EQUIVARIANT SPLITTINGS ASSOCIATED WITH SMOOTH TORAL ACTIONS

Michał Sadowski
Department of Mathematics, The University of Gdańsk
80–952 Gdańsk, Wita Stwosza 57, Poland

0. Introduction. Let M be a connected manifold and let T be a torus acting smoothly on M. We consider the question when a given splitting $\Gamma_0 \cong \mathrm{im}\,\mathrm{ev}_* \times \Delta$ of a finite index subqroup Γ_0 of $\Pi_1(M)$ is induced by an equivariant diffeomorphism $\phi : \hat{M} \to T \times \hat{M}/T$, where $\hat{M}$ is an appropriate finite covering space of M. Here $\mathrm{ev}_* : \Pi_1(T) \to \Pi_1(M)$ is the homomorphism induced by $\mathrm{ev} : T \ni t \to tx \in M$. Our first result is the following.

THEOREM 1. *Let M be a connected paracompact manifold and let T be a torus acting smoothly on M. A continuous map $f : M \to T$ is homotopic to a transversally equivariant fibration $p : M \to T$ iff $f_* \circ \mathrm{ev}_*$ is a monomorphism.*

Here by a *transversally equivariant fibration* (*t-e fibration*) we mean a smooth fibration $p : M \to T = \mathbb{R}^k/\mathbb{Z}^k$ such that the orbits of the action of T are transversal to the fibers of p and $p(tx) - p(x)$ depends on $t \in T$ only. Note the a t-e fibration has always finite structure group so that M is finitely covered by $T \times M/T$ (see Proposition 2.1). Theorem 1 is a generalization of the smooth case of the main result of [6] (see [6, Theorem 4.2], [7, Theorem 7.5] see also Remark 1.2).

Applying Theorem 1 we show in particular that a smooth T-manifold M is T-equivariantly diffeomorphic to $T \times M/T$ iff there is an appropriate decomposition of $\Pi_1(M)$ (see § 1, Theorem 2). This result reduces the classification of all smooth actions of T on M such that $\mathrm{im}\,\mathrm{ev}_*$ is an appropriate subgroup of $\Pi_1(M)$ to the classification of all manifolds F such that $F \times T$ is diffeomorphic to M. It seems that Theorem 2 cannot be derived from [6, Theorem 4.2]. The topological version of Theorem 2 was proved for hypertoral manifolds only (see [19] see also [2, ch. 4, Th. 9.5]). It is false in general, because there are nonmanifolds whose Cartesian products by S^1 are topological manifolds (see [1, 17]).

Some applications of Theorem 1 and Theorem 2 are discussed in § 3 and § 4. We give a topological characterization of group actions generated by parallel fields and we extend a classic result of P. E. Conner and D. Montgomery (see e.g. [2, ch. 4, Th. 9.5]) about toral actions on tori to smooth toral actions on nilmanifolds. We also consider when a continuous map $f : M \to T$ is homotopic to a fibration whose structure group is finite.

The proof of Theorem 1 is quite different then the proof of the related result of P. E. Conner and F. Raymond ([9, 10]). We use a variant of Tischler's theorem showing that every T^k-invariant 1-form $\omega : TM \to \mathbb{R}^k$ such that the bundle $\ker\omega$ is transversal to the orbits determines a t-e fibration over T^k. This kind of argument also works for holomorphic complex torus actions on some complex manifolds. The details will appear elsewhere.

Throughout this paper the following notation will be used. By $\widetilde{Z}$ we shall denote the universal covering space of a given topological space Z. The symbol I will denote the canonical identification of $\Pi_1(Z, z_0)$ with the deck group Γ of Z. By definition $I^{-1}(\gamma)$ is the homotopy class of $\Pi \circ c$, where c is any curve in $\widetilde{Z}$ joining z_0 with $\gamma(z_0)$ and $\Pi : \widetilde{Z} \to Z$ is the canonical projection. If G is a group, $g_1, \ldots, g_k \in G$, then $\langle g_1, \ldots, g_k \rangle$ wil denote the subgroup of G generated by $g_1, \ldots, g_k$, and $Z(G)$ will denote the center of G.

I am particulary grateful to Dr J. Popko for many useful remarks which clarified the arguments. I would like to express my gratitude to Professor F. T. Farrell and to Professor R. Schultz for helpful remarks. It is a pleasure to thank to Dr Z. Szafraniec for his careful reading of the earlier version of the paper and for pointing out some errors.

1. Equivariant fibrations associated with smooth toral actions. The aim of this section is to prove Theorem 1. First we need some definitions.

DEFINITION 1.1. Let $\Phi_t : M \to M$, $t \in T^k$, be a smooth T^k action on a connected manifold M. A fibration $p : M \to T^k$ is T^k-equivariant if $p(\Phi_t(x)) - p(x) \in \mathbb{R}^k/\mathbb{Z}^k = T^k$ depends on t only.

REMARK 1.1.

a) A fibration $p : M \to T^k$ will be T^k-equivariant in the usual meaning if we take an appropriate T^k-action on T^k. To be more specific the formula

$$\rho(t)(s) = p(\Phi_t(p^{-1}(s))), \tag{1}$$

where $t \in T^k$, $s \in T^k$, defines a T^k action ρ on T^k such that

$$p(\Phi_t(x)) = \rho(t)(p(x)). \tag{1'}$$

b) As the difference $p(\Phi_t(x)) - p(x)$ depends on t only we have

$$\rho(t)(s) = p(\Phi_t(p^{-1}(s))) = p(\Phi_t(p^{-1}(s))) - p(p^{-1}(s)) + p(p^{-1}(s)) = \rho'(t) + s$$

for some homomorphism $\rho' : T^k \to T^k$.

DEFINITION 1.2. Let W be a finite dimensional vector space. A closed 1-form $\omega : TM \to W$ is regular if $\omega(x) : TM_x \to W$ is a surjection for every $x \in M$.

The foliation associated to a regular 1-form ω is a foliation $\mathfrak{F}(\omega)$, whose tangent bundle $T\mathfrak{F}(\omega)$ is equal to $\ker\omega$. The form ω determines the set $P_\omega = \{\int_u \omega : u \in H_1(M; Z)\} \subset W$ of periods of ω. Recall that a discrete subgroup B of W is a lattice if $B \approx \mathbb{Z}^k$, where $k = \dim W$. Note that $W/_B$ is a k-dimensional torus T^k and we have a canonical isomorphism $I : \Pi_1(W/_B) \approx B$.

We shall need the following variant of Tischler's theorem (compare [4], [21]).

LEMMA 1.1. *Let M be a connected manifold, let W be a k-dimensional vector space, let B be a lattice in W, let x_0 be a chosen base point of M, and let $\omega : TM \to W$ be a regular 1-form. Assumme that $P_\omega \subseteq B$. Then the map $p : M \to W/_B = T^k$, given by*

$$p(x) = \int_{x_0}^{x} \omega \bmod B \tag{2}$$

is a submersion. The leaves of $\mathfrak{F}(\omega)$ are connected components of the inverse images of p.

PROOF. The map p is well defined because $P_\omega \subseteq B$. In order to check that p is a submersion consider $\widetilde{p} : \widetilde{M} \to W$ given by $\widetilde{p}(y) = \int_{y_0}^{y} \widetilde{\omega}$, where $y_0 \in \widetilde{M}$ is a chosen point above x_0 and $\widetilde{\omega}$ is a lift of ω to $\widetilde{M}$. Then $\widetilde{p}$ is a lift of p and for every smooth curve $c : [0,1] \to \widetilde{M}$ we have

$$\begin{aligned} d\widetilde{p}\left(\frac{dc}{dt}(0)\right) &= \frac{d}{dt}\left(t \to \int_{c(0)}^{c(t)} \widetilde{\omega}\right)\Big|_{t=0} \qquad (@)\\ &= \frac{d}{dt}\left(t \to \int_0^t \widetilde{\omega}\left(\frac{dc}{dt}(s)\right) ds\right)\Big|_{t=0} = \widetilde{\omega}\left(\frac{dc}{dt}(0)\right). \end{aligned}$$

This shows that $\widetilde{p}$ is a submersion.

The bundle $\ker\omega$ is the bundle tangent to the inverse images of p so that each connected component of a given fiber of p is contained in a leaf of $\mathfrak{F}(\omega)$. If L is a leaf of $\mathfrak{F}(\omega)$ and $x, y \in L$, then $p(x) = p(y)$ so that L is a connected submanifold of $p^{-1}(p(x))$. This completes the proof of Lemma 1.1. ∎

Let $h : \Pi_1(M) \to \Pi_1(W/_B)$ be a given homomorphism. Then h determines a unique homomorphism $h_1 : H_1(M;Z) \to \Pi_1(W/_B) \xrightarrow[\approx]{I} B \subset W$ and then an element $[h]$ of $H^1(M;W)$. Note that $[h]$ is a unique element of $H^1(M;W)$ such that

$$\int_u [h] = Ih(u) \qquad \text{for every} \qquad u \in \Pi_1(M). \tag{3}$$

LEMMA 1.2. *Let M be a connected smooth T^k-manifold, let $h : \Pi_1(M) \to \Pi_1(T^k)$ be a given homomorphism such that $h \circ \mathrm{ev}_*$ is a monomorphism, and let $[h]$ be the element of $H^1(M;W)$ determined by h. Let $\omega : TM \to W$ be any T^k-invariant form representing $[h]$. Then the map $p : M \to T^k$ defined in Lemma 1.1 is a T^k-equivariant fibration such that $p_* = h$.*

PROOF. Fix $x \in M$. Write T^k as $S^1_1 \times \cdots \times S^1_k$. Let X_j denote the vector field generating the action of S^1_j on M, let $c : [0,1] \to M$ be the S^1_j orbit of x, and let $\sigma_j \in \Pi_1(M,x)$ be the homotopy class of c. Since $\frac{dc}{dt}(c(t)) = X_j(c(t))$ we have

$$Ih(\sigma_j) = \int_c \omega = \int_0^1 \omega(X_j) = \omega(X_j(x)), \tag{*}$$

because X_j and ω are S^1_j-invariant.

As rank im $h \circ \mathrm{ev}_* = k$, the vectors $\omega(X_1(x)), \ldots, \omega(X_k(x))$ are linearly independent. Hence $\dim \omega(TM_x) \geq k$ for every $x \in M$. Let $p : M \to T^k$ be the submersion defined by Lemma 1.1. If F is a connected component of $p^{-1}(1)$, $x, y \in M$, then $x \in \phi_t(F)$, $y \in \phi_s(F)$ for some $t, s \in T^k$. By the T^k-invariance of ω we have $Cd(t,s) \leq d(p(x),p(y)) = d(p(\phi_t(F)), p(\phi_s(F))) \leq C'd(t,s)$ for some positive constants C, C'. Here d denotes an euclidian distance in T^k. Using this it is easy to see that p is a smooth fibration. The T^k-equivariance of p follows form the T^k-invariance of ω.

We show that $p_* = h$. Take $\gamma \in \Pi_1(M)$. Let $c : [0,1] \to M$ be any curve representing γ and let $\tilde{c} : [0,1] \to \widetilde{M}$ denote the lift of c to the universal covering space. As in the proof of Lemma 1.1 consider the fibration $\tilde{p} : \widetilde{M} \to \mathsf{R}$ covering p and the form $\tilde{\omega}$ covering ω. Observe that $\tilde{c}(1) = \gamma''(\tilde{c}(0))$, where γ'' is the deck transformation corresponding to γ. Hence

$$Ip_*(\gamma) = \tilde{p}(\tilde{c}(1)) - \tilde{p}(\tilde{c}(0)) = \int_0^1 \tilde{\omega}\left(\frac{d\tilde{c}}{dt}\right) dt = \int_0^1 \omega\left(\frac{dc}{dt}\right) dt = \int_c \omega = \int_\gamma [h] = Ih(\gamma)$$

and $p_* = h$ as claimed. This completes the proof of Lemma 1.2. ∎

LEMMA 1.3. *Let $p : M \to T^k$ be a smooth T^k-equvariant fibration. The following conditions are equivalent:*

(i) *The orbits of the T^k action are transversal to the fibers of p.*

(ii) *$p_* \circ \mathrm{ev}_* : \Pi_1(T^k) \to \Pi_1(T^k)$ is a monomorphism.*

(iii) *One orbit of the T^k action is transversal to a fiber of p at a point $x_0 \in M$.*

PROOF. (iii)⇒(ii). Let $\sigma \in \mathrm{im}\,[\mathrm{ev}_* : \Pi_1(T^k, 1) \to \Pi_1(M, x_0)] - \{1\}$. Then σ can be represented by the orbit $c(t)$, $t \in [0,1]$, $c(0) = x_0$ of an S^1 action embedded into the T^k action. Let Θ be an invariant 1-form on T^k. Then by the S^1-invariance of dc/dt and $p^*\Theta$,

$$\int_{p_*(\sigma)} \Theta = \int_c p^*\Theta = \int_0^1 p^*\Theta\left(\frac{dc}{dt}\right) dt = p^*\Theta\left(\frac{dc}{dt}(0)\right) = \Theta\left(dp\left(\frac{dc}{dt}(0)\right)\right).$$

As $c(0) = x_0$ we have $dp(\frac{dc}{dt}(0)) \neq 0$ so that $\Theta\big(dp(\frac{dc}{dt}(0))\big) \neq 0$ for some invariant 1-form Θ. Hence $p_*(\sigma) \neq 0$ as claimed.

(ii)⇒(i). Assume the contrary that there is a vector $v \in TM_x$ that is tangent to the orbit and to the fiber simultanously. Let $\lambda_t : M \to M$, $t \in \mathsf{R}$, be the flow embedded into the T^k action that is tangent to v at x. By the λ_t-invariance of $d\lambda_t(v)$ and the bundle T_F tangent to the fibers of p we have $d\lambda_t(v) \in T_F$ for $t \in \mathsf{R}$. As the orbit $\lambda_t(x)$, $t \in \mathsf{R}$, is the integral curve of $d\lambda_t(v)$, $t \in \mathsf{R}$, this orbit is contained in $p^{-1}(p(x))$.

Let $G(x) = \{t \in T^k : \Phi_t(p^{-1}(p(x))) = p^{-1}(p(x))\}$. Then $\dim G(x) \geq 1$. Since $G(x)$ is a compact subgroup of T^k, $G(x)$ contains a subgroup isomorphic to S^1 so that $p_* \circ \mathrm{ev}_*$ cannot be a monomorphism. This is a contradiction.

PROOF of Theorem 1. Assume that $f : M \to T$ is a continuous map such that $f_* \circ \mathrm{ev}_*$ is a monomorphism. By Lemma 1.2 there is a T-equivariant fibration $p : M \to T$ such that $p_* = f_*$. The last equality shows that p and f are homotopic (compare [20, ch. 8, § 2, Theorem 15]). According to Lemma 1.3 the fibers of p are transversal to the orbits of the action of T. The reverse implication follows form Lemma 1.3. This finishes the proof of Theorem 1. ∎

REMARK 1.2. If the group $H_1(M;Z)$ is finitely generated and the action of T on M is homologically injective then there is a map $f : M \to T$ such that $f_* \circ \mathrm{ev}_*$ is a monomorphism so that Theorem 1 implies the smooth variant of the Conner Raymond theorem ([6, Theorem 4.2]). Recall that an action of T on M is *homologically injective* if $\mathrm{ev}_* : H_1(T;\mathbb{Z}) \to H_1(M;\mathbb{Z})$ is a monomorphism.

REMARK 1.3. The fibrations constructed in [6, 7] are not associated with the homomorphisms form $\Pi_1(M)$ to $\mathbb{Z}$ but with the elements of $H^1(G\,;\,\mathrm{Maps}(Y,T))$, where $G \to Y \times T \to M$ is a suitably chosen covering of M. This looks complicated and the results of [6, 7] assert the existence of a T-equivariant fibration on a given homologically injective T-manifold only. These results does not allow to derive Theorem 2 and Proposition 2.1 directly from them.

2. Cartesian product decompositions induced by smooth toral actions.

First we prove the following .

PROPOSITION 2.1. *If $p : M \to T$ is a t-e fibration then the structure group of p can be reduced to the finite group $A = \Pi_1(T)/\mathrm{im}\,(p_* \circ \mathrm{ev}_*)$.*

PROOF. Let $\Phi_t : M \to M$, $t \in T$, be our T action on M and let F be a fiber of p. Consider $G = \{g \in T : \Phi_g(F) = F\}$. Then G is a closed discrete subgroup of T. It is easy to see that $\Psi : F \times T \to M$, given by $\Psi(u,t) = \Phi_t(u)$ is a covering and the group of covering transformations of Ψ is isomorphic to G.

We show that there is a monomorphism $\lambda : G \to A$. Fix $x_0 \in F$. Let $g \in G$ and let c_g be the piece of the orbit $t \to \Phi_t(x_0)$ joining x_0 with $\Phi_g(x_0)$, let $[p \circ c_g]$ be the homotopy class of $p \circ c_g$ in $\Pi_1(T)$ and let $\lambda(g)$ be the class of $[p \circ c_g]$ in A. It is easy to see that $[p \circ c_g] \in \mathrm{im}\,(p_* \circ \mathrm{ev}_*)$ iff $g = 1$. The group A is finite by Lemma 1.3. ∎

THEOREM 2. *Let M be a smooth T^k manifold.*

a) *The manifold M is T^k-equivariantly diffeomorphic to $(M/T^k) \times T^k$ (where the T^k action on $(M/T^k) \times T^k$ is given by $(t,v,s) \to (v,st)$ for every $v \in M/T^k$, $t,s \in T^k$) if and only if* $\mathrm{im}\,\mathrm{ev}_* \approx Z^k$ *and there is a direct sum decomposition* $\Pi_1(M) \approx \Delta \times \mathrm{im}\,\mathrm{ev}_*$.

b) *If M is an aspherical manifold, then M is T^k-equivariantly diffomorphic to $(M/T^k) \times T^k$ (where T^k acts on T^k by translations) iff* $\mathrm{im}\,\mathrm{ev}_* \approx \mathbb{Z}^k$ *and there is a direct sum decomposition* $\Pi_1(M) = \Delta \times \Lambda$ *such that* $\mathrm{im}\,\mathrm{ev}_*$ *is a finite index subgroup of* Λ.

We say that a torus T^k acts on T^k by translations if there is a homomorphism $\lambda : T^k \to T^k$ such that the T^k action can be written as $(t,x) \to \lambda(t)x$. A topological space is aspherical if its universal covering space is contractible.

PROOF of Theorem 2. We show a) first. The necessity is clear. Assume that $\pi_1(M) = \mathrm{im}\,\mathrm{ev}_* \times \Delta$. Let $\pi : \Pi_1(M) \to \mathrm{im}\,\mathrm{ev}_*$ be the associated projection and let $h = \mathrm{ev}_*^{-1} \circ \pi : \Pi_1(M) \to \Pi_1(T^k)$ (note that $\mathrm{ev}_*^{-1} : \mathrm{im}\,\mathrm{ev}_* \to \Pi_1(T^k)$ is well defined because we have assumed that ev_* is a monomorphism). By Theorem 1 there is a t-e fibration p such that $p_* = h$. We have $\Pi_1(T^k)/\mathrm{im}\,(p_* \circ \mathrm{ev}_*) = \{1\}$ so that the structure group of p is trivial and $M = F \times T^k$, where F is a fiber of p.

b) If M is an aspherical manifold and im $\mathrm{ev}_* \not\approx \Lambda$ then the action of T^k is not effective (see [8, Lemma 1]). Hence there is a new action of T^k having the same orbits and satisfying im $\mathrm{ev}_* = \Lambda$. Using Theorem 2.a), the conclusion follows. ∎

REMARK 2.1. The assumption that M is an aspherical manifold can be replaced by the assumption that M is a K-manifold because Lemma 1 from [8] can be extended to K-manifolds (see [18]). A closed manifold M is a K-manifold if there is a torsion free group H and a map $f : M \to K(H,1)$ such that $f^* : H^m(K(H,1);\mathbf{Z}) \to H^m(M;\mathbf{Z})$ is onto, where $m = \dim M$.

Proposition 2.1. implies the following.

COROLLARY 2.1. *Let $p : M \to T^k$ be a fibration. The following two conditions are equivalent.*

(i) *The structure group of p can be reduced to a finite group.*
(ii) *The fibration p is transversally equivariant with respect to some T^k action on M.*

PROOF. By Proposition 2.1 it suffices to show that (i)⇒(ii). We can assume that the fibers of p are connected. If the structure group of p can be reduced to a finite group G, then this group can be reduced to $\mathbf{Z}_{r_1} \oplus \cdots \oplus \mathbf{Z}_{r_k}$ for some $r_1, \ldots, r_k$, because the image of every homomorphism $\Pi_1(T^k) \to \Pi_1(BG) \cong G$ is a finite abelian group. Let $k = 1$. By the above M can be written as $(F \times I)/\big((x,0) \sim (\alpha(x),1)\big)$, for some diffeomorphism α of a finite order r. Under this identification $p([u,t]) = [t]$, where $[t]$ is the class of $t \in I$ in S^1 and $[u,t]$ is the class of $(u,t) \in F \times I$ in M (compare [10, ch. 12, p. 121]). The formula $\Phi_t([u,s]) = [u, s+rt]$, defines the desired S^1 action on M. The proof of the general case is similar. The details are left to the reader. ∎

3. Some applications. First consider some special properties of toral actions on nilmanifolds. *A nilmanifold* is an orbit space $M = G/_\Gamma$, where G is a nilpotent, connected, simply connected Lie group and Γ is a discrete subgroup of G acting on G by right translations.

The following result describes smooth homologically injective toral actions on nilmanifolds.

PROPOSITION 3.1. *Let M be a closed nilmanifold. Assume that the T^k action on M is smooth and homologically injective. Then M is T^k equivariantly diffeomorphic to $(M/T^k) \times T^k$.*

Before proving Proposition 3.1 let us recall some special properties of nilpotent Lie groups. If G is a nilpotent, connected, simply connected Lie group and K is a subgroup of G, then K generates a Lie subgroup $G(K)$ of G. By definition $G(K)$ is the set of all $\gamma_1^{t_1}, \ldots, \gamma_r^{t_r}$, where $\gamma_1, \ldots, \gamma_r \in K$, $t_1, \ldots, t_r \in \mathbf{R}$, and where $t \to \gamma^t$ is the one-parameter subgroup of G such that $\gamma^1 = \gamma$. If A is a discrete subgroup of G such that $G/_A$ is a compact manifold, then there is a basis $\alpha_1, \ldots, \alpha_n$ of A such that every element $x \in G$ can be written uniquely as $\alpha_1^{t_1} \ldots \alpha_n^{t_n}$ for some $t_1, \ldots, t_n \in \mathbf{R}$ ([3, Proposition 5.1.8], [15, § 2, Lemma 4]). In particular $G(A) = G(A') = G$ if A' is a finite index subgroup of A.

PROOF. It is enough to consider effective S^1 actions only. Assume that $M = G/_\Gamma$ and that $\Phi_t : M \to M$, $t \in [0,1]$, is a smooth homologically injective S^1 action on M. Let $H_1(\Gamma) = \Gamma/[\Gamma,\Gamma] \approx H_1(M)$, let $\Pi : \Gamma \to H_1(\Gamma)$ be the canonical projection, let $\sigma = \widetilde{\Phi}_1$, and let $\sigma_0 = \Pi(\sigma)$. Take a direct sum decomposition $H_1(\Gamma) = H \oplus \langle \delta_0 \rangle$

such that $\sigma_0 = \delta_0^r$ for some $r \in Z$. Let $\Delta = \Pi^{-1}(H)$ and let δ be a fixed element of $\Pi^{-1}(\delta_0)$. The subgroup of Γ generated by σ and Δ is isomorphic to $\Delta \times \langle\sigma\rangle$ and $G = G(\Delta) \times G(\langle\sigma\rangle)$, because $\Delta \times \langle\sigma\rangle$ is a finite index subgroup of Γ.

Let $P : G \to G(\Delta)$, $Q : G \to G(\langle\sigma\rangle)$ by the projections, let $\tau = Q(\delta)$, and let $g = \tau\delta^{-1}$. Note that $g \in \ker Q = G(\Delta)$. Let $\hat{\Delta}$ be the subgroup of $G(\Delta)$ generated by Δ and g and let $\Gamma_1 = \hat{\Delta}\cdot\langle\tau\rangle$. Then $\tau \in G(\langle\sigma\rangle) \subseteq Z(G)$, $\Gamma \subseteq \Gamma_1$, and $d^r \in \langle\sigma\rangle \times \Delta \subseteq \Gamma$ for every $d \in \Gamma_1$. Every element γ of Γ_1 can be written as $\tau^k u$ (where $k \in Z$, $u \in \hat{\Delta}$) and it is easy to check that this representation is unique. Since $\tau \in Z(G)$, $\Gamma_1 = \hat{\Delta} \times \langle\tau\rangle$.

Let $\widehat{M} = G/_{\Gamma_1}$. Then $\widehat{M}$ is finitely covred by M, the S^1 action on M induces an S^1 action on $\widehat{M}$ and, according to Theorem 2, $\widehat{M}$ is S^1-equivariantly diffeomorphic to $F \times S^1$, where F is a submanifold of $\widehat{M}$. If F_0 is a connected submanifold of M covering F, then every orbit of our S^1 action intersects F_0 at exactly one point. This completes the proof of Proposition 3.1. ∎

REMARK 3.1. A weaker version of Proposition 3.1, asserting that every S^1 action on a nilmanifold is free, can be found in [14, section 5.7].

REMARK 3.2a) Generally nilmanifolds admit many injective but not homologically injective S^1 actions. To be more specific *any nilmanifold* $M = G/_\Gamma$ *not homeomorphic to a torus admits an injective but not homologically injective* S^1 *action.* This can be seen as follows. By the nilpotency of Γ, $Z(\Gamma) \neq \{1\}$ and, by [23, Theorem 1.3], $Z(\Gamma) \cap [\Gamma,\Gamma] \neq \{1\}$. Take $\alpha \in Z(\Gamma) - \{1\}$ whose image in $H_1(\Gamma)$ vanishes. By [11, section 4.3] there is an S^1 action $\Phi_t : M \to M$, $t \in [0,1]$, such that $\widetilde{\Phi}_1 = \alpha$.

REMARK 3.2b) The simplest example of a compact nilmanifold having nonabelian fundamental group is the quotient $M = G/_\Gamma$ of the Heisenberg group of all real upper triangular matrices having ones on the diagonal by its subgroup of integer matrices. Then $Z(G) = [G,G] \approx \mathbb{R}$, $Z(\Gamma) \approx [\Gamma,\Gamma] \approx Z$ and accordingly there is an injective but not homologically injective S^1 action on M.

Let M be a closed manifold. As another application of Theorem 1 consider the problem when a given continuous map $f : M \to T^k$ is homotopic to a smooth fibration p whose structure group is finite. By Corollary 2.1 the condition that rank $f_*(Z(\Pi_1(M))) = k$ is necessary.

The problem when f is homotopic to a fibration was completely solved by F. T. Farrell (see [9]) and the related problem when two smooth fibrations are isotopic was investigated in [12]. The arising obstructions described in [9, 12] belong to appropriate Whithead's groups, to appropriate K groups, and to appropriate pseudo isotopy groups. Our problem is connected with the problem of the existence of a periodic diffeomorphism that is homotopic to a given diffeomorphism so that it does not seem likely that in general the solution of it can be decribed in terms of some obstructions belonging to appropriate pseudo isotopy classes or belonging to some groups depending on $\Pi_1(M)$ only.

Let M be a closed manifold. Recall that *a* T^k *action on* M *is maximal* if $\operatorname{im} \mathrm{ev}_* = Z(\Pi_1(M))$. Our next result is an immediate consequence of Corollary 2.1 and Theorem 1.

PROPOSITION 3.2. *The following conditions are equivalent.*

(i) *Any map* $f : M \to T^k$ *such that* rank $f_*(Z(\Pi_1(M))) = k$ *is homotopic to a smooth fibration whose structure group is finite.*
(ii) *There is a smooth maximal toral action on* M.

REMARK 3.3. a) The condition (ii) is always satisfied if M is a nonpositive curvature manifold (see e.g. [13]) or if M is an infranilmanifold ([11, section 4.3]).

b) Note that any manifold $\mathcal{T}$ homeomorphic but not diffeomorphic to a torus does not admit a smooth maximal toral action, because the existence of such an action would imply that $\mathcal{T}$ was diffeomorphic to a torus.

4. A topological characterization of group actions generated by parallel fields. Let G be a compact connected Lie group. In this section we give a topological characterization of closed G manifolds admitting G-parallel metrics (a Riemannian metric is G-parallel if the action of G is generated by parallel fields). Note that parallel fields commute so that it suffices to consider the case when G is a torus T^k only. The simplest examples of T^k parallel metrics are given by Riemannian products of Riemannian manifolds by flat tori. By Bochner's theorem isometric actions of connected groups on closed nonpositive curvature manifolds are generated by parallel fields. Our result can be stated as follows.

THEOREM 3. *Let G be a compact connected Lie group acting smoothly and effectively on a closed manifold M. Then there is a G-parallel metric on M if and only if G is a torus and the action of G is homologically injective.*

Theorem 3 is related to one result of D. J. Welsh [24], who solved a problem posed by S. S. Chern. Welsh's theorem and our Theorem 3 are simple consequences of Theorem 1 (see Remark 4.1).

PROOF of Theorem 3. Assume that an action of a torus T^k on M is homologically injective (see Remark 2.1 for a definition). By Theorem 1, by Remark 2.1, and by Proposition 2.1 (see also [6, Theorem 4.2]), the manifold M is T^k-equivariantly diffeomorphic to $T^k \times_H F$, where the structure group H is finite and for every $h \in H$ we have $h = h_T \times h_F$ for some translation h_T of T^k and for some diffeomorphism h_F of F. Now (as in [24]) it suffices to take a flat metric on T^k, an H-invariant metric on F, the corresponding product metric on $T^k \times F$, and then arising metric on M.

Assume that there is a T^k-parallel metric on M. Let X be a parallel vector field which generates an S^1 action embedded into the torus action. Consider the 1-form ω given by $\omega(X) = 1$, $\omega(v) = 0$ if v is perpendicular to X. It is not difficult to check that this form is parallel in particular harmonic. Let $c : [0,1] \to M$ be the orbit of a given point $x \in M$ under the action of S^1. Then

$$\int_c \omega = \int_0^1 \omega\left(\frac{dc}{dt}\right) dt = \int_0^1 \omega(X)\, dt = 1$$

so that $\mathrm{ev}_* : H_1(T^k) \to H_1(M)$ is a monomorphism. This finishes the proof of Theorem 3. ∎

REMARK 4.1. The main result of [24] shows that a closed manifold M has a vector field which is parallel with respect to some Riemannian metric iff M is a fibre bundle over a torus with finite structure group. It can be derived from Theorem 1 as follows. Let X be a parallel vector field on M and let ω be the parallel 1-form such that $\omega(X) = 1$, $\omega(v) = 0$ if v is perpendicular to X. Then X generates an one-parameter subgroup P of $I(M)$, the isometry group of M. The closure $\overline{P}$ of P in $I(M)$ is a torus. Take $x_0 \in M$. There is a vector field Y generating a subgroup of $\overline{P}$ isomorphic to S^1 and such that $\omega(Y(x_0)) \neq 0$. As ω is a harmonic 1-form and as Y is a Killing vector field, $\omega(Y(x)) = \omega(Y(x_0)) \neq 0$ for every $x \in M$ (see e.g. [24, p. 10]). If c is an orbit of the S^1 action generated by Y, then $\int_c \omega \neq 0$ (compare the proof of Theorem 3) so that the S^1 action is homologically injective and there is a fibration $p : M \to S^1$ whose structure group can be reduced to a finite group (see Remark 1.2, [6, Theorem 4.2]). The reverse implication is easy (see [24]). It follows from the argument that is given in the proof of Theorem 3.

Let M be a closed S^1 manifold. By [22], there is an S^1-invariant Riemannian metric on M such that the orbits of the S^1 action are geodesics iff the S^1 action is fixed point free. A related question is the question of the existence of an S^1-invariant Riemannian metric on M such that the orbits of the action are geodesics and at every $x \in M$ there is a totally geodesic slice V that is perpendicular to the orbits. It turns out that the existence of such a metric is equivalent to the homological injectivity of the S^1 action as shows the following.

PROPOSITION 4.1. *Let M be a closed Riemannian S^1 manifold. The following two conditions are equivalent.*
(a) *At every point $x \in M$ there is a slice V that is perpendicular to the orbits.*
(b) *The S^1 action is generated by a parallel vector field X.*

PROOF. If (a) holds, then the map $\Phi : V \times S^1 \ni (v,t) \to tv \in M$ is a local isometry so that the S^1 action is a generated by a parallel field. In order to prove that (b)$\Rightarrow$(a) consider the orthogonal complement E of X in the tangent bundle of M. The vector bundle E is integrable because $E = \ker\omega$, where ω is a parallel 1-form (compare the proof of Theorem 3).

REFERENCES.

[1] R. H. Bing, *The cartesian product of a certain nonmanifold and a line is E^4*. Ann. of Math. 70(1959), 399–412.

[2] G. B. Bredon, *Introduction to compact transformation groups*. Academic Press, Now York–London 1972.

[3] P. Buser, H. Karcher, *Gromov's almost flat manifolds*. Asterisque No. 81, Soc. Math. France 1981.

[4] J. Cerf, 1*-formes fermées non singulieres sur les variétés compactes de dimension 3*. Séminaire Bourbaki No. 574.

[5] P. E. Conner, F. Raymond, *Actions of compact Lie groups on aspherical manifolds*. Topology of Manifolds, Markham, Chicago, 1970, 227–264.

[6] P. E. Conner, F. Raymond, *Injective operations of the toral groups*. Topology 10(1971), 283–296.

[7] P. E. Conner, F. Raymond, *Holomorphic Seifert fiberings*. Proc. Second Conference on Compact Transformation Groups, Part 2, Springer Lecture Notes in Math. 299(1972), 124–204.

[8] P. E. Conner, F. Raymond, *Realising finite groups of homeomorphisms from homotopy classes of self-homotopy equivalences*. Manifolds, Tokyo 1973, University of Tokyo Press, Tokyo 1975, 231–238.
[9] F. T. Farrell, *The obstruction to fibering a manifold over a circle*. Indiana Univ. Math. Journal 21(1971), 315–346 .
[10] J. Hempel, *3-Manifolds*. Priceton University Press 1976.
[11] Y. Kamishima, K. B. Lee, F. Raymond, *The Seifert construction and its applications to infranilmanifolds*. Quart. J. Math. Oxford 34(1983), 433–452.
[12] C. Kinsey, *Pseudoisotopies and submersions of a compact manifold to the circle*. Topology 26(1988), 433–452.
[13] H. B. Lawson, S. T. Yau, *Compact manifolds of nonpositive curvature*. J. Differential Geometry, 7(1972), 211–228.
[14] K. B. Lee, F. Raymond, *Geometric realisation of group extensions by the Seifert construction*. Contemporary Math. AMS, vol. 33(1984), 353–411.
[15] A. I. Malcev, *On a class of homogeneous spaces*. AMS Transl. 39(1951), 1–33.
[16] J. J. Rotman, *The theory of groups*. An introduction, second edition, Allyn and Bacon, Inc, Boston 1973.
[17] T. B. Rushing, *Topological Embeddings*. Academic Press, New York-London, 1973.
[18] M. Sadowski, *Freeness and effectiveness of some toral actions*, to appear.
[19] R. Schultz, *Group actions on hypertoral manifolds 1*, Topology Symp. Siegen 1979, Springer Lecture Notes in Math. 788(1980), 364–377.
[20] E. H. Spanier, *Algebraic topology*. Mc Graw Hill, Berkeley 1967.
[21] D. Tischler, *On fibering certain foliated manifolds over S^1*. Topology 9(1970), 153–154.
[22] A. W. Wadsley, *Geodesic foliations by circles*. J. Differential Geometry, 10(1975), 541–549.
[23] R. B. Warfield, *Nilpotent groups*. Springer Lecture Notes in Math. 513, 1976.
[24] D. J. Welsh, *Manifolds that admit parallel vector fields*. Illinois Journ. of Math. 30(1986).

LEFSCHETZ NUMBERS OF C^*-COMPLEXES

E. V. Troitsky
Chair of Higher Geometry and Topology, Dept. of Mech. and Math.,
Moscov State University, Moscov, 119899, USSR

1. Introduction. In this paper based on the results of [10], [11] (see §2) we obtain a formula (§3) of the Atiyah-Segal type for a G-invariant complex E of C^*-eliptic pseudodifferential operators over an unital C^*-algebra A (see 4.4 of [9]), where G is a compact Lie group. If $A = C$ this theorem is one of [1]. In general the Lefschetz numbers of this (the first) type take values in $K_0(A) \bigoplus C$. If we have a single endomorphism T of E, it turns out to be possible to define (§4) the Lefschetz number of the second type valued in the cyclic homology $HC_0(A) = A/[A,A]$. We show in §5 that $\widetilde{Ch}_0^0$, $\widetilde{Ch}_0^0(a \otimes z) = Ch_0^0(a)z$, transforms the Lefschetz number of the first type into the Lefschetz number of the second type in the case $T = g_*$, $g \in G$, where Ch_0^0 is defined, for example, in [4].

With the help of [8] we can define traces valued in $HC_{2l}(A)$ for every l and correspondent Lefschetz numbers of the second type. In this case the Lefschetz numbers are connected via Ch_0^1 of [4]. We shall discuss in detail this theorem in the next peper. Among the other prospects of development of the field of [10], [11] and the present paper J. Rosenberg has purposed to investigate G-algebras and crossed products. It seems to be of some interest to work with algebras with a countable approximate unity.

I wish to express gratitude to A. S. Mishchenko, V. Ya. Pidstrigach and J. Rosenberg for stimulating discussions.

2. Some information about the exact C-index formula. For the basic definitions and demonstrations see [9], [10], [11].

Definition 2.1. Let $p : F \to X$ be a G-C-bundle over a compact Hausdorff G-space X. Let $\Lambda(p^*F, s_F)$ be the well known complex of G-C-bundles (see [2]) with, in general, non-compact support. Let a complex (E, α) represent an element $a \in K_G(X; A)$ (see [9], sect. 1.3), then $(p^*E, p^*\alpha) \otimes \Lambda(p^*F, s_F)$ has a compact support and defines an element of $K_G(F; A)$. We get the *Thom homomorphism* of $R(G)$-modules

$$\varphi = \varphi_A^F : K_G(X; A) \to K_G(F; A).$$

If we pass to K_G^1 by the Bott periodicity ([9], 1.2.4), we can define

$$\varphi : K_G^*(X; A) \to K_G^*(F; A).$$

Theorem 2.2. *If X is separable and metrizable, then φ is an isomorphism.*

With the help of 2.2 we can define the *Gysin homomorphism* $i_! : K_G(TX;A) \to K_G(TY;A)$ and the *topological index* $t\text{-}ind_G^X = t\text{-}ind_{G,A}^X : K_G(TX;A) \to K^G(A)$ in a way similar to the case $A = C$ [2]. Here $i : X \to Y$ is a G-inclusion of smooth manifolds and TX, TY are (co)tangent bundles.

We need the following property of the Gysin homomorphism.

LEMMA 2.3. *Let $i : Z \to X$ be a G-inclusion of smooth manifolds, N its normal bundle. Then the homomorphism*

$$(di)^* i_! : K_G(TZ;A) \to K_G(TZ;A)$$

is the multiplication by

$$[\lambda_{-1}(N \otimes_R C)] = \sum (-1)^i [\Lambda^i (N \otimes_R C)] \in K_G(Z),$$

where Λ^i are the exterior powers, and we consider $K_G(TZ;A)$ as a $K_G(Z)$-module in the usual way.

THEOREM 2.4. *Let $a\text{-}ind\, D \in K^G(A)$ be the analytic index of a pseudodifferential equivariant C^*-elliptic operator* [9], *$\sigma(D) \in K_G(TX;A)$ its symbol's class. Then*

$$t\text{-}ind_{G,A}^X\, \sigma(D) = a\text{-}ind\, D.$$

3. An Atiyah-Segal type formula valued in $K_0(A)$. In this section we generalize the result of [1]. Let, as above, G be a compact Lie group, X a G-space, γ a classs of conjugate elements of G, X^g the set of fixed points of $g : X \to X$. Then $X = \bigcup_{g \in \gamma} X^g$ is a closed G-subspace of X. The class γ defines an ideal in $R(G) = K^G(C)$ with localization $R(G)_\gamma$, and localizations of $R(G)$-modules.

LEMMA 3.1. *Let X be a compact G-space with G-map into G/H, where H is a closed subgroup of G, $H \cap \gamma = \emptyset$. Then $K_G^*(X;A)\gamma = 0$.*

PROOF. $K_G(X;A)$ is a $R(H)$-module, and $R(H)_\gamma = 0$ (see [1]). ∎

LEMMA 3.2. *Let X be a locally compact G-space, γ has no fixed points in X. Then $K_G^*(X;\, A)_\gamma = 0$.*

THEOREM 3.3. *Let G be an abelian group, X a smooth G-manifold, $\gamma = g \in G$, $i : X^g \to X$ the inclusion. Then*

$$i^* : K_G(X;A) \to K_G(X^g;A)$$

becomes an isomorphism after localization

$$i_g^* : K_G(X;A)_g \to K_G(X^g;A)_g.$$

Demonstration of 3.2 and 3.3 is similar to that in [1].

LEMMA 3.4. *Let G be a topologically cyclic group with a generator g, X a compact G-manifold. Then*

$$i_! : K_G(TX^g;A) \to K_G(TX;A)$$

becomes an isomorphism after localization

$$(i_!)_g : K_G(TX^g; A_g) \to K_G(TX; A_g)$$

with the inverse

$$\frac{i_g^*}{\lambda_{-1}(N^g \otimes_{\mathbb{R}} C)},$$

where i_g^* *is the localization of the restricion homomorphism*

$$i^* : K_G(TX; A) \to K_G(TX^g; A)$$

and $\lambda_{-1}(N^g \otimes_{\mathbb{R}} C)$ *denotes* $[\lambda_{-1}(N^g \otimes_{\mathbb{R}} C)] \in K_G(X^g)$ *as well as its class in* $K_G(X^g)_g$.

PROOF. The results is a consequence of 3.3, 2.3 and [1]. ∎

LEMMA 3.5. *Let* G *be a topologically cyclic group with a generator* g, X *a compact* G*-manifold,* $u \in K_G(TX; A)$*. Then*

$$(ind^X_{G,A}(u))_g = (ind^{X^g}_{G,A})_g\Big(\frac{i^* u}{\lambda_{-1}(N^g \otimes_{\mathbb{R}} C)}\Big). \tag{1}$$

PROOF. By triviality of G-action on TX^g

$$K_G(TX^g; A) \cong K(TX^g; A) \otimes R(G)$$

(see [9]), and

$$ind^{X^g}_{G,A} \cong ind^{X^g}_{1,A} \otimes id,$$

since the Thom homomorphism is induced by a G-trivial complex. Let us localize the following commutative diagram from [10]

$$\begin{array}{ccc} K_G(TX^g; A) & \xrightarrow{i_!} & K_G(TX; A) \\ & ind^{X^g}_{G,A} \searrow \quad \swarrow ind^X_{G,A} & \\ & K^G(A) & \end{array}$$

We obtain the following commutative diagram

$$\begin{array}{ccc} K(TX^g; A) \otimes R(G)_g = K_G(TX^g; A)_g & \xrightarrow{(i_!)_g} & K_G(TX; A)_g \\ ind^{X^g}_{1,A} \otimes id \searrow & & \swarrow (ind^X_{G,A})_g \\ & K_0(A) \otimes R(G)_g & \end{array}$$

Lemma 3.4 completes the proof. ∎

Let us define the evaluation map $\alpha \mapsto \alpha(g)$ for a G-trivial space Y

$$K_G(Y; A) \cong K(Y; A) \otimes R(G) \longrightarrow K(Y; A) \otimes C,$$
$$K_G(Y; A)_g \cong K(Y; A) \otimes R(G)_g \longrightarrow K(Y;) \otimes C$$

by

$$u \otimes \chi \longrightarrow u \otimes \chi(g),$$
$$u \otimes \chi/\psi \longrightarrow u \otimes \chi(g)/\psi(g).$$

If we take the value of (1) on g, then we get

$$(ind^X_{G,A}\, u)(g) = (ind^{X^g}_{1,A} \otimes \mathrm{Id})\Big(\frac{i_* u(g)}{\lambda_{-1}(N^g \otimes_R C)(g)}\Big), \tag{2}$$

where $ind^{X^g}_{1,A} \otimes \mathrm{Id} : K(TX^g; A) \otimes C \to K_0(A) \otimes C$, since the map of evaluation can be passed throught the local ring.

Let G be an arbitrary compact Lie group, H its subgroup, generated by $g, j : H \to G$ the inclusion, $u \in K_G(TX; A)$. Then (see [10])

$$j^* ind^X_{G,A}(u) = ind^X_{H,A}\, j^*(u)$$

and from (2), when $G = H$, we get

$$\begin{aligned} ind^X_{G,A}(u)(g) &= (j^* ind^X_{G,A}(u))(g) = (ind^X_{H,A}\, j^*(u))(g) = \\ &= (ind^{X^g}_{1,A} \otimes \mathrm{Id})\Big(\frac{i_* j^* u(g)}{\lambda_{-1}(N^g \otimes_R C)(g)}\Big) = \\ &= (ind^{X^g}_{1,A} \otimes \mathrm{Id})\Big(\frac{i_* u(g)}{\lambda_{-1}(N^g \otimes_R C)(g)}\Big). \end{aligned}$$

So we can operate now with an arbitrary G.

DEFINITION 3.6. Let E be a G-invariant A-complex on E, $\sigma(E)$ its sequence of symbols (see [9]), $u = [\sigma(E)] \in K_G(TX; A)$, $ind^X_{G,A}(u) \in K_0(A) \otimes R(G)$. The *Lefschetz number of the first type* is

$$L_1(g, E) = ind^X_{G,A}(u)(g) \in K_0(A) \otimes C.$$

The consideration above is the proof of the following assertion.

THEOREM 3.7. *Using the notation as above we have*

$$L_1(g, E) = (ind^{X^g}_{1,A} \otimes 1)\Big(\frac{i_* u(g)}{\lambda_{-1}(N^g \otimes_R C)(g)}\Big).$$

4. Traces and Lefschetz numbers valued in $HC_0(A)$.

DEFINITION 4.1. Let $\{e_1, e_2, \ldots\}$ be an A-orthobasis of H_A (Hilbert module over A) with A-inner product $(\,,\,)$ (see [5], [7], [12]). Let $S \in \mathrm{End}^*_A H_A$ (see [12]) and $S(e_i) = 0\,(i > k)$. We define the trace of S by

$$t(S, \{e_i\}, k) = \sum_{i=1}^{\infty} f((Se_i, e_i)) = \sum_{i=1}^{k} f(S^i_i),$$

where $f : A \to A/[A,A] = HC_0(A)$, $\|S^i_j\|$ is the matrix of S with respect to $\{e_i\}$, $S^i_j \in A$.

Lemma 4.2. $t(S,\{e_i\},k) = t(S,\{e_i\},l) := t(S,\{e_i\})$ *for* $l \geq k$.

Proof. Proof is evident. ∎

Lemma 4.3. *Let $S, \{e_i\}, k$ be as in 4.1 and $\{h_j\}$ a new A-basis of H_A (in general non-orthogonal). Then the series*

$$\sum_{r=1}^{\infty} f((S_h)^r_r)$$

converges to $t(S,\{e_i\})$, where $(S_h)^p_r$ are the matrix elements of S with respect to $\{h_i\}$.

Proof. Let $\{h_j\}$ be an A-orthogonal basis. Then

$$\sum_{i=1}^{\infty} f(Se_i,e_i) = \sum_{i=1}^{k} f\Big(\sum_{j=1}^{\infty}(Se_i,h_j)^*(e_i,h_j)\Big) =$$
$$= \sum_{j=1}^{\infty}\sum_{i=1}^{k} f((Se_i,h_j)^*(e_i,h_j)) = \sum_{j=1}^{\infty}\sum_{i=1}^{k} f((h_j,e_i)^*(Se_i,h_j)^*) =$$
$$= \sum_{j=1}^{\infty} f\Big(\sum_{i=1}^{k}(h_j,e_i)^*(Se_i,h_j)^*\Big) = \sum_{j=1}^{\infty} f\Big(\sum_{i=1}^{\infty}(h_j,e_i)^*(Se_i,h_j)^*\Big) =$$
$$= \sum_{j=1}^{\infty} f\Big(\sum_{i=1}^{\infty}(h_j,e_i)^*(S^*h_j,e_i)\Big) = \sum_{j=1}^{\infty} f((h_j,S^*h_j)) = \sum_{j=1}^{\infty} f((Sh_j,h_j)). \blacksquare$$

The trace of S does not depend on the choice of basis, so we put $t(S) := t(S,\{e_i\})$. If $V \in \mathrm{End}^*_A H_A$ is an A-unitary operator, then $\{Ve_j\}$ is an A-orthogonal basis, so

$$\sum_{i=1}^{\infty} f((Se_i,e_i)) = \sum_{j=1}^{\infty} f((SVe_j,Ve_j)) = \sum_{j=1}^{\infty} f(((V^{-1}SV)e_j,e_j)),$$

and the trace of V^*SV is equal to the trace of S.

Let now $B \in \mathrm{End}_A H_A$ be an arbitrary isomorphism and S as in 4.1. Then BS, SB and consequenlty $B^{-1}SB$ are of the form as in 4.1. Then

$$\sum_{i=1}^{\infty} f((BSe_i,e_i)) = \sum_{i=1}^{\infty} f((SBe_i,e_i))$$

Indeed, B is a C-linear combination of four A-unitary operators, but for an A-unitary operator V

$$V^{-1}(VS)V = SV,$$

so, as it is proved above,

$$\sum_{i=1}^{\infty} f((SVe_i, e_i)) = \sum_{i=1}^{\infty} f((VSe_i, e_i)).$$

If we take the C-linear combination, then we obtain the desired result.

Hence, if $\{h_i\}$ is an arbitrary basis, $Be_i = h_i$, then

$$\sum_{i=1}^{\infty} f((Se_i, e_i)) = \sum_{i=1}^{\infty} f((((SB)B^{-1}e_i, e_i)) = \sum_{i=1}^{\infty} f((B^{-1}(SB)e_i, e_i)) =$$

$$= \sum_{i=1}^{\infty} f((B^{-1}SB)_i^i) = \sum_{i=1}^{\infty} (f(S_h)_i^i). \blacksquare$$

So we can take instead of 4.1 the following correct definition.

Definition 4.4.Let $S \in \mathrm{End}_A^* H_A$, M and N Hilbert submodules of H_A, N finitely generated, $H_A = M \oplus N$, $S|_M = 0$. For an arbitrary basis $\{e_i\}$

$$t(S) = \sum_{i=1}^{\infty} f(S_i^i).$$

Remark. Here the matrix of S with respect to an A-orthogonal basis $\{h_i\}$ is $\|(Sh_i, h_j)\|$, and with respect to an arbitrary basis $\{\overline{h}_i\}$, $\overline{h}_i = Bh_i$, $B \in \mathrm{End}_A^* H_A$, is $\|(B^{-1}SBh_i, h_j)\|$. We obtain the correctness of this definition immedietly.

Lemma 4.5. *Let M, N, S be as in 4.4, and N acountably generated Hilbert A-module, $\tilde{H}_A = H_A \hat{\oplus} \tilde{N} \cong H_A$ (see [5]),*

$$\tilde{S} = \begin{pmatrix} S & 0 \\ 0 & 0 \end{pmatrix} : H_A \hat{\oplus} \tilde{N} \to H_A \hat{\oplus} \tilde{N},$$

where $\hat{\oplus}$ denotes A-orthogonal sum. Then $t(S) = t(\tilde{S})$.

Proof. If $\tilde{N} \cong H_A$, then the result is evident. So by the stabilization theorem of [5] we can take $M \cong H_A$ or $M = H_A^1 \hat{\oplus} H_A^2$ from the very begining. The end of the proof is clear. $\blacksquare$

Lemma 4.6. *Let M, N, S be as in 4.4, $M \cong H_A$, $N = \overline{N} \oplus \overline{\overline{N}}$, $S|_{\overline{\overline{N}}} = 0$. Then*

$$t(S) = t(pSp),$$

where $p : M \oplus \overline{N} \oplus \overline{\overline{N}} \to M \oplus \overline{N}$ is a projector, and the trace on the right is in the space $M \oplus \overline{N}$.

Proof. Let a projective module $\tilde{N}$ be added to H_A, as in 4.5 and $\overline{\overline{N}} \hat{\oplus} \tilde{N} = A^m$ with orthobasis $h_0, \ldots, h_{-m+1}$. Let $\{h_1, h_2, \ldots\}$ be an A-orthobasis of $M \hat{\oplus} \overline{N} \cong H_A$.

Then, as in 4.5,

$$t(S) = t(\tilde{S}) = \sum_{i=-m+1}^{\infty} f(B^{-1}\tilde{S}Be_i, e_i),$$

where $\{e_i\}$ is an A-orthobasis of $H_A \hat{\oplus} \tilde{N}$ with $h_i = Be_i$, $e_i = h_i \, (i \geq 1)$. We have $\tilde{S}Be_i = \tilde{S}h_i = 0$, when $i = 0, \dots, -m+1$, so

$$t(S) = \sum_{i=1}^{\infty} f(B^{-1}\tilde{S}Be_i, e_i).$$

When $i \geq 1$, we obtain, that $Be_i = e_i \in M \oplus \overline{N}$,

$$t(S) = \sum_{i=1}^{\infty} f(B^{-1}SBe_i, e_i) =$$

$$= \sum_{i=1}^{\infty} f((B^{-1}pSe_i, e_i) + (B^{-1}(1-p)Se_i, e_i)).$$

$(1-p)Se_i \in \overline{\overline{N}} \oplus \tilde{N} = span_A\{h_0, \dots, h_{-m+1}\}$, hence $B^{-1}(1-p)Se_i \in span_A\{e_0, \dots, e_{-m+1}\}$ and the second summand vanishes. Also $B^{-1}|_{Im\, p} = id$, so

$$t(S) = \sum_{i=1}^{\infty} f(pSe_i, e_i) = t(pSp). \ \blacksquare$$

Corollary 4.6. *If in 4.5 $M \oplus \overline{N}$ is orthogonal to N, and $\{h_i\}$ is an A-orthobasis of $M \oplus \overline{N}$, then*

$$t(S) = \sum_{i=1}^{\infty} f(Sh_i, h_i).$$

Definition 4.7. Let $F : H_A \to H_A$ be an A-Fredholm operator admitting an adjoint,

$$H_A = M_0 \oplus N_0 \xrightarrow{\begin{pmatrix} F_1 & 0 \\ 0 & F_2 \end{pmatrix}} M_1 \oplus N_1 = H_A \qquad (D)$$

a corespondent decomposition (see [7]), S_0, S_1 are from 3xEnd$^*_A H_A$ such that the diagram

$$\begin{array}{ccc} H_A & \xrightarrow{F} & H_A \\ \downarrow S_0 & & \downarrow S_1 \\ H_A & \xrightarrow{F} & H_A \end{array}$$

commutes. Let us define

$$\tilde{S}_0 = \begin{cases} 0 & \text{on } M_0 \\ S_0 & \text{on } N_0 \end{cases}, \qquad \tilde{S}_1 = \begin{cases} 0 & \text{on } M_1 \\ S_1 & \text{on } N_1 \end{cases}$$

and

$$L(F,S,D) = t(\tilde{S}_0) - t(\tilde{S}_1).$$

LEMMA 4.8. *Let*

$$H_A = M_0 \oplus N_0 \longrightarrow M_1 \oplus N_1 = H_A, \tag{D}$$
$$H_A = \widetilde{M}_0 \oplus N_0 \longrightarrow \widetilde{M}_1 \oplus N_1 = H_A \tag{$\tilde{D}$}$$

be two decompositions for F. Then

$$L(F,S,D) = L(F,S,\tilde{D}).$$

PROOF. 1Choose such A-inner products, that M_0 and N_0, M_1 and N_1 are orthogonal, $F : M_0 \cong M_1$ preserves product. Let $p_0 : M_0 \oplus N_0 \to N_0$, $\tilde{p}_0 : \widetilde{M}_0 \oplus N_0 \to N_0$, $p_1 : M_1 \oplus N_1 \to N_1$, $\tilde{p}_1 : \widetilde{M}_1 \oplus N_1 \to N_1$ be projectors. Then

$$\begin{aligned} L(F,S,D) &= t(S_0 p_0) - t(S_1 p_1), \\ L(F,S,\tilde{D}) &= t(S_0 \tilde{p}_0) - t(S_1 \tilde{p}_1) = \\ &= t(S_0 \tilde{p}_0 p_0) + t(S_0 \tilde{p}_0 (1-p_0)) - t(S_1 \tilde{p}_1 p_1) - t(S_1 \tilde{p}_1 (1-p_1)) = \\ &= t(S_0 p_0) + t(S_0 \tilde{p}_0 (1-p_0)) - t(S_1 p_1) - t(S_1 \tilde{p}_1 (1-p_1)), \end{aligned}$$

since $\tilde{p}_0$ on $N_0 = im p_0$ is the indentity operator, as well as $\tilde{p}_1$ on N_1. So

$$L(F,S,\tilde{D}) = L(F,S,D) + t(S_0 \tilde{p}_0 (1-p_0)) - t(S_1 \tilde{p}_1 (1-p_1)).$$

Without loss of generality (see the proof of 4.5) we can take $M_0 \cong H_A$ and let $\{e_1, e_2, \ldots\}$ be its A-orthobasis. Then $\{Fe_1, Fe_2, \ldots\}$ is an A-orthobasis of M_1 and by 4.6

$$\begin{aligned} t(S_1 \tilde{p}_1 (1-p_1)) &= \sum_{i=1}^{\infty} f(S_1 \tilde{p}_1 F e_i, F e_i) = \sum_{i=1}^{\infty} f(S_1 F \tilde{p}_0 e_i, F e_i) = \\ &= \sum_{i=1}^{\infty} f(F S_0 \tilde{p}_0 e_i, F e_i) = \\ &= \sum_{i=1}^{\infty} (f(F p_0 S_0 \tilde{p}_0 e_i, F e_i) + f(F(1-p_0) S_0 \tilde{p}_0 e_i, F e_i)) = \\ &= \sum_{i=1}^{\infty} (f(p_1 F S_0 \tilde{p}_0 e_i, F e_i) + f((1-p_0) S_0 \tilde{p}_0 e_i, e_i)) = \\ &= \sum_{i=1}^{\infty} f(S_0 \tilde{p}_0 e_i, e_i) = t(S_0 \tilde{p}_0 (1-p_0)). \ \blacksquare \end{aligned}$$

Lemma 4.9. *Let*

$$H_A = (M_0 \oplus N_0) \oplus K_0 \longrightarrow (M_1 \oplus N_1) \oplus K_1 = H_A \qquad (D_1)$$

and

$$H_A = M_0 \oplus (N_0 \oplus K_0) \longrightarrow M_1 \oplus (N_1 \oplus K_1) = H_A \qquad (D_2)$$

be two decompositions for F. *Then* $L(F,S,D_1) = L(F,S,D_2)$.

Proof. We have to verify, that $t(S_0p_0) = t(S_1p_1)$, where $p_0 : H_A \to N_0$ and $p_1 : H_A \to N_1$ are the projectors. Choose the metrics such that $M_j \perp N_j \perp K_j$ $(j = 1,2)$ and $F : M_0 \oplus N_0 \to M_1 \oplus N_1$ preserves A-inner product. Let $N_0 \subset span_A(e_1, \ldots, e_m)$, then $N_1 \subset span_A(Fe_1, \ldots, Fe_m)$, where $\{e_i\}$ and $\{Fe_i\}$ are A-orthobases of $M_0 \oplus N_0$ and $M_1 \oplus N_1$ (we assume them to be isomorphic to H_A as above). Let $q_0 : H_A \to M_0 \oplus N_0$ and $q_1 : H_A \to M_1 \oplus N_1$ be projectors. Then

$$\begin{aligned}
t(S_1p_1) &= \sum_{i=1}^{m} f(S_1p_1Fe_i, Fe_i) = \sum_{i=1}^{m} f(S_1Fp_0e_i, Fe_i) = \\
&= \sum_{i=1}^{m} f(FS_0p_0e_i, Fe_i) = \sum_{i=1}^{m} (f(Fq_0S_0p_0e_i, Fe_i) + f(F(1-q_0)S_0p_0e_i, Fe_i)) = \\
&= \sum_{i=1}^{m} (f(q_0S_0p_0e_i, e_i) + f((1-q_1)FS_0p_0e_i, Fe_i) = \sum_{i=1}^{m} f(S_0p_0e_i, e_i) = \\
&= t(S_0p_0). \blacksquare
\end{aligned}$$

Lemma 4.10. *Let*

$$H_A = M_0 \oplus N_0 \longrightarrow M_1 \oplus N_1 = H_A \qquad (D)$$

and

$$H_A = \overline{M}_0 \oplus \overline{N}_0 \longrightarrow \overline{M}_1 \oplus \overline{N}_1 = H_A \qquad (\overline{D})$$

be two decompositions for F. *Then* $L(F,S,D) = L(F,S,\overline{D})$. *So* L *does not depend on* D *and we denote it by* $L(F,S)$.

Proof. Choose a free module V_0 with A-orthobasis $e_1, \ldots, e_r$, $\overline{N}_0 \subset V_0$, and a free V_1 with A-orthobasis $h_1, \ldots, h_m$, $\overline{N}_1 \subset V_1$. Choose an arbitrary $\varepsilon > 0$ and a projective finitely generated module $K_0 \subset M_0$, $M_0 = L_0 \oplus K_0$, such that

$$\|(1 - p_{K_0 \oplus N_0})|_{V_0}\| < \delta_0 = \frac{\varepsilon}{2r\|S_0\|\,\|f\|},$$
$$\|(1 - p_{K_1 \oplus N_1})|_{V_1}\| < \delta_1 = \frac{\varepsilon}{2m\|S_1\|\,\|f\|},$$

where $K_1 = F(K_0)$, $L_1 = F(L_0)$,

$$p_{K_0 \oplus N_0} : L_0 \oplus K_0 \oplus N_0 \to K_0 \oplus N_0,$$
$$p_{K_1 \oplus N_1} : L_1 \oplus K_1 \oplus N_1 \to K_1 \oplus N_1$$

are projectors.

Consider the decomposition

$$L_0 \oplus (K_0 \oplus N_0) \longrightarrow L_1 \oplus (K_1 \oplus N_1). \qquad (D_K)$$

By Lemma 4.5 $L(F,S,D) = L(F,S,D_K)$. Take

$$Q_0 = p_{K_0\oplus N_0}(\overline{N}_0), \qquad Q_1 = p_{K_1\oplus N_1}(\overline{N}_1),$$
$$R_0 = \overline{M}_0 \cap (K_0 \oplus N_0), \qquad R_1 = \overline{M}_1 \cap (K_1 \oplus N_1).$$

When δ is small (and K_0 is "large")

$$Q_0 \cong \overline{N}_0, \qquad Q_1 \cong \overline{N}_1,$$
$$H_A = \overline{M}_0 \oplus Q_0, \qquad H_A = \overline{M}_1 \oplus Q_1,$$
$$K_0 \oplus N_0 = R_0 \oplus Q_0, \qquad K_1 \oplus N_1 = R_1 \oplus Q_1,$$

where R_0 and R_1 are projective finitely generated modules. We have

$$F: R_0 \xrightarrow{\cong} R_1, \qquad F: Q_0 \to Q_1.$$

Indeed, $F: \overline{M}_0 \cong \overline{M}_1$ and $F: K_0 \oplus N_0 \to K_1 \oplus N_1$. So

$$F: R_0 = \overline{M}_0 \cap (K_0 \oplus N_0) \xrightarrow{\cong} \overline{M}_1 \cap (K_1 \oplus N_1) = R_1.$$

Let $x \in Q_0$, then $x = p_{K_0\oplus N_0}(y)$, $y \in \overline{N}_0$, and

$$F(x) = Fp_{K_0\oplus N_0}(y) = p_{K_1\oplus N_1}F(y) \in Q_1,$$

since $F(y) \in \overline{N}_1$. Let

$$(L_0 \oplus R_0) \oplus Q_0 \longrightarrow (L_1 \oplus R_1) \oplus Q_1 \qquad (D_1)$$

and

$$\overline{M}_0 \oplus Q_0 \longrightarrow \overline{M}_1 \oplus Q_1. \qquad (D_2)$$

Then by 4.9 $L(F,S,D_1) = L(F,S,D)$, and by Lemma 4.8

$$L(F,S,D_2) = L(F,S,D_1) = L(F,S,D).$$

Let

$$p_0: \overline{M}_0 \oplus \overline{N}_0 \to \overline{N}_0, \qquad p_1: \overline{M}_1 \oplus \overline{N}_1 \to \overline{N}_1,$$
$$q_0: \overline{M}_0 \oplus Q_0 \to Q_0, \qquad q_1: \overline{M}_1 \oplus Q_1 \to Q_1$$

be projectors. We can assume that sums on the first line are orthogonal.

$$L(F,S,D_2) - L(F,S,\bar{D}) = t(S_0p_0) - t(S_0q_0) - t(S_1p_1) + t(S_1q_1).$$

Since $V_0^\perp \subset \overline{M}_0$, $V_1^\perp \subset \overline{M}_1$, we have $p_0|_{V_0^\perp} = q_0|_{V_0^\perp} = 0$, $p_1|_{V_1^\perp} = q_1|_{V_1^\perp} = 0$ and

$$\begin{aligned}
L(F,S,D_2)-&L(F,S,\tilde{D}) = \\
&= \sum_{i=1}^{r} f(S_0p_0e_i,e_i) - \sum_{i=1}^{r} f(S_0q_0e_i,e_i) - \sum_{j=1}^{m} f(S_1p_1h_j,h_j) + \\
&\quad + \sum_{j=1}^{m} f(S_1q_1h_j,h_j) = \\
&= \sum_{i=1}^{r} f(S_0(p_0-q_0)e_i,e_i) - \sum_{j=1}^{m} f(S_1(p_1-q_1)h_j,h_j) = \\
&= \sum_{i=1}^{r} (f(S_0(p_0-q_0)p_0e_i,e_i) + f(S_0(p_0-q_0)(1-p_0)e_i,e_i)) + \\
&\quad - \sum_{j=1}^{m} (f(S_1(p_1-q_1)p_1h_j,h_j) + f(S_1(p_1-q_1)(1-p_1)h_j,h_j)).
\end{aligned}$$

Since $(1-p_0)e_i \in \overline{M}_0$, $(1-p_1)h_j \in \overline{M}_1$ and $q_0|_{\overline{M}_0} = p_0|_{\overline{M}_0} = 0$, $q_1|_{\overline{M}_1} = p_1|_{\overline{M}_1} = 0$, then the second term in each brackets vanishes. Also

$$\begin{aligned}
(p_0-q_0)p_0 &= (1-q_0)p_0, \\
(p_1-q_1)p_1 &= (1-q_1)p_1.
\end{aligned}$$

By the estimate in the begining of the proof we have

$$\begin{aligned}
\|L(F,S,D_2) - L(F,S,\tilde{D})\| \le& \\
&\le r\|f\|\cdot\|S_0\|\cdot\|(1-q_0)|_{\overline{N}_0}\| + m\|f\|\cdot\|S_1\|\cdot\|(1-q_1)|_{\overline{N}_1}\| \le \\
&\le r\|f\|\cdot\|S_0\|\cdot\|(1-p_{K_0\oplus N_0})|_{V_0}\| + m\|f\|\cdot\|S_1\|(1-p_{K_1\oplus N_1})|_{V_1}\| < \varepsilon,
\end{aligned}$$

since $Q_0 \subset K_0 \oplus N_0$, $Q_1 \subset K_1 \oplus N_1$, $\overline{N}_0 \subset V_0$, and $\overline{N}_1 \subset V_1$. Since ε is arbitrary, we obtain

$$L(F,S,D_2) = L(F,S,\tilde{D}). \blacksquare$$

REMARK 4.11. By the stabilization theorem and Lemma 4.5 we can define $L(F,S)$ for any countably generatted Hilbert A-module instead of H_A.

DEFINITION 4.12. Let $T = \{T_i\}$ be an endomorphism of an A-elliptic complex E :

$$\begin{array}{ccccccc}
0 & \longrightarrow & \Gamma(E_0) & \xrightarrow{d_0} & \Gamma(E_1) & \longrightarrow & \dots \\
 & & \downarrow T_0 & & \downarrow T_1 & & , \\
0 & \longrightarrow & \Gamma(E_0) & \xrightarrow{d_0} & \Gamma(E_1) & \longrightarrow & \dots
\end{array}$$

$$T_{i+1}d_i = d_iT_i, \qquad T_i \in \mathrm{End}_A^*\Gamma(E_i).$$

Let be fulfilled

CONDITION 4.13. Sobolev products in $\Gamma(E_i)$ can be choosen in such a way, that

$$T_i d_i = d_i T_{i+1}.$$

We take $E_{ev} = \bigoplus E_{2i}$, $E_{od} = \bigoplus E_{2i+1}$,

$$F = d + d^* : \Gamma(E_{ev}) \to \Gamma(E_{od}),$$

then F is an A-Fredholm operator and the diagram, stated below, commutes, where

$$S_0 = \bigoplus T_{2i}, \qquad S_1 = \bigoplus T_{2i+1}.$$

$$\begin{array}{ccc} \Gamma(E_{ev}) & \xrightarrow{F} & \Gamma(E_{od}) \\ \downarrow S_0 & & \downarrow S_1 \\ \Gamma(E_{ev}) & \xrightarrow{F} & \Gamma(E_{od}). \end{array}$$

We define the Lefschetz number of the second type as

$$L_2(E, T, m) = L(F, S) \in HC_0(A),$$

where m denotes the dependence on inner products (via d^*).

5. A connection between Lefschetz numbers of the two types via the Chern character. In the situation of §3 we shall denote the endomorphism of complex by T_g.

LEMMA 5.1. *Let $T = T_g$, $g \in G$. Then Condition 4.13 is fulfilled.*

PROOF. Average A-inner product via the action of G (see [9], 3.1). The new product will be denoted by m_G. Then $T = T_g$ is a unitary operator and from $Td_i = d_i T$ we get

$$\begin{aligned} d_i^* T^* &= T^* d_i^*, \\ d_i^* T^{-1} &= T^{-1} d_i^*, \\ T(d_i^* T^{-1})T &= T(T^{-1} d_i^*)T, \\ T d_i^* &= d_i^* T. \end{aligned}$$

THEOREM 5.2. *If $T = T_g$, $g \in G$, then*

$$L_2(E, T_g, m_G) = \widetilde{Ch}_0^0(L_1(g, E)),$$

where Ch_0^0 is the Chern character

$$Ch_0^0 : K_0(A) \to HC_0(A)$$

(see [4]), and

$$\widetilde{Ch}_0^0(a \otimes z) = Ch_0^0(a) z,$$

where $z \in C$. In particular L_2 does not depend on m_G.

PROOF. We have

$$L_1(g, E) = ind_{G,A}^X([\sigma(E)])(g) = ind_{G,A}^X(F)(g).$$

Let (D) $M_0 \oplus N_0 \longrightarrow M_1 \oplus N_1$ be a decomposition for F. Then (see 2.1 of [9] and [6])

$$N_0 = \bigoplus_{k=1}^{K} V_k \otimes P_k, \qquad N_1 = \bigoplus_{l=1}^{L} W_l \otimes Q_l,$$

where V_k and W_l are C-vector spaces of irreducible representations of G, P_k and Q_l are G-trivial projective finitely generated A-modules. Then (representations are unitary)

$$ind_{G,A}^X(F) = \sum_{k=1}^{K} [P_k] \otimes \chi(V_k) - \sum_{l=1}^{L} [Q_l] \otimes \chi(W_l)$$

and

$$L(g, E) = \sum_{k=1}^{K} [P_k] \otimes \text{Trace}(g|V_k) - \sum_{l=1}^{L} [Q_l] \otimes \text{Trace}(g|W_l).$$

If $[P_k] = [p_k]$, where $p_k : A^{n(k)} \to A^{n(k)}$ is a projector, $M(p_k) \in M(A, n(k))$ is its matrix in the standard basis, $[Q_l] = [q_l]$, $M(q_l) \in M(A, r(l))$, then

$$\widetilde{Ch}_0^0([P_k] \otimes \text{Trace}(g|V_k)) = Tr_*^{n(k)}([M(p_k)]) \cdot \text{Trace}(g|V_k),$$

$$\widetilde{Ch}_0^0([Q_l] \otimes \text{Trace}(g|W_l)) = Tr_*^{r(l)}([M(q_l)]) \cdot \text{Trace}(g|W_l),$$

where $Tr^n : M(A, n) \to A$ is the trace and

$$Tr_*^n : HC_0(M(A, n)) \to HC_0(A).$$

Let $\{e_1^k, \ldots, e_{n(k)}^k\}$ be the standard basis of $A^{n(k)}$, $\{v_1^k, \ldots, v_{m(k)}^k\}$ an orthobasis of V_k, $\{h_1^l, \ldots, h_{r(l)}^l\}$ the standard basis of $A^{r(l)}$, $\{w_1^l, \ldots, w_{s(l)}^l\}$ an orthobasis of W_l. We have

$$L_2(E, T_g, m_G) = \sum_{k=1}^{K} \sum_{i,j} f(T_g(1 \otimes p_k)(v_j^k \otimes e_i^k), v_j^k \otimes e_i^k) +$$

$$- \sum_{l=1}^{L} \sum_{c,d} f(T_g(1 \otimes q_l)(w_c^l \otimes h_d^l), w_c^l \otimes h_d^l) =$$

$$= \sum_{k} \sum_{i,j} f((g(v_j^k), v_j^k)(p_k e_i^k, e_i^k)) - \sum_{l} \sum_{c,d} f((g(w_c^l), w_c^l)(q_l h_d^l, h_d^l)) =$$

$$= \sum_{k} \left(\sum_{l} f(p_k e_i^k, e_i) \text{Trace}(g|V_k) \right) - \sum_{l} \left(\sum_{d} f(q_l h_d^l, h_d^l) \text{Trace}(g|W_l) \right).$$

Here we denote by (,) the C-inner product as well as the A-inner one. The identification $HC_0(A) = A/[A,A]$ (see [3]) is the "identity" map : $[a] \mapsto [a]$. So

$$Tr_*[M(p_k)] = Tr_*[\|(p_k)^i_j\|] = [(p_k)^1_1 + \cdots + (p_k)^{n(k)}_{n(k)}],$$
$$\sum_i f(p_k e^k_i, e^k_i) = [(p_k)^1_1 + \cdots + (p_k)^{n(k)}_{n(k)}].$$

In a similar way we get

$$Tr_*[M(q_l)] = \sum_d f(q_l h^l_d, h^l_d). \blacksquare$$

REFERENCES

[1] Atiyah, M. F. and Segal, G. B. *The index of elliptic operators. II.* Ann. of Math. (2) 87 (1968), 531-545.

[2] Friedrich, T. *Vorlesungen über K-Theorie.* Leipzig: Teubner, 1987.

[3] Karoubi, M. *Homologie cyclique des groupes et des algébres.* C. R. Ac. Sci. Paris, Série 1, 297 (1983), 381-384.

[4] Karoubi, M. *Homologie cyclique et K-théorie algébrique. I.* C. R. Ac. Sci. Paris, Série 1, 297 (1983), 447-450.

[5] Kasparov, G. G. *Hilbert C*-modules: theorems of Stinespring and Voiculescu.* J. Operator Theory 4 (1980), 133-150.

[6] Mischenko, A. S. *Representations of compact groups in Hilbert modules over C*-algebras.* Trudy Mat. Inst. Steklov. 166 (1984), 161-176 (in Russian). English transl. in Proc. Steklov Inst. Math., 1986, No.1 (166).

[7] Mischenko, A. S. and Fomenko, A. T. *The index of elliptic operators over C*-algebras.* Izv. Akad. Nauk SSSR. Ser. Mat., 43 (1979), 831-859 (in Russian). English transl. in Math. USSR Izv., 15 (1980).

[8] Mischenko, A. S. and Pidstrigach, V. Ya. *On non-commutative local index formula.* in Baku International Topological Conference. Abstracts (Part II). Baku 1987, 244. (in Russian).

[9] Troitsky, E. V. *The index of equivariant elliptic operators over C*-algebras.* Ann. Global Anal. Geom., 5, No.1 (1987), 3-22.

[10] Troitsky, E. V. *An exact K-cohomological C*-index formula. I. The Thom isomorphism and the topological index.* Vestnik Mosk. Univ. Ser. 1. Mat. Meh., No.2 (1988), 83-85 (in Russian). English transl. in Moscow Univ. Math. Bull., to appear.

[11] Troitsky, E. V. *An exact K-cohomological C*-index formula. II. The index theorem and its applications.* Usp. Mat. Nauk, 44 (1989), No. 1, 213-214 (in Russian).

[12] Troitsky, E. V. *The contractibility of the full general linear group of the C*-Hilbert module* $l_2(A)$. Funkz. Anal. i Priloz., 20 (1986), No.4, 58-64 (in Russian). English transl. in Funct. Anal. Appl., to appear.

On the homotopy category of Moore spaces and an old result of Barratt

Hans Joachim Baues

Let $n \geq 1$, in this paper we describe a minimal algebraic model for the homotopy category $\underline{\underline{P}}_n/\simeq$ of Moores spaces $M(\mathbb{Z}/f,n)$ of cyclic groups $\mathbb{Z}/f$, $f \in \mathbb{N}$. For $n = 1$ we obtain the isomorphism of categories

(1)
$$\underline{\underline{P}}_1/\simeq \xrightarrow{\cong} \underline{\underline{R}}/\simeq$$

where $\underline{\underline{R}}$ is a category derived from group rings of cyclic groups, see (1.6) and (1.7). This seems to be the most elegant description of the category $\underline{\underline{P}}_1/\simeq$; results of Rutter [10] are immediate consequences of the isomorphism (1).

For $n \geq 2$ we show that the category $\underline{\underline{P}}_n/\simeq$ is a split linear extension of the category $\underline{\underline{F\ Cyc}}$ of cyclic groups, see (2.5). Moreover we compute the suspension functors

(2)
$$\underline{\underline{P}}_1/\simeq \xrightarrow{\Sigma} \underline{\underline{P}}_2/\simeq \xrightarrow{\Sigma} \underline{\underline{P}}_3/\simeq \xrightarrow{\Sigma} \dots$$

on these categories, see (2.7) and (2.5). Using these functors Σ we obtain a canonical splitting functor B_n of the homology functor H_n:

(3)
$$\underline{\underline{P}}_n/\simeq \underset{B_n}{\overset{H_n}{\rightleftarrows}} \underline{\underline{F\ Cyc}}, \quad n \geq 2,$$

compare (2.3). We determine the additive structure of $P_n/\simeq$, $n \geq 2$, by computing the term

(4)
$$\Delta(\varphi,\varphi') = B_n(\varphi + \varphi') - B_n(\varphi) - B_n(\varphi')$$

for $\varphi,\varphi' \in \mathrm{Hom}(\mathbb{Z}/f,\mathbb{Z}/g)$, see (2.10). Using this formula we obtain a new proof of an old result of Barratt [2] on the homotopy groups

(5)
$$[M(\mathbb{Z}/f,n),M(\mathbb{Z}/g,n)], \quad n \geq 2,$$

see (2.13) and (2.14). Our method for the computation of the group (5) is algebraic and very different from Barratt's highly involved geometrical techniques, compare the remark following (2.14). We also derive from (1) and (3) an algebraic description of the group of homotopy equivalences

(6) $$\mathrm{Aut}(M(\mathbb{Z}/f,n))^*, \quad n \geq 1 .$$

For $n = 1$ this yields an easy proof of a result of Olum [9], see (1.15). For $n \geq 2$ the description of the group (6) gives us the result of Sieradski [11], see (2.8). Our computation of the homotopy category $P_n/\simeq$, $n \geq 2$, also solves a problem of Barratt [1], compare the remark following (2.14).

In the first two sections § 1, § 2 we describe the main results of this paper. In section § 3 we recall some basic facts on crossed chain complexes which are the crucial tools in our proofs in section § 4. In particular we derive from the tensor product for crossed chain complexes (due to Brown–Higgins [6]) a formula for the crossed chain complex of the James construction $J(X)$ of a CW–complex X, see (3.5). This formula is essential in our computation of (4) and (5) above, see (4.5) and (4.9).

The author would like to acknowledge the support of the Max–Planck–Institut für Mathematik in Bonn.

§ 1. The homotopy category of pseudo projective planes

Pseudo projective planes, $P_f = M(\mathbb{Z}/f,1)$, are the most elementary 2–dimensional CW–complexes. They are obtained by attaching a 2–cell e^2 to a 1–sphere S^1 by an attaching map $f: S^1 \to S^1$ of degree $f \geq 1$, that is

(1.1) $$P_f = S^1 \cup_f e^2 = D/\sim_f .$$

Here D is the unit disk of complex numbers with boundary $S^1 = \partial D$ and with basepoint $* = 1$. The equivalence relation $\sim_f$ is generated by the relations $x \sim_f y \Leftrightarrow x^f = y^f$ with $x,y \in S^1$. Clearly $P_2 = \mathbb{R}P_2$ is the real projective plane. Let $\underline{\underline{P}}$ be the category consisting of pseudo projective planes P_f and of cellular maps. We consider the quotient functors

(1.2) $$\underline{\underline{P}} \longrightarrow \underline{\underline{P}}/\overset{0}{\simeq} \longrightarrow \underline{\underline{P}}/\simeq$$

where we use 0–homotopies ($\overset{0}{\simeq}$) running through cellular maps and homotopies ($\simeq$) relative $*$. Moreover, there is a canonical functor

$$\tau : \underline{\underline{\mathrm{Pair}}}(\mathbb{N}) \longrightarrow \underline{\underline{P}} \tag{1.3}$$

where $\underline{\underline{\mathrm{Pair}}}(\mathbb{N})$ is the category of pairs in the monoid $\mathbb{N}$ of natural numbers. Objects are elements $f \in \mathbb{N}$ and morphisms $f \longrightarrow g$ are pairs $(\xi,\eta) \in \mathbb{N} \times \mathbb{N}$ with $g\xi = \eta f$. Let $[f,g]$ be the set of such morphisms $(\xi,\eta) : f \longrightarrow g$. The functor τ carries f to P_f and (ξ,η) to the map $\tau_\xi : P_f \longrightarrow P_g$ with $\tau_\xi\{x\} = \{x^\xi\}$ for $x \in D$, see (1.1). The induced homomorphism

$$\pi_1(\xi,\eta) = \pi_1(\tau_\xi) : \pi_1(P_f) = \mathbb{Z}/f \longrightarrow \pi_1(P_g) = \mathbb{Z}/g \tag{1.4}$$

on fundamental groups is given by the number $\eta = g\xi/f$ which carries the generator $1 \in \mathbb{Z}/f$ to $\eta \cdot 1 \in \mathbb{Z}/g$. Clearly τ above is a faithful functor. We now introduce the natural equivalence relation $\simeq$ on $\underline{\underline{\mathrm{Pair}}}(\mathbb{N})$ which is generated by the relations

$$\begin{aligned} (\xi,\eta) \simeq (\xi',\eta') \qquad & \Leftrightarrow \eta,\eta' \equiv 0 \bmod g\,, \\ & \Leftrightarrow \pi_1(\xi,\eta) = \pi_1(\xi',\eta') = 0. \end{aligned}$$

(1.5) Theorem: The functor τ induces faithful functors

$$\tau : \underline{\underline{\mathrm{Pair}}}(\mathbb{N}) \rightarrowtail \underline{\underline{P}}/\overset{0}{\simeq}\,, \quad \text{and} \quad \tau : \underline{\underline{\mathrm{Pair}}}(\mathbb{N})/\simeq \; \rightarrowtail \underline{\underline{P}}/\simeq .$$

The image category of τ in $\underline{\underline{P}}/\simeq$ is the subcategory of principal maps in the sense of (V.§3) in Baues [3]. We now define a category $\underline{\underline{R}}$ which is actually a simple algebraic model of the category $\underline{\underline{P}}/\overset{0}{\simeq}$.

(1.6) Definition: The objects of the category $\underline{\underline{R}}$ are the elements $f \in \mathbb{N}$. A morphism $\lambda \in \underline{\underline{R}}(f,g)$ is an element $\lambda \in \mathbb{Z}[\mathbb{Z}/g]$ for which there is $\eta \in \mathbb{Z}$ with $g \cdot \epsilon(\lambda) = f \cdot \eta$. Here $\epsilon : \mathbb{Z}[\mathbb{Z}/g] \longrightarrow \mathbb{Z}$ is the augmentation of the group ring. Composition $\lambda \circ \mu$ for $\mu \in \underline{\underline{R}}(h,f)$ is defined by

$$\lambda \circ \mu = \lambda \cdot \lambda_\#(\mu) \tag{1}$$

where the right hand side is a product in the group ring $\mathbb{Z}[\mathbb{Z}/g]$. The homomorphism $\lambda_\# : \mathbb{Z}[\mathbb{Z}/f] \longrightarrow \mathbb{Z}[\mathbb{Z}/g]$ with $\lambda_\#[x] = [\eta x]$ is induced by the homomorphism $\pi_1(\lambda) = \eta : \mathbb{Z}/f \longrightarrow \mathbb{Z}/g$. Let

$$\partial_f = \sum_{x \in \mathbb{Z}/f} [x] \tag{2}$$

be the norm element in $\mathbb{Z}[\mathbb{Z}/f]$. We introduce a natural equivalence relation $\simeq$ on the category $\underline{\underline{R}}$ as follows $(\lambda,\mu \in \underline{\underline{R}}(f,g))$:

(3) $$\lambda \simeq \mu \Leftrightarrow \pi_1(\lambda) = \pi_1(\mu) \text{ and } \exists\ \beta \in \mathbb{Z}[\mathbb{Z}/g] \text{ with } \lambda - \mu = \lambda_{\#}(\partial_f) \cdot \beta ,$$

(1.7) Theorem: There are isomorphisms of categories

$$\rho : \underline{\underline{P}}/\underline{\underline{Q}} \xrightarrow{\sim} \underline{\underline{R}} \text{ , and } \rho : \underline{\underline{P}}\simeq \xrightarrow{\sim} \underline{\underline{R}} /\simeq .$$

Various results of Olum [9] and Rutter [10] are immediate consequences of this theorem. For $\varphi \in \mathrm{Hom}(\mathbb{Z}/f,\mathbb{Z}/g)$ let $[f,g]_\varphi$ and $[P_f,P_g]_\varphi$ be the set of all morphisms in $[f,g]$ and $[P_f,P_g]$ respectively which induce φ on fundamental groups, see (1.4). By (1.5) the function

(1.8) $$\tau : [f,g]_\varphi \longrightarrow [P_f,P_g]_\varphi$$

is injective for $\varphi \neq 0$ and is identically 0 if $\varphi = 0$. The group of integers $\mathbb{Z}$ acts freely on $[f,g]$ by $(\xi,\eta) + k = (\xi + kf,\ \eta + kg)$ and $[f,g]_\varphi$ is the orbit of (ξ,η) with $\pi_1(\xi,\eta) = \varphi$. On the other hand the coaction $P_f \longrightarrow P_f \vee S^2$ induces an action $+$ of the cohomology group

(1.9) $$E_\varphi = \hat{H}^2(P_f,\varphi^*\pi_2 P_g) = \pi_2 P_g/(\pi_2 P_g)\cdot \varphi_{\#}\partial_f ,$$

on the set $[P_f,P_g]_\varphi$ which is transitive and effective. The group $\pi_2 P_g$ can be described by each of the following equations

(1.10) $$\begin{aligned} \pi_2 P_g = H_2 \hat{P}_g &= \{x \in \mathbb{Z}[\mathbb{Z}/g] \mid \partial_g \cdot x = 0\} \\ &= \text{kernel } (\epsilon : \mathbb{Z}[\mathbb{Z}/g] \longrightarrow \mathbb{Z}) \\ &= ([0] - [1]) \cdot \mathbb{Z}[\mathbb{Z}/g] . \end{aligned}$$

Let $t : \mathbb{Z} \longrightarrow E_\varphi$ be the homomorphism mapping 1 to the class of

$$t_\varphi = f\cdot[0] - \varphi_{\#}\partial_f \in \text{kernel } (\epsilon) = \pi_2 P_g .$$

(1.11) Proposition: τ in (1.8) is t–equivariant or equivalently

$$\tau(\xi,\eta) + k\cdot t_\varphi = \tau(\xi + kf,\ \eta + kg) \text{ in } [P_f,P_g].$$

This result follows easily from (1.7).

We next derive from (1.7) a result on the group of homotopy equivalences $\mathrm{Aut}(P_f)^*$, in the category $\underline{P}/\simeq$. Let I be the ideal generated by the norm element ∂_f in $\mathbb{Z}[\mathbb{Z}/f]$ and let U_f be the group of units in the quotient ring $\mathbb{Z}[\mathbb{Z}/f]/I$. Moreover let $\tilde{U}_f$ be the group whose elements are those of U_f but with a multiplication

$$\{\lambda\} \circ \{\mu\} = \{\lambda \cdot \lambda_{\#}(\mu)\} .$$

Here $\{\lambda\}$ denotes the class of $\lambda \in \mathbb{Z}[\mathbb{Z}/f]$ modulo I .

(1.12) Proposition: There is an isomorphism of groups

$$\mathrm{Aut}(P_f)^* \xrightarrow{\cong} \tilde{U}_f .$$

Proof: Let E(f) be the group of equivalences of the object f in $\underline{R}/\simeq$. Then we have $\{\lambda\} \in E(f)$ iff there is μ with $\lambda\mu \simeq [0]$, $\mu\lambda \simeq [0]$. This is equivalent to $\mu \cdot (\mu_{\#}\lambda) \simeq [0]$ and $\pi_1\mu = (\pi_1\lambda)^{-1}$. This is the case iff

$$\begin{aligned} & \exists \beta \text{ with } \mu \cdot (\mu_{\#}\lambda) = [0] + \beta \cdot \mu_{\#}\partial_f \\ \Leftrightarrow\ & \exists \beta \text{ with } (\lambda_{\#}\mu) \cdot \lambda = [0] + (\lambda_{\#}\beta) \cdot \partial_f \\ \Leftrightarrow\ & \{\lambda\} \in U_f . \end{aligned}$$

Since the composition in $\mathrm{Aut}(P_f)^*$ corresponds to the composition in $\underline{R}$, we get the isomorphism for $\mathrm{Aut}(P_f)^*$.

//

We do not know whether the functor

$$\pi_1 : \underline{P}/\simeq \xrightarrow{\sim} \underline{R}/\simeq \twoheadrightarrow \underline{FCyc}$$

admits a splitting where $\underline{FCyc}$ is the category of finite cyclic groups $\mathbb{Z}/f$, $f \in \mathbb{N}$. However a splitting of the homomorphism

$$\pi_1 : \mathrm{Aut}(P_f)^* \cong \tilde{U}_f \twoheadrightarrow \mathrm{Aut}(\mathbb{Z}/f)$$

can be constructed as follows. For this we consider the commutative diagram

$$
(1.13) \qquad \begin{array}{ccc}
[f,g] & \xrightarrow{\quad\Gamma\quad} & \underline{\underline{R}}\,[f,g] \\
\downarrow \pi_1 & & \downarrow q \\
\mathrm{Hom}(\mathbb{Z}/f,\mathbb{Z}/g) & \overset{\tilde{\Gamma}}{--\to} & \underline{\underline{R}}\,[f,g]/\simeq
\end{array}
$$

Here Γ is defined for $(\xi,\eta) \in [f,g]$ by

$$\Gamma(\xi,\eta) = \sum_{j=0}^{\xi-1} [j\cdot\varphi 1] \in \mathbb{Z}[\mathbb{Z}/g]$$

with $\varphi = \pi_1(\xi,\eta)$ and q is the quotient map.

(1.14) Lemma: The function Γ induces a function $\tilde{\Gamma}$ such that (1.13) commutes.

Proof: We have to check that $\pi_1(\xi,\eta) = \pi_1(\xi',\eta') = \varphi$ implies $\Gamma(\xi,\eta) \simeq \Gamma(\xi',\eta')$:

$$
\begin{aligned}
\Gamma(\xi+f,\eta+g) &= \sum_{j=0}^{\xi+f-1} [j\cdot\varphi 1] = \Gamma(\xi,\eta) + \sum_{j=\xi}^{\xi+f-1} [j\cdot\varphi 1] \\
&= \Gamma(\xi,\eta) + \varphi_{\#}([\xi\cdot 1] \cdot \sum_{j=0}^{f-1} [j\cdot 1] \\
&= \Gamma(\xi,\eta) + \varphi_{\#}([\xi\cdot 1] \cdot \varphi_{\#}\partial_f
\end{aligned}
$$

//

(1.15) Proposition: Let U_f^1 be the group of units x in the quotient ring $\mathbb{Z}[\mathbb{Z}/f]/I$ with $\epsilon(x) = 1$. Then we have the split short exact sequence of groups

$$0 \to U_f^1 \to \mho_f \xrightarrow{\pi_1} \mathrm{Aut}(\mathbb{Z}/f) \to 0$$

where $\mho_f \cong \mathrm{Aut}(P_f)^*$ by (1.12). The splitting carries $\varphi \in \mathrm{Aut}(\mathbb{Z}/f)$ to $(\varphi^{-1})_{\#}\tilde{\Gamma}(\varphi)$.

(1.16) Remark: Proposition (1.15) is proved by different methods in (3.5) of Olum. It is known that there is an isomorphism of abelian groups

$$U_f^1 \cong \mathbb{Z}^{\chi} \oplus \mathbb{Z}/f\,,$$

where χ is the number of all $i \in \mathbb{N}$, $1 \leq i \leq f/2$, for which i is not a divisor of f. It is, however, a deep number theoretic problem to determine the action of $\mathrm{Aut}(\mathbb{Z}/f)$ on $\mathbb{Z}^{\chi} \oplus \mathbb{Z}/f$ in terms of basis elements. This action is defined by the split exact sequence (1.15).

//

§ 2. The homotopy category of suspended pseudo projective planes

We consider the suspensions

$$\Sigma^{n-1}P_f = M(\mathbb{Z}/f,n) = S^n \cup_f e^n \tag{2.1}$$

of pseudo projective planes, $n \geq 1$, which are Moore spaces of cyclic groups. Let $\underline{\underline{P}}_n$ be the category consisting of the spaces $\Sigma^{n-1}P_f$, $f \geq 1$, and of cellular maps. In section 1 above we studied the category $\underline{\underline{P}} = \underline{\underline{P}}_1$ of pseudo projective planes and its homotopy category $\underline{\underline{P}}/\simeq$. We here compute the suspension functor $\Sigma : \underline{\underline{P}}_n/\simeq \longrightarrow \underline{\underline{P}}_{n+1}/\simeq$, $n \geq 1$, which is an isomorphism of categories for $n \geq 3$. For this we consider the commutative diagram of functors

$$\begin{array}{ccccccc} \underline{\underline{\mathrm{Pair}}}(\mathbb{N}) & \xrightarrow{\tau} & \underline{\underline{P}}/\simeq & \xrightarrow{\Sigma} & \underline{\underline{P}}_2/\simeq & \xrightarrow{\Sigma} & \underline{\underline{P}}_3/\simeq \\ & & \downarrow \pi_1 & & H_2 & & H_3 \\ & & \underline{\underline{\mathrm{FCyc}}} & & & & \end{array} \tag{2.2}$$

where H_2 and H_3 are the homology functors. The next result seems to be new; recall that $[f,g]$ is the set of morphisms $f \longrightarrow g$ in $\underline{\underline{\mathrm{Pair}}}(\mathbb{N})$, $f,g, \in \mathbb{N}$, see (1.3).

(2.3) Theorem: Let $\varphi \in \mathrm{Hom}(\mathbb{Z}/f,\mathbb{Z}/g)$. Then there is a unique element $\overline{\varphi} = B_2(\varphi)$ in the image of

$$\Sigma\tau : [f,g] \longrightarrow [\Sigma P_f,\Sigma P_g]$$

with $H_2\overline{\varphi} = \varphi$. Moreover there is a unique element $\overline{\overline{\varphi}} = B_3(\varphi)$ in the image of

$$\Sigma\Sigma : [P_f,P_g] \longrightarrow [\Sigma^2 P_f,\Sigma^2 P_g]$$

with $H_3\overline{\overline{\varphi}} = \varphi$.

(2.4) Corollary: The functors H_n $(n = 2,3)$ in (2.2) admit a splitting functor

$$B_n : \underline{\mathrm{FCyc}} \longrightarrow \underline{\underline{P}}_n/\simeq$$

with $H_nB_n = 1$

This follows immediately from (2.3) since the definition of $B_n(\varphi)$ is compatible with compositions. The splitting functor B_n, however, is not additive; below we describe the distributivity law for $B_n(\varphi + \varphi')$. The functors H_n in (2.2) are part of the following commutative diagram in which the rows are split linear extensions of categories (compare IV. § 3 and V § 3a in [3])

(2.5)
$$\begin{array}{ccccc} E^2 & \rightarrowtail & \underline{\underline{P}}_2/\simeq & \xrightarrow{H_2} & \underline{\mathrm{FCyc}} \\ \downarrow{\sigma_*} & & \downarrow{\Sigma} & & \| \\ E^3 & \rightarrowtail & \underline{\underline{P}}_3/\simeq & \xrightarrow[H_3]{} & \underline{\mathrm{FCyc}} \end{array}$$

Here E^n is the bifunctor on $\underline{\mathrm{FCyc}}$ given by

(1)
$$E^n(\mathbb{Z}/f,\mathbb{Z}/g) = \mathrm{Ext}(\mathbb{Z}/f,\Gamma^1_n\mathbb{Z}/g)$$

where Γ^1_n is Whitehead's functor Γ for $n = 2$ and the functor $-\otimes\mathbb{Z}/2$ for $n \geq 3$. The group (1) is a cyclic group of order $(f,2g,g^2)$ for $n = 2$ and $(f,g,2)$ for $n \geq 3$ where the bracket $(\ldots)$ denotes the greatest common divisor. The natural transformation σ_* in (2.5) is induced by the surjection

(2)
$$\sigma : \Gamma(\mathbb{Z}/g) \longrightarrow \mathbb{Z}/g \otimes \mathbb{Z}/2$$

compare (IX. 4.4) [3]. The action of E^n on $\underline{\underline{P}}_n/\simeq$ is given by the well known central extension of groups

(3)
$$\mathrm{Ext}(\mathbb{Z}/f,\pi_{n+1}U) \overset{i}{\rightarrowtail} [\Sigma^{n-1}P_f,U] \xrightarrow{\pi_n} \mathrm{Hom}(\mathbb{Z}/f,\pi_nU)$$

which is known as the '*universal coefficient sequence*' compare Hilton [7] or (V.3a) in Baues [3]. For $U = \Sigma^{n-1}P_g$ we have $\pi_{n+1}U = \Gamma^1_n(\mathbb{Z}/g)$. The splitting B_n gives us an identification $(n \geq 2)$

(4)
$$\mathrm{Hom}(\mathbb{Z}/f,\mathbb{Z}/g) \times E^n(\mathbb{Z}/f,\mathbb{Z}/g) = [\Sigma^{n-1}P_f,\Sigma^{n-1}P_g]$$

which carries (φ,α) to $B_n(\varphi)+i(\alpha)$. The composition in $\underline{P}_n/\simeq$ then satisfies the simple formula

(5) $$(\varphi,\alpha)\circ(\Psi,\beta)=(\varphi\Psi,\varphi_*\alpha+\Psi^*\beta).$$

This indeed yields a very simple algebraic description of the category $\underline{P}_n/\simeq$. The suspension functor in (2.5) is given by $\Sigma(\varphi,\alpha)=(\varphi,\sigma_*\alpha)$. We now consider the image category of the functor $\Sigma:\underline{P}/\simeq\longrightarrow\underline{P}_2/\simeq$. Recall that $1\in\mathbb{Z}/f$ denotes the canonical generator.

(2.6) Definition: For maps $u,v:P_f\longrightarrow P_g$ in $\underline{P}$ we set $u\equiv v$ if $\Sigma f\simeq\Sigma g$. Whence the quotient category $\underline{P}/\equiv$ is the same as the image category $\Sigma(\underline{P}/\simeq)$. For morphisms $\lambda,\mu\in\underline{R}(f,g)$, see (1.6), we set $\lambda\equiv\mu$ if $\pi_1(\lambda)=\pi_1(\mu)$ and if for some $\beta\in\mathbb{Z}[\mathbb{Z}/g]$ with

$$\lambda-\mu-\epsilon(\lambda-\mu)[0]=([0]-[1])\cdot\beta$$

the greatest common divisor $(f,g^2,2g)$ divides $g\cdot\epsilon(\beta)$.

//

The following result shows that the image category $\Sigma(\underline{P}/\simeq)$ is surprisingly small. By (2.3) we know that the image category $\Sigma\Sigma(\underline{P}/\simeq)$ is isomorphic to $\underline{FCyc}$.

(2.7) Theorem: The isomorphism ρ in (1.7) induces an isomorphism of categories

$$\rho:\Sigma(\underline{P}/\simeq)=\underline{P}/\equiv\xrightarrow{\cong}\underline{R}/\equiv .$$

Moreover one has a split linear extension of categories

$$\tilde{E}+\!\!>\!\longrightarrow\underline{P}/\equiv\xrightarrow[\pi_1]{}\!\!\!\rightarrow\underline{FCyc}$$

where $\tilde{E}$ is the quotient of E^2 above with $\tilde{E}(\mathbb{Z}/f,\mathbb{Z}/g)=g\cdot\mathrm{Ext}(\mathbb{Z}/f,\Gamma(\mathbb{Z}/g))$. This group is $\mathbb{Z}/2$ if $(f,g^2,2g)=2g$ and in 0 otherwise. The splitting is given by B_2 in (2.3).

We derive from (2.6) and (2.7) the following commutative diagram in which the rows are split extensions of groups. Here $\mathrm{Aut}(X)^*$ denotes the group of homotopy classes of basepoint preserving homotopy equivalences of X.

$$(2.8)\qquad \begin{array}{ccccc} & & \Sigma Aut(P_f)^* & \cong & Aut(\mathbb{Z}/f) \\ & & \downarrow & & \| \\ \mathbb{Z}/f & \rightarrowtail & Aut(\Sigma P_f)^* & \twoheadrightarrow & Aut(\mathbb{Z}/f) \\ \downarrow & & \downarrow & & \| \\ \mathbb{Z}/(f,2) & \rightarrowtail & Aut(\Sigma^2 P_f)^* & \twoheadrightarrow & Aut(\mathbb{Z}/f) \end{array}$$

Using different methods the split extension for $Aut(\Sigma P_f)^*$ was obtained by Sieradski [11].

The morphism sets $[\Sigma P_f, \Sigma P_g]$ in $\underline{P}_2/\simeq$ are groups since the suspension ΣP_f is a co-H-group. As pointed out in (2.4) the splitting

$$(2.9)\qquad B_2 : Hom(\mathbb{Z}/f, \mathbb{Z}/g) \longrightarrow [\Sigma P_f, \Sigma P_g]$$

is not additive. We now describe the distributivity law for $B_2(\varphi + \varphi')$. Let

$$\Delta : Hom(\mathbb{Z}/f, \mathbb{Z}/g) \times Hom(\mathbb{Z}/f, \mathbb{Z}/g) \longrightarrow Ext(\mathbb{Z}/f, \Gamma\mathbb{Z}/g) \cong \mathbb{Z}/(f, 2g, g^2)$$

be the linear map which carries the pair (φ, φ') to the element

$$\Delta(\varphi, \varphi') = (f(f-1)/2)\varphi_1 \cdot \varphi_1' \cdot 1$$

where $\varphi(1) = \varphi_1 1$, $\varphi'(1) = \varphi_1' 1$. Then we get

(2.10) <u>Theorem</u>: $B_2(\varphi + \varphi') = B_2(\varphi) + B_2(\varphi') + \Delta(\varphi, \varphi')$

The splitting B_n, $n \geq 3$, satisfies the addition law $B_n(\varphi, \varphi') = B_n(\varphi) + B_n(\varphi') + \sigma_* \Delta(\varphi, \varphi')$. This follows from (2.5). The formula for Δ yields the following property.

(2.11) <u>Lemma</u>: Let $f = 2^a f_0$, $g = 2^b g_0$ where f_0 and g_0 are odd. Then we have $\Delta \neq 0$ iff $a = b \geq 1$ or $a = b + 1 \geq 2$ and we have $\sigma_* \Delta \neq 0$ iff $a = b = 1$.

Using the identification (2.5)(4) we can describe the group structure $+$ of the group $[\Sigma^{n-1} P_f, \Sigma^{n-1} P_g]$, $n \geq 2$, by the formula

$$(2.12)\qquad (\varphi, \alpha) + (\varphi', \alpha') = (\varphi + \varphi', \alpha + \alpha' + \Delta_n(\varphi, \varphi'))$$

where $\Delta_n = \Delta$ for $n = 2$ and $\Delta_n = \sigma_*\Delta$ for $n \geq 3$. This formula describes completely the additive structure of the category $\underline{\underline{P}}_n/\simeq$. Since $\Delta(\varphi,\varphi') = \Delta(\varphi',\varphi)$ we see that also the group $[\Sigma P_f,\Sigma P_g]$ is abelian for all $f,g \in \mathbb{N}$. The cyclic summands and explicit generators of this group are described in the next result.

(2.13) Corollary: Let $f = 2^a f_0$ and $g = 2^b g_0$ where f_0 and g_0 are odd. Then the homomorphism

$$H_2 : [\Sigma P_f,\Sigma P_g] \longrightarrow \mathrm{Hom}(\mathbb{Z}/f,\mathbb{Z}/g) \cong \mathbb{Z}/d$$

has an additive splitting of abelian groups if and only if $(a,b) \neq (1,1)$. Moreover for the greatest common divisors $d = (f,g)$ and $c = (f,g^2,2g)$ one has

$$[\Sigma P_f,\Sigma P_g] = \begin{cases} \mathbb{Z}/d \oplus \mathbb{Z}/c & \text{for } (a,b) \neq (1,1), \\ \\ \mathbb{Z}/2d \oplus \mathbb{Z}/(c/2) & \text{for } (a,b) = (1,1). \end{cases}$$

The generator of the first summand is $(\varphi_0,(f/4)1)$ if $a > b = 1$ and $(\varphi_0,0)$ otherwise where φ_0 is a generator of $\mathrm{Hom}(\mathbb{Z}/f,\mathbb{Z}/g)$. The generator of the second summand is $(0,1)$ if $(a,b) \neq (1,1)$ and is $(0,2\cdot 1)$ if $(a,b) = (1,1)$. Here we use again the identification in (2.5)(4).

(2.14) Addendum: The homomorphism

$$H_3 : [\Sigma^2 P_f,\Sigma^2 P_g] \longrightarrow \mathrm{Hom}(\mathbb{Z}/f,\mathbb{Z}/g) = \mathbb{Z}/d$$

has an additive splitting if and only if $(a,b) \neq (1,1)$. Moreover for $e = (f,g^2,2)$ one has

$$[\Sigma^2 P_f,\Sigma^2 P_g] = \begin{cases} \mathbb{Z}/d \oplus \mathbb{Z}/e & \text{for } (a,b) \neq (1,1), \\ \\ \mathbb{Z}/2d & \text{for } (a,b) = (1,1). \end{cases}$$

The generator of the first summand is $(\varphi_0,0)$ and the generator of the second summand $\mathbb{Z}/e$ is $(0,1)$.

Remark: The result in (2.13), (2.14) is due to Barratt [2], (table 2 in 10.6). Barratt uses Whitney's tube system for proving this result; his arguments are highly geometrical and totally different from our method. Hilton (p. 125) presents a different approach for the stable groups $[\Sigma^2 P_f,\Sigma^2 P_g]$ and points out that a more simple minded proof of Barratt's result is needed. A further improvement in the results above is the fact that we describe explicitly generators of the cyclic summands. The algebraic description of the category $\underline{\underline{P}}_n/\simeq$ by (2.5)(5) and (2.12) solves a problem of Barratt [1] who used generators and relations for the description of $\underline{\underline{P}}_n/\simeq$, $n \geq 3$. Our algebraic model of $\underline{\underline{P}}_n/\simeq$ is simpler and also available for $n = 2$.

//

Proof of (2.13): H_2 has a splitting if and only if there is α such that (φ_0,α) has order d. By the group law in (2.12) we obtain the formula

(1) $$(\varphi_0,\alpha)\cdot d = (0,\alpha\cdot d + \sum_{t=1}^{d-1} \Delta(\varphi_0,t\varphi_0)) .$$

We choose the generator $\varphi_0 = \pi_1(\xi,\eta)$ with $\eta = g/d$, see (1.4). Then we have

(2) $$\Delta(\varphi_0,t\varphi_0) = (f(f-1)/2)\eta\cdot t\eta\cdot 1$$

and therefore we have $(\varphi_0,\alpha)\cdot d = 0$ iff

(3) $$\alpha d = (f(f-1)/2)\eta\eta(d(d-1)/2)\cdot 1 .$$

If $a > b = 1$ we see that η and $d/2$ are odd. Thus $\alpha = (f/4)1$ satisfies the equation (3). Otherwise $\alpha = 0$ satisfies (3) for $(a,b) \neq (1,1)$. For $(a,b) = (1,1)$ we have $\alpha d = 0$ for all α, however, the right hand side of (3) is a non trivial element of order 2 in this case. This proves the proposition. If we reduce equation (3) modulo 2 then both sides of (3) are zero for $a > b = 1$. This shows that $B_3(\varphi)$ yields an additive splitting for if $(a,b) \neq (1,1)$.

//

Finally we consider the group structure of the homotopy groups with coefficients in $\mathbb{Z}/f$. As in (2.5)(3) we have the central extension of groups

(2.14) $$\mathbb{Z}/f \otimes \pi_{n+1}U \rightarrowtail [\Sigma^{n-1}P_f,U] \xrightarrow{\pi_n} Hom(\mathbb{Z}/f,\pi_n U)$$

where we identify $Ext(\mathbb{Z}/f,\pi_{n+1}U) = \mathbb{Z}/f \otimes \pi_{n+1}U$. This extension is completely determined by the following proposition which completes the partial results on the extension (2.14) in Hilton [7] (page 125–128).

(2.15) Proposition: For $x,y \in [\Sigma P_f,U]$ we have the commutator rule $(v = f(f-1)/2)$

$$-x-y+x+y = v1 \otimes [i^*x,i^*y]$$

where $i : S^2 \subset \Sigma P_f$ is the inclusion and where $[i^*x, i^*y] \in \pi_3 U$ is the Whitehead product. Moreover let $\mathbb{Z}\varphi$ be the subgroup of $\mathrm{Hom}(\mathbb{Z}/f, \pi_n U)$ generated by an element φ. Then there is a function $T : \mathbb{Z}\varphi \longrightarrow [\Sigma^{n-1}P_f, U]$ with $\pi_n T(x) = x$ for $x \in \mathbb{Z}\varphi$ and with $(r,s \in \mathbb{Z})$

$$-T((r+s)\varphi) + T(r\varphi) + T(s\varphi) = rtv1 \otimes (\eta^* \varphi(1))$$

where $\eta : S^{n+1} \longrightarrow S^n$ is the Hopf element.

Proof: The property of the commutator follows from the definition of the Whitehead product and the lemma on the reduced diagonal $\Delta : P_f \longrightarrow P_f \wedge P_f$ in (4.10) below, see for example II.1.12 in Baues [4]. Next let $\mathbb{Z}/g$ be the cyclic group generated by $\varphi(1)$ in $\pi_n U$. Then we can choose a map $F : \Sigma^{n-1}P_g \longrightarrow U$ with $i^*F = \varphi(1)$. Moreover an element $t\varphi \in \mathbb{Z}\varphi$ corresponds to a homomorphism $t\varphi : \mathbb{Z}/f \longrightarrow \mathbb{Z}/g$. Now we define T in (D.22) by $T(t\varphi) = F_* B_n(t\varphi)$ where B_n is the splitting in (2.4) and (2.10).

//

§ 3 Crossed chain complexes

Let $\underline{\underline{CW}}$ be the category of CW–complexes X with $X^0 = *$ and of cellular maps. Our main tool for the proofs of the results in § 1 and § 2 is the functor

$$\rho : \underline{\underline{CW}} \longrightarrow \underline{\underline{H}} \tag{3.1}$$

which carries X to the crossed chain complex $\rho(X)$. Here $\underline{\underline{H}}$ is the category of totally free crossed chain complexes which are called homotopy systems in Whitehead [12], compare also (VI. § 1) [3] where we set $D = *$ and $G = 0$. The crossed chain complex $\rho(X)$ is given by the sequence of boundary homomorphisms

$$\cdots \xrightarrow{d_4} \pi_3(X^3,X^2) \xrightarrow{d_3} \pi_2(X^2,X^1) \xrightarrow{d_2} \pi_1(X^1)$$

with $d_{n-1}d_n = 0$. The cells of X form a basis of the totally free crossed chain complex ρX, that is $\pi_1(X^1)$ is a free group generated by the 1–cells of X and d_2 is a free crossed module generated by the 2–cells of X, moreover $\pi_n(X^n,X^{n-1})$, $n \geq 3$, is a free $\pi_1(X)$–module generated by the n–cells of X. There is a notion of homotopy $\simeq$ for morphisms in $\underline{\underline{H}}$ such that ρ induces a functor $\rho : \underline{\underline{CW}}/\simeq \longrightarrow \underline{\underline{H}}/\simeq$ between homotopy categories. Here we use basepoint preserving homotopies for maps in $\underline{\underline{CW}}$ denoted by $f \simeq g$. Let $f \overset{0}{\simeq} g$ be a homotopy running through cellular maps.

(3.2) Theorem: The functor ρ induces equivalences of categories

$$\begin{cases} \rho : \underline{\underline{CW}}^2/\overset{0}{\simeq} \xrightarrow{\sim} \underline{\underline{H}}^2 , \\ \rho : \underline{\underline{CW}}^2/\simeq \xrightarrow{\sim} \underline{\underline{H}}^2/\simeq \end{cases}$$

where $\underline{\underline{CW}}^2$ and $\underline{\underline{H}}^2$ denote the full subcategories of 2–dimensional objects. Moreover $\rho : \underline{\underline{CW}}/\simeq \longrightarrow \underline{\underline{H}}/\simeq$ induces the map

$$\rho : [X,Y] \longrightarrow [\rho X, \rho Y]$$

between homotopy sets which is a bijection if $\dim X \leq 2$, $X,Y \in \underline{\underline{CW}}$.

This theorem is an old result of J.H.C. Whitehead [12], it is as well proved in chapter VI of [3], (compare (VI.3.5) and (VI.6.5)). We use the theorem as the main tool in the proofs of § 4. We shall also use the tensor product of Brown–Higgins [6] which gives us a functor $\otimes : \underline{\underline{H}} \times \underline{\underline{H}} \longrightarrow \underline{\underline{H}}$ such that there is a natural isomorphism

$$\rho(X \times Y) = \rho(X) \otimes \rho(Y) . \tag{3.3}$$

Here $X \times Y$ is the product with the CW–topology given by product cells $e \times f$. The crossed chain complex $A \otimes B$ is generated by elements $a \otimes b$, $a \otimes *$, $* \otimes b$ where $a \in A$, $b \in B$ with the following defining relations (plus, of course, the laws for crossed chain complexes):

(1) $|a \otimes b| = |a| + |b|$, $|a \otimes *| = |a|$, $|* \otimes b| = |b|$.

(2) $(*\otimes b)^{*\otimes t} = *\otimes(b^t)$ for $|t|=1$ and $(a\otimes b)^{*\otimes t} = a\otimes(b^t)$ for $|t|=1$, $|b|\geq 2$,
$(a\otimes *)^{s\otimes *} = (a^s)\otimes *$ for $|s|=1$ and $(a\otimes b)^{s\otimes *} = (a^s)\otimes b$ for $|s|=1$, $|a|\geq 2$.

(3) $(a + a') \otimes * = a \otimes * + a' \otimes *$,
$(a + a') \otimes b = a \otimes b + a' \otimes b$ for $|a| \geq 2$,
$(a + a') \otimes b = (a \otimes b)^{a' \otimes *} + a' \otimes b$ for $|a| = 1$.

(4) $* \otimes (b + b') = * \otimes b + * \otimes b'$,
$a \otimes (b + b') = a \otimes b + a \otimes b'$ for $|b| \geq 2$,
$a \otimes (b + b') = a \otimes b' + (a \otimes b)^{* \otimes b'}$ for $|b| = 1$.

(5) $d(a \otimes *) = (da) \otimes *$, $d(* \otimes b) = * \otimes (db)$ and $d(a \otimes b) =$

$$\begin{cases} -a \otimes * - * \otimes b + a \otimes * + * \otimes b & \text{for } |a| = |b| = 1 \quad , \\ [-(* \otimes b)^{a \otimes *} + * \otimes b] - a \otimes db & \text{for } |a| = 1 \ , \quad |b| \geq 2 \, , \\ (da) \otimes b + (-1)^{|a|}[-(a \otimes *)^{* \otimes b} + a \otimes *] & \text{for } |a| \geq 2 \ , \quad |b| = 1 \, , \\ (da) \otimes b + (-1)^{|a|} a \otimes (db) & \text{for } |a| \geq 2 \ , \quad |b| \geq 2 \, . \end{cases}$$

Moreover $A \otimes B$ is totally free if A and B are totally free, a basis of $A \otimes B$ is given by the elements $* \otimes b$, $a \otimes *$, $a * b$ where a and b are basis elements of A and B respectively. The isomorphism (3.3) carries $e \times f$ to $e \otimes f$.

Next we consider the James construction which is a functor $J : \underline{\underline{CW}} \longrightarrow \underline{\underline{CW}}$ given by the direct limit $JX = \varinjlim J_n X$ where $J_n X = (X \times \ldots \times X)/\sim$ is given by the relations $(x_1,\ldots,x_{n-1},*) \sim (x_1,\ldots,x_{t-1},*,x_t,\ldots,x_{n-1})$ for $t = 1,\ldots,n$. It is a classical result of James [8] that there is a natural homotopy equivalence

(3.4) $$J(X) \simeq \Omega\Sigma X$$

for X in $\underline{\underline{CW}}$. Using (3.3) we obtain a functor $J : \underline{\underline{H}} \longrightarrow \underline{\underline{H}}$ together with a natural isomorphism

(3.5) $$\rho J(X) = J\rho(X) \, .$$

Let A be a crossed chain complex. The crossed chain complex JA is generated by all words $a_1 \ldots a_n$ $(a_i \in A, i = 1,\ldots,n$ and $n \geq 1)$ with the following defining relations (plus, of course, the laws of crossed chain complexes). Let u,v be such words or empty words ϕ and let $a,a',t \in A$. Then (a) denotes the word given by $a \in A$.

(1) $$|a_1 \ldots a_n| = |a_1| + \ldots + |a_n| \, .$$

(2) $$(uav)^t = u(a^t)v \ \text{ for } |t| = 1, \ |a| \geq 2 \ \text{ and } (a)^t = (a^t) \ \text{ for } |t| = 1 \, .$$

(3) $$u(a + a')v = \begin{cases} uav + ua'v & \text{for } |a| \geq 2 \\ ua'v + (uav)^{a'} & \text{for } |a| = 1, \ |u| \geq 1 \\ (uav)^{a'} + ua'v & \text{for } |a| = 1, \ |v| \geq 1 \\ (a) + (a') & \text{for } u = \phi = v \end{cases}$$

(4) $$d(a) = (da) \text{ and } d(uv) = \begin{cases} -u-v+u+v & \text{for } |u| = |v| = 1, \\ -v^u+v-u(dv) & \text{for } |u| = 1, |v| \geq 2, \\ (du)v + (-1)^{|u|}(-u^v+u) & \text{for } |u| \geq 2, |v| = 1, \\ (du)v + (-1)^{|u|}u(dv) & \text{for } |u| \geq 2, |v| \geq 2. \end{cases}$$

For a map $F : A \longrightarrow B$ in $\underline{\underline{H}}$ the induced map $JF : JA \longrightarrow JB$ is defined by $(JF)(a_1...a_n) = (Fa_1)...(Fa_n)$. There is a well defined natural map

(5) $$\mu : JA \otimes JA \longrightarrow JA$$

given by $\mu(u \otimes v) = uv$, $\mu(* \otimes v) = v$, $\mu(u \otimes *) = u$. Therefore JA is an example of a '*crossed chain algebra*'.

//

One can check that JA is totally free if A is totally free. In fact, if Z is a basis of A then $\text{Mon}(Z) - *$ is a basis of JA: Here Mon(Z) is the free monoid generated by Z. As an application of (3.2) we get:

(3.6) Corollary: Let X,Y be CW–complexes in $\underline{CW}$ with $\dim(X) \leq 2$. Then one has the binatural isomorphism of groups

$$[\Sigma X, \Sigma Y] \cong [X, JY] \cong [\rho X, J\rho Y]$$

where the group structure in $[\rho X, J\rho Y]$ is induced by μ in (3.5) (5). As a special case one gets $\pi_3(\Sigma Y) = \pi_2(J\rho Y)$.

A more detailed study of the James construction of crossed chain complexes can be found in Baues [5].

§ 4. Proofs

We here prove the main results of § 1 and § 2. Using (3.2) and (3.6) these proofs turn out to be purely algebraic. This indeed is an advantage compared with the longwinded sequence of geometric arguments of Barratt [2]. For a more detailed discussion of the following proofs see Baues [5].

We first observe that the group $\pi_2(P_f,S^1)$ is abelian. Let e_2 be the 2–cell of P_f and let e be the 1–cell of P_f. Then the elements e_2^{ne}, $n \in \mathbb{Z}$, generate the group $\pi_2(P_f,S^1)$. The commutators satisfy the formula

$$(4.1)\qquad -e_2^{ne} - e_2^{me} + e_2^{ne} + e_2^{me} = \langle e_2^{ne}, e_2^{me} \rangle - \langle e_2^{me}, e_2^{me} \rangle .$$

Here $\langle x,y \rangle = -x-y+x+y^{\partial x}$ is the Peiffer commutator which is trivial in the crossed module $d_2 : \pi_2(P_f,S^1) \longrightarrow \pi_1(S^1)$.

(4.2) Proof of (1.7): Since $\pi_2(P_f,S^1)$ is abelian we have an isomorphism

$$h_2 : \pi_2(P_f,S^1) \cong \mathbb{Z}[\mathbb{Z}/f] ,$$

this follows from a result of J.H.C. Whitehead [12], compare for example (VI.1.12) in Baues [3]. As a special case of diagram (3) in (VI.1.14) [3] we obtain the commutative diagram

$$\begin{array}{ccc}
\pi_2(P_f,S^1) & \xrightarrow{\ d_2\ } & \pi_1(S^1) = \mathbb{Z} \\
{\scriptstyle h_2}\downarrow & f\cdot\epsilon & \downarrow{\scriptstyle h_1} \\
\mathbb{Z}[\mathbb{Z}/f] & \xrightarrow[\ d\ ]{} & \mathbb{Z}[\mathbb{Z}/f]
\end{array}$$

where $d(x) = x \cdot \partial_f$ is given by the norm element ∂_f in (1.6). The boundary d describes the cellular chain complex of the universal covering of P_f and h_1 is a $(\mathbb{Z} \longrightarrow \mathbb{Z}/f)$–crossed homomorphism. Using the isomorphism h_2 we can identify the crossed module $d_2 = \rho(P_f)$ with the map $f\cdot\epsilon$ where ϵ is the augmentation of the group ring $\mathbb{Z}[\mathbb{Z}/f]$. We now restrict the functor ρ in (3.1) to the subcategory $\underline{\underline{P}} \subset \underline{\underline{CW}}$. The functor ρ carries P_f to the totally free crossed module $f\cdot\epsilon$ and carries a map $F : P_f \longrightarrow P_g$ to a map $(\xi,\eta) : f\cdot\epsilon \longrightarrow g\cdot\epsilon$ which is given by a commutative diagram

$$\begin{array}{ccc}
\mathbb{Z}[\mathbb{Z}/f] & \xrightarrow{\ f\cdot\epsilon\ } & \mathbb{Z} \\
\downarrow{\scriptstyle \xi} & & \downarrow{\scriptstyle \eta} \\
\mathbb{Z}[\mathbb{Z}/g] & \xrightarrow[\ g\cdot\epsilon\ ]{} & \mathbb{Z}
\end{array}$$

We identify the full subcategory of $\underline{\underline{H}}^2$ consisting of the objects $f\cdot\epsilon$, $f \in \mathbb{N}$, with the category $\underline{\underline{R}}$ defined in (1.6). The identification carries the morphism (ξ,η) in $\underline{\underline{H}}^2$ to the morphism

$\lambda = \xi[1]$ in $\underline{\underline{R}}$. This proves that $\rho : \underline{P}/\underline{\underline{Q}} \xrightarrow{\sim} \underline{\underline{R}}$ is the restriction of the first equivalence in (3.2). A homotopy $\alpha : (\xi,\eta) \simeq (\xi',\eta')$ in $\underline{\underline{H}}^2$ is an η-crossed homomorphism $\alpha : \mathbb{Z} \longrightarrow \mathbb{Z}[\mathbb{Z}/g]$ which is determined by an element $\alpha(1) = \beta$ as in (1.6) (3). The equation $-\xi + \xi' = \alpha(f\epsilon)$ is equivalent to the equation

$$-\lambda + \lambda' = -\xi[1] + \xi'[1] = \alpha(f) = \alpha(1) \cdot (\eta_{\#}\partial_f)$$

which is equivalent to the equation in (1.6) (3).

//

(4.3) <u>Proof of</u> (1.5): The functor ρ in (1.7) carries τ_ξ to the element $\xi \cdot [0] \in \underline{R}(f,g) \subset [\mathbb{Z}/g]$ where $[0]$ is the unit of the ring $\mathbb{Z}[\mathbb{Z}/g]$. By (1.7) we know

$$\tau(\xi,\eta) \simeq \tau(\xi',\eta') \Leftrightarrow \varphi = \pi_1(\xi,\eta) = \pi_1(\xi',\eta') \text{ and}$$
$$\exists \beta \in \mathbb{Z}[\mathbb{Z}/g] \text{ with } (\xi-\xi')[0] = \beta \cdot \varphi_{\#}(\partial_f).$$

This implies $\eta - \eta' = \epsilon(\beta) \cdot g$. We now observe

(1)
$$\varphi_{\#}\partial_f = \sum_{x\in\mathbb{Z}/f} [\varphi x] = t \cdot \sum_{y\in\varphi\mathbb{Z}/f} [y]$$

where t is the number of elements in the kernel of φ. For $\beta = \sum_{y\in\mathbb{Z}/g} a_y[y]$ in $\mathbb{Z}[\mathbb{Z}/g]$ we have

(2)
$$\beta \cdot \varphi_{\#}\partial_f = t \cdot \sum_{y\in\mathbb{Z}/g,\ v\in\varphi\mathbb{Z}/f} a_y[y+v],$$
$$= t \cdot \sum_{u\in\mathbb{Z}/g} \Big(\sum_{y\in u+\varphi\mathbb{Z}/f} a_y \Big)[u].$$

Now $\beta \cdot \varphi_{\#}\partial_f = (\xi'-\xi)[0]$ with $\varphi = \pi_1\tau_\xi = \pi_1\tau_{\xi'}$ implies

(3)
$$0 = \sum_{y\in u+\varphi\mathbb{Z}/f} a_y \quad \text{for } u \in \mathbb{Z}/g,\ u \neq 0.$$

The number of elements in $\varphi\mathbb{Z}/f$ is $g/\gcd(\eta,g)$. If $\gcd(\eta,g) < g$ we add up the equations in (3) for $u \in U = \{x\cdot 1; 1 \leq y \leq \gcd(\eta,g)\} \subset \mathbb{Z}/g$. Since the union of all $u + \varphi\mathbb{Z}/f$, $u \in U$, is $\mathbb{Z}/g$ we get $\epsilon(\beta) = \sum_y a_y = 0$ for $\gcd(\eta,g) < g$. Whence in this case $\eta = \eta'$ and thus $(\xi,\eta) = (\xi',\eta')$. If $\gcd(\eta,g) = g$, that is, if $\varphi = 0$ we see by (3) that $a_y = 0$ for $y \neq 0$ and $a_0 = \xi' - \xi$, so that in this case β exists.

//

The crucial step for the proof of (2.3) and (2.7) is the computation of the suspension homomorphism Σ on $\pi_2 P_g$. For this we consider the diagram

(4.4)
$$\begin{array}{ccc} \pi_2(P_g) & \xrightarrow{\quad\Sigma\quad} & \pi_3 \Sigma P_g \\ \cong\Vert & & \cong\Vert \\ ([0]-[1])\mathbb{Z}[\mathbb{Z}/g] & \xrightarrow{\quad\Sigma'\quad} & \mathbb{Z}/(g^2,2g) \end{array}$$

where Σ' is defined by the formula $\Sigma'(([0]-[1])\beta) = g\cdot\epsilon(\beta)\cdot 1$ for $\beta \in \mathbb{Z}[\mathbb{Z}/g]$. Here ϵ is the augmentation. The right hand isomorphism carries the generator 1 to the Hopf element $S^3 \longrightarrow S^2 \subset \Sigma P_g$. The left hand side isomorphism is described in (1.12).

(4.5) <u>Lemma</u>: Diagram (4.4) commutes. This shows that $\Sigma\pi_2(P_g)$ is of order 2 if g is even, and is trivial if g is odd. Moreover the subgroup $\Sigma^2\pi_2(P_g) = 0$ is trivial in $\pi_4(\Sigma^2 P_g)$ for all g.

<u>Proof</u>: We use (3.6) so that Σ in (4.4) corresponds to the map

(1)
$$i_* : \pi_2(\rho) \longrightarrow \pi_2 J(\rho) = H_2 CJ(\rho)$$

where $\rho = \rho(P_g)$ and where $i : \rho \subset J(\rho)$ is the inclusion. Let $e = e_1$ and e_2 be the cells of P_g which are the generators of ρ with $d(e_2) = g\cdot e = e+\ldots+e$. The boundary $d : J(\rho)_3 \longrightarrow J(\rho)_2$ is given on generators by the following formulas.

(2)
$$\begin{aligned} d(ee_2) &= -e_2^e + e_2 - e(g\cdot e) \\ &= -e_2^e + e_2 - \sum_{i=0}^{g-1} (ee)^{ie} \end{aligned}$$

(3)
$$\begin{aligned} d(e_2 e) &= (g\cdot e)e - e_2^e + e_2 \\ &= \Big(\sum_{i=1}^{g} (ee)^{(g-i)e}\Big) - e_2^e + e_2 \end{aligned}$$

(4)
$$d(eee) = -(ee)^e + ee$$

These equations are simple applications of (3.5) (4) since $d(ee) = 0$. The same equation hold in $CJ(\rho)$, this is the cellular chain complex of the universal covering of $J(P_g)$, compare the definition of the functor C in (VI.1.2) of Baues [3]. Let $B = d(CJ\rho)_3$ be the group of boundaries. Then we get modulo B the congruences (see (1.6)(2))

(5) $$e_2[0]-[1]) \equiv (ee)\cdot \partial_g \equiv (ee)(g\cdot[0])\,.$$

The Hopf element in (2.9) corresponds to the cycle ee in $CJ(\rho)$. Whence (5) shows that Σ' is defined correctly for $\beta=[0]$. For general β we can choose $\gamma \in \mathbb{Z}[\mathbb{Z}/g]$ such that $\beta=\epsilon(\beta)[0]+\gamma([0]-[1])$. Then we get

(6) $$e_2([0]-[1])\beta \equiv (ee)\partial_g\beta = (ee)\partial_g\epsilon(\beta) \equiv (ee)(g\cdot\epsilon(\beta)[0])$$

since $([0]-[1])\partial_g = 0$ for the norm element ∂_g. This proves (2.10).

//

(4.6) Proof of (2.3) and (2.7): The second part of (2.3) and also (2.7) follow immediately from lemma (2.10) since Σ is compatible with the coaction on P_f. For the first part of (2.3) we have to check that $\Sigma' t_\varphi$ is trivial if considered as an element in the group

(1) $$\mathrm{Ext}(\mathbb{Z}/f,\Gamma(\mathbb{Z}/g)) = \mathbb{Z}/(f,2g,g^2)\,.$$

Here we use (1.13). By (1.9)(1) we know

(2) $$t_\varphi = f\cdot[0]-\varphi_\#\partial_f = t\cdot(v[0]-\sum_{x\in V}[x])$$

where $t=|\ker\varphi|$, $t\cdot v=f$, $V=\mathrm{image}(\varphi)$ with $|V|=v$. Since $t_\varphi \in \ker(\epsilon)$ there is β with

(3) $$v[0]-\sum_{x\in V}[x] = ([0]-[1])\beta\,.$$

This shows that there is an integer b with

(4) $$\epsilon(\beta) = g\cdot b - u(v(v-1))/2$$

where $u\cdot v=g$. Whence $\Sigma'(t_\varphi)$ is given by the element

(5) $$\Sigma'(t_\varphi) = t\cdot g\cdot\epsilon(\beta)\cdot 1 = -(tg^2(v-1)/2)1$$

in $\mathbb{Z}/(g^2,2g)$. Now it is clear that (5) represents the trivial element in the group (1).

//

The proof of (2.10) is based on the following lemma on commutators $(a,b)=-a-b+a+b$.

(4.7) Lemma: Let $G = <a,b>$ be the free group generated by elements a and b and let $f \in \mathbb{N}$. Then there exists elements $\xi_i \in G$ $(i = 1,2,\dots,f(f-1)/2)$ such that

$$(a+b)\cdot f = a\cdot f + b\cdot f - \sum_{i=1}^{f(f-1)/2} (a,b)^{\xi_i}.$$

Here we set $x\cdot f = x+\dots+x$ (f–times x) and we set $x^y = -y + x + y$. The sum is the ordered sum in the non commutative group G.

Proof: We show inductively that there are $\alpha_i \in G$ with

(1) $$(a+b)\cdot f = a\cdot f + b\cdot f - \sum_{i=1}^{f-1} (a,b\cdot i)^{\alpha_i}.$$

This is true for $f = 1$. Now we get for $(a+b)(f+1)$:

$$\begin{aligned}(a+b)f+(a+b) &= a\cdot f+b\cdot f - \sum_{i=1}^{f-1} (a,bi)^{\alpha_i} + (a+b)\\ &= a\cdot f+b\cdot f+(a+b) - \sum_{i=1}^{f-1} (a,bi)^{\alpha_i+a+b}\\ &= a\cdot(f+1)+b(f+1) - (a,bf)^b - \sum_{i=1}^{f-1} (a,bi)^{\alpha_i+a+b}.\end{aligned}$$

This proves (1). Moreover there are $\beta_j \in G$ with

(2) $$(a,b\cdot i) = \sum_{j=1}^{i} (a,b)^{\beta_j}.$$

This follows inductively from $(a,y+z) = (a,z) + (a,y)^z$ where we set $y = b\cdot i$ and $z = b$. From (1) and (2) we derive the proposition.

//

We derive from (4.7) the following algebraic description of the diagonal $\Delta : P_f \longrightarrow P_f \times P_f$.

(4.8) Corollary: Let e, e_2 be the generators of $\rho = \rho(P_f)$. Then $a = e\otimes *$, $b = *\otimes e$ are generators of $\rho\otimes\rho$ so that $(\rho\otimes\rho)_1 = <a,b>$. Moreover we obtain a map $\Delta : \rho \longrightarrow \rho\otimes\rho$ by $\Delta(e) = a+b$ and

$$\Delta(e_2) = e_2 \otimes * + *\otimes e_2 - \sum_{i=1}^{f(f-1)/2} (e\otimes e)^{\xi_i}$$

where the ξ_i are the elements in (4.7). The map Δ satisfies $p_1\Delta = 1$ and $p_2\Delta = 1$ where p_1,p_2 are the projections of the tensor product $\rho \otimes \rho$.

Since we have $[P_f,P_f{\times}P_f] = [\rho,\rho\otimes\rho]$ by (3.2) we see that Δ in (4.8) represents the homotopy class of the topological diagonal of P_f.

(4.9) Proof of (2.10): The group addition in the group

(1)
$$[\Sigma P_f,\Sigma P_g] = [\rho P_f,J\rho P_g]$$

can be described by the composition $F + F' = G$,

(2)
$$G : \rho P_f \xrightarrow{\Delta} \rho P_f \otimes \rho P_f \xrightarrow{F\otimes F'} J\rho P_g \otimes J\rho P_g \xrightarrow{\mu} J\rho P_g\,,$$

where Δ is the diagonal in (4.8). Now we choose (ξ,η), resp. $(\xi',\eta') \in [f,g]$ which induces φ, resp. φ', in (2.10). We may set $\eta = \varphi_1, \eta' = \varphi_1'$. Moreover let F, resp. F', be given by

(3)
$$\begin{cases} F(e) = \eta e,\ F(e_2) = \xi e_2\,, \\ \\ F'(e) = \eta' e,\ F'(e_2) = \xi' e_2\,. \end{cases}$$

Then the composition G in (2) represents $B_2(\varphi) + B_2(\varphi')$. Explicitly we get G by the formulas:

(4)
$$\begin{cases} G(e) = (\eta + \eta')e\ , \\ G(e_2) = (\xi + \xi')e_2 - \sum_{i=1}^{v} (\eta e)\cdot(\eta' e)^{\mu(\xi_i)} \end{cases}$$

where $v = f(f-1)/2$ and $\eta e = e+\dots e = \eta$-fold sum of e. On the other hand $B_2(\varphi + \varphi')$ is represented by the map $G' : \rho P_f \longrightarrow J\rho P_g$ with $G'(e) = (\eta + \eta')e$ and $G'(e_2) = (\xi + \xi')e_2$. This shows that $\Delta(\varphi,\varphi')$ is represented by the element

(5)
$$\alpha = \sum_{i=1}^{v} (\eta e)(\eta' e)^{\mu(\xi_i)} \in \text{kernel}(d_2)$$

where d_2 is the boundary of $J\rho(Pg)$. We know that $\pi_2 J\rho P_g = \mathrm{kernel}(d_2)/\mathrm{image}(d_3)$ is generated by the cycle ee. Whence we have to show

(6) $$\alpha \equiv v \cdot \eta \cdot \eta'(ee) \ \text{ modulo image } (d_3)\,.$$

This is easily checked by use of (4.5) (4) and (3.5) (3). Therefore the proof of (2.19) is complete. //

We also derive from (4.8) the following well known result on the reduced diagonal of P_f.

(4.10) Corollary: The following diagram homotopy commutes.

$$\begin{array}{ccc} P_f & \xrightarrow{\ \Delta\ } & P_f \times P_f \\ \downarrow q & & \downarrow q \\ S^2 \xrightarrow{\ v\ } S^1 \wedge S^2 & \subset & P_f \wedge P_j \end{array}$$

Here v is a map of degree $f(f-1)/2$ and q denotes the quotient maps. Recall that the smash product $A \wedge B$ is defined by the quotient $A \wedge B = A \times B/(A \vee B)$. We obtain (4.10) directly from (4.8) since $\rho(P_f \wedge P_f) = p(P_f) \otimes \rho(P_f)/\rho(P_f \vee P_f)$.

Literature

[1] Barratt, M.G.: (HR) Homotopy ringoids and homotopy groups. Q. J. Math. Ox (2) 5 (1954) 271–290.

[2] Barratt, M.G.: (T) Track groups (II). Proc. London Math. Soc. (3) 5 (1955) 285–329.

[3] Baues, H.J.: Algebraic Homotopy, Cambridge Studies in Advanced Mathematics 15. Cambridge University Press (1988) 450 pages.

[4] Baues, H.J.: Commutator calculus and groups of homotopy classes. London Math. Soc. LNS 50 (1981).

[5] Baues, H.J.: Combinatorial homotopy and 4–dimensional complexes, to appear in Walter de Gruyter Verlag (350 pages).

[6] Brown, R. and Higgins, P.J.: Tensor products and homotopies for ω–groupoids and crossed complexes. Journal of Pure and Applied Algebra 47 (1987) 1–33.

[7] Hilton, P.: Homotopy theory and duality. Nelson (1965) Gordon Breach.

[8] James, I.M.: Reduced product spaces, Ann. of Math. 62 (1955), 170–97.

[9] Olum, P.: Self–equivalences of pseudo–projective planes. Topology 4 (1965) 109–127.

[10] Rutter, J.W.: Homotopy classification of maps between pseudo–projective planes. Quaestiones Math. 11 (1988) 409–422.

[11] Sieradski, A.: A semigroup of simple homotopy types. Math. Zeit. 153 (1977) 135–148.

[12] Whitehead, J.H.C.: Combinatorial homotopy II. Bull. AMS 55 (1949) 213–245.

H.J. Baues
Max–Planck–Institut für Mathematik
Gottfried–Claren–Straße 26
5300 Bonn 3

BRD

AN ADDITIVE BASIS FOR THE COHOMOLOGY OF REAL GRASSMANNIANS

Jan Jaworowski
Department of Mathematics, Indiana University
Bloomington, Indiana 47405 EMail: JAWOROWS@IUBACS

1. Introduction

Let $G_m(\mathbb{R}^{m+n})$ be the Grassmann manifold of m-dimensional subspaces of $\mathbb{R}^{m+n}$; and let $G_m = \varinjlim G_m(\mathbb{R}^{m+n})$ be the corresponding infinite Grassmannian. There are at least two descriptions of the cohomology ring H^*G_m (with coefficients in Z_2): Ehresmann [4] constructed an explicit and, in a sense, minimal, cell decomposition of the Grassmann manifolds based on ideas dating back to Schubert [8]. Using this decomposition, Chern gave a description of the ring structure in Grassmannians in terms of the so-called Schubert cycles. On the other hand, Borel [1] showed that H^*G_m is isomorphic to the polynomial algebra $Z_2[w_1,\cdots,w_m]$, where $w_1,\cdots,w_m$ are the universal Stiefel-Whitney classes, $w_i \in H^iG_m$. An excellent account of the cohomology of Grassmannians is given in [6]. Borel also showed that $H^*G_m(R^{m+n})$ is the quotient of this algebra by the ideal generated by the relation $w\overline{w} = 1$, where $w = 1 + w_1 +\cdots+ w_m$ is the total Stiefel-Whitney class of the canonical m-plane bundle over $G_m(\mathbb{R}^{m+n})$, and $\overline{w} = 1 + \overline{w}_1 + \cdots + \overline{w}_n$ is its dual. This quotient algebra can also be described as follows. Let $J(m,n)$ be the ideal in $Z_2[w_1,\cdots,w_m]$ generated by $\overline{w}_{1+n},\dots,\overline{w}_{m+n}$ expressed as polynomials in $w_1,\dots,w_m$ by using the relation $w\overline{w} = 1$. Then

$$H^*G_m(\mathbb{R}^{m+n}) \cong Z_2[w_1,\cdots,w_m]/J(m,n).$$

The Chern description, extended to G_m, is as follows. A <u>Schubert symbol</u> (of length m) is a sequence $(a_1,\dots,a_m)$ of integers satisfying $0 \le a_1 \le \cdots \le a_m$. There is a bijection between the set of Schubert symbols $(a_1,\cdots,a_m)$ satisfying $a_1 + \dots + a_m = d$ and the set of d-cells of the Chern-Ehresmann decomposition of G_m. We will also say that $d = a_1 + \cdots + a_m$ is the dimension of $(a_1,\cdots,a_m)$. Note that a Schubert symbol $(a_1,\cdots,a_m)$ with $a_1 + \cdots + a_m = d$ corresponds to a partition of d into at most m positive integers. The Schubert symbols of length m and dimension d form an additive basis for the vector space H^dG_m. This is a consequence of the fact that there is a bijection between the set of Schubert symbols $(a_1,\cdots,a_m)$ of dimension d and the set of monomials $w_1^{r_1}\cdots w_m^{r_m}$ of a total degree $d = \sum_{i=1}^{m} ir_i$ (see [6], p. 85). As for $G_m(\mathbb{R}^{m+n})$, it follows that the Schubert symbols $(a_1,\cdots a_m)$ of dimension d satisfying $a_m \le n$ correspond to d-cells of $G_m(\mathbb{R}^{m+n})$ and thus they form an additive basis for $H^dG_m(\mathbb{R}^{m+n})$. However, it is much less clear what monomials in the Stiefel-Whitney classes may form a basis for this vector space. The difficulty is caused by the fact that the multiplication rule for Schubert symbols is quite complicated. We propose to overcome this difficulty and prove the following theorem.

Theorem. The set of monomials $w_1^{r_1}\cdots w_m^{r_m}$ of a total degree $d = \sum_{i=1}^{m} ir_i$ satisfying the condition $\sum_{i=1}^{m} r_i \le n$ forms an additive basis for $H^dG_m(\mathbb{R}^{m+n})$.

2. Multiplication rules and the lexicographic order

In terms of Schubert symbols, the j-th Stiefel-Whitney class of G_m is $w_j = (0,\cdots,0,\underbrace{1,\cdots,1}_{j})$. The dual classes are given by $\overline{w}_k = (0,\cdots,0,k)$. The multiplication is first defined for products of a class by a Schubert symbol according to the Pieri formula:

$$\overline{w}_k(a_1,\cdots,a_m) = \Sigma(b_1,\cdots,b_m)$$

where the summation on the right hand side extends over all Schubert symbols $(b_1,\cdots,b_m)$ satisfying $a_i \le b_i \le a_{i+1}$ for $i = 1,\cdots,m-1$, $a_m \le b_m$, and $a_1 + \cdots + a_m + k = b_1 + \cdots + b_m$.

A consequence of this rule is that every Schubert symbol $(a_1,\cdots,a_m)$, and hence every cohomology class in H^dG_m, can be expressed as a polynomial in the dual classes $\overline{w}_1,\overline{w}_2,\cdots$, by using the formula $(a_1,\cdots,a_m) = \det(\overline{w}_{a_i+i-j})$, where $\overline{w}_k = 0$ if $k < 0$. This formula can also be used to compute the product of arbitrary Schubert symbols, although such computations tend to be quite complicated and quickly get out of hand. For references, see [2], [5] and [7].

To obtain a multiplication rule in $H^*G_m(\mathbb{R}^{m+n})$ we simply restrict the results to the terms $(b_1,\cdots,b_m)$ such that $b_m \le n$.

To prove the theorem, we will establish some auxiliary multiplication rules. First of all, there is a natural lexicographic order among the Schubert symbols.

(2.1) Definition. Let $(a_1,\cdots,a_m)$, $(b_1,\cdots,b_m)$ be Schubert symbols. We say that $(a_1,\cdots,a_m) < (b_1,\cdots,b_m)$ if $a_i = b_i$ for $i<j$ and $a_j < b_j$.

If $x \in H^dG_m$ then $\lambda(x)$ will denote the lowest term in the expansion of x with respect to the Schubert basis; $\lambda(x)$ will also be called the leading term of x.

We will always write Schubert expansions in this lexicographic order.

(2.1) Notation. If $a = (a_1,\cdots,a_m)$ is a Schubert symbol in G_m, we write f_ra for the Schubert symbol $(a_1 + r,\cdots,a_m + r)$. The operator f_r can then be extended to every cohomology class $x \in H^q$.

(2.2) Lemma. For each $x \in H^q$, $x(f_ra) = f_r(xa)$.

Proof. It is clear that the assertion is valid if x is a dual Stiefel-Whitney class, $x = \overline{w}_k = (0,\cdots,k)$. Therefore it is also valid for every monomial in dual Stiefel-Whitney classes:

$$(\overline{w}_k\overline{w}_j)(f_ra) = \overline{w}_k(\overline{w}_jf_ra) = \overline{w}_k(f_r(\overline{w}_ja)) = f_r(\overline{w}_k\overline{w}_ja).$$

It follows that the formula is valid for every polynomial in $\overline{w}_k$; and thus is true in general since every $x \in H^q$ can be written as such a polynomial.

(2.3) Lemma. For any $x \in H^q$, $w_m^rx = f_rx$.

Proof. It suffices to prove this if $x = a$ is a Schubert symbol. First, if $r = 1$, we have by (2.2);

$$w_m s = s w_m = s(1,\cdots,1) = s f_1(0,\cdots,0) = f_1(s(0,\cdots,0)) = f_1 s .$$

Then we apply an induction on r.

(2.4) Corollary. $w_m^r = (r,\cdots,r)$ (Compare [5], p. 523 and [7], p. 180).

(2.5) Lemma. Let $w_j = (0,\cdots,0,\underbrace{1,\cdots,1}_{j})$ be the j-th Stiefel-Whitney class. Then the leading term of $w_j s$ is

$$\lambda(w_j s) = (s_1,\cdots,s_{m-j},s_{m-j+1} + 1,\cdots,s_m + 1) .$$

More generally,

$$\lambda(w_j^r s) = (s_1,\cdots,s_{m-j},s_{m-j+1} + r,\cdots,s_m + r) .$$

Proof. Write w_j in the determinant form

$$w_j = \det\begin{bmatrix} 1 & 0 & .. & 0 & 0 & \cdots & 0 \\ * & 1 & .. & .. & . & \cdots & . \\ .. & .. & .. & .. & .. & \cdots & . \\ * & * & .. & 1 & 0 & 0\ \cdots & 0 \\ * & * & .. & * & \overline{w}_1 & 1\ \cdots & 0 \\ * & * & .. & * & \overline{w}_2 & \overline{w}_1\ \cdots & 0 \\ .. & .. & .. & .. & .. & \cdots & . \\ * & * & .. & * & \overline{w}_j & \overline{w}_{j-1}\ \cdots & \overline{w}_1 \end{bmatrix}$$

$$= \det\begin{bmatrix} \overline{w}_1 & 1 & 0 & \cdots & 0 \\ \overline{w}_2 & \overline{w}_1 & 1 & \cdots & 0 \\ . & \cdots & . & \cdots & \cdots \\ \overline{w}_{j-1} & . & \cdots & \cdots & 1 \\ \overline{w}_j & \overline{w}_{j-1} & . & \cdots & \overline{w}_1 \end{bmatrix}$$

Consider the initial terms of the product $w_j s$, those in which the first $m - j$ entries will remain $s_1,\cdots,s_{m-j}$ (all the other terms will be higher). The remaining j entries will be affected by multiplying with w_j in the same way as the Schubert symbol $(s_{m-j+1},\cdots,s_m)$ in G_j is affected by multiplying it with the top Stiefel-Whitney class w_j in G_j; that is, they will become $s_{m-j+1} + 1,\cdots,s_m + 1$, by Lemma (2.2). It follows that the leading term of the product $w_j s$ will be $(s_1,\cdots,s_{m-j},s_{m-j+1} + 1,\cdots,s_m + 1)$. The more general case follows by induction.

(2.6) Corollary. For every monomial $w_1^{r_1}\cdot \cdots \cdot w_m^{r_m}$ in $H^d G_m$,

$$\lambda\left(w_1^{r_1}\cdot \cdots \cdot w_m^{r_m}\right) = (a_1,\cdots,a_m) ,$$

where $a_j = \sum_{k=0}^{j-1} r_{m-k}$, $j = 1,\cdots,m$.

The proof uses Lemma (2.5) and induction, beginning with w_m and going down to w_1.

3. Proof of the Theorem

The function which associates to the monomial $w_1^{r_1} \cdot \ldots \cdot w_m^{r_m} \in H^d G_m$, the Schubert symbol $(a_1, \cdots, a_m)$ where $a_j = \sum_{k=0}^{j-1} r_{m-k}$ is a bijection from the set of such monomials to the set of Schubert symbols in G_m of dimension d; its inverse is obtained by assigning to the Schubert symbol $(a_1, \cdots, a_m)$, the monomial $w_1^{r_1} \cdot \ldots \cdot w_m^{r_m}$, where $r_1 = a_{m-i-1} - a_{m-i}$ (compare [6], p. 85). Corollary (2.6) says that this bijection is given by λ, the leading term of the Schubert expansion. This means that, with respect to this bijection, the matrix of the Schubert expansions of the monomials is a triangular matrix.

In $H^d G_m(\mathbb{R}^{m+n})$, with respect to this bijection, the Schubert symbols $(a_1, \cdots, a_m)$ with $a_m \leq n$ correspond exactly to the monomials $w_1^{r_1} \cdot \ldots \cdot w_1^{r_m} \in H^d G_m(\mathbb{R}^{m+n})$ satisfying $r_1 + \cdots + r_m \leq n$; and the Schubert expansion matrix remains triangular.

This completes the proof.

REFERENCES

[1] A. Borel: La cohomologie mod 2 de certains espaces homogènes, Comm. Math. Helv. 27 (1953),165-197.

[2] I. Berstein: On the Lusternik-Schnirelmann category of Grassmannians, Math. Proc. Camb. Phil.Soc. 79 (1976), 129-134.

[3] S. Chern: On the multiplication in the characteristic ring of a sphere bundle, Ann. of Math. 49 (1948), 362-372.

[4] C. Ehresmann: Sur la topologie de certains espaces homogènes, Ann. of Math. 35 (1934), 396-443.

[5] H.L. Hiller: On the cohomology of real Grassmannians, Trans. Amer. Math. Soc. 257 (1980), 521-533.

[6] J.W. Milnor, J.D. Stasheff: Characteristic Classes, Ann. of Math. Studies, Princeton, 1974.

[7] V. Oproiu, Some non-embedding theorems for the Grassmann manifolds $G_{2,n}$ and $G_{3,n}$, Proc.Edinburgh Math. Soc. 20 (1976-77), 177-185.

[8] H. Schubert: Kalkül der abzählenden Geometrie, Teubner, Leipzig, 1879.

ON THE TOPOLOGY OF THE SPACE OF REACHABLE SYMMETRIC LINEAR SYSTEMS

Nguyễn Huỳnh Phàn
Department of Mathematics
Pedagogical Institute of VINH, Viet Nam.

Introduction. Let us consider the reachable linear system in the state space F^n described by a following equation

$$\dot{x}(t) = Ax(t) + Bu(t) \tag{1.1}$$

where $x(t) \in F^n$ and $u(t) \in F^m$ are the values of state and control respectively, at time $t \in \mathbb{R}$, $(A, B) \in F^{n\times n} \times F^{n\times m}$. F denotes the field of real numbers $\mathbb{R}$ or complex numbers $\mathbb{C}$.

Recall that a linear system which is desribed by the equation (1.1) is called reachable if and only if

$$\operatorname{rank}[B, AB, \ldots, A^{n-1}B] = n.$$

We denote by $\widetilde{\Sigma}_{n,m}(F)$ the subspace of $F^{n\times m} \times F^{n\times m}$ consisting of all reachable couples (A, B). An invertible linear transformation $T : F^n \to F^n$ transforms the couple (A, B), which represents the system by equation (1.1) to the couple (TAT^{-1}, TB). Hence if we identify the equation (1.1) with a couple (A, B) then the similar class $[A, B] := \{(TAT^{-1}, TB), T \in GL(n, F)\}$ is identified with the system. Therefore via the similar action of the general linear group $GL(n, F)$ on $\widetilde{\Sigma}_{n,m}(F)$ given by $T(A, B) = (TAT^{-1}, TB)$, the space of reachable linear systems in the state space F^n is identified with the orbit space

$$\Sigma_{n,m} = \widetilde{\Sigma}_{n,m} \big/ _{GL(n,F)}.$$

Let $U(n, F)$ be the unitary group of degree n over F. We now consider linear systems which are described by equation (1.1) with $A = A^*$, where A^* denotes the conjugate of the matrix A. Those linear systems are called reachable symmetric linear systems in the state space F^n.

Consider the $U(n, F)$-subspace

$$\widetilde{S}_{n,m}(F) = \left\{ (A, B) \in \widetilde{\Sigma}_{n,m}(F),\ A = A^* \right\} \text{ of the space } \widetilde{\Sigma}_{n,m}(F).$$

Since $\widetilde{S}_{n,m}(F)$ is an open (and dence) subset of the vector space C consisting of all couples (A, B) with $A = A^*$, it is an analytic manifold.

By an argument analogous to the previous one, we get that the space of reachable symmetric linear systems in the state space F^n is identified with the orbit space

$$S_{n,m}(F) := \widetilde{S}_{n,m}(F) \,/_{U(n,F)}.$$

Here the unitary group $U(n,F)$ acts on $\widetilde{S}_{n,m}(F)$ as the restriction of $Gl(n,F)$.

Topological and geometrical properties of $\Sigma_{n,m}(F)$ and canonical forms of reachable linear systems have been studied by various authors, see e.g. Popov [17](1972), Brockett [4](1976), Helmke [7](1982), Hinrichsen and Prätzel-Wolters [10](1983), Helmke and Hinrichsen [8](1986), Hinrichsen [9](1987).

As far as we know, using the methods of cell decompositions, Helmke [7] for the originally obtained the integral (resp. modulo 2) homology group of $S_{n,m}(F)$ for $F = \mathbf{C}$ (resp. $\mathbf{R}$).

Now consider the following commutative diagram where maps are canonical embeddings

$$\begin{array}{ccc} & \widetilde{\Sigma}_{n,m}(F) \,/_{U(n,F)} & \\ \nearrow & & \overset{\simeq}{\searrow} \Psi \\ S_{n,m}(F) & \xrightarrow{J} & \Sigma_{n,m}(F) . \end{array}$$

The map Ψ is a fibretion with the fibre isomorphic to the homogeneous space $Gl(n, F)/_{U(n,F)}$ (see Dieudonné [6]). Moreover, since the orbit spaces $\widetilde{\Sigma}_{n,m}(F)/_{U(n,F)}$ and $\Sigma_{n,m}(F)$ are separable analytic manifolds, they have the homotopy type of a countable CW-complex (see McCleary [14], Thm. 4.4 or Spanier [19]).

In 1985 Huynh Mui conjectured that J is a homotopy equivalence. In other words, on the level of the homotopy theory of the space of reachable linear systems, the topological invariants of the space $S_{n,m}(F)$ are "good enough" in a certain sence. Furthermore, the study of $S_{n,m}(F)$ is not only more convenient, but also (as can be seen later) gives us a simple canonical form.

The first purpose of this paper is study the topology of the space $S_{n,m}(F)$ and the second one is to prove the conjecture of Huynh Mui in certain cases. To study the topology of $S_{n,m}(F)$, we use the method of cell decompositions.

The paper is divided into 4 sections. Section 1 deals with a canonical form and complete invariants of reachable symmetric linear systems. In section 2 we prove that the canonical projection $P : \widetilde{S}_{n,m}(F) \to S_{n,m}(F)$ is a principal $U(n,F)$-bundle. In section 3 we give a cell decompositions of $S_{n,m}(F)$ and show that the integral (resp. modulo 2) homology group of $S_{n,m}(F)$ is isomorphic to those of the Grassmann manifold $G_{n,n+m-1}(F)$ for $F = \mathbf{C}$ (resp. $\mathbf{R}$). Finally, for the case $F = \mathbf{C}$ we prove in section 4 that the conjecture of Huynh Mui is true.

For convenience, we write $\widetilde{S}_{n,m}$, $S_{n,m}$, $\widetilde{\Sigma}_{n,m}$, $\Sigma_{n,m}$ and $GL(n)$ instead of $\widetilde{S}_{n,m}(F)$, $S_{n,m}(F)$, $\widetilde{\Sigma}_{n,m}(F)$, $\Sigma_{n,m}(F)$ and $GL(n,F)$.

Acknowledgements.
I wish to acknowledge my deep gratitude to my teacher, Prof. Dr. Huynh Mui for his inspiring guidance and constant encouragements. I would like to express my warm thanks

to DR U. Helmke for his helpful suggestions. It is the author's pleasure to acknowledge Prof. Dr. D. Hinrichsen for his many valuable comments and suggestions. Finally, very special thanks are due to the referee and for helpful suggestions for correcting my English.

1. Invariants and canonical forms of symmetric linear systems. First we recall that the scalar product on the vector space F^n is defined by

$$(x,y) := \sum_{i=1}^{n} x_i \overline{y}_i$$

where $x = (x_1, \dots, x_n) \in F^n$, $y = (y_1, \dots, x_n) \in F^n$ and $\overline{y}_i$ denotes the conjugate of y_i.

The Gram-Schmidt orthogonalization process defines an analytic map from $GL(n)$ into $U(n)$,

$$(w_1, \dots, x_n) \longrightarrow (v_1, \dots, v_n) \quad \text{where}$$

$$v_1 := \frac{w_1}{\|w_1\|}, \quad v_j := \frac{w_j - \sum_{i<j}(w_j, v_i)v_i}{\|w_j - \sum_{i<j}(w_j, v_i)v_i\|}.$$

Hinrichsen and Prätzel-Wolters gave in [10] the Hermite canonical form for reachable linear systems under a similar action of the general linear group $GL(n)$. Using their method, we introduce in this section a canonical form, which will be called the Hermite canonical form, for reachable symmetric linear systems under a similar action of the unitary group $U(n)$.

Let $(A,B) \in \widetilde{S}_{n,m}$ be a reachable symmetric linear system. Let $B = [b_1, b_2, \dots, b_m]$. Since rank $[B, AB, \dots, A^{n-1}B] = n$ hence in the set of nm vectors $A^i b_j$ $(0 \le i \le n-1$, $1 \le j \le m)$ we can choose n linearly independent vectors by the following elimination procedure:

Going from the left to the right through the $n \times m$-matrix

$$[b_1, Ab_1, \dots, A^{n-1}b_1, b_2, Ab_2, \dots, A^{n-1}b_2, \dots, b_m, Ab_m, \dots, A^{n-1}b_m]$$

delete all the column vectors which are linearly dependent upon their predecessors.

By reachability, the remaining vectors from a basis of F^n. They can be ordered in the following way

$$H = H(A,B) := [b_1, Ab_1, \dots, A^{k_1-1}b_1, \dots, b_m, Ab_m, \dots, A^{k_m-1}b_m]$$

where $k_1, k_2, \dots, k_m$ are non-negative integers satisfying $k_1 + k_2 + \dots + k_m = n$ and the vector $A^{k_j} b_j$ is lineary dependent on the system $\{b_1, Ab_1, \dots, A^{k_1-1}b_1, \dots, b_j, Ab_j, \dots, A^{k_j-1}b_j\}$ for $1 \le j \le m$.

As in [10], $k(A,B) := (k_1, \dots, k_m)$ is called the list of Hermite indices of (A,B). Obviously $k(A,B) := k(TAT^{-1}, TB)$ for each $(A,B) \in \widetilde{S}_{n,m}$ and for every $T \in U(n)$. So we have

DEFINITION 1.1. Let $K_{n,m} := \{(k_1, \dots, k_m) \in Z^m, k_j \ge 0 \text{ and } \sum_{j=1}^{m} k_j = n\}$. For each $k \in K_{n,m}$, the set $\widetilde{H}(k) := \{(A,B) \in \widetilde{S}_{n,m} k(A,B) = k\}$ is called the Hermite stratum of the space $\widetilde{S}_{n,m}$ corresponding to k.

In the control theoretic literature the following definition taken from Birkhoff and Maclane [2](1977) are standard.

Definition 1.2. Let α be a group action of a group G on a space X, i.e. $\alpha : G \times X \to X$, $(g,x) \mapsto g \cdot x$ such that $e \cdot x = x$ and $g(g' \cdot x) = gg' \cdot x$ for all $x \in X$ and $g, g' \in G$, here e denotes the neutral element of G.

Group action α induces an equivalence relation on X by

$$x \sim y \Leftrightarrow \exists g \in G \qquad \text{such that} \qquad y = g \cdot x.$$

(i) A map $f : X \to Y$ (Y is a set) is called an invariant for α if $x \sim y$ implies $f(x) = f(y)$ for all $x, y \in X$. In the case of $Y = F$ then f is called a scalar invariant.

(ii) An invariant f is called complete if it separates the orbits of α, i.e.

$$x \sim y \Leftrightarrow f(x) = f(y) \qquad \text{for all} \qquad x, y \in X.$$

Definition 1.3. A canonical form for a group action α on X is a map $c : X \to X$ which satisfies for all $x, y \in X$:

$$c(x) \sim x \text{ and } x \sim y \Leftrightarrow c(x) = c(y).$$

The following theorem gives the canonical form and the complete invariant family for the similar action of $U(n)$ on a Hermite stratum.

Theorem 1.4. *Suppose $(A,B) \in \widetilde{H}(k)$, $k = (k_1, \ldots, k_m) \in K_{n,m}$. Then (A,B) is similar by the unitary group to exactly one couple (A_k, B_k) of the form*

$$A_k = \begin{pmatrix} N_1 & & & \\ & N_2 & & \\ & & \ddots & \\ & & & \ddots \\ & & & & N_m \end{pmatrix} \qquad B_k = \begin{pmatrix} a_{11} & * & \cdots & * \\ 0 & \vdots & & \vdots \\ \vdots & a_{21} & & \vdots \\ \vdots & 0 & & * \\ \vdots & \vdots & & a_{m1} \\ \vdots & \vdots & & 0 \\ \vdots & \vdots & & \vdots \\ 0 & 0 & \cdots & 0 \end{pmatrix}$$

where

$$N_j = \begin{pmatrix} x_{j1} & a_{j2} & & & \\ a_{j2} & \ddots & \ddots & & \\ & \ddots & \ddots & \ddots & \\ & & \ddots & \ddots & a_{jk_j} \\ & & & a_{jk_j} & x_{jk_j} \end{pmatrix} \qquad b_j^k = \begin{pmatrix} * \\ \vdots \\ a_{jl} \\ 0 \\ \vdots \\ 0 \end{pmatrix} \leftarrow \textit{row } k_1 + k_2 + \cdots + k_{j-1} + 1$$

$x_{j1},\dots,x_{jk_j}$ are real numbers, $a_{j2},\dots,a_{jk_j}$ are positive real numbers. Moreover, a_{j1} is positive if $k_j \geq 1$, $a_{j1} = 0$ if $k_j = 0$ and the entries $*$ in B_k are elements of F. Here we set $k_0 := 0$.

The couple (A_k, B_k) is called the Hermite canonical form of the couple (A, B).

PROOF. First we prove the uniqueness of (A_k, B_k). Suppose (A_k, B_k) and (A'_k, B'_k) are of the form as in Theorem 1.4 and they are similar to another, i.e. $(SA_kS^{-1}, SB_k) = (A'_k, B'_k)$ for some $S \in U(n)$. Let

$$H(A_k, B_k) := [b_1^k, A_k b_1^k, \dots, A_k^{k_1-1} b_1^k, \dots, b_m^k, A_k b_m^k, \dots, A_k^{k_m-1} b_m^k].$$

It can be verified that $H(A_k, B_k)$ and $H(A'_k, B'_k)$ are the $n \times n$-upper triangular matrices, whose entries on the diagonal are positive real numbers, and $H(SA_kS^{-1}, SB_k) = S\,H(A'_k, B'_k)$ for every $S \in U(n)$.

Since $(SA_kS^{-1}, SB_k) = (A'_k, B'_k)$, so we have $H(SA_kS^{-1}, SB_k) = S\,H(A'_k, B'_k)$. Because the set of all $n \times n$-upper triangular matrices whose entries on the diagonal are positive real numbers form a subgroup of the general linear group $GL(n)$, hence S is an upper triangular matrix with positive real numbers at diagonal entries, too. Furthermore, $S \in U(n)$, we get $S = I_n$. The uniqueness is proved.

Now we will show that (A, B) is similar to a couple (A_k, B_k). Let

$$X := [X_1^0, \dots, X_1^{k_1-1}, \dots, X_m^0, \dots, X_m^{k_m-1}]$$

be the unitary matrix obtained by the Gram-Schmidt ortogonalization of the system

$$H(A, B) = [b_1, Ab_1, \dots, A^{k_1-1} b_1, \dots, b_m, Ab_m, \dots, A^{k_m-1} b_m].$$

That means

$$X = H(A, B)\,C \tag{1.4.1}$$

where C is the upper triangular matrix with positive real numbers on diagonal. From (1.4.1) we have

$$XC^{-1} = [b_1, Ab_1, \dots, A^{k_1-1} b_1, \dots, b_m, Ab_m, \dots, A^{k_m-1} b_m] \tag{1.4.2}$$

where C^{-1} has the form like C.

We consider the following cases:

(i) If $k_j > 0$ $(1 \leq j \leq m)$, then by (1.4.2), b_j is of the form

$$b_j = X \begin{pmatrix} * \\ \vdots \\ * \\ a_{j1} \\ 0 \\ \vdots \\ 0 \end{pmatrix} \quad \leftarrow \text{row } k_1 + k_2 + \cdots + k_{j-1} + 1$$

where a_{j1} is the positive real number and the entries $*$ are elements of F.

(ii) If $k_j = 0$, then

$$\begin{aligned} b_j &\in \langle b_1, Ab_1, \ldots, A^{k_1-1}b_1, \ldots, b_{j-1}, Ab_{j-1}, \ldots, A^{k_{j-1}-1}b_{j-1}\rangle \\ &= \langle X_1^0, \ldots, X_1^{k_1-1}, \ldots, X_{j-1}^0, \ldots, X_{j-1}^{k_{j-1}-1}\rangle \end{aligned}$$

where the notation $\langle v_1, \ldots, v_t\rangle$ indicates the vector space generated by vectors $v_1, \ldots, v_t$. Hence b_j is of the following form

$$b_j = X \begin{pmatrix} * \\ \vdots \\ * \\ 0 \\ \vdots \\ 0 \end{pmatrix} \quad \leftarrow \text{row } k_1 + k_2 + \cdots + k_{j-1}$$

where the entries $*$ are elements of F. Thus, it follows from (i) and (ii) that B has the form

$$B = XB_k \tag{1.4.3}$$

Also, it follows readily from (1.4.1) that

$$AX = [Ab_1, A^2b_1, \ldots, A^{k_1}b_1, \ldots, Ab_m, A^2b_m, \ldots, A^{k_m}b_m]C. \tag{1.4.4}$$

We will show that the right side of (1.4.4) has the form XA_k. Since

$$[b_1, Ab_1, \ldots, A^{k_1-1}b_1, \ldots, b_m, Ab_m, \ldots, A^{k_m-1}b_m] = XC^{-1},$$

it follows that $[Ab_1, A^2b_1, \ldots, A^{k_1}b_1, \ldots, Ab_m, A^2b_m, \ldots, A^{k_m}b_m]$ has the form

$$X \begin{pmatrix} Y_{11} & Y_{12} & \cdots & Y_{1m} \\ & Y_{22} & & \vdots \\ & 0 & \ddots & \vdots \\ & & & Y_{mm} \end{pmatrix} =: XY \tag{1.4.5}$$

where Y_{tj} $(1 \le t, j \le m)$ are excluded if either $k_t = 0$ or $k_j = 0$ and Y_{tj} is the $k_t \times k_j$-matrix if both $k_t > 0$ and $k_j > 0$.
Furthermore, if $k_j > 0$ then Y_{jj} has the form

$$Y_{jj} = \begin{pmatrix} * & \cdots & \cdots & * & * \\ x_{j2} & \ddots & & \vdots & \vdots \\ & 0 & \ddots & * & \vdots \\ & & & x_{jk_j} & * \end{pmatrix}$$

where $x_{j2}, \ldots, x_{jk_j}$ are positive real numbers. Thus, it follows from (1.4.4) and (1.4.5) that

$$AX = XYC.$$

Since C is the upper triangular matrix with positive real numbers at diagonal entries, we conclude that YC is of the form like Y. Let

$$YC = \begin{pmatrix} A_{11} & \cdots & A_{1m} \\ & \ddots & \vdots \\ 0 & & A_{mm} \end{pmatrix} \tag{1.4.6}$$

where A_{jj} has the following form

$$A_{jj} = \begin{pmatrix} * & \cdots & \cdots & * & * \\ a_{j2} & \ddots & & \vdots & \vdots \\ & 0 & \ddots & * & \vdots \\ & & & a_{jk_j} & * \end{pmatrix} \tag{1.4.7}$$

where $a_{j2}, \ldots, a_{jk_j}$ are positive real numbers.

Since $A^* = A$ and X is the unitary matrix, it follows that

$$(YC)^* = YC.$$

Combining (1.4.6), (1.4.7) and the fact $(YC)^* = YC$, we see that YC has the form A_k. Thus

$$AX = XA_k \tag{1.4.8}$$

From (1.4.3) and (1.4.8) we get that (A, B) is similar to (A_k, B_k). The proof of Theorem 1.4 is complete. ∎

It follows from the Theorem 1.4 that $\widetilde{H}(k) \neq \emptyset$ for every $k \in K_{n,m}$. Hence the family $\{\widetilde{H}(k),\ k \in K_{n,m}\}$ is a partition of the space $\widetilde{S}_{n,m}$; $\widetilde{S}_{n,m} = \bigcup_{k \in K_{n,m}} \widetilde{H}(k)$.

By this Theorem we get the following:

COROLLARY 1.5. *For every $k \in K_{n,m}$, $\widetilde{H}(k)$ is an analytic submanifold of the analytic manifold $\widetilde{S}_{n,m}$ and the map*

$$h : \widetilde{H}(k) \to U(n) \times H(k),$$

$$(A, B) \mapsto (X, a_{12}, \ldots, a_{1k_1}, \ldots, a_{m1}, \ldots, a_{mk_m}, b_1^k, \ldots, b_m^k)$$

is an analytic homeomorphism.

Here $H(k) := \mathsf{R}_+^n \times \mathsf{R}^n \times F^{g(k)} \cong \mathsf{R}^{2n+dg(k)}$,

$$g(k) := k_1 + (k_1 + k_2) + \cdots + (k_1 + k_2 + \cdots + k_{m-1}),$$

$$d := \dim_{\mathsf{R}} F = 1, 2, \qquad \mathsf{R}_+ := \{x \in \mathsf{R},\ x > 0\}, \qquad k_0 := 0.$$

Furthermore, the map $(A, B) \to (A_k, B_k)$ is the analytic canonical form on $\widetilde{H}(k)$ and the orbit space $\widetilde{H}(k)/_{U(n)}$ is analytic isomorphic to $\mathsf{R}^{2n+dg(k)}$.

Particularly, we have:

The case $m = 1 \Rightarrow \widetilde{S}_{n,1} \cong U(n) \times \mathsf{R}^{2n}$ and $S_{n,1} \cong \mathsf{R}^{2n}$.

The case $n=1 \Rightarrow \widetilde{S}_{1,m} \cong F \times (F^m \setminus \{0\})$ and $S_{1,m} \cong F \times (F^m \setminus \{0\})/_{U(1)}$.
Hence $S_{1,m}(F)$ is homotopy equivalent to the projective space $P_{m-1}(F)$.

2. The principal $U(n)$-bundle structure on $S_{n,m}$. In this section we shall prove the following:

THEOREM 2.1. *The space of reachable symmetric linear systems $S_{n,m}$ is an analytic manifold. The canonical projection $p : \widetilde{S}_{n,m} \to S_{n,m}$ is a principal $U(n)$-bundle. It is trivial if and only if $m = 1$.*

PROOF. Let the Lie group acts analytic and freely on an analytic manifold M so that the graph of the action (i.e. the set $Q := \{(x, g \cdot x),\ x \in M$ and $g \in G\})$ is a closed analytic submanifold in $M \times M$. Then the orbit space $M/_G$ is a principal G-bundle (see e.g. Dieudonné [6], 16.10, 16.14).

Clearly that $U(n)$ acts analytically and freely on $\widetilde{S}_{n,m}$. So it sufficies to prove the following two lemmas: ∎

LEMMA 2.2. *The set*

$$Q := \{((A,B), S \cdot (A,B)),\ S \in U(n) \text{ and } (A,B) \in \widetilde{S}_{n,m}\}$$

is closed analytic submanifold of $\widetilde{S}_{n,m} \times \widetilde{S}_{n,m}$.

PROOF. Since $U(n)$ acts analytically and freely on $\widetilde{S}_{n,m}$ hence the map

$$q : U(n) \times \widetilde{S}_{n,m} \to \widetilde{S}_{n,m} \times \widetilde{S}_{n,m},$$

$$(S,(A,B)) \mapsto ((A,B), S \cdot (A,B))$$

is an analytic embedding and im $q \cong U(n) \times \widetilde{S}_{n,m}$. Hence Q is an analytic submanifold in $\widetilde{S}_{n,m} \times \widetilde{S}_{n,m}$ (see Dieudonné [6]).

Now suppose $((A_t, B_t), (\widetilde{A}_t, \widetilde{B}_t))$ is a convergent sequence in Q. Since $U(n)$ acts analytically and freely, this sequence is of the form $((A_t, B_t), S_t \cdot (A_t, B_t))$; $\lim_{t \to \infty}(A_t, B_t) = (A, B)$ and $\lim_{t\to\infty} S_t \cdot (A_t, B_t) = (\widetilde{A}, \widetilde{B})$.

We have to prove that $((A,B), (\widetilde{A}, \widetilde{B})) \in Q$. We shall prove that the sequence S_t has a limit in $U(n)$.

Denote $R(A_t, B_t) := [B_t, A_t B_t, \ldots, A_t^{n-1} B_t]$. Since rank $R(A_t, B_t) = n$ hence $R(A_t, B_t) R(A_t, B_t)^*$ is an $n \times n$-nondegenerate matrix. So we have

$$\begin{aligned} S_t &= S_t R(A_t, B_t) R(A_t, B_t)^* \left(R(A_t, B_t) R(A_t, B_t)^*\right)^{-1} \\ &= R(S_t, A_t S_t^{-1}, S_t, B_t) R(A_t, B_t)^* \left(R(A_t, B_t) R(A_t, B_t)^*\right)^{-1}. \end{aligned}$$

Hence $\lim_{t\to\infty} S_t = R(\widetilde{A}, \widetilde{B}) R(A,B)^* \left(R(A,B), R(A,B)^*\right)^{-1}$. Since $U(n)$ is closed in $GL(n)$, we have $(\lim_{t\to\infty} S_t) \in U(n)$.

The proof of the Lemma is completed. ∎

LEMMA 2.3. *$P : \widetilde{S}_{n,m} \to S_{n,m}$ is trivial principal $U(n)$-bundle if and only if $m = 1$.*

PROOF. The "if" part follows from the comments after Corollary 1.5.

The "only if": If $n = 1$, we have $\widetilde{S}_{1,m} \cong \mathbb{R} \times (F^m \setminus \{0\})$ and $S_{1,m} \simeq P_{m-1}(F)$ ($S_{1,m}$ is homotopy equivalent to $P_{m-1}(F)$). If P is trivial then $\widetilde{S}_{1,m} \cong S_{1,m} \times U(1)$. Hence $F^m \setminus \{0\} \simeq U(1) \times P_{m-1}(F)$. This is a contradiction if $m > 1$.

If $n > 1$ and $m > 1$: Let $1 \le r < \min\{n, m\}$ and let $V_{r,m}(F)$ be the Stiefel manifold consisting of all orthogonal vector r-frames in the vector space F^m. Let $G_{r,m}(F)$ be the Grassmann manifold consisting of all r-planes in F^m.

For any $(\widehat{A}, \widehat{B}) \in \widetilde{S}_{n-r,m}$ with $\det \widehat{A} \neq 0$ we consider the following monomorphism of principal bundles

$$V_{r,m} \longrightarrow \widetilde{S}_{n,m}$$

$$Y \longrightarrow \left(\begin{pmatrix} \widehat{A} & 0 \\ 0 & 0 \end{pmatrix}, \begin{pmatrix} \widehat{B} \\ Y \end{pmatrix} \right).$$

Since $(V_{r,m}(F), P, G_{r,m}(F))$ is a nontrivial principal $U(n)$-bundle for $m > 1$, the principal $U(n)$-bundle $(\widetilde{S}_{n,m}, P, S_{n,m})$ is also nontrivial for $m > 1$.

The proof of the Lemma is completed. ■

3. An analytic cell decomposition of the space $S_{n,m}$.

3.1. We equip $K_{n,m}$ with the order given by

$$k \preceq 1 :\Leftrightarrow \sum_{t=1}^{j} k_t \le \sum_{t=1}^{j} l_t \quad \text{for} \quad 1 \le j \le m$$

and $k = (k_1, \ldots, k_m)$, $l = (l_1, \ldots, l_m) \in K_{n,m}$. Obiously $(K_{n,m}, \preceq)$ is a lattice with the minimum element $(0, \ldots, 0, n)$ and the maximum element $(n, 0, \ldots, 0)$.

Let l and k be two elements of $K_{n,m}$. We say that k is covered by l if and only if $k \prec l$ and there does not exist $x \in K_{n,m}$ such that $k \prec x \prec l$. By Helmke [7], Lemma 5, p. 67, we see that k is covered by l if and only if l is of the form

$$l = (k_1, \ldots, k_{i-1}, k_i + 1, k_{i+1} - 1, k_{i+2}, \ldots, k_m).$$

Here i is some integer less than m such that $k_{i+1} \ge 1$.

Consider now the set

$$Y_{m,n} := \{(a_1, \ldots, a_m) \in Z^m, \; 0 \le a_1 \le \ldots \le a_m \le n\}$$

equipped with the order defined by

$$a \preceq b :=\Leftrightarrow a_j \le b_j \qquad \text{for} \qquad 1 \le j \le m.$$

Then $(Y_{n,m}, \preceq)$ is also a lattice.

We define the map $\widetilde{g} : K_{n,m} \to Y_{m-1,n}$

$$\widetilde{g}(k_1, \ldots, k_m) = (k_1, k_1 + k_2, \ldots, k_1 + \cdots + k_{m-1}).$$

It is easily seen that $\widetilde{g}$ is a lattice isomorphism.

The following Theorem is main result of this section.

THEOREM 3.2. *For every $k \in K_{n,m}$ let $\overline{\widetilde{H}}(k)$ be the closure of $\widetilde{H}(k)$ in $\widetilde{S}_{n,m}$. Then the following statements are equivalent*
(i) $\widetilde{H}(k) \subset \overline{\widetilde{H}}(l)$,
(ii) $\widetilde{H}(k) \cap \overline{\widetilde{H}}(l) \neq \emptyset$,
(iii) $k \preceq l$.

PROOF. (i)⇒(ii) is trivial.
(ii)⇒(iii): take $(A, B) \in \widetilde{H}(k) \cap \overline{\widetilde{H}}(k)$. Denote by b_j the j-th column of B. We have

$$\text{rank}\,[b_1, Ab_1, \ldots, A^{n-1}b_1, \ldots, b_j, Ab_j, \ldots, A^{n-1}b_j] = k_1 + k_2 + \cdots + k_j.$$

Suppose that $(\widetilde{A}, \widetilde{B}) \in \widetilde{H}(l)$ converges to (A, B). Since the rank function is upper semicontinuous, it follows that

$$\text{rank}\,[\widetilde{b}_1, \widetilde{Ab}_1, \ldots, \widetilde{A}^{n-1}\widetilde{b}_1, \ldots, \widetilde{b}_j, \widetilde{Ab}_j, \ldots, \widetilde{A}^{n-1}\widetilde{b}_j] = l_1 + l_2 + \cdots + l_j \geq k_1 + k_2 + \cdots + k_j.$$

Hence $k \preceq l$.

Now in order to show (iii)⇒(i) we need two following Lemmata.

LEMMA 3.3 *Let (A_t, B_t) be a sequence $\widetilde{S}_{n,m}$ which converges to $(A, B) \in \widetilde{S}_{n,m}$. Then for every $S \in U(n)$, the sequence (SA_tS^{-1}, SB_t) converges to (SAS^{-1}, SB).*

PROOF. It is well-known that for $S \in U(n)$, then the map $(A, B) \to (SAS^{-1}, SB)$ is an analytic map from $\widetilde{S}_{n,m}$ into $\widetilde{S}_{n,m}$. Hence the Lemma follows. ∎

LEMMA 3.4. *For $1 \leq j \leq m$, the eigenvalues of the $k_j \times k_j$-matrix N_j are pairwise different, where N_j is indicated as in Theorem 1.4.*

PROOF. Let $x = (1, 0, \ldots, 0)$ be a k_j-vector. It is clear that the couple (N_j, x) is reachable. Hence

$$\text{rank}\,[x, N_jx, \ldots, N_j^{k_j-1}x] = k_j.$$

Since N_j is symmetric, there exists a unitary matrix $S_j \in U(k_j)$ such that

$$S_jN_jS_j^{-1} = \text{diag}\,(y_1, \ldots, y_{k_j}) =: D_j.$$

We put $S_jx = (x_1, \ldots, x_{k_j}) =: \widetilde{x}$. Since the couple $(S_jN_jS_j^{-1}, S_jN_j)$ is also reachable, it follows that

$$0 \neq \det\,[\widetilde{x}, D_j\widetilde{x}, \ldots, D_j^{k_j-1}\widetilde{x}] = \det\begin{pmatrix} x_1 & & \\ 0 & \ddots & 0 \\ & & x_{k_j}\end{pmatrix} \det\begin{pmatrix} 1 & y_1 & \cdots & y_1^{k_j-1} \\ 1 & \vdots & & \vdots \\ 1 & y_{k_j} & \cdots & y_{k_j}^{k_j-1}\end{pmatrix}$$

The second factor on he right side is a Vandermonde determinant. Hence $x_t \neq 0$ for $1 \leq t \leq k_j$ and $y_t \neq y_l$ if $t \neq l$ for $1 \leq t, l \leq k_j$.

The Lemma is proved. ∎

Now we prove (iii)⇒(i) by showing that every $(A, B) \in \widetilde{H}(k)$ is the limit of some sequence in $\widetilde{H}(1)$. By means of Lemma 3.3 we need only to consider the couple (A, B) which is of the Hermite canonical form.

For $m = 1$, it is obvious that the assertion is trivial. Hence we only consider the case $m \geq 2$.

Since $k \prec l$ and $K_{n,m}$ is a lattice, there exist elements $d_1, d_2, \ldots, d_s \in K_{n,m}$ such that $k = d_1 \prec d_2 \prec \cdots \prec d_{s-1} \prec d_s = l$ and d_q is covered by d_{q-1}, $1 \leq q \leq s-1$. (see Aigner [1]).

Hence we need only to prove the case where k is covered by l i.e. l is of the form $l = (k_1, \ldots, k_{i-1}, k_i + 1, k_{i+1} - 1, k_{i+2}, \ldots, k_m)$, $k_{i+1} \geq 1$.

We first consider the case $m = 2$. Suppose that $(A, B) \in \widetilde{H}(k)$, $k = (k_1, k_2) \in K_{n,2}$ and (A, B) is of the Hermite canonical form as in Theorem 1.2:

$$A_k = \begin{pmatrix} N_1 & \\ & N_2 \end{pmatrix} \qquad B_k = \begin{pmatrix} a_{11} & * \\ 0 & \vdots \\ \vdots & * \\ \vdots & a_{21} \\ \vdots & 0 \\ \vdots & \vdots \\ 0 & 0 \end{pmatrix} \leftarrow \text{row } k_1 - 1.$$

Since $k \prec l$ hence $k_2 \geq 1$.

Suppose that $S = \begin{pmatrix} S_1 & \\ & S_2 \end{pmatrix} \in U(k_1 + k_2)$ is a unitary matrix such that $SA_kS^{-1} = \operatorname{diag}(w_1, \ldots, w_{k_1}, z_1, \ldots, z_{k_2})$, where $y_t \neq y_l$ if $t \neq l$ and $x_u \neq x_v$ if $u \neq v$ by Lemma 3.4 and

$$SB_k = \begin{pmatrix} x_1 & * \\ \vdots & \vdots \\ x_{k_1} & * \\ 0 & y_1 \\ \vdots & \vdots \\ 0 & y_{k_2} \end{pmatrix},$$

where $x_t \neq 0$ for $1 \leq t \leq k_1$ and $y_u \neq 0$ for $1 \leq u \leq k_2$ by Lemma 3.4.

Put $\widetilde{A} := \operatorname{diag}(w_1, \ldots, w_{k_1}, \widetilde{z}, z_2, \ldots, z_{k_2})$ where $\widetilde{z} \to z_1$ and $\widetilde{z} \neq w_t$ for $1 \leq t \leq k_1$.

Put

$$\widetilde{B} = \begin{pmatrix} x_1 & * \\ \vdots & \vdots \\ x_{k_1} & * \\ x_{k_1+1} & y_1 \\ 0 & \vdots \\ \vdots & \vdots \\ 0 & y_{k_2} \end{pmatrix}$$

where $x_{k_1+1} \neq 0$ and $x_{k_1+1} \to 0$. We can check that $(\widetilde{A}, \widetilde{B})$ converges to the couple (SA_kS^{-1}, SB_k) and (applying Lemma 3.4) we have $(\widetilde{A}, \widetilde{B}) \in \widetilde{H}(l)$ where $l = (k_1 + 1, k_2 - 1)$.

Finally, we consider the case $m > 2$. Suppose that (A, B) is of the Hermite canonical form:

$$A_k = \text{diag}\,[N_1, \ldots, N_i, N_{i+1}, \ldots, N_m], \qquad B_k = [b_1, \ldots, b_i, b_{i+1}, \ldots, b_m]$$

and suppose that the $n \times 2$-matrix $[b_i, b_{i+1}]$ is of the following form

$$[b_i, b_{i+1}] = \begin{pmatrix} B_{i-1} \\ B_i \\ 0 \end{pmatrix} \begin{matrix} \leftarrow \text{ row } k_1 + \cdots + k_{i-1} + 1 \\ \leftarrow \text{ row } k_1 + \cdots + k_{i-1} + k_i + 1 \\ \end{matrix},$$

where the $(k_i + k_{i+1}) \times 2$-matrix B_i has the form

$$B_i = \begin{pmatrix} a_{i1} & * \\ 0 & \vdots \\ \vdots & * \\ \vdots & a_{i+1,1} \\ \vdots & 0 \\ \vdots & \vdots \\ 0 & 0 \end{pmatrix} \begin{matrix} \\ \\ \\ \leftarrow \text{ row } k_i + 1 \\ \\ \\ \\ \end{matrix}$$

By an argument analogous to the previous one for the couple $\left(\begin{pmatrix} N_i & \\ & N_{i+1} \end{pmatrix}, B_i\right)$, we can construct a sequence $(\widetilde{A}_t, \widetilde{B}_t) \in \widetilde{H}(1)$ such that $(\widetilde{A}_t, \widetilde{B}_t)$ converges to (A, B). The proof of Theorem is completed. ■

Corollary 3.5. (i) *For $k \in K_{n,m}$, $\overline{\widetilde{H}}(k) = \bigcup\{\widetilde{H}(l), l \preceq k\}$ and $\overline{\widetilde{H}}(k)$ is an analytic subset of $\widetilde{S}_{n,m}$, i.e. for $x \in \overline{\widetilde{H}}(k)$, there exists a neighbourhood W in $\widetilde{S}_{n,m}$ of x and there exist analytic functions $f_1, \ldots, f_s : W \to \mathrm{R}$ such that $\overline{\widetilde{H}}(k) \cap W = \bigcap_{u=1}^{s} f_u^{-1}(0)$. In particular, $\widetilde{H}((n, 0, \ldots, 0))$ is open and dense in $\widetilde{S}_{n,m}$ and $\widetilde{H}((0, \ldots, 0, n))$ is closed analytic submanifold in $\widetilde{S}_{n,m}$.*

(ii) *The boundary in $\widetilde{S}_{n,m}$ of each $\widetilde{H}(k)$ is the union of some $\widetilde{H}(l)$ with $\dim \widetilde{H}(l) < \dim \widetilde{H}(k)$.*

Proof. (i) By Theorem 3.4 we have $\overline{\widetilde{H}}(k) = \bigcup\{\widetilde{H}(l); l \preceq k\}$. It follows that

$$(A, B) \in \overline{\widetilde{H}}(k) \Leftrightarrow \text{rank}\,[b_1, Ab_1, \ldots, A^{n_1}b_1, \ldots, b_j, Ab_j, \ldots, A^{n-1}b_j] \leq k_1 + k_2 + \cdots + k_j$$

for every $1 \leq j \leq m$. Hence $\overline{\widetilde{H}}(k)$ is an algebraic subset of $\widetilde{S}_{n,m}$. So it is an analytic subset of $\widetilde{S}_{n,m}$.

Since $(n, 0, \ldots, 0)$ is the maximum element of the lattice $K_{n,m}$, we have

$$\begin{aligned} \overline{\widetilde{H}}((n, 0, \ldots, 0)) &= \bigcup\{\widetilde{H}(l), l \preceq (n, 0, \ldots, 0)\} \\ &= \bigcup\{\widetilde{H}(k), k \in K_{n,m}\} = \widetilde{S}_{n,m}. \end{aligned}$$

And since $(0,\dots,0,n)$ is the minimum element of the lattice $K_{n,m}$ hence we have $\overline{\widetilde{H}}((0,\dots,0,n)) = \widetilde{H}((0,\dots,0,n))$, i.e. $\widetilde{H}((0,\dots,0,n))$ is closed in $\widetilde{S}_{n,m}$.

Finally, $(A,B) \in \widetilde{H}((0,\dots,0,n))$ if and only if rank $[b_1, Ab_1, \dots, A^{n-1}b_1] = n$ where b_1 is the first column of B. Hence $\widetilde{H}((n,0,\dots,0))$ is open in $\widetilde{S}_{n,m}$. The statement (i) is proved.

(ii) We have $\overline{\widetilde{H}}(k)\setminus\widetilde{H}(k) = \bigcup\{\widetilde{H}(l), l \prec k\}$ and $\dim \widetilde{H}(k) = \dim_{\mathbb{R}} U(n)+2n+dg(k)$, where d and $g(k)$ defined as in Corollary 1.5. Hence if $l \prec k$ then $\dim \widetilde{H}(l) < \dim \widetilde{H}(k)$ by the Definition of the $\prec$ in 3.1. Statement (ii) is proved. The proof of the Corollary is complete. ∎

Recall that $H(k) \simeq \mathbb{R}^{2n+dg(k)}$ and the orbit space $\widetilde{H}(k)/_{U(n)}$ is analytically homeomorphic to $H(k)$. Now we have:

PRPOPOSITION 3.6. *Let* $\overline{H}(k)$ *be closure of* $H(k)$ *in* $\widetilde{S}_{n,m}$ *and* $P : \widetilde{S}_{n,m} \to S_{n,m}$ *be the canonical projection. Then for* $k \in K_{n,m}$ *we have:*

(i) $P(\overline{\widetilde{H}}(k)) = \overline{H}(k)$.

(ii) $\overline{H}(k)$ *is an analytic subset of* $S_{n,m}$ *and* $H(k)$ *is an analytic submanifold of* $S_{n,m}$. $H(k)$ *is called a symmetric Hermite cell.*

(iii) $\overline{H}(k) = \bigcup\{H(l), l \preceq k\}$. *So the boundary in* $\widetilde{S}_{n,m}$ *of each cell is the union of all cells of strictly lower dimensions than* $\dim H(k)$: $\partial H(k) = \bigcup\{H(l), \dim H(l) < \dim H(k)\}$.

PROOF. (i): Follows from the fact that $P : \widetilde{S}_{n,m} \to S_{n,m}$ is a principal $U(n)$-bundle.

(ii): Since $P : \widetilde{S}_{n,m} \to S_{n,m}$ is a principal bundle, for every $[A,B] \in H(k)$ there exists a neighbourhood W of $[A,B]$ in $S_{n,m}$ and a locally trivial analytic section s : $W \to \widetilde{S}_{n,m}$ satisfying

$$s([A,B]) = (A,B) \in \widetilde{H}(k).$$

Since $\overline{\widetilde{H}}(k)$ is an anlytic subset of $\widetilde{S}_{n,m}$, there exist a neighbourhood V of (A,B) in $\widetilde{S}_{n,m}$ and analytic functions

$$g_1,\dots,g_u : V \to \mathbb{R} \qquad \text{such that} \qquad V \cap \overline{\widetilde{H}}(k) = \bigcap_{i=1}^{u} g_i^{-1}(0).$$

We can choose W such that $s(W) \subset V$ and put

$$f_i = g_i s : W \to \mathbb{R}, \qquad 1 \le i \le u.$$

Clearly that f_i are analytic functions and $W \cap \overline{H}(k) = \bigcap_{i=1}^{u} f_i^{-1}(0)$. This proves that $\overline{H}(k)$ is an anlytic subset of $S_{n,m}$.

Finally, since $\widetilde{H}(k)$ is an analytic submanifold of $\widetilde{S}_{n,m}$ and $P_k := P \mid \widetilde{H}(k) : \widetilde{H}(k) \to H(k)$ is the principal $U(n)$-bundle hence $H(k)$ is an analytic submanifold of $S_{n,m}$.

(iii): Follows from Statement (i) and Corollary 3.5 (ii).

The proof of the Proposition is complete. ∎

We now recall the definition of an analytic cell decomposition

DEFINITION 3.7 (see Helmke [7], Massey [13]). Let X be an analytic manifold. By an analytic cell decomposition of X we mean a locally finite partition $\{X_i, i \in I\}$ of X satisfying the following conditions

(i) Each X_i is an analytic submanifold of X and it is analytic homeomorphic to some $\mathbf{R}^{n_i}$, with $n_i \in \mathbf{N}$, i.e. X_i is an n_i-cell.

(ii) The closure $\overline{X}_i$ of X_i in X is an analytic subset of X and the boundary $\partial X_i = \overline{X}_i \setminus X_i$ of each cell X_i is the union of some cells X_j of strictly lower dimension than X_i.

We now shall give an analytic cell decomposition of $S_{n,m}$.

THEOREM 3.8. (i) *The manifold $S_{n,m}$ is connected and not compact. Further* $\dim_{\mathbf{R}} S_{n,m} = 2n + dn(m-1)$ *where* $d := \dim_{\mathbf{R}} F$.

(ii) *The set of all symmetric hermite cells $\{H(k), k \in K_{n,m}\}$ forms a finite analytic cell decomposition of $S_{n,m}$.*

PROOF. (i) Recall that $H((n,0,\dots,0))$ is connected and dence in $S_{n,m}$ by Corollary 3.5 (i) and by Proposition 3.6 (i). Hence $S_{n,m}$ is connected. Since $H((0,\dots,0,n))$ is closed in $S_{n,m}$ and it is not compact hence $S_{n,m}$ is not compact, either. Since $P : \widetilde{S}_{n,m} \to S_{n,m}$ is an open map (see Dieudonné [6], 12.10.6), it follows that $H((n,0,\dots,0)) = P(\widetilde{H}((n,0,\dots,0)))$ is open. Hence $\dim_{\mathbf{R}} S_{n,m} = \dim_{\mathbf{R}} H((n,0,\dots,0)) = 2n + dn(m-1)$.

Part (ii) follows from the Definition of analytic cell decomposition and from Proposition 3.6. The proof of Theorem 3.8 is complete. ∎

Before stating the result about the homology group of $S_{n,m}$ we first recall that the following beautiful and deep result is due to Borel and Haefliger [3](1961).

THEOREM 3.9. *Suppose $\{X_i, i \in I\}$ is an analytic cell decomposition of an analytic n-manifold X. Let* $h_q = \text{card}\{i \in I, \text{codim}\, X_i = q\}$, *we have:*

(i) $H_q(X; Z_2) \simeq Z_2^{h_q}$,

(ii) $H_q(X; Z) \simeq Z^{h_q}$ *if X is an orientable differentiable manifold and each X_i has an even dimension.*

PROOF. See Borel and Haefliger [3] or Helmke [7] or Massey [13]. ∎

THEOREM 3.10. *For every q we have:*

(i) $H_q(S_{n,m}(\mathbf{R}); Z_2) \simeq H_q(G_{n,n+m-1}(\mathbf{R}); Z_2)$,

(ii) $H_q(S_{n,m}(\mathbf{C}); Z) \simeq H_q(G_{n,n+m-1}(\mathbf{C}); Z)$.

Here $G_{n,n+m-1}(F)$ is the Grassmann manifold consisting of all n-planes in F^{n+m-1}.

PROOF. We put $h_q := \text{card}\{k \in K_{n,m}, \text{codim}\, H(k) = q\}$. Then $h_q = \text{card}\{k \in K_{n,m}, dn(m-1) - dg(k) = q\}$, where d and $g(k)$ are defined as in 1.5, i.e.

$$g(k) = k_1 + (k_1 + k_2) + \dots + (k_1 + k_2 + \dots + k_{m-1}) \quad \text{and} \quad d = \dim_{\mathbf{R}} F.$$

We get $H_q(S_{n,m}; Z_2) \simeq Z^{h_q}$ by Theorem 3.8 and by Theorem 3.9.

Recall that for element $a \in Y_{m-1,n}$, $a = (a_1, \dots, a_{m-1})$ the Schubert cells

$$S(a) := \left\{ V \in G_{m-1,n+m-1}(F) \;\middle|\; \begin{array}{ll} \dim V \cap F^{a_i+i-1} = i-1 & \text{and} \\ \dim V \cap F^{a_i+i} = i & \text{for } 1 \le i \le m-1 \end{array} \right\}$$

form a cell decomposition of $G_{m-1,n+m-1}(F)$ (see e.g. Milnor and Stasheff [15]). Here we have identified F^r with $F^r \times \{0\}$. Furthermore, $\dim_{\mathbf{R}} S(a) = d(a_1 + a_2 + \cdots + a_{m-1})$. Hence the Schubert cells $S(a)$ of $G_{m-1,n-m-1}(F)$ with codimension q are characterized by following:

(i) $0 \leq a_1 \leq a_2 \leq \cdots \leq a_{m-1} \leq n$;

(ii) $d(a_1 + a_2 + \cdots + a_{m-1}) = dn(m-1) - q$.

Let $b_q := \text{card}\{a \in Y_{m-1,n}, \text{codim}_{\mathbf{R}} S(a) = q\}$. We have $H_q(G_{m-1,n+m-1}(F); Z_2) \simeq Z_2^{b_q}$ by Theorem 3.9. Since the map $\widetilde{g} : K_{n,m} \to Y_{m-1,n}$ defined by $(k_1, \ldots, k_m) \to (k_1, k_1 + k_2, \ldots, k_1 + k_2 + \cdots + k_{m-1})$ is lattice isomorphism, we have $b_q = h_q$. It follows

$$H_q(S_{n,m}(F); Z_2) \simeq H_q(G_{m-1,n+m-1}(F); Z_2) \text{ for every } q.$$

Furthermore, $G_{m-1,n+m-1}(F) \simeq G_{n,n+m-1}(F)$. Hence $H_q(S_{n,m}; Z_2) \simeq H_q(G_{n,n+m-1}; Z_2)$ for every q.

(ii) If $F = \mathbf{C}$ then $S_{n,m}(F)$ is a simply-connected (see N.H.Phan and L.C.Dung [17]). Hence $S_{n,m}(F)$ is an orientable manifold (see Dold [5], VIII, 2, 12 or Spanier [19]). Moreover, all symmetric Hermite cells are of even dimensions, by mean of Theorem 3.9 (ii) we get

$$H_q(S_{n,m}(\mathbf{C}); Z) \simeq H_q(G_{n,n+m-1}(\mathbf{C}); Z) \text{ for every } q.$$

The proof of Theorem is complete. ∎

From Theorem 3.10 we obtain the following result:

COROLLARY 3.11. *There does not exists a continuous canonical form on the space* $\widetilde{S}_{n,m}$ *under similar action of* $U(n)$ *if and only if* $m > 1$.

PROOF. If $m = 1$ then the set $K_{n,1} = \{(n)\}$ (the set $K_{n,1}$ has one element) hence the canonical form in Theorem 1.4 is the continuous canonical form.

Now we consider case $m > 1$. Assume the contrary that $c : \widetilde{S}_{n,m} \to \widetilde{S}_{n,m}$ is a continuous canonical form. We shall prove that $\widetilde{S}_{n,m}$ is homeomorphic to the product $U(n) \times \widetilde{S}_{n,m}/_{U(n)}$.

Denote by $Q := \{((A,B), S\cdot(A,B)),\ S \in U(n),\ (A,B) \in \widetilde{S}_{n,m}\}$ the graph of action of $U(n)$ and denote by

$$R(A,B) := [B, AB, \ldots, A^{n-1}B].$$

Since $U(n)$ acts analytically and freely on $\widetilde{S}_{n,m}$, the map

$$f : Q \to U(n); \quad ((A,B),(A',B')) \to R(A',B')R(A,B)^* \left(R(A,B)R(A,B)^*\right)^{-1}$$

is the continuous map. Hence the map

$$g : U(n) \times S_{n,m} \to \widetilde{S}_{n,m}, (S, [A,B]) \to S \cdot c(A,B)$$

is the continuous bijection. The inverse is

$$(A,B) \to \left(f(c(A,B),(A,B)), [A,B]\right).$$

Therefore, we have $H_*(U(n) \times S_{n,m}) \simeq H_*(\widetilde{S}_{n,m})$. But this is incompatible with $H_*(S_{n,m}; G) \simeq H_*(G_{n,n+m-1}; G)$ ($G = Z_2$ or Z) by Theorem 3.10 and $\pi_i(\widetilde{S}_{n,m}) = 0$ for $i \leq dm - 2$ by N.H.Phan and L.C.Dung [17], Thm. 2.1.

The proof of the Corollary is complete. ∎

4. About the conjecture of Huynh Mui. In this section we study the homology and homotopy relation between the space of reachable symmetric linear systems $S_{n,m}(F)$ and the space of reachable linear systems $\Sigma_{n,m}(F)$ and prove that the conjecture of Huynh Mui is true in case $F = C$.

Recall that

$$\widetilde{\Sigma}_{n,m}(F) = \{(A, B) \in F^{n\times n} \times F^{n\times m}, \quad (A, B) \text{ is reachable}\}.$$

The general linear group $GL(n)$ acts on $\widetilde{\Sigma}_{n,m}(F)$ by

$$T(A, B) = (TAT^{-1}, TB).$$

The orbit space $\Sigma_{n,m} = \widetilde{\Sigma}_{n,m}(F)/_{GL(n)}$ is called the space of reachable linear systems.

Suppose that $inc : \widetilde{S}_{n,m} \to \widetilde{\Sigma}_{n,m}$ is the canonical embedding and $J : S_{n,m} \to \Sigma_{n,m}$ is the map which makes the following diagram commutative

$$\begin{array}{ccc} \widetilde{S}_{n,m} & \xrightarrow{inc} & \widetilde{\Sigma}_{n,m} \\ P\downarrow & & \downarrow P \\ S_{n,m} & \xrightarrow{J} & \Sigma_{n,m} \end{array} \tag{4.0}$$

The following Theorem is main result of this section:

THEOREM 4.1. *If $F = C$ then $J : S_{n,m} \to \Sigma_{n,m}$ is a homotopy equivalence.*

First we consider some topological properties of the space $\Sigma_{n,m}(F)$. The following Theorem has been proved by Helmke [7] in the complex case (see Helmke [7], Thm 2., p. 68 and Thm. 4, p. 135).

THEOREM 4.2. *If $F = C$ then:* (i) *$\Sigma_{n,m}(F)$ is an analytic manifold of dimension* $2nm$.

(ii) *For $k \in K_{n,m}$, let $\widetilde{H}er(k) := \{(A, B) \in \widetilde{\Sigma}_{n,m}(F),\ k(A, B) = k\}$ and let $Her(k) = \widetilde{H}er(k)/_{GL(n)}$. Then the family $\{Her(k), k \in K_{n,m}\}$ is a finite analytic cell decomposition of $\Sigma_{n,m}(F)$ and $\dim_{\mathbb{R}} Her(k) = 2n + 2g(k)$, where $g(k)$ is defined as in 1.5. i.e.*

$$g(k) = k_1 + (k_1 + k_2) + \cdots + (k_1 + k_2 + \cdots + k_{m-1}).$$

PROOF. See Helmke [7]. ∎

We now prove Theorem 4.1.

PROOF of Theorem 4.1. Since $\widetilde{S}_{n,m}(C)$ and $\widetilde{\Sigma}_{n,m}(C)$ are separable manifolds hence orbit spaces $S_{n,m}(C)$ and $\Sigma_{n,m}(C)$ are separable manifolds, too (see Dieudonné [6], 12.10.9). It follows that they have the homotopy type of a (countable) CW-complex (see McCleary [14]). Hence by Whitehead Theorem (see McCleary [14] or Spanier [19]), we need only to prove that J induces an isomorphism on integral homology.

Let $H^q_c(\Sigma_{n,m}(\mathbf{C}); Z)$ and $H^q_c(S_{n,m}(\mathbf{C}); Z)$ be respectively the q-th Alexander-Spanier cohomology groups with compact support of $\Sigma_{n,m}(\mathbf{C})$ and $S_{n,m}(\mathbf{C})$, where the coefficients are taken in Z (see Massey [13]). Denote by

$$L^q = \bigcup\{H(k), \dim_{\mathbf{R}} H(k) \le q\} \qquad \text{and}$$
$$Q^q = \bigcup\{Her(k), \dim_{\mathbf{R}} Her(k) \le q\}$$

the q-th skeletons of $S_{n,m}(\mathbf{C})$ and $\Sigma_{n,m}(\mathbf{C})$ respectively.

Let

$$C^q(S_{n,m}(\mathbf{C})) := H^q_c(L^q \setminus L^{q-1}; Z)$$
$$C^q(\Sigma_{n,m}(\mathbf{C})) := H^q_c(Q^q \setminus Q^{q-1}; Z).$$

By Massey [13] Thm. 3.3, $H^q_c(S_{n,m}(\mathbf{C}); Z)$ and $H^q_c(\Sigma_{n,m}(\mathbf{C}); Z)$ are respectively the q-th cohomology groups of cochain complex with trivial boundary operations

$$C^*(S_{n,m}(\mathbf{C})) = \{C^q(S_{n,m}(\mathbf{C})), 0\}$$
$$\text{and} \qquad C^*(\Sigma_{n,m}(\mathbf{C})) = \{C^q(\Sigma_{n,m}(\mathbf{C})), 0\}$$

(note that all cells of $S_{n,m}(\mathbf{C})$ and of $\Sigma_{n,m}(\mathbf{C})$ have even dimensions). Hence we have

$$\begin{aligned} H^q_c(S_{n,m}(\mathbf{C}); Z) = C^q(S_{n,m}(\mathbf{C}); Z) &= \bigoplus_k \{H^q_c(H(k); Z), \dim_{\mathbf{R}} H(k) = q\} \\ &= \bigoplus_k \{Z, \dim_{\mathbf{R}} H(k) = q\}, \end{aligned}$$

and

$$\begin{aligned} H^q_c\Big(\Sigma_{n,m}(\mathbf{C}); Z\Big) = C^q\Big(\Sigma_{n,m}(\mathbf{C})\Big) &= \bigoplus_k \{H^q_c(Her(k); Z), \dim_{\mathbf{R}} Her(k) = q\} \\ &= \bigoplus_k \{Z, \dim_{\mathbf{R}} Her(k) = q\}. \end{aligned}$$

For every $k \in K_{n,m}$, we have $J(H(k)) \subset Her(k)$ and

$$\begin{aligned} \dim_{\mathbf{R}} H(k) &= \dim_{\mathbf{R}} Her(k) = 2n + 2g(k) = \\ &= 2n + 2\big(k_1 + (k_1 + k_2) + \cdots + (k_1 + k_2 + \cdots + k_{m-1})\big) \end{aligned}$$

by Corollary 1.5 and by Theorem 4.2. Hence we get

$$J_k := J \mid_{H(k)} : H(k) \to Her(k)$$

induces an isomorphism on cohomology groups. It follows that

$$J^* : H^q_c(\Sigma_{n,m}(\mathbf{C}); Z) \to H^q_c(S_{n,m}(\mathbf{C}); Z)$$

is an isomorphism for every q.

Since $\Sigma_{n,m}(\mathbf{C})$ and $S_{n,m}(\mathbf{C})$ are simply-connected (see Helmke [7] and N.H.Phan and L.C.Dung [17], Thm. 2.1 and Thm. A') hence they are orientable manifolds (see Dold [5], VIII, 2.12 or Spanier [19]). Furthermore, $\dim_{\mathbf{R}} S_{n,m}(\mathbf{C}) = \dim_{\mathbf{R}} \Sigma_{n,m}(\mathbf{C}) = 2nm$. Hence applying the Poincare duality Theorem (see Massey [13]), Thm. 11.2) we have

$$H_q(\Sigma_{n,m}(\mathbf{C}); Z) \simeq H_c^{2nm-q}(\Sigma_{n,m}(\mathbf{C}); Z) \qquad \text{and}$$
$$H_q(S_{n,m}(\mathbf{C}); Z) \simeq H_c^{2nm-q}(S_{n,m}(\mathbf{C}); Z).$$

Thus

$$J_q : H_q(S_{n,m}(\mathbf{C}); Z) \to H_q(\Sigma_{n,m}(\mathbf{C}); Z)$$

is an isomorphism for every q.

The proof of Theorem 4.1 is completed. ∎

REMARK 4.4. By the diagram (4.0) and by Theorem 4.1, the inclusion map $inc : \widetilde{S}_{n,m}(\mathbf{C}) \to \widetilde{\Sigma}_{n,m}(\mathbf{C})$ induces the isomorphism of homotopy groups. Further, since $\widetilde{S}_{n,m}(\mathbf{C})$ and $\widetilde{\Sigma}_{n,m}(\mathbf{C})$ have the homotopy type of (countable) CW-complexes (see McCleary [14]) hence the map inc is a homotopy equivalence by Whitehead Theorem.

REFERENCES

[1] M. Aigner: *Combinatorial theory*. Grundlehren der Math. Wissenschaften 234, Springer 1979.

[2] G. Birkhoff and Maclane : *A survey of modern algebra*. Macmillan, New York, 4th edition 1977.

[3] A. Borel and A. Haefliger: *La classe d'homologie fondamentale d'un espace analytique*. Bull. Soc. Math. France, 89, 461-513 1961.

[4] R. Brockett: *Some geometric questions in theory of linear systems*. IEEE Trans. Autom. Control AC-21, 449-455, 1976.

[5] A. Dold: *Lectures on algebraic topology*. Springer-Verlage, Berlin Heidelberg, New York, 1972.

[6] J. Dieudonné: *Foundations of modern analysis*. Vol. 3, Academic Press, New York, 1972.

[7] U. Helmke: *Zur topologie des Raumes linearer Kontrollsysteme*. Ph. D. Thesis, Report 100, Forschungsschwerpunkt Dynamische Systeme, University of Bremen, West Germany, 1982.

[8] U. Helmke and D. Hinrichsen: *Canonical forms and orbit spaces of linear systems*. IMA Journal of Mathematical Control and Information, 3, 167-184, 1986.

[9] D. Hinrichsen: *Metrical and topological aspects of linear control theory*. Syst. Anal. Model. Simul. 4, 13-36, 1987.

[10] D. Hinrichsen and D. Prätzel-Wolters: *Generalized Hermite matrices and complete invariants of (strict) system equivalence*. SIAM J. Control and Optimazation 21, 289-305, 1983.

[11] D. Hinrichsen, D. Salamon, A. J. Pritchard, P. E. Crouch and e.a.: *Introduction to mathematical system theory*. Lecture Notes for a Join Cource at the Univerities of Warwick and Bremen, 1980.

[12] R. E. Kalman, P. L. Falb and M. A. Arbib: *Topics in mathematical system theory*. McGraw-Hill, New York, 1969.
[13] W. S. Massey: *Homology and cohomology theory*. Marcel Dekker, New York, 1978.
[14] J. McCleary: *User's guide to spectral sequences*. Publish or Perish, Inc. Wilming, Delaware (U.S.A.), 1985.
[15] J. Milnor and J. Stasheff : *Characteristic classes*. Princeton University Press, 1974.
[16] N. H. Phan: *Topo cùa không gian các hê thõng tuyẽn tính dói xúng*. TAP CHI TOAN HOC, Vol XV, No 1, 26-31, 1987, (in Vietnamese).
[17] N. H. Phan and L. C. Dung: *On the topology of the space of reachable observable symmetric linear systems*. To appear in the Report Series of Forschungsschwerpunkt Dynamische Systeme, University of Bremen, West Germany.
[18] V. M. Popov: *Invariant description of linear time-invariant controllable systems*. SIAM J. Control, 10, 252-264, 1972.
[19] E. H. Spanier: *Algebraic topology*. McGraw-Hill, New York, 1966.

Homotopy Ring Spaces and Their Matrix Rings

R. Schwänzl and R. M. Vogt

0. Introduction

This paper is an account of methods and (elementary) results we have used without giving details since 1984 (e.g. see [FSSV], [FSV1], [FSV2]). Utilizing a result of R. Steiner [St2; Lemma 1.7] we in [SV1] altered P. May's definitions of A_∞ and E_∞ ring spaces [M2], [M3] to a homotopy invariant one, and could show that a ring space in this new sense can be converted up to coherent homotopy in one of May's sense without changing its homotopy type. This new definition is flexible enough to allow to carry out a large number of classical constructions of basic ring theory in the A_∞ and E_∞ setting in a fairly straight-forward way, a point which May discussed rather pessimistically [M3; Introduction, Remarks 10.3 and 12.4]. One aim of this paper is to demonstrate this transfer from classical ring theory to the A_∞ world for constructions involving matrices.

In Section 1 we give definitions of A_∞ and E_∞ monoids and rings which are most suited for our purpose and compare them with previous definitions in the literature. Homomorphisms between homotopy ring spaces are too rigid for our theory; in particular, the notion of homomorphism is not homotopy invariant. There are two possible substitutes introduced and shortly discussed in Section 2. In Section 3 we construct the A_∞ rings M_nX of n-squared matrices over an A_∞ ring X. This cannot be done in May's original setting (see [M3; § 4]). Special cases have been treated previously by K. Igusa [I]. For the generalization of triangular or diagonal matrices we have two options which we introduce and compare. Finally we study the multiplicative A_∞ submonoid Gl_nX of homotopy invertible n-squared matrices over X. The methods of Section 3 are applied to give a plus-construction definition of the algebraic K-theory KX of an A_∞ ring X using a suitable stabilization sequence of Gl_nX. This definition is in the spirit of [M3] but can do without technical arguments. The alternative approach to KX due to Steiner [St2] will be the subject of [SV6] (see also [SV2]). There we among other things prove Morita-invariance for KX and show that KX of an E_∞ ring X is itself an E_∞ ring, extending a result of May [M3; Prop. 10.12] who proved this to be true for genuine commutative topological rings.

Acknowledgement: The work on this paper has partly been supported by the DFG.

1. Homotopy monoids and homotopy rings

As pointed out in the introduction, many of the concepts of classical monoid and ring theory carry over to the homotopy coherent case if they can be expressed in universal terms, i.e. if they are independent of the particular monoid or ring under consideration. It is the purpose of this section to set up this "universal" language.

Throughout this paper we work in the category *Top* of compactly generated spaces in the sense of [V]. Products, subspaces, function spaces etc. are taken in this category. The corresponding homotopy category is denoted by Top_h.

The language we are going to use is the one of universal algebra, i.e. we are going to codify (topological) algebraic structures universally by exhibiting the category of all operations which can be written down in the given algebraic structure. Such categories are called theories (see definition below). We always have some operations for free, called *set operations*: Let S be the category of finite sets $\underline{n} = \{1, 2, 3, ...\}, n \geq 0$, with $\underline{0} = \emptyset$ and all maps. For each space X and each $\alpha \in S(\underline{m}, \underline{n})$ we have the set operation

$$(1.1) \qquad \alpha^* : X^n \to X^m, \quad (x_1, ..., x_n) \mapsto (x_{\alpha 1}, ..., x_{\alpha m})$$

1.2 Definition: A *theory* is a category Θ with objects 0,1,2,..., topologized morphism sets $\Theta(m, n)$, and products, together with a faithful functor $S^{op} \subset \Theta$ preserving objects and products. Composition is continuous, and the canonical map $\Theta(m, n) \to \Theta(m, 1)^n$ is a homeomorphism.
A *Θ-space* is a continuous functor $X : \Theta \to Top$ such that $S^{op} \subset \Theta \to Top$ is product preserving. The space $X(1)$ is called the *underlying space* of X.
A *homomorphism* of Θ-spaces is a natural transformation of such functors.
A *theory functor* $\Theta_1 \to \Theta_2$ is a continuous functor of theories such that

$$\begin{array}{ccc} & S^{op} & \\ \swarrow & & \searrow \\ \Theta_1 & \longrightarrow & \Theta_2 \end{array}$$

commutes.

We will often find it convenient to overlook the distinction between X and $X(1)$ and refer to X when we mean $X(1)$ and vice versa.

1.3 Examples: By definition, S^{op} is initial in the category of theories, and each topological space is an S^{op}-space by (1.1).

The first non-trivial example is the theory Θ_m of monoids: $\Theta_m(n, 1)$ is the free monoid on n generators $x_1, ..., x_n$, and $\Theta_m(n, k) = (\Theta_m(n, 1))^k$. Composition is defined by substitution;

e.g.

$$(x_1x_2x_1)\circ(x_1x_3,x_2x_3)=(x_1x_3x_2x_3x_1x_3):3\to 2\to 1.$$

A monoid X determines and is determined by a Θ_m-space $\Theta_m \to Top$, given by sending a word $w(x_1,...,x_n)$ in n generators to the operation

$$X^n \to X, \qquad (x_1,...,x_n)\mapsto w(x_1,...,x_n)$$

where the word is evaluated in X.

Θ_m and the theories Θ_{cm} of commutative monoids, Θ_r of semirings, and Θ_{cr} of commutative semirings, which are defined similarly, will play the central role in this paper.

We are now going to define the homotopy coherent analogues of monoids and semirings. In Θ_m a morphism from n to k is a k-tuple of monomials

(1.4) $$x_{i_1}^{r_1}\cdot\ldots\cdot x_{i_p}^{r_p}$$

in n non-commuting variables, in Θ_{cm} it is a k-tuple of monomials

(1.5) $$x_1^{r_1}\cdot\ldots\cdot x_n^{r_n}$$

in n commuting variables, in Θ_r it is a k-tuple of finite sums of monomials

(1.6) $$l\cdot x_{i_1}^{r_1}\cdot\ldots\cdot x_{i_p}^{r_p}$$

of type (1.4) with coefficients $l\in I\!N$ (0 is contained in $I\!N$), and in Θ_{cr} it is a k-tuple of finite sums of monomials

(1.7) $$l\cdot x_1^{r_1}\cdot\ldots\cdot x_n^{r_n}$$

of type (1.5) with coefficients $l\in I\!N$. Such a morphism is called *simple* if all coefficients l are 0 or 1 and if in the commutative cases all exponents r_i are 0 or 1. In particular, all morphisms of Θ_m are simple.

In the following definition Θ_* is $\Theta_m, \Theta_{cm}, \Theta_r, \Theta_{cr}$, whatever is appropriate, and $s\Theta_*$ its subset of simple morphisms.

1.8 Definition An A_∞ or E_∞ *monoid* or *ring theory* is a theory Θ together with a theory functor $F=F_\Theta:\Theta\to\Theta_*$ such that
(1) $ob\Theta\subset mor\Theta$ is a closed cofibration
(2) $F: mor\Theta\to mor\Theta_*$ is bijective on path components and a homotopy equivalence over $s\Theta_*$.
A Θ-space is called an A_∞ or E_∞ *monoid* or *ring*.

1.9 Comparison of definitions: The definitions of A_∞ or E_∞ monoids in [BV1], [BV2] and of A_∞ and E_∞ rings in [SV1] require less stringent conditions of the augmentation functor. As theories Θ_m and Θ_{cm} are generated by the simple morphisms

(1.10) $\mu_n = x_1 \cdot x_2 \cdot \ldots \cdot x_n, \quad n > 0, \text{ and } \mu_0 = 1$

subject to the obvious relations. Let $A \subset \Theta_m$ and $CA \subset \Theta_{cm}$ be the subcategories generated under composition and taking products by the μ_n and, in the commutative case, by the permutation set operations. In [BV1], [BV2] a theory augmented over $\Theta_* = \Theta_m$ or Θ_{cm}, $F : \Theta \to \Theta_*$ is called an A_∞ resp. E_∞ monoid theory if F is a homotopy equivalence over A resp. CA. This condition is weaker than (1.8) because A and CA consist of simple morphisms only. The subcategory $B = F^{-1}(A)$ resp. $F^{-1}(CA)$ of Θ is the spine of a split theory Θ_N over Θ_m resp. Θ_{cm} in the sense of [BV2; p. 58 ff.]. The identity on B extends to a theory functor $G : \Theta_N \to \Theta$. It is easy to show that $F_\Theta \circ G$ is a homotopy euqivalence over simple morphisms. Hence any A_∞ or E_∞ monoid in the sense of [BV1], [BV2] is in a canonical way an A_∞ or E_∞ monoid in the sense of (1.8) by substituting Θ by Θ_N via G.

The theories Θ_r and Θ_{cr} are generated by the simple multiplicative operations μ_n of (1.10) and the simple additive relations

(1.11) $\lambda_n = x_1 + \ldots + x_n, \quad n > 0, \quad \lambda_0 = 0.$

The subcategories generated by the μ_n and λ_n under composition and pro- duct contain non-simple morphisms, which complicates the picture. Let $R \subset \Theta_r$ and $CR \subset \Theta_{cr}$ be the subsets consisting of μ_0 and the morphisms of the subcategories generated under composition and product by the $\lambda_n, n \geq 0$, the $\mu_k, k \geq 1$, and the permutation set operations. In [SV1] a theory augmented over $\Theta_* = \Theta_r$ or $\Theta_{cr}, F_\Theta : \Theta \to \Theta_*$ is called an A_∞ resp. E_∞ ring theory if F_Θ is surjective and a homotopy equivalence over R resp. CR. We could show [SV1; Thm. 5.1] that there is a universal A_∞ resp. E_∞ theory $\mathcal{U}$ derived from Steiner's canonical operad pair [St1] and a functor $G : \mathcal{U} \to \Theta$ of A_∞ resp. E_∞ ring theories. This G is unique up to contractible choice; i.e. there is a canonical contractible space of A_∞ resp. E_∞ theory functors $\mathcal{U} \to \Theta$. Since $\mathcal{U}$ is an A_∞ resp. E_∞ theory in the sense of (1.8) ([St2; Lemma 1.7] and [SV1; Prop. 2.5]) any A_∞ or E_∞ ring in the sense of [SV1] is one in the sense of (1.8), canonically up to contractible choice. Moreover, by [SV1; Cor. 5.2], any $\mathcal{U}$-space X is homotopy equivalent to a space Y on which Steiner's operad pair acts in the sense of May [M2]. Since any A_∞ or E_∞ operad pair in the sense of [M3] or [M2] gives rise to an A_∞ or E_∞ ring theory by [St2], any A_∞ or E_∞ ring in the sense of May is one in the sense of (1.8), while any A_∞ or E_∞ ring in our sense is homotopy equivalent to one in May's sense.

There are more combinatorial definitions of A_∞ or E_∞ monoids and rings, the Δ-spaces and Γ-spaces of Thomason [Th] and Segal [Se] in the monoid cases and Woolfson's hyper-Γ-spaces in the E_∞ ring case [Wo]. By various replacement procedures ([M4], [MT], [SV3]) these functors are homotopy equivalent to A_∞ resp. E_∞ monoids and rings, and vice versa.

1.12 Remarks: One may think that in Definition 1.8 it would be more convenient to postulate that F_Θ be a global equivalence. Unfortunately this cuts down the range of examples in the commutative monoid case and hence also in both ring cases basicly to products of Eilenberg-MacLane spaces, because by [BV2; Thm. 4.58] a Θ-space would then be weakly equivalent to a topological abelian monoid resp. a (commutative) topological semiring, which excludes important examples such as stable homotopy, complex bordism [M2], algebraic K-theory of E_∞ rings [SV2] etc. For similar reasons we have to augment our theories over commutative monoids and semirings rather than abelian groups and rings: Let Θ_{cg} be the theory of abelian groups. Then $\Theta_{cm} \subset \Theta_{cg}$. In [SV4] we showed

(1.13) Let $F : \Theta \to \Theta_{cg}$ be a theory functor which is bijective on path components and defines an E_∞ monoid theory upon restriction to Θ_{cm}. Then every Θ-space is weakly equivalent to a weak product of Eilenberg-MacLane spaces.

1.14 Remark: Since the standard CW-approximation functor preserves products and contractability, we may assume that our A_∞ and E_∞ monoid and ring spaces have structures codified by CW-theories which are contractible over simple morphisms. Moreover, up to weak equivalence, we can substitute our A_∞ and E_∞ monoids and rings by their CW-approximations. The passage to CW-theories takes automatically care of the technical requirement that $ob\Theta \subset mor\Theta$ be a closed cofibration.

2. Homotopy homomorphisms and hammocks

As map between A_∞ and E_∞ monoids and rings we have two options: Homotopy homomorphisms, studied in detail in [BV2] in the monoid cases and in [SV1] in the ring cases, and hammocks, introduced in [DK]. Our definition of homotopy homomorphisms below is equivalent to the ones in [BV2] and [SV1] but avoid the use of universal constructions.

A homotopy homomorphism can be replaced by a hammock (see [SV5]). While homotopy homomorphisms arise naturally when one translates standard linear maps such as the stabilization $Gl_nR \to Gl_{n+1}R$ of the general linear group to the A_∞ and E_∞ world, functoriality of our constructions can be described more easily in terms of hammocks.

Homotopy homomorphisms: In general we have a family $\{X_k; k \in K\}$ of A_∞ or E_∞ monoids or rings. Let S_k^{op} be the category of set operations on X_k. Since $X_k^0 = \star$ for all $k \in K$, the category of set operations of the whole family is

$$S_K^{op} = \coprod_{k\in K} S_k^{op}/\sim$$

with the objects $\underline{0} \in S_k^{op}, k \in K$, all identified to a single terminal object 0. We denote the object $\underline{n} \in S_k$ by (n, k).

2.1 Definition: A *K-coloured theory* is a category Θ with $ob\Theta = obS_K$ together with a faithful functor $S_K^{op} \subset \Theta$ preserving objects and products. The morphism sets of Θ are topologized, composition is continuous and the canonical map $\Theta((m,k),(n,l)) \to [\Theta((m,k),(1,l))]^n$ is a homeomorphism.
A *theory functor* from a K-coloured theory Θ_1 to an L-coloured theory Θ_2 consists of a map $f : K \to L$ and a continuous functor $G : \Theta_1 \to \Theta_2$ such that

$$\begin{array}{ccc} S_K^{op} & \xrightarrow{f_*} & S_L^{op} \\ \cap & & \cap \\ \Theta_1 & \xrightarrow{F} & \Theta_2 \end{array}$$

commutes.
The definitions of Θ-spaces and homomorphisms extend to the K-coloured case in the obvious way.

2.2 Example: Let Θ be a monochrome theory and D a small category. A D-diagram of Θ-spaces is codified by an obD-coloured theory $\Theta \diamond D$ defined as follows:

$$(\Theta \diamond D)((m,d_1),(1,d_2)) = \Theta(m,1) \times D(d_1,d_2)^m.$$

An injection $\sigma : \underline{1} \to \underline{m}$ in $S_d \subset S_{obD}$ is mapped to $(\sigma^*; id_d, ..., id_d)$ which specifies the functor $S_{obD}^{op} \to \Theta \diamond D$. Composition of $c = (a; f_1, ..., f_m) \in (\Theta \diamond D)((m,d_1),(1,d_2))$ with $(c_1, ..., c_m) \in (\Theta \diamond D)((n,d_0)(m,d_1))$ is defined by

$$(a \circ (b_1, ..., b_m); f_1 \circ g_{11}, ..., f_1 \circ g_{1n}, ..., f_m \circ g_{m1}, ..., f_m \circ g_{mn})$$

where $c_i = (b_i; g_{i1}, ..., g_{in})$. This determines $\Theta \diamond D$ completely.
It is easy to check that a $(\Theta \diamond D)$-space is a D-diagram of Θ-spaces and homomorphisms and vice versa.

2.3 Remark: In [SV1] we used the ambiguous symbol $\Theta \times D$ for $\Theta \diamond D$. Note that $\Theta \diamond D$ is the quotient of the usual product $\Theta \times D$ of categories, obtained by identifying all objects $(0,d)$ to a single terminal object 0. The results of [SV1] apply to $\Theta \diamond D$.

In the following definition Θ_* is $\Theta_m, \Theta_{cm}, \Theta_r$ or Θ_{cr}, and $s\Theta_*$ its subset of simple morphisms.

2.4 Definition: Let D be a small category. A *D-indexed* A_∞ or E_∞ *monoid* or *ring theory* is an obD-coloured theory Θ with an augmentation functor $F = F_\Theta : \Theta \to \Theta_* \diamond D$ of obD-coloured theories such that

(1) $ob\Theta \subset mor\Theta$ is a closed cofibration
(2) F preserves objects
(3) F is bijective on path components and a homotopy equivalence over $s\Theta_* \diamond D$.
A Θ-space is called a *D-indexed* A_∞ or E_∞ monoid or ring. To subsume the four definitions into one, we call Θ a *theory over* $\Theta_* \diamond D$.

Categories D of special interest for us are

$$L_n : 0 \to 1 \to 2 \ldots \to n.$$

For constructions generalizing the bar construction $D = \Delta^{op}$, the simplicial index category, is of importance (see [FSSV]). Since a homomorphism of Θ-spaces, Θ a monochrome theory, is simply a $\Theta \diamond L_1$-space, we are led to the following definition.

2.5 Definition: Let Θ_i be theories over Θ_* and X_i a Θ_i-space, $i = 0, 1$. An *h-morphisms* (or *homotopy homomorphism*) from X_0 to X_1 consists of theories Φ_i over $\Theta_*, i = 0, 1$, a theory Ψ over $\Theta_* \diamond L_1$, theory functors

$$\Theta_i \xleftarrow{p_i} \Phi_i \xrightarrow{q_i} \Psi_i = F_\Psi^{-1}(\Theta_* \diamond \{i\}) \qquad i = 0, 1$$

over Θ_*, and a Ψ-space $G : \Psi \to Top$ such that

$$\begin{array}{ccc} \Phi_i & \xrightarrow{q_i} \Psi_i \subset & \Psi \\ \downarrow{p_i} & & \downarrow{G} \\ \Theta_i & \xrightarrow{X_i} & Top \end{array} \qquad i = 0, 1$$

commutes. Any map $H(a) : X_0 \to X_1$ with $a \in F_\Psi^{-1}(((id_1; 0 \to 1))$ is called an *underlying map* of the h-morphism.

Although the theories Φ_i, Ψ_i and Θ_i might have very little to do with each other, the functors p_i and q_i are homotopy equivalences over $s\Theta_*$. This definition allows to compare spaces structured by different theories without the need to introduce the universal theory $\mathcal{U}$. Using the universality of $\mathcal{U}$ it is immediate to relate this definition to the ones of [BV2; 4.2] and [SV1; 3.1]. We recall one important fact from [SV1].

2.6 Proposition: Let $\Theta_i, i = 0, 1$ be theories over Θ_*, let X_i be a Θ_i-space and $H : X_0 \to X_1$ an h-morphisms. Let $h : X_0(1) \to X_1(1)$ be an underlying map and a homotopy equivalence. Then any homotopy inverse of h is the underlying map of an h-morphism $X_1 \to X_0$. We say that X_0 and X_1 are homotopy equivalent as A_∞ or E_∞ monoids or rings (whatever applies).

Hammocks: Fix a theory Θ over Θ_*. A homomorphism $f : X \to Y$ of Θ-spaces will be called a weak equivalence if $f(1) : X(1) \to Y(1)$ is a homotopy equivalence.

2.7 Definition: A *hammock* of length $n \geq 0$ and width $k \geq 0$ is a commutative diagram of Θ-spaces and homomorphisms

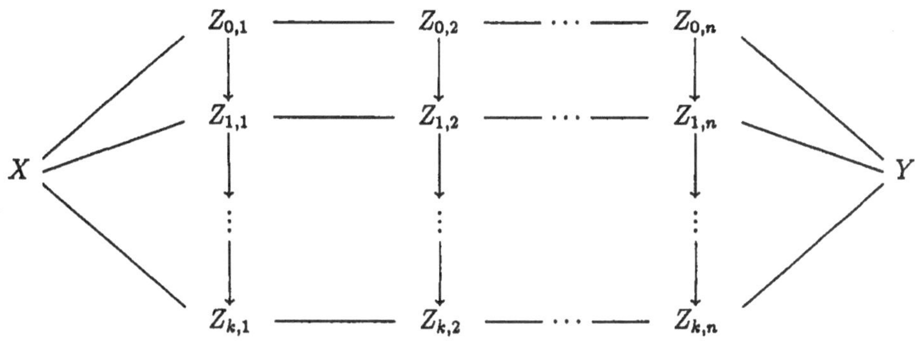

such that
(1) all vertical maps are weak equivalences
(2) in each column all maps go in the same direction. If they go to the left, they are weak equivalences
(3) the maps in two adjacent columns go in different directions
(4) no column contains only identity maps.

After restriction to a sufficiently large set of Θ-spaces the hammocks form a simplicial category. Its k-simplices are the hammocks of width k. The i-th degeneracy repeats the i-th row, the i-th face omits the i-th row. It can happen that the resulting "hammock" fails to satisfy (4), but we can reduce it to a genuine hammock by omitting a column if it contains only identities and then compose columns to establish property (3), and then carry on if necessary.

Hammocks are composed in the obvious way, and the composition is associative and commutative.

3. Matrices over A_∞ ring spaces

Throughout this section let Θ be an A_∞ ring theory and X a Θ-space. Since $F_\Theta : \Theta \to \Theta_r$ is bijective on path components, X defines a semiring in the homotopy category

$$[X] : \Theta_r \longrightarrow Top_h$$

such that

$$\begin{array}{ccc} \Theta & \xrightarrow{X} & Top \\ \big\downarrow F_\Theta & & \big\downarrow \text{proj.} \\ \Theta_r & \xrightarrow{[X]} & Top_h \end{array}$$

commutes.

Let M_nX denote the space of all n-squared matrices with entries in X. We define a functor

$$M_n : \Theta_r \to \Theta_r$$

such that $[X] \circ M_n$ is the semiring of n-squared matrices over $[X]$ in Top_h: The category S of sets $\underline{n}$ comes equipped with a canonical sum and a canoni- cal product. We identify $\underline{m} \sqcup \underline{n}$ with $\underline{m+n}$ in blocks and $\underline{m} \times \underline{n}$ with $\underline{m \cdot n}$ via lexicographical ordering of pairs. The functor M_n sends $1 \in \Theta_r$ to $n^2 = n \times n$. Think of n^2 as the entries of an $(n \times n)$-matrix in lexicographical ordering. A set operation $\alpha^* : q \to p$ is sent to the set operation $\sigma^* = M_n(\alpha^*)$, where

$$\sigma : \underline{pn^2} = (\underline{n} \times \underline{n}) \sqcup ... \sqcup (\underline{n} \times \underline{n}) \to (\underline{n} \times \underline{n}) \sqcup ... \sqcup (\underline{n} \times \underline{n}) = \underline{qn^2}$$

maps the i-th summand $\underline{n} \times \underline{n}$ identically to the $\alpha(i)$-th summand. In matrix terms, σ^* is given by the set operation α^* applied to a q-tuple of $(n \times n)$-matrices. The operation $x_1 + ... + x_p$ is mapped to the p-fold matrix addition $pn^2 \to n^2$ whose (i,j)-th component is

$$x^1_{ij} + ... + x^p_{ij}$$

where the superscript k stands for the k-th summand $n \times n$ in pn^2. Similarly $x_1 \cdot ... \cdot x_p$ is sent to the p-th fold matrix multiplication. This determines the functor M_n completely. Now form the pullback theory $M_n\Theta$:

$$\begin{array}{ccc} M_n\Theta & \xrightarrow{\tilde{M}_n} & \Theta \\ \big\downarrow F_{M_n\Theta} & & \big\downarrow F_\Theta \\ \Theta_r & \xrightarrow{M_n} & \Theta_r \end{array}$$

Since M_n maps simple operations to simple operations, $F : M_n\Theta \to \Theta_r$ is an A_∞ ring theory, and we obtain

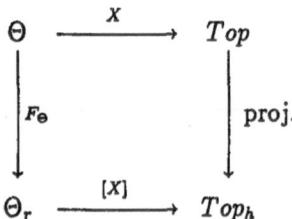

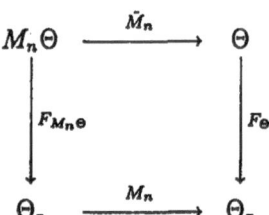

3.1 Theorem: Let Θ be an A_∞ ring theory and $X : \Theta \to Top$ an A_∞ ring. Then $M_n\Theta$ is an A_∞ ring theory and $M_nX = X \circ \tilde{M}_n : M_n\Theta \to Top$ endows the space of $(n \times n)$-matrices over X with an A_∞ ring structure such that $[M_nX] : \Theta_r \to Top_h$ is the usual matrix semiring over $[X]$. The correspondence $X \mapsto M_n X$ is functorial with respect to hammocks.

To treat the A_∞ analogues of upper or lower triangular or of diagonal matrices let X_z be the space $X(F_\Theta^{-1}(\lambda_0)) \subset X(1)$, i.e. the space of zeroes in X (which is not the full path component of 0 in the semiring $\pi_0(X)$). Let $U_n^z X, L_n^z X$ and $D_n^z X$ be the subspaces of all matrices in $M_n X$ having elements in X_z below, above, respectively below and above the diagonal. We have

(3.2) The subspaces $U_n^z X, L_n^z X$, and$D_n^z X$ of $M_n X$ are $M_n\Theta$-subspaces of $M_n X$. In particular, they are A_∞ rings.

3.3 Remarks: Although $F_\Theta^{-1}(\lambda_0)$ is contractible, X_z need not be, so that $U_n^z X, L_n^z X$, and $D_n^z X$ may not have the desired homotopy type of $X^{\frac{1}{2}n(n+1)}$ resp. X^n. We are going to construct a functor $U_n : \Theta_r \to \Theta_r$ similar to M_n, such that $[X] \circ U_n$ codifies the upper triangular matrix semiring with the zeroes below the diagonal ignored. By [SV2; Thm. 4.12], any Θ-space X is equivalent as A_∞ ring to a Θ-space Y such that

$$Y : \Theta(0,1) \to Top(*, Y(1)) = Y(1)$$

maps $F_\Theta^{-1}(\lambda_0)$ and $F_\Theta^{-1}(\mu_0)$ homeomorphically onto their images. The homotopy equivalence is given by a Θ-homomorphism $Y \to X$ and a homotopy homomorphism $X \to Y$. So up to homotopy, we can always arrange that $X_z \simeq *$. In this case $U_n^z X$ is equivalent as A_∞ ring to $U_n X$ to be constructed.

The functor $U_n : \Theta_r \to \Theta_r$ maps is given on objects by

$$p \mapsto p \cdot \frac{1}{2} n \cdot (n+1) = \frac{1}{2} n(n+1) \sqcup \ldots \sqcup \frac{1}{2} n(n+1).$$

Each summand should be considered as the $\frac{1}{2}n(n+1)$ interesting entries of an upper triangular $(n \times n)$-matrix. Set operations are mapped in the analogous way to the M_n-case. Addition $x_1 + \ldots + x_p$ is mapped to componentwise addition again, but the multiplication $x_1 \cdot \ldots \cdot x_p$ is mapped to the operation whose (ij)-th component, $i \le j$, is

$$\sum_{i \le r_1 \le \ldots \le r_{p-1} \le j} x^1_{ir_1} x^2_{r_1 r_2} \ldots x^p_{r_{p-1} j}.$$

As in the M_n-case we form the pullback theory $U_n\Theta$, which is an A_∞ ring theory. By construction $[X] \circ U_n$ is isomorphic to the semiring of upper triangular matrices over $[X]$.

Proceeding analogously in the L_n and D_n cases we obtain

3.4 Proposition: $U_n\Theta$, $L_n\Theta$ and $D_n\Theta$ are A_∞ ring theories and hence $U_nX = X \circ \tilde{U}_n$ and similarly L_nX and D_nX are A_∞ rings such that $[U_nX]$, $[L_nX]$ and $[D_nX]$ are isomorphic to the upper resp. lower triangular resp. diagonal matrix semirings over $[X]$. The correspondences $X \mapsto U_nX, L_nX, D_nX$ are functional with respect to hammocks. Moreover $U_nX(1) = L_nX(1) = X^{\frac{1}{2}n(n+1)}$ and $D_nX(1) = X^n$.

Since $U_n\Theta$, $L_n\Theta$ and $D_n\Theta$ differ from $M_n\Theta$, the A_∞ rings U_nX, L_nX and D_nX are not subsemirings of M_nX, but there are h-morphisms into M_nX.

3.5 Proposition: There are h-morphisms

$$u_nX : U_nX \to U_n^zX, \quad l_nX : L_nX \to L_n^zX, \quad d_nX : D_nX \to D_n^zX$$

which are homotopy equivalences of A_∞ rings if X_z is contractible. Their underlying maps include all maps of the form

$$(x_{ij})_{i\le j\le n} \mapsto \begin{pmatrix} x_{11} & & x_{ij} \\ & \ddots & \\ 0 & & x_{nn} \end{pmatrix}, \qquad (x_{ij})_{j\le i\le n} \mapsto \begin{pmatrix} x_{11} & & 0 \\ & \ddots & \\ x_{ij} & & x_{nn} \end{pmatrix}$$

and $(x_1, ..., x_n) \mapsto \begin{pmatrix} x_1 & & 0 \\ & \ddots & \\ 0 & & x_n \end{pmatrix}$ respectively, where 0 stands for any collection of zeroes in X_z. All three h-morphisms are natural with respect to hammocks.

The proof of 3.5 is typical for the passage from strict data to A_∞ data: Since M_n and U_n are injective, so are $\tilde{M}_n$ and $\tilde{U}_n$. Hence the following definition of a theory Ψ over $\Theta_r \diamond L_1$ is legitimate:

$$\begin{aligned} \Psi((p,0),(1,0)) &= U_n\Theta(p,1) \subset \Theta\left(\frac{1}{2}pn(n+1), \frac{1}{2}n(n+1)\right) \\ \Psi((p,1),(1,1)) &= M_n\Theta(p,1) \subset \Theta(pn^2, n^2) \\ \Psi((p,0),(1,1)) &\subset \Theta\left(\frac{1}{2}pn(n+1), n^2\right). \end{aligned}$$

The last morphism space is the subspace of all operations c such that $F_\Theta(c)$ is a composite of the form

$$\left((x_{ij}^1)_{i\le j}, \ldots, (x_{ij}^p)_{i\le j}\right) \xrightarrow{a} (y_{ij})_{i\le j} \longrightarrow \begin{pmatrix} \ddots & & y_{ij} \\ & \ddots & \\ 0 & & \ddots \end{pmatrix} \qquad (*)$$

with $a = U_n(a')$ and $a' \in \Theta_r(p, 1)$.

Composition in Ψ is induced by the composition in Θ. The augmentation F_Ψ is the obvious one with $F_{u_n\Theta}$ and $F_{M_n\Theta}$ on colours 0 and 1 and on $\Psi((p,0),(1,1))$ by $F_\Psi(c) = (a', 0 \to 1)$ in the notation of $(*)$.
$X : \Theta \to Top$ induces a Ψ-space $H : \Psi \to Top$ which defines the required h-morphism with q_i and p_i in Def. 2.5 being identities.
The functionality with respect to hammocks is obvious.

□

A short investigation of the construction shows that $D_n\Theta \cong \Theta$ and that $D_nX \cong X^n$ as A_∞ ring with the diagonal action of Θ on X^n. The diagonal

$$\Delta : X \to X^n$$

is a Θ homomorphism. Hence as a corollary to 3.5 we obtain

3.6 Proposition: There are h-morphisms

$$X \to M_nX \text{ and } X^n \to M_nX$$

functorial with respect to hammocks, whose underlying maps include all maps of the form

$$x \mapsto \begin{pmatrix} x & & 0 \\ & \ddots & \\ 0 & & x \end{pmatrix} \text{ resp. } (x_1, \ldots, x_n) \mapsto \begin{pmatrix} x_1 & & 0 \\ & \ddots & \\ 0 & & x_n \end{pmatrix}$$

for any collection of zeroes in X_z.

Other subspaces of M_nX enherit the multiplicative structure only. Let $\Theta_m \subset \Theta_r$ be the embedding of the multiplicative monoid structure of semi- rings. The pullback

$$\begin{array}{ccc} A_n\Theta & \subset & M_n\Theta \\ \downarrow F_{A_n\Theta} & & \downarrow F_{M_n\Theta} \\ \Theta_m & \subset & \Theta_r \end{array}$$

is an A_∞ monoid theory. We consider it as the multiplicative part of $M_n\Theta$. Restriction to $A_n\Theta$ makes M_nX an A_∞ monoid. Interesting $A_n\Theta$-subspaces of M_nX are the monomial matrices Mon_nX and the homotopy invertible matrices Gl_nX. A matrix is called monomial if each row and column has at most one element which is not in X_z. The subspace Gl_nX is by definition the pullback

(3.7)

$$\begin{array}{ccc} Gl_nX & \subset & M_nX \\ \downarrow & & \downarrow \pi_0 \\ Gl_n(\pi_0X) & \subset & M_n(\pi_0X) \end{array}$$

where π_0 is the discretization map. So we have

(3.8) $Mon_n X \cap Gl_n X$, $Gl_n X$ and $Mon_n X$ are $A_n\Theta$-submonoids of $M_n X$. The correspondences $X \mapsto Mon_n X \cap Gl_n X$, $Gl_n X$, $Mon_n X$ are functorial with respect to hammocks.

As with upper triangular matrices there are versions of $Mon_n X \cap Gl_n X$ and $Mon_n X$ which disregard the zeroes and the analogues of 3.4 and 3.5 hold. These versions are of importance in the set up of the so-called monomial map in K-theory (see § 4), we denote them by $\Sigma_n \int X^*$ and $\Sigma_n \int X$ respectively (May uses $\Sigma_n \int FX$ for $\Sigma_n \int X^*$ in [M3]). The treatment of the monomial map in [M3] is incorrect as pointed out in [M4; Appendix B]. We sketch a correction in § 4.

3.9 Definition: An A_∞ or E_∞ ring X is called *ringlike* if the semiring $\pi_0 X$ is a genuine ring.

If X is ringlike, π_0 in (3.7) is a fibration with fiber $X_0^{n^2}$, where X_0 is the path component of $0 \in \pi_0 X$. Since (3.7) is a pullback, we obtain

(3.10) If X is ringlike, then for any choice of base points

$$\pi_i(Gl_n X) \cong \begin{cases} Gl_n(\pi_0 X) & i = 0 \\ \pi_i(X_0)^{n^2} & i > 0 \end{cases}$$

Let $(Gl_n X)_E \subset Gl_n X$ be the component of the unit matrix E in $Gl_n(\pi_0 X)$. For various applications (e.g. see [FSSV]) it is useful to identify $(Gl_n X)_E$ with $M_n(X_0)$: If X is a genuine topological ring, there is an isomorphism of groups

$$M_n(X_0) \cong Gl_n(X)_E, \quad A \mapsto E + A$$

with the group structure $A * B = A + B + AB$ on $M_n(X_0)$. We will establish the A_∞ analogue as an additional example of the method used in this section. We first compare the $*$-multiplication on $M_n X$ with the usual one, and then obtain the result by restriction to subfunctors $M_n(X_0)$ and $(Gl_n X)_E$.

Define $G_n\Theta$ to be the pullback

$$\begin{array}{ccc} G_n\Theta & \xrightarrow{\bar{G}_n} & M_n\Theta \\ \Big\downarrow{\scriptstyle F_{G_n\Theta}} & & \Big\downarrow{\scriptstyle F_{M_n\Theta}} \\ \Theta_m & \xrightarrow{G_n} & \Theta_r \end{array}$$

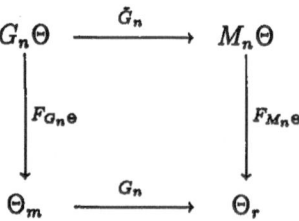

where G_n is determined by mapping the operation $\mu_p = x_1 \cdot \ldots \cdot x_p$ to

$$\sum_{k=1}^{p} \sum_{1 \leq i_1 < \ldots < i_k \leq p} x_{i_1} \cdot \ldots \cdot x_{i_k} \quad \text{for } p > 0$$

and the operation $\mu_0 = 1$ to $\lambda_0 = 0$. This G_n is a functor of theories, and $G_n\Theta$ is an A_∞ monoid theory.

By construction $M_nX \circ \tilde{G}_n$ defines an A_∞-monoid structure on the space M_nX sucht that $[M_nX \circ \tilde{G}_N]$ has the structure $*$ defined above. To distinguish the structures we denote $M_nX \circ \tilde{G}$ by G_nX and the usual multiplicative structure on M_nX, which is codified by $A_n\Theta$, by A_nX.

We next define an h-morphism $G_nX \to A_nX$: Since $\tilde{G}_n$ is injective we may consider $A_n\Theta$ and $G_n\Theta$ as embedded in $M_n\Theta$. Set

$$\begin{aligned}
\Psi((p,0),(1,1) &= G_n\Theta(p,1) \subset M_n\Theta(p,1)\\
\Psi((p,1),(1,1) &= A_n\Theta(p,1) \subset M_n\Theta(p,1)\\
\Psi((p,0),(1,1) &\subset M_n\Theta(p,1)
\end{aligned}$$

The last space consists of all operations c such that $F_{M_n\Theta}(c)$ is a composite of the form

$$(x_1, \ldots, x_p) \to (1 + x_1, \ldots, 1 + x_p) \xrightarrow{a} y$$

where $a \in \Theta_m(p,1) \subset \Theta_r(p,1)$. Composition is induced by the composition in $M_n\Theta$.

3.11 Proposition: $G_nX = M_nX \circ \tilde{G}_n$ is an A_∞ monoid such that $[G_nX]$ has the monoid structure $*$ defined above. There is an h-morphism $H : G_nX \to A_nX$ from G_nX to the multiplicative A_∞ monoid structure A_nX of M_nX. Its set of underlying maps contains all maps of the form $A \mapsto E + A$, where $+$ is any matrix addition and E any unit matrix in M_nX. If X is ringlike, H is an equivalence of A_∞ monoids.

We prove the last part: The space of unit elements in $X = X(1)$ is contained in the path component 1 of the semiring π_0X. If X is ringlike, there is an additive inverse component -1. Take any element e in that component, and let E' be the matrix with entries e in the diagonal and any collection of zeroes at the other entries. Since $\pi_0(M_nX) = M_n(\pi_0X)$, we have $[E] + [E'] = [E'] + [E] =$ 0-matrix, where $[\ ]$ denotes path components. Hence any addition with E' in M_nX defines a homotopy inverse of the map $A \mapsto E + A$.

□

Let $(M_nX)_0$ be the component of the 0-matrix. Since $\pi_0(M_nX) = M_n(\pi_0X)$, we have $(M_nX)_0 = M_n(X_0)$. The structure on G_nX preserves the component $(M_nX)_0$, and the structure on A_nX the component $(M_nX)_E = (Gl_nX)_E$.

3.12 Corollary: The h-morphism H of 3.11 restricts to an h-morphism of A_∞ monoids $M_n(X_0) \to (Gl_nX)_E$. If X is ringlike, $M_n(X_0) \simeq (Gl_nX)_E$ as A_∞ monoids.

§ 4 Algebraic K-theory via matrices

One of Quillen's definitions of algebraic K-groups $K_iR, i > 0$, of a genuine ring R is $K_iR = \pi_i(BGl_\infty R)^+$, where $BGl_\infty R = colim_n BGl_n R$ and $(-)^+$ is Quillen's plus construction. May and Steinberger ([M3], [Stb]) extended this definition to the A_∞ ring case. Since the stabilization $Gl_nR \to Gl_{n+1}R$ does not translate to an honest homomorphism in the A_∞ case, their construction uses technical ad hoc arguments. The methods of § 3 allow a smoother treatment.

Let X_0 and X_1 denote the subspaces of zeroes and ones in $X = X(1)$, i.e. the images of $F_\Theta^{-1}(\lambda_0)$ and $F_\Theta^{-1}(\mu_0)$.

Since composites of h-morphisms are not well-defined we construct the stabilization sequence of the Gl_nX as a whole. It is a sequence of h-morphisms with well-defined composites.

Let $\mathbb{N}^*$ be the infinite linear category

$$1 \to 2 \to 3 \to \cdots$$

4.1 Proposition: There is an $\mathbb{N}^*$-indexed A_∞ monoid theory $A_*\Theta$ and a $A_*\Theta$-space Gl_*X. The colour $n \in \mathbb{N}^*$ of the pair $(Gl_*X, A_*\Theta)$ is $(Gl_nX, A_n\Theta)$. The underlying maps from Gl_nX to $Gl_{n+k}X$ contain all maps of the form

$$A \mapsto \begin{pmatrix} A & 0 \\ 0 & E_k \end{pmatrix}$$

where E_k is some unit matrix and the 0's some zero matrices. $\pi_0(Gl_*X)$ is the classical sequence of stabilizations

$$Gl_1(\pi_0X) \to Gl_2(\pi_0X) \to \ldots$$

(here we identify $\pi_0(Gl_n)$ with $Gl_n(\pi_0X)$). The correspondence $X \mapsto Gl_*X$ is functorial with respect to hammocks between ringlike A_∞ rings.

Proof: We are forced to set $A_*\Theta((p,n),(q,n)) = A_n\Theta(p,q) \subset \Theta(pn^2, qn^2)$. We define $A_*\Theta((p,m),(1,n)) \subset \Theta(pm^2,n^2), m \leq n$, to be the subspace of all morphisms $c \in \Theta(pm^2,n^2)$ such that $F_\Theta(c)$ decomposes into

$$pm^2 \xrightarrow{M_m(a)} m^2 \xrightarrow{b} n^2$$

with $a \in \Theta_m(p,1)$, so that $M_m(a)$ is contained in $A_m\Theta(p,1)$, and

$$b : A \mapsto \begin{pmatrix} A & 0 \\ 0 & E_{n-m} \end{pmatrix}.$$

The augmentation $A_*\Theta \to \Theta_m \diamond \mathbb{N}^*$ maps this particular c to $(a, m \to n)$ and is given by $F_{A_n\Theta}$ on the colour n.

The composite

$$A_*\Theta \to \Theta \xrightarrow{X} Top$$

defines an $\mathbb{N}^*$-indexed A_∞ monoid, whose colour n is the multiplicative A_∞ monoid A_nX. Restriction to the invertible components defines Gl_*X. Clearly, this construction is functorial with respect to hammocks.

□

We now apply the rectification construction of [BV2; Thm. 4.49], which associates with each D-indexed A_∞ monoid a D-diagram of strict monoids. This construction is functorial with respect to hammocks. Application of this process to Gl_*X yields a sequence

$$R_1X \to R_2X \to R_3X \to \ldots$$

of monoids and homomorphisms, such that Gl_nX is a strong deformation retract of R_nX and the diagram

$$\begin{array}{ccc} Gl_nX & \xrightarrow{\sigma_n} & Gl_{n+1}X \\ \cap & & \cap \\ R_nX & \longrightarrow & R_{n+1}X \end{array}$$

commutes up to homotopy for any choice of underlying map σ_n from Gl_*X. The inclusions $Gl_nX \subset R_nX$ are h-morphisms and the maps $R_nX \to R_{n+1}X$ cofibrations.

We define $Gl_\infty X$ to be the telescope of the σ_n and $R_\infty X$ to be the union of the R_nX. Since the telescope of the R_nX is equivalent to $R_\infty X$, the space $Gl_\infty X$ is equivalent to $R_\infty X$. Though $Gl_\infty X$ depends on the choice of the σ_n, $R_\infty X$ only depends on X and the homotopy type of $Gl_\infty X$ is represented by $R_\infty X$. Let B be Milgram's classifying space functor [Mi]. We define

$$BGlX = colim_n BR_nX = B(R_\infty X).$$

By construction, the correspondence $X \mapsto BGlX$ is functorial with respect to hammocks, and so is $X \mapsto (BGlX)^+$ if we use a functorial version of Quillen's plus construction (e.g. [HH]).

4.2 Definition: The algebraic K-theory of a ringlike A_∞ ring $X : \Theta \to Top$ is defined to be

$$KX = K_0(\pi_0X) \times (BGlX)^+.$$

The correspondence $X \mapsto KX$ is functorial with respect to hammocks.

It is an easy excercise to show that this definition coincides with the one of May and Steinberger.

Let $f : X \to Y$ be a k-connected homomorphism of A_∞ rings. Then the induced map $Gl_nX \to Gl_nY$ is k-connected by (3.10). Hence $R_\infty X \to R_\infty Y$ is k-connected. Since $R_\infty X$ is a grouplike monoid, $BGlX \to BGlY$ is $(k+1)$-connected. It can be shown that KX is simple (in fact it is an infinite loop space [St2; Thm. 3.3]). Hence passage to homology yields

4.3 Proposition: If $f : X \to Y$ is a k-connected homomorphism of A_∞ rings, the induced map $KX \to KY$ is $(k+1)$-connected.

The monomial map: Steinberger proved [Stb; Thm. 2] that Waldhausen's algebraic K-theory $A(Y)$ of a topological space Y is homotopy equivalent to the algebraic K-theory of the A_∞ ring $\Omega^\infty S^\infty(\Omega Y_+)$ where $\Omega Y_+ = \Omega Y \cup \{*\}$. In [W2] Waldhausen constructed a splitting

$$A(Y) \simeq \Omega^\infty S^\infty Y \times Wh\, Y \times \mu Y$$

where WhY is the differentiable Whitehead space and μY a homology theory, which later was proved to be trivial. The matrix analogue of Waldhausen's splitting map $\Omega^\infty S^\infty Y \to A(Y)$ is the monomial map. Our machinery allows us to correct the treatment of this map in [M3]:

Let $\sum_n \int X^*$ be the A_∞ monoid introduced in § 3. Apply the stabilization process to the various $\sum_n \int X^*$ to obtain a $\mathbb{N}^*$-indexed A_∞ monoid $\sum_* \int X^*$. The h-morphisms $\sum_n \int X^* \to Gl_nX$ combine to give an h-morphism $\sum_* \int X^* \to Gl_*X$ of $\mathbb{N}^*$-indexed A_∞ monoids. Rectification and the application of Milgram's classifying space functor yields a map

$$B\left(\sum_\infty \int X^*\right) \to BGlX$$

where $B\left(\sum_\infty \int X^*\right)$ is the colimit of the sequence of maps obtained from $\sum_* \int X^*$ by this process. For non-pathological spaces the Barratt-Priddy-Quillen theorem provides a homotopy equivalence

$$\Omega^\infty S^\infty((BX^*)_+) \to \mathbb{Z} \times B(\sum \int X^*)^+$$

where $X^* = Gl_1X \subset X(1)$ is the space of homotopy units. Hence we obtain a map

$$\Omega^\infty S^\infty((BX^*)_+) \to KX.$$

If X is $\Omega^\infty S^\infty(\Omega Y_+)$, we can compose this map with

$$\Omega^\infty S^\infty Y \to \Omega^\infty S^\infty(B(\Omega^\infty S^\infty(\Omega Y_+)^*)$$

induced by the stabilization $\Omega Y \to \Omega^\infty S^\infty(\Omega Y_+)$, to obtain Waldhausen's splitting map. On π_0 the latter map is the inclusion of the trivial units

$$\pi_0(\Omega Y) = \pi_1(Y) \to \pi_0(\Omega^\infty S^\infty(\Omega Y_+)) = \mathbb{Z}[\pi_1 Y].$$

Hence Waldhausen's splitting map is induced by the A_∞ version of the inclusion of trivial units.
For a detailed treatment of this summary on the mononial map see [SV2].

References

[BV1] J.M. Boardman and R.M. Vogt, Homotopy everything H-spaces, Bull. Amer. Math. Soc. 74 (1968), 1117-1122.

[BV2] J.M. Boardman and R.M. Vogt, Homotopy invariant structures on topological spaces, Lecture Notes in Mathematics No. 347 (Springer, Berlin-New York, 1973).

[DK] W.G. Dwyer and D.M. Kan, Function complexes in homotopical algebra, Topology 19 (1980), 427-440.

[FSSV] Z. Fiedorowicz, R. Schwänzl, R. Steiner, and R.M. Vogt, Non-connective delooping of K-theory of an A_∞ ring space, Math. Z. 203 (1990), 43-57.

[HH] J.C. Hausmann and D. Husemoller, Acyclic maps, Enseignement Math. 25 (1979), 53-75.

[I] K. Igusa, On the algebraic K-theory of A_∞ ring spaces, Lecture Notes in Mathematics No. 967 (Springer, Berlin-New York, 1982), 146-194.

[M1] J.P. May, The geometry of iterated loop spaces, Lecture Notes in Mathematics No. 171 (Springer, Berlin-New York, 1972).

[M2] J.P. May (with contributions by N. Ray, F. Quinn, and J. Tornehave), E_∞ ring spaces and E_∞ ring spectra, Lecture Notes in Mathematics No. 577 (Springer, Berlin-New York, 1977).

[M3] J.P. May, A_∞ ring spaces and algebraic K-theory, Lecture Notes in Mathematics No. 658 (Springer, Berlin-New York, 1978), 240-315.

[M4] J.P. May, Mulitplicative infinite loop space theory, J. Pure Appl. Algebra 26 (1982), 1-69.

[MT] J.P. May and R. Thomason, The uniqueness of infinite loop space machines, Topology 17 (1978), 205-224.

[Mi] R.J. Milgram, The bar construction and abelian H-spaces, Illinois J. Math. 11 (1967), 242-250.

[SV1] R. Schwänzl and R.M. Vogt, Homotopy Invariance of A_∞ and E_∞ ring spaces, Proc. Conf. Alg. Topology, Aarhus 1982, Lecture in Mathematics No. 1051 (Springer, Berlin-New York, 1984), 442-481.

[SV2] R. Schwänzl and R.M. Vogt, Matrices over homotopy ring spaces and algebraic K-theory, OSM, P73 (1984), Universität Osnabrück (preprint).

[SV3] R. Schwänzl and R.M. Vogt, E_∞ spaces and injective Γ-spaces, manuscripta math. 61 (1988), 203-214.

[SV4] R. Schwänzl and R.M. Vogt, E_∞ monoids with coherent homotopy inverses are abelian groups, Topology 28 (1989), 481-484.

[SV5] R. Schwänzl and R.M. Vogt, Homotopy homomorphisms versa hammocks, to appear.

[SV6] R. Schwänzl and R.M. Vogt, Basic constructions in the algebraic K-theory of homotopy ring spaces, to appear.

[Se] G. Segal, Categories and cohomology theories, Topology 13 (1974), 293-312.

[Stb] M. Steinberger, On the equivalence of the two definitions of the algebraic K-theory of a topological space, Lecture Notes in Mathematics No, 763 (Springer, Berlin-New York, 1979), 317-331.

[ST1] R. Steiner, A canonical operad pair, Math. Proc. Camb. Phil. Soc. 86 (1979), 443-449.

[St2] R. Steiner, Infinite loop structures on the algebraic K-theory of spaces, Math. Proc. Camb. Phil. Soc. 90 (1981), 85-111.

[Th] R. Thomason, Uniqueness of delooping machines, Duke, Math. J. 46 (1979), 217-256.

[V] R.M. Vogt, Convenient categories of topological spaces for homotopy theory, Arch. Math. 22 (1971), 545-555.

[W1] F. Waldhausen, Algebraic K-theory of topological spaces I, Proc. Symp. Pure Math. 32 (1978), 35-60.

[W2] F. Waldhausen, Algebraic K-theory of topological spaces II, Lecture Notes in Mathematics No. 763 (Springer, Berlin-New York, 1979), 356-394.

[Wo] R. Woolfson, Hyper Γ-spaces and hyperspectra, Quart. J. Math. 30 (1979), 229-255.

HOMOTOPY COLIMITS ON E-I-CATEGORIES

Jolanta Słomińska
Instytut Matematyki, Uniwersytet M. Kopernika
87-100 Toruń, Poland

Introduction.

By an E-I-category we mean a small category whose all endomorphisms are isomorphisms. The posets, groupoids and certain orbit categories are special cases of E-I-categories. Homotopy colimits on such categories appear in the transformation group theory and its applications to the homotopy theory ([8],[9]). The aim of this paper is to express homotopy colimits on E-I-categories by homotopy colimits on groupoids and posets.

Assume that $\mathcal{A}$ is an E-I-category and that $\mathrm{Is}\mathcal{A}$ is the poset of isomorphism classes of objects of $\mathcal{A}$ ([10]B). Let $s_0(\mathcal{A})$ be the opposite poset to the simplicial complex associated to $\mathrm{Is}\mathcal{A}$.

In Section 1 we construct a functor $\mathcal{M}_{\mathcal{A}} : s_0(\mathcal{A}) \longrightarrow \mathrm{Grp}$, where Grp is the category of groupoids. The category Grp is a subcategory of the category Cat of small categories. Hence we can construct the "semi-direct product" $s(\mathcal{A}) = s_0(\mathcal{A}) \int \mathcal{M}_{\mathcal{A}}$, where $\int$ is the Grothendieck construction described, for instance, in [12],1.1. We recall the definition of $\int$ and some of its properties in Section 0 of this paper. In Section 1 we show that there exists a right cofinal functor $p_{\mathcal{A}} : s(\mathcal{A}) \longrightarrow \mathcal{A}$.

Let Top be a category of topological spaces satisfying the conditions described in ([4]). It can be, for example, the Steenrod category of compactly generated Hausdorff spaces. From the general homotopy colimit theory it follows that, for any functor $X : \mathcal{A} \longrightarrow \mathrm{Top}$, there is a homotopy equivalence:

$$\operatorname*{hocolim}_{a \in s_0(\mathcal{A})} \operatorname*{hocolim}_{\mathcal{M}(a)} X(a) \cong \operatorname*{hocolim}_{s(\mathcal{A})} X p_{\mathcal{A}} ,$$

where $X(a)$ is the restriction of $Xp_{\mathcal{A}}$ to $\mathcal{M}(a) = \mathcal{M}_{\mathcal{A}}(a)$. The fact that $p_{\mathcal{A}}$ is a right cofinal functor implies that there is a homotopy equivalence

$$\operatorname*{hocolim}_{a\in s_o(\mathcal{A})} \operatorname*{hocolim}_{\mathcal{M}(a)} X(a) \cong \operatorname*{hocolim}_{\mathcal{A}} X.$$

This result can be also obtained as a consequence of Theorem 3.4 in Dwyer and Kan's paper [13]. It can be regarded as a generalization of the result described in [8], where the case of a category with two objects is considered.

There is another consequence of the construction described above. Let $N:\mathcal{A}^{op}\longrightarrow Ab$ be a contravariant functor from $\mathcal{A}$ to the category Ab of abelian groups. Then our approach gives a simple proof of the unpublished result of Jackowski and Segal which asserts that there exists a spectral sequence:

$$E_2^{p,q} = \lim_{a\in s_o(\mathcal{A})}{}^{p} \prod_{[m]\in Is\mathcal{M}(a)} H^q(Aut(m), Np_{\mathcal{A}}(m)) \Rightarrow \lim_{\mathcal{A}}{}^{p+q} N\ ,$$

where Aut(m) is the automorphism group of the object m of $\mathcal{M}(a)$.

In the case where $\mathcal{A}$ is an orbit category, a similar spectral sequence was considered by the author in [11]. If, for any two objects x, y of $\mathcal{A}$, the group Aut(y) acts freely on the morphism set $\mathcal{A}(x,y)$, then the spectral sequence described in 17.18 of [10] has the same $E_2^{p,q}$ -groups.

In Section 2 of this paper we investigate an example of an E-I-category. Let G be a finite group and let Sub-G be the poset of all subgroups of G. Assume that W is a poset and that G acts on W preserving order. Let $d:W\longrightarrow$Sub-G be an equivariant poset map such that, for each element w of W, d(w) is a subgroup of the isotropy group $G_w=\{g\in G:\ gw=w\}$. Pairs of the form (W, d) are considered for example in [3] and are called there G-posets.

We associate to each pair (W, d) a certain E-I-category W(d). We show that if $\mathcal{A}$ is an EI-category, then there exists a group $G(\mathcal{A})$, a $G(\mathcal{A})$-poset $W_{\mathcal{A}}$ and an equivariant map $d_{\mathcal{A}}:W_{\mathcal{A}}\longrightarrow$Sub-$G(\mathcal{A})$ such that $W_{\mathcal{A}}(d_{\mathcal{A}})=s(\mathcal{A})$. In Section 2 we also study properties of functors between categories of the form W(d), which are induced by G-posets maps. As corollaries we obtain results, which can be considered as special cases of the results of Section 1. We also apply these results to the case of orbit categories.

0. Preliminaries.

Let $\mathcal{C}$ be a small category and let $\mathcal{F}:\mathcal{C}\longrightarrow Cat$ be a functor. The Grothendieck construction on $\mathcal{F}$, $\mathcal{C}\int\mathcal{F}$, is the category whose objects are the pairs (c,F), where c is an object of $\mathcal{C}$ and F is an object of $\mathcal{F}(c)$. The morphisms of $\mathcal{C}\int\mathcal{F}$ are given by the pairs $(\gamma,f):(c',F')\longrightarrow(c,F)$, where $\gamma:c'\longrightarrow c$ is a morphism of $\mathcal{C}$ and $f:\mathcal{F}(\gamma)(F')\longrightarrow F$ is a morphism of $\mathcal{F}(c)$. Composition is defined by

$$(\gamma,f)(\gamma',f')=(\gamma\gamma',f\mathcal{F}(\gamma)(f')).$$

By $\pi_{\mathcal{F}}:\mathcal{C}\int\mathcal{F}\longrightarrow\mathcal{C}$, we shall denote the functor such that $\pi_{\mathcal{F}}(c,F)=c$, $\pi_{\mathcal{F}}(\gamma,f)=\gamma$.

Assume now that $\mathcal{R}$ is a subcategory of Cat and that $\mathcal{F}$ is a functor from $\mathcal{C}$ to $\mathcal{R}$. Let $i_{\mathcal{R}}:\mathcal{R}\longrightarrow Cat$ denote the natural inclusion. We shall use the notation $\mathcal{C}\int\mathcal{F}=\mathcal{C}\int i_{\mathcal{R}}\mathcal{F}$. In particular, we shall consider the following subcategories of Cat. Let Gr be the category of groups. Assume that G is a group. By $\mathcal{G}$ we shall denote the category with one object $*_G$ whose endomorphism monoid is equal to G. This correspondence gives us the natural inclusion of categories $i_{Gr}:Gr\longrightarrow Cat$. We can also consider any set T as a discrete category with object set equal to T. This correspondence gives us the natural inclusion $i_{Set}:Set\longrightarrow Cat$. We shall also use the natural inclusion $i_{Poset}:Poset\longrightarrow Cat$, where Poset is the category of all posets.

The construction $\int$ can be regarded as a generalization of the semi-direct product of groups. If G is a group, then for any functor $\mathcal{F}:\mathcal{G}\longrightarrow Gr$, the construction $\mathcal{G}\int\mathcal{F}$ is equal to the semi-direct product of G and $\mathcal{F}(*_G)$.

Assume now that $\mathcal{C}$ and $\mathcal{D}$ are small categories and that $\phi:\mathcal{D}\longrightarrow\mathcal{C}$ is a functor. Let c be an object of $\mathcal{C}$. Let $\phi_c=\mathcal{C}(c,\phi(-)):\mathcal{D}\longrightarrow Set$ and $\phi^c=\mathcal{C}(\phi(-),c):\mathcal{D}^{op}\longrightarrow Set$. It is easy to check that the following two categories $c\backslash\phi=\mathcal{D}\int\phi_c$ and $\phi/c=(\mathcal{D}^{op}\int\phi^c)^{op}$ are the usual "over" and "under" categories. In the case where $\mathcal{C}=\mathcal{D}$ and $\phi=id_{\mathcal{C}}$, $c\backslash\phi=c\backslash\mathcal{C}$ and $\mathcal{C}/c=\phi/c$.

For any small category $\mathcal{C}$ and two functors $Y:\mathcal{C}\longrightarrow Top$ and $Y':\mathcal{C}^{op}\longrightarrow Top$, one can define the topological space $Y\times_{\mathcal{C}}Y'$. This construction is described, for example, in 2.16. of [4]. By $N_*\mathcal{C}$ we shall denote the simplicial nerve of the category $\mathcal{C}$. Its realization will be denoted by $B\mathcal{C}$. We shall consider the functor $E\mathcal{C}=B(-\backslash\mathcal{C}):\mathcal{C}^{op}\longrightarrow Top$. If $Y:\mathcal{C}\longrightarrow Top$, then

$$\operatorname*{hocolim}_{\mathcal{C}} Y = Y\times_{\mathcal{C}}E\mathcal{C} = |Y\times_{\mathcal{C}}N_*(-\backslash\mathcal{C})|,$$

where symbol $||$ denotes the realization of the simplicial space.

The homotopy colimits on the following two categories appear in our further considerations.

0.1. Examples. (i) Let S be a simplicial complex and let K(S) be the polyhedron obtained as the geometrical realization of S. The category associated to the poset defined by S will be denoted by the same letter. Let $\mathcal{K}_S: S \longrightarrow Top$ be the functor such that, for every element s of S, the space $\mathcal{K}_S(s)$ is equal to the closed simplex K(s) of K(S) determined by s. If $s \subseteq s'$, then $\mathcal{K}(s \subseteq s')$ is equal to the inclusion $K(s) \subseteq K(s')$. It follows from the results of [4], that $\mathcal{K}_S$ is a free, locally contractible S-CW-complex and that for any functor $X: S^{op} \longrightarrow Top$, there is a homotopy equivalence

$$\operatorname{hocolim}_{S^{op}} X \cong \mathcal{K}_S \times_S X.$$

(ii) Let G be a group. We shall use the notation $EG = B(*_G \backslash \mathcal{G}) = E\mathcal{G}(*_G)$. It is obvious that EG is a free universal G-CW-complex. A functor $T: \mathcal{G} \longrightarrow Top$ can be considered as a topological G-space and

$$\operatorname{hocolim}_{\mathcal{G}} T = EG \times_G T. \blacksquare$$

The following fact is a generalization of the main Theorem of Thomason's paper [12]. The method of proof is essentially the same as that of [12].

0.2. Proposition. Let $Y: \mathcal{C} \int \mathcal{F} \longrightarrow Top$ be a functor. Then there is a homotopy equivalence

$$\varphi: \operatorname{hocolim}_{c \in \mathcal{C}} \operatorname{hocolim}_{\mathcal{F}(c)} Y(c) \longrightarrow \operatorname{hocolim}_{\mathcal{C} \int \mathcal{F}} Y,$$

where Y(c) is the restriction of Y to $\mathcal{F}(c)$.

Proof. For any functor $\phi: \mathcal{D} \longrightarrow \mathcal{C}$ of small categories and for any functor $Y: \mathcal{D} \longrightarrow Top$, there is a homotopy equivalence

$$\nu: \operatorname{hocolim}_{c \in \mathcal{C}} \operatorname{hocolim}_{\phi/c} Y/c \longrightarrow \operatorname{hocolim}_{\mathcal{D}} Y.$$

The fact above follows from the results of [1],[5],9.8, and [4]. Let $\nu_c: \phi/c \longrightarrow \mathcal{D}$ be the functor equal to $\pi_\phi c$, where $\phi^c: \mathcal{D}^{op} \longrightarrow Set$. It is clear that ν_c is a usual forgetful functor. Then $Y/c = Y\nu_c$ and the homotopy equivalence ν is induced by the maps hocolim ν_c.

Assume now that $\mathcal{D} = \mathcal{C} \int \mathcal{F}$ and that $\phi = \pi_{\mathcal{F}}$. If c is an object of $\mathcal{C}$, then by $i(c): \mathcal{F}(c) \longrightarrow \phi/c$ we shall denote the inclusion of categories such that

$i(c)(F)=((c,F),id_c)$. This functor has a left adjoint functor $p(c):\phi/c \longrightarrow \mathcal{F}(c)$ such that $p(c)((c',F'),\gamma)=\mathcal{F}(\gamma)F'$. In fact, in this case, we have the natural transformation p from $\phi/-$ to $\mathcal{F}$. This natural transformation induces the map

$$\lambda: \underset{c\in\mathcal{C}}{\text{hocolim}}\ \underset{\phi/c}{\text{hocolim}}\ Y/c \longrightarrow \underset{c\in\mathcal{C}}{\text{hocolim}}\ \underset{\mathcal{F}(c)}{\text{hocolim}}\ Y(c).$$

From the fact that $i(c)$ is a right adjoint to $p(c)$ and that the composition $p(c)i(c):\mathcal{F}(c)\longrightarrow\mathcal{F}(c)$ is the identity map it follows that, for every object c of $\mathcal{C}$, the natural map

$$\lambda(c): \underset{\phi/c}{\text{hocolim}}\ Y/c \longrightarrow \underset{\mathcal{F}(c)}{\text{hocolim}}\ Y(c)$$

induced by $p(c)$, is a homotopy equivalence. Thus λ is a homotopy equivalence, too. ([1],[5],[4].) This implies that

$$\underset{c\in\mathcal{C}}{\text{hocolim}}\ \underset{\mathcal{F}(c)}{\text{hocolim}}\ Y(c) \cong \underset{\mathcal{C}\int\mathcal{F}}{\text{hocolim}}\ Y.$$

In fact, using the same method as in [12], one can construct a canonical map φ such that $\varphi\lambda$ is homotopic to ν.∎

The following fact, which belongs to homological algebra, can be proved using arguments like those in the proof of 0.2. (See, for example [7],IX,6.) Let us consider a functor $N:(\mathcal{C}\int\mathcal{F})^{op}\longrightarrow Ab$.

0.3. Proposition. There is a spectral sequence

$$\underset{c\in\mathcal{C}}{\lim}^p\ \underset{\mathcal{F}(c)}{\lim}^q\ N(c) \Rightarrow \underset{\mathcal{C}\int\mathcal{F}}{\lim}^{p+q}\ N,$$

where $N(c)$ is the restriction of N to $\mathcal{F}(c)$.∎

We shall now describe a category, which will be used in the next Section in the definition of the functor $\mathcal{M}_{\mathcal{A}}$.

Let (Gr, Set) be the category whose objects are the pairs (G,T), where G is a group and T is a G-set, and whose morphisms are the pairs (γ,η), where $\gamma:G\longrightarrow G'$ is a group homomorphism and $\eta:T\longrightarrow T'$ is an equivariant map. The one point trivial G-set will be denoted by $*_G$. Any G-set T can be considered as a functor $T:\mathcal{G}\longrightarrow Set$. By $\iota_{(Gr.Set)}:(Gr,Set)\longrightarrow Cat$ we shall denote the natural inclusion such that

$$\iota_{(Gr,Set)}(G,T)=\mathcal{G}\int T\ .$$

It is easy to check that the object set of $\mathcal{G}\int T$ is equal to $T(*_G)$ and that, for any two elements t, t' of $T(*_G)$, we have:

$$\mathcal{G}\int T(t,t')=\{g\in G: gt=t'\}.$$

The composition of morphisms of $\mathcal{G}\int T$ is defined by multiplication of G. It is obvious that $\iota_{(Gr,Set)}(G,*_G)=\mathcal{G}$.

<u>0.4. Corollary.</u> Suppose that G is a group and T is a G-set. Let $\pi_T:\mathcal{G}\int T\longrightarrow\mathcal{G}$ be the natural projection.

(i) Let X be a G-topological space. Then there is a homotopy equivalence

$$\varphi(G,T,X): EG\times_G(X\times T)\longrightarrow \operatorname{hocolim}_{\mathcal{G}\int T} X\pi_T .$$

(ii) Let N be a $\mathbb{Z}(G)$-module. Then

$$\lim_{\mathcal{G}\int T}{}^{*} N\pi_T = H^*(G,\mathrm{Hom}_{\mathbb{Z}}(\mathbb{Z}(T),N)),$$

where $\mathbb{Z}(T)$ is the permutation $\mathbb{Z}(G)$-module with the basis equal to T.

<u>Proof.</u> (i) By 0.1.(ii), we obtain that

$$EG\times_G(X\times T)=\operatorname{hocolim}_{\mathcal{G}} X\times T=\operatorname{hocolim}_{\mathcal{G}}\ \operatorname{hocolim}_{T(*_G)} (X\pi_T)(*_G).$$

Now, it is sufficient to apply 0.2.

(ii) This fact is a consequence of 0.3.■

1. Main results.

In this section we shall assume that $\mathcal{A}$ is an E-I-category satisfying the additional condition that all isomorphisms of $\mathcal{A}$ are automorphisms. We shall say that $\mathcal{A}$ is an E-I-A-category .

Let $\tilde{\mathcal{A}}$ be the set of all objects of $\mathcal{A}$. This set can be ordered in such a way that, for any two objects x, y of $\mathcal{A}$, we have $x\le y$, if and only if the morphism set $\mathcal{A}(x,y)$ is not empty. The symbol $x<y$ means that $x\le y$ and $x\ne y$. The category defined by the poset $\tilde{\mathcal{A}}$ will be denoted by the same letter.

It is clear that the simplicial subdivision $sd\tilde{\mathcal{A}}$ of $\tilde{A}$, in the sense of 5.2.[6], is equal to the simplicial complex associated to $\tilde{\mathcal{A}}$. Let $s_o(\mathcal{A})$ be the poset (and category) opposite (dual) to $sd\tilde{\mathcal{A}}$. The objects of $s_o(\mathcal{A})$ are the strictly increasing sequences of elements of $\tilde{\mathcal{A}}$. For any two such sequences $a=(a_o,\dots,a_n)$ and $a'=(a',\dots,a')$, we have $a\le a'$, if and only if the set $\{a',\dots,a'\}$ is a subset of $\{a_o,\dots,a_n\}$.

Let $T_{\mathcal{A}}:s_o(\mathcal{A})\longrightarrow Set$ be the functor such that, for each object a of $s_o(\mathcal{A})$,

$$T_{\mathcal{A}}(a)=\mathcal{A}(a_o,a_1)\times\dots\times\mathcal{A}(a_{n-1},a_n),$$

if $n\neq o$, and $T_{\mathcal{A}}(a_o)$ is equal to the one point set $\{*_{a_o}\}$, if $n=o$. For each morphism $a\le a'$ of $s_o(\mathcal{A})$, $T_{\mathcal{A}}(a\le a')$ is defined by the appropriate projections and compositions of morphisms of $\mathcal{A}$.

Let $G_{\mathcal{A}}:s_o(\mathcal{A})\longrightarrow Gr$ be a functor such that, for each object a of $s_o(\mathcal{A})$,

$$G_{\mathcal{A}}(a)=Aut(a_o)\times\dots\times Aut(a_n),$$

where $Aut(a_i)$ denotes the automorphism group of the object a_i of $\mathcal{A}$, and for each morphism $a\le a'$ of $s_o(\mathcal{A})$, $G_{\mathcal{A}}(a\le a')$ is the appropriate projection map.

We shall also use the notation $T(a)=T_{\mathcal{A}}(a)$, $G(a)=G_{\mathcal{A}}(a)$. The group $G(a)$ acts on $T(a)$ in such a way that, for any two elements $g=(g_o,\dots,g_n)$ of $G(a)$ and $f=(f_1,\dots,f_n)$ of $T(a)$, we have

$$gf=(g_1f_1g_o^{-1},\dots,g_nf_ng_{n-1}^{-1}) \quad .$$

Hence, for any element a of $s_o(\mathcal{A})$, we can consider the pair $(G(a),T(a))$, which is an object of the category (Gr,Set). One can easily check that there is a functor

$$\mathcal{T}_{\mathcal{A}}:s_o(\mathcal{A})\longrightarrow(Gr,Set)$$

such that, for every $a\in s_o(\mathcal{A})$,

$$\mathcal{T}_{\mathcal{A}}(a)=(G(a),T(a)),$$

$\rho_{Gr}\mathcal{T}_{\mathcal{A}}=G_{\mathcal{A}}$ and $\rho_{Set}\mathcal{T}_{\mathcal{A}}=T_{\mathcal{A}}$, where $\rho_{Gr}:(Gr,Set)\longrightarrow Gr$ and $\rho_{Set}:(Gr,Set)\longrightarrow Set$ are the natural projections.

<u>1.1. Definition</u>. We define

$$\mathcal{M}_{\mathcal{A}}=\iota_{(Gr,Set)}\mathcal{T}_{\mathcal{A}} \quad \text{and} \quad s(\mathcal{A})=s_o(\mathcal{A})\textstyle\int\mathcal{M}_{\mathcal{A}} \ .\blacksquare$$

We shall also use the notation $\mathcal{M}=\mathcal{M}_{\mathcal{A}}$, $\mathcal{T}=\mathcal{T}_{\mathcal{A}}$. It follows from the definition that $s(\tilde{\mathcal{A}})=s_o(\mathcal{A})$. The following facts are immediate consequences of definitions.

1.2. Corollary. We have

$$\mathcal{M}(a)=\mathcal{G}(a)\int T(a)$$

whenever $a\in s_o(\mathcal{A})$. ∎

1.3. Proposition. (i) The object set of $s(\mathcal{A})$ is equal to the set

$$\bigcup_{a\in s_o(\mathcal{A})} T(a) = \bigcup_{a_o<a_1<\ldots<a_n} \mathcal{A}(a_o,a_1)\times\ldots\times\mathcal{A}(a_{n-1},a_n).$$

(ii) For any two elements f of T(a) and f' of T(a'),

$$s(\mathcal{A})(f,f')=\{g\in G(a'):T(a\le a')f=g^{-1}f'\}.$$

(iii) The construction of $s(\mathcal{A})$ is natural in $\mathcal{A}$. ∎

1.4. Definition. Let

$$p_{\mathcal{A}}:s(\mathcal{A})\longrightarrow\mathcal{A}$$

be the functor defined in the following way. If $f=(f_1,\ldots,f_n)$ is an element of T(a), then we define $p_{\mathcal{A}}(f)=a_o$. Suppose that $a\le a'$ and that $a'_o=a_r$. If $g\in G(a')$ defines a certain element of $s(\mathcal{A})(f,f')$, then

$$p_{\mathcal{A}}(g)=g_of_r\ldots f_1. ∎$$

It is easy to check that this definition is correct. If $\rho:\mathcal{A}\longrightarrow\mathcal{A}'$ is a functor between E-I-A-categories, then $\rho p_{\mathcal{A}}=p_{\mathcal{A}'}s(\rho)$.

1.5. Proposition. For any E-I-A-category $\mathcal{A}$, the functor

$$p_{\mathcal{A}}:s(\mathcal{A})\longrightarrow\mathcal{A}$$

is right cofinal.

Proof. We have to prove that, for each object x of $\mathcal{A}$ the classifying space (i.e. the geometrical realization of the nerve) of the category $x\backslash p_{\mathcal{A}}$, is contractible. The objects of $x\backslash p_{\mathcal{A}}$ are all morphisms of $\mathcal{A}$ of the form $h:x\longrightarrow p_{\mathcal{A}}(f)$, where f is an object of $s(\mathcal{A})$. The morphisms $h\longrightarrow h'$ are all morphisms $\alpha:f\longrightarrow f'$ satisfying condition $p_{\mathcal{A}}(\alpha)h=h'$.

Let $p_{\mathcal{A}}^{-1}(x)$ be the subcategory of $s(\mathcal{A})$ whose objects are all f such that $p_{\mathcal{A}}(f)=x$. If f, f' are objects of $p_{\mathcal{A}}^{-1}(x)$ then

$$p_{\mathcal{A}}^{-1}(x)(f,f')=\{\ g\in s(\mathcal{A})(f,f')\ :\ id_{a_o}=g_o=p_{\mathcal{A}}(g)\ \}.$$

The classifying space of the category $p_{\mathcal{A}}^{-1}(x)$ is contractible because $*_x\in T(x)$ is a final object of this category.

We shall prove that there exists a right adjoint functor to the natural inclusion $\omega:p_{\mathcal{A}}^{-1}(x)\longrightarrow x\backslash p_{\mathcal{A}}$. This fact implies that the classifying space of $x\backslash p_{\mathcal{A}}$ is contractible.

We define a functor $R: x\backslash p_{\mathcal{A}} \longrightarrow p_{\mathcal{A}}^{-1}(x)$ in the following way. Let $f=(f_1,\ldots,f_n)$ belong to $T(a_o,\ldots,a_n)$ and let $h: x \longrightarrow p_{\mathcal{A}}(f)=a_o$ be an object of $x\backslash p_{\mathcal{A}}$. Then we take $R(h)=(f_1h,\ldots,f_n)$.

Now, let $\alpha': h \longrightarrow h'$ be the morphism of $x\backslash p_{\mathcal{A}}$ induced by the morphism $\alpha: f \longrightarrow f'$ of $s(\mathcal{A})$. Then $p_{\mathcal{A}}(f')=a_o'=a_r$. Suppose that α is defined by the element $g=(g_o,\ldots,g_k)$ of $G(a')$. Hence $h'=f_r\ldots f_1g_oh$, because $h'=p_{\mathcal{A}}(\alpha)h$. We can now define $R(\alpha')$ to be equal to the morphism $(f_1h,\ldots,f_n) \longrightarrow (f_1'h',\ldots,f_k')$ given by $(id,g_1,\ldots,g_k) \in G(x,a_1',\ldots,a_k')$.

It is easy to check that R is a right adjoint functor to ω. Hence the classifying spaces of the categories $x\backslash p_{\mathcal{A}}$ and $p_{\mathcal{A}}^{-1}(x)$ are homotopy equivalent and this ends the proof.∎

For every functor $X: \mathcal{A} \longrightarrow Top$ there exists a functor

$$X_E: s_o(\mathcal{A}) \longrightarrow (Gr, Top)$$

such that

$$X_E(a)=(G(a), EG(a) \times T(a) \times X(a_o)),$$

where $T(a)$ is considered as a discrete topological space.

The map $X_E(a \leq a')$ is the product of the map induced by the projection $G(a) \longrightarrow G(a')$ and the map

$\kappa: T(a) \times X(a_o) \longrightarrow T(a') \times X(a_r)$ such that $\kappa(f,x)=(T(a \leq a')f, X(f_r\ldots f_1)x)$.

The functor X_E induces the functor $X_o: s_o(\mathcal{A}) \longrightarrow Top$ such that

$$X_o(a)=X_E(a)/G(a)=EG(a) \times_{G(a)} (T(a) \times X(a_o)).$$

<u>1.6. Proposition.</u> Let $\mathcal{A}$ be an E-I-A-category. For any functor $X: \mathcal{A} \longrightarrow Top$, there are homotopy equivalences

$$\operatorname*{hocolim}_{a \in s_o(\mathcal{A})} EG(a) \times_{G(a)} (T(a) \times X(a_o)) \cong \operatorname*{hocolim}_{a \in s_o(\mathcal{A})} \operatorname*{hocolim}_{\mathcal{G}(a)} (T(a) \times X(a_o)) \cong \operatorname*{hocolim}_{\mathcal{A}} X .$$

<u>Proof.</u> The general homotopy colimit theory ([1],[4],[5]) and 1.5. imply that there is a homotopy equivalence

$$\operatorname*{hocolim}_{s(\mathcal{A})} Xp_{\mathcal{A}} \longrightarrow \operatorname*{hocolim}_{\mathcal{A}} X.$$

From Proposition 0.2, we obtain that, for any functor $Y: s(\mathcal{A}) \longrightarrow Top$, there is a homotopy equivalence

$$\varphi: \operatorname*{hocolim}_{a \in s_o(\mathcal{A})} \operatorname*{hocolim}_{\mathcal{M}(a)} Y(a) \longrightarrow \operatorname*{hocolim}_{s(\mathcal{A})} Y$$

where $Y(a)$ is the restriction of Y to $\mathcal{M}(a)$. We shall use the fact that $\mathcal{M}(a)=\mathcal{G}(a)\int T(a)$. If $Y=Xp_{\mathcal{A}}$, then $Y(a)(f)=X(a_o)$ and $Y(a)(g)=X(g_o)$. In this case, by 0.3, we have homotopy equivalences

$$\sigma_{\mathcal{A}}(a):X_o(a)=EG(a)\times_{G(a)}(T(a)\times X(a_o))\cong \operatorname*{hocolim}_{\mathcal{G}(a)}\ \operatorname*{hocolim}_{T(a)} X(a_o)\cong \operatorname*{hocolim}_{\mathcal{M}(a)} Y(a).$$

The family $\{\sigma_{\mathcal{A}}(a)\}$ can be constructed in such a way that we obtain a natural transformation of functors

$$\sigma_{\mathcal{A}}:X_o \longrightarrow \operatorname*{hocolim}_{\mathcal{M}(-)} Y(-),$$

which induces a homotopy equivalence between the homotopy colimits of the above functors. This fact ends the proof.■

Let us now consider the simplicial complex $sd\tilde{\mathcal{A}}=s_o(A)^{op}$ associated to the poset $\tilde{\mathcal{A}}$. From 0.1.(i), we obtain the following fact.

<u>1.7. Corollary.</u> There is a homotopy equivalence

$$X_o\times_{s_o(\mathcal{A})}\mathcal{K}_{sd\tilde{\mathcal{A}}} \cong \operatorname*{hocolim}_{\mathcal{A}} X\ .■$$

Assume now that the category $\mathcal{A}$ has only two objects. This is the example, which was described in [8]. Suppose that a_o and a_1 are the objects of $\mathcal{A}$ and that that the set $\mathcal{A}(a_1,a_o)$ is empty. If the set $\mathcal{A}(a_o,a_1)$ is not empty, then $\tilde{\mathcal{A}}$ is equal to the poset $\{a_o,a_1\}$ ordered by the relation $a_o\le a_1$. In this case

$$X_o(a_i)=E(G(a_i))\times_{G(a_i)}X(a_i),\ \text{for } i=o,1$$

and

$$X_o(a_o,a_1)=E(G(a_o)\times G(a_1))\times_{G(a_o)\times G(a_1)}(\mathcal{A}(a_o,a_1)\times X(a_o)).$$

The poset $sd\tilde{\mathcal{A}}$ is isomorphic to the poset of all non-empty subsets of the set $\{o,1\}$ and $\mathcal{K}_{sd\tilde{\mathcal{A}}}$ is the one dimensional simplex. Thus $\operatorname*{hocolim}_{\mathcal{A}} X$ is homotopy equivalent to the homotopy push out of the diagram consisting of the two maps $X_o(a_o,a_1)\longrightarrow X_o(a_i)$, for $i=o,1$, which are induced by the morphisms $(a_o,a_1)\le(a_i)$ of $s_o(\mathcal{A})$.

Consider now a contravariant functor $N:\mathcal{A}^{op}\longrightarrow Ab$. The following result is a further consequence of 1.5.

<u>1.8. Proposition.</u> There exists a spectral sequence

$$\lim_{a\in s_o(\mathcal{A})}{}^{p}\ H^q(G(a),Hom_{\mathbb{Z}}(\mathbb{Z}(T(a)),N(a_o))) \Rightarrow \lim_{\mathcal{A}}{}^{p+q} N\ .$$

Proof. It follows from 1.5 that there exists a group isomorphism

$$\lim_{\mathcal{A}}{}^{*} N = \lim_{s(\mathcal{A})}{}^{*} Np_{\mathcal{A}}.$$

Now it is sufficient to apply 0.3 and 0.4.(ii), because $s(\mathcal{A})=s_o(\mathcal{A})\int\mathcal{M}$ and $\mathcal{M}(a)=\mathcal{G}(a)\int T(a)$. ■

Let $\mathcal{A}_{Gr}$ be the category with the same objects as $\mathcal{A}$ and with the morphism sets defined, for any two objects x, y of $\mathcal{A}$, as follows:

$$\mathcal{A}_{Gr}(x,y)=\tilde{\mathcal{A}}(x,y) \text{ for } x\neq y, \quad \mathcal{A}_{Gr}(x,x)=\mathcal{A}(x,x)=G(x).$$

There exists an isomorphism of categories

$$s(\mathcal{A}_{Gr}) = s_o(\mathcal{A})\int G_{\mathcal{A}}.$$

The category $s_o(\mathcal{A})$ can be considered as a subcategory of $s(\mathcal{A}_{Gr})$. There exists the natural extension of the functor $T_{\mathcal{A}}$ to the functor from $s(\mathcal{A}_{Gr})$ to Set. This extension will be also denoted by $T_{\mathcal{A}}$. It is easy to check that

$$s(\mathcal{A}_{Gr})\int T_{\mathcal{A}}=s(\mathcal{A}).$$

1.9. Corollary. Let $\mathcal{A}$ and X be the same as in 1.6. Then there exists a homotopy equivalence

$$\operatorname{hocolim}_{a\in s(\mathcal{A}_{Gr})} (T(a)\times X(a_o)) \cong \operatorname{hocolim}_{\mathcal{A}} X .$$

Proof. This result is a consequence of 1.5 and 0.2. ■

1.10. Corollary. There exists a group isomorphism

$$\lim_{\mathcal{A}}{}^{*} N = \lim_{a\in s(\mathcal{A}_{Gr})}{}^{*} \operatorname{Hom}_{\mathbb{Z}}(\mathbb{Z}(T(a)),N(a_o))$$

whenever $N:\mathcal{A}^{op}\longrightarrow Ab$.

Proof. This result is a consequence of 1.5 and 0.3. ■

Assume now that $\mathcal{F}:\mathcal{C}\longrightarrow(Gr,Set)$ is a functor such that, for every object c of $\mathcal{C}$, $\mathcal{F}(c)=(\mathcal{F}_1(c),\mathcal{F}_2(c))$. By $\mathcal{F}/\mathcal{F}_1$ we shall denote the canonical functor from $\mathcal{C}$ to Set such that, for every object c of $\mathcal{C}$,

$$(\mathcal{F}/\mathcal{F}_1)(c)=\mathcal{F}_2(c)/\mathcal{F}_1(c).$$

The category $\mathcal{C}\int(\mathcal{F}/\mathcal{F}_1)$ will be also denoted by $[\mathcal{C}\int\mathcal{F}]$. A proof of the following facts is easy and will be left to the reader.

1.11. Lemma. Let $\mathcal{P}$ be a poset.

(i) For any functor $\mathcal{L}:\mathcal{P}\longrightarrow Set$, the category $\mathcal{P}\int\mathcal{L}$ is the poset such that $(p,f)\le(p',f')$ if and only if $p\le p'$ and $\mathcal{L}(p\le p')f=f'$.

(ii) Let $\mathcal{F}:\mathcal{P}\longrightarrow(Gr,Set)$ be a functor. Then there exists a functor

$$\mathcal{F}_t:[\mathcal{P}\int\mathcal{F}]\longrightarrow(Gr,Set)$$

such that,

$$\mathcal{P}\int\mathcal{F} = [\mathcal{P}\int\mathcal{F}]\int\mathcal{F}_t$$

and

$$\mathcal{F}_t(p,[f])=(\mathcal{F}_1(p),\mathcal{F}_1(p)f)$$

whenever $(p,[f])\in\mathcal{P}\int(\mathcal{F}/\mathcal{F}_1)$. ∎

Consider now the case where $\mathcal{P}=s_o(\mathcal{A})$ and $\mathcal{F}=\mathcal{M}=(\mathcal{T}_{\mathcal{A}},\mathcal{G}_{\mathcal{A}})$. Then

$$s_o(\mathcal{A})\int\mathcal{T}_{\mathcal{A}}/\mathcal{G}_{\mathcal{A}} = [s(\mathcal{A})]$$

and

$$s(\mathcal{A})=s_o(\mathcal{A})\int\mathcal{M} =[s(\mathcal{A})]\int\mathcal{M}_t,$$

where $\mathcal{M}_t:[s(\mathcal{A})]\longrightarrow(Gr,Set)$ is the functor such that

$$\mathcal{M}_t(a,[f])=(G(a),G(a)f).$$

1.12. Corollary. Let $\mathcal{A}$ and X be the same as in 1.6. Then there exists a homotopy equivalence

$$\operatorname*{hocolim}_{(a,[f])\in[s(\mathcal{A})]} EG(a)\times_{G(a)}(G(a)f\times X(a_o)) \cong \operatorname*{hocolim}_{\mathcal{A}} X.$$

Proof. This result is a consequence of 1.5, 0.2 and 0.4.(i). ∎

Consider now a functor $N:\mathcal{A}^{op}\longrightarrow Ab$.

1.13. Corollary. There exists a spectral sequence

$$\lim_{(a,[f])\in[s(\mathcal{A})]}{}^{p} H^q(G(a),Hom_{\mathbb{Z}}(\mathbb{Z}(G(a)f),N(a_o))) \Rightarrow \lim_{\mathcal{A}}{}^{p+q} N .$$

Proof. This result follows from 1.5, 0.3 and 0.4.(ii). ∎

The referee of this paper pointed out to the author that the existence of a homotopy equivalence

$$\psi:\operatorname*{hocolim}_{a\in s_o(\mathcal{A})}\ \operatorname*{hocolim}_{\mathcal{M}(a)} X(a)\longrightarrow \operatorname*{hocolim}_{\mathcal{A}} X$$

is a consequence of Theorem 3.4 of [13]. The Dwyer and Kan's condition "locally grouplike and reduced" is equivalent to the E-I-A condition. Their category $\pi.\underset{\sim}{C}$ is equal to $\tilde{\mathcal{A}}$. Their category $\underset{\sim}{D}$ is the opposite of $s_o(\mathcal{A})$. In the definition of the functor h, the index k can be ignored in this situation. The category $p^{-1}\underset{\sim}{D}$ is our category $\mathcal{M}(a)$, the functor j^*X is equal to X(a) and the functor $h(X):\underset{\sim}{D}\longrightarrow\underset{\sim}{S}$ is our functor

$$a\longrightarrow\operatorname{hocolim}_{\mathcal{M}(a)} X(a)$$

(except that the variance is reversed). To deduce the existence of a homotopy equivalence ψ from Theorem 3.4 of [13] one uses the following facts

1. For any category $\underset{\sim}{A}$ the functor

$$\operatorname{hocolim}_{\underset{\sim}{A}} : Ho(\underset{\sim}{S}^{\underset{\sim}{A}})\longrightarrow Ho(\underset{\sim}{S})$$

is the left adjoint of the "constant" functor (see [1], XII.2.4.)

2. The functor $Ho(\underset{\sim}{S})\longrightarrow\underset{\sim}{K}$ obtained by composing the constant functor $Ho(\underset{\sim}{S})\longrightarrow Ho(\underset{\sim}{S}^{\underset{\sim}{C}})$ with $h_{\#}$ is equal to the composite of the constant functor $Ho(\underset{\sim}{S})\longrightarrow Ho(\underset{\sim}{S}^{\underset{\sim}{D}})$ with the functor $xH:\underset{\sim}{S}^{\underset{\sim}{D}}\longrightarrow\underset{\sim}{S}^{\underset{\sim}{D}}/H$ which takes an object Y of $\underset{\sim}{S}^{\underset{\sim}{D}}$ to the projection map $YxH\longrightarrow H$.

3. The left adjoint of xH is the forgetful functor $\underset{\sim}{S}^{\underset{\sim}{D}}/H\longrightarrow\underset{\sim}{S}^{\underset{\sim}{D}}$.

2. Categories associated to G-posets.

Let G be a finite group. By a G-poset we mean a poset with an order preserving left action of G on it. A G-poset W can be regarded as a functor $\mathcal{G}\longrightarrow$Poset. The poset of the orbits of G-action on W will be denoted by W/G. A G-set T will be considered as a G-poset ordered by the identity relation. The category of all G-sets (G-posets) will be denoted by G-Set (G-Poset). We shall consider the natural inclusion $\iota_{G\text{-Poset}}$: G-Poset$\longrightarrow$Cat such that $\iota_{G\text{-Poset}}(W)=\mathcal{G}\int W$.

The underlying functor G-Poset$\longrightarrow$Poset will be denoted by ζ. Thus $W(*_G)=\zeta(W)$. The G-poset of all subgroups of G will be denoted by Sub-G.

2.1. Definition. A G-poset map $d: W \longrightarrow \text{Sub-}G$ is called admissible if, for every element w of W, dw is a subgroup of the isotropy group

$$G_w = \{g \in G: gw = w\} = \mathcal{G}\!\int\! W(w,w) \;.$$

If d is an admissible map, then $W(d)$ is the category whose objects are the elements of $\zeta(W)$ and whose morphism sets are defined as follows:

$$W(d)(w,w') = \mathcal{G}\!\int\! W(w,w')/dw' = \{g \in G: gw \le w'\}/dw' =$$

$$= \bigcup_{[g] \in G/dw'} \zeta(W)(w, gw') = \bigcup_{[g] \in G/dw'} \zeta(W)(g^{-1}w, w'). \blacksquare$$

The natural inclusion $\zeta(W) \longrightarrow W(d)$ will be denoted by $i(d)$. We shall say that the morphism of $W(d)$ which corresponds to $[g] \in G/dw'$ is determined by g. An immediate consequence of this definition is the following fact.

2.2. Lemma. Suppose that $d: W \longrightarrow \text{Sub-}G$ is an admissible map. Then dw is a normal subgroup of G_w. ■

Next we consider some examples of categories of the form $W(d)$ and describe homotopy colimits of functors $X: W(d) \longrightarrow \text{Top}$.

2.3. Examples. (i) Let d be the admissible map such that, for every element w of W, dw is the trivial subgroup of G. Then $W(d) = \mathcal{G}\!\int\! W$ and

$$EG \times_G \operatorname*{hocolim}_{\zeta(W)} Xi(d) \cong \operatorname*{hocolim}_{W(d)} X \;.$$

(ii) If the action of G on W is trivial, then $W(d) = W\!\int\!\alpha(d)$, where $\alpha(d): W \longrightarrow \text{Cat}$ is a functor such that, for every element w of W, $\alpha(d)w = \mathcal{G}/dw = \iota_{Gr}(G/dw)$ and

$$\operatorname*{hocolim}_{w \in \zeta(W)} E(G/dw) \cong \operatorname*{hocolim}_{W(d)} X \;.$$

If $dw = G$ whenever w belongs to W, then $W(d) = \zeta(W) = W/G$.

(iii) Suppose that H is a subgroup of G and that H' is a normal subgroup of H. Let $d: G/H \longrightarrow \text{Sub-}G$ be the admissible map such that $d(gH) = gH'g^{-1}$. The category $(G/H)(d)$ will be denoted by $(G/H)(H')$. By $E(G/H')$ we shall denote the classifying space of the category whose object set is equal to G/H' and whose morphism sets contain exactly one element. This category is isomorphic to $H\backslash(G/H)(H')$. The space $E(G/H')$ can be considered as a free right H/H' space.

There are homotopy equivalences

$$\operatorname{hocolim}_{G/H)(H')} X \cong E(G/H')\times_{H/H'}X(H) \cong E(H/H')\times_{H/H'}X(H)\ .■$$

Let W_o be a G-subposet of W. The restriction of an admissible map $d: W \longrightarrow \text{Sub-}G$ to W_o will be also denoted by d. If w belongs to W, then the category (Gw)(d) is isomorphic to $(G/G_w)(dw)$.

2.4. Example. Let $d: \text{Sub-}G \longrightarrow \text{Sub-}G$ be the identity map. Then $(\text{Sub-}G)(d)=O_G$ is the orbit category of the group G. ([2], I.3.) The objects of O_G are the G-sets of the form G/H, where H is a subgroup of G. The morphisms of O_G are the G-maps. For a G-subposet F of Sub-G the subcategory F(d) of O_G will be denoted by O(F). ■

Let W and V be G-posets and let $d: W \longrightarrow \text{Sub-}G$, $d': V \longrightarrow \text{Sub-}G$ be admissible maps. Suppose that $\vartheta: W \longrightarrow V$ is a G-poset map such that $d \le d'\vartheta$. Then ϑ induces the functor $\vartheta(d): W(d) \longrightarrow V(d')$ such that $\vartheta(d)w = \vartheta w$. The projection $W(d) \longrightarrow W/G$ will be denoted by p(d). We need the following notation.

2.4. Definition. (i) A G-poset W is called normal if, for any two elements w, w' of W, the condition $w \le w'$ implies that G_w is a subgroup of $G_{w'}$.

(ii) A G-poset W is called regular if, for any two elements w, w' of W, the conditions $w \le w'$ and $w \le gw'$ imply that g belongs to $G_{w'}$. ■

It is obvious that every regular G-poset is normal.

2.5. Examples. (i) Assume that W is a normal G-poset. Then the map $d_G: W \longrightarrow \text{Sub-}G$ such that $d_G w = G_w$, whenever w belongs to W, is admissible. In this case, the category $W(d_G)$ will be denoted by W(G). If d is an arbitrary admissible map defined on W, then there is a functor $\pi(d): W(d) \longrightarrow W(G)$ such that $\pi(d)(w)=w$.

Assume now that K is a G-complex in the sense of [2]. The G-poset of all finite subcomplexes of K of the form K(s), where s is a cell of K and K(s) is the smallest subcomplex of K containing s, will be denoted by SK. This G-poset is normal and the category $SK(G)^{op}$ is a full subcategory of the category $\mathcal{K}$ described in the section I-5 of [2].

(ii) Let W be an regular G-poset. Then the projection $p(G): W(G) \longrightarrow W/G$ is a natural equivalence of categories. It follows from the fact that, in this case, p(G) is a full and faithful functor .

(iii) Let W be a G-poset. By sdW^{op} we shall denote the G-poset such that $\zeta(sdW^{op})=sd\zeta(W)^{op}$. The action of G on this poset is defined by the action on each coordinate. It is easy to check that sdW^{op} is a regular G-poset. There is a G-poset map $q_W: sdW^{op} \longrightarrow W$, which after restriction to the underlying posets is a right cofinal functor. (See[6].)■

Assume now, that P is a poset and that τ is a functor $P \longrightarrow G\text{-Set}$. Then by $P\int\tau$ we shall denote the G-poset such that

$$\iota_{G\text{-Poset}}(P\int\tau)=P\int\iota_{G\text{-Set}}\tau.$$

The elements of $P\int\tau$ are the pairs (p,t) , where $p\in P, t\in\tau(p)$, and

$$(p,t)\le(p',t') \text{ iff } p\le p' \text{ and } \tau(p\le p')t=t'.$$

If $g\in G$, then $g(p,t)=(p,gt)$. The following fact is an immediate consequence of the definition.

<u>2.6. Proposition.</u> For every poset P and every functor $\tau: P \longrightarrow G\text{-Set}$, the G-poset $P\int\tau$ is regular. There is an isomorphism of posets

$$(P\int\tau)/G \longrightarrow P\int(\tau/G)$$

where $\tau/G=\tau(-)/G$.■

The reverse result is also true.

<u>2.7. Proposition.</u> Let W be a regular G-poset. Then there exists a functor $\tau: W/G \longrightarrow G\text{-Set}$ such that W is isomorphic to $(W/G)\int\tau$. For every element w of W, $\tau([w])=Gw$.

<u>Proof</u>. Let w and w′ be such elements of W that $[w]\le[w']$. There exists exactly one element $gG_{w'}$ of $G/G_{w'}$ such that $w\le gw'$. Hence, for every element g′ of G, we can define a G-map $\tau([w]\le[w']): Gw \longrightarrow Gw'$ in such a way that $\tau([w]\le[w'])(g'w)=g'gw'$. It is easy to check that this definition is correct and that $W=(W/G)\int\tau$.■

The following result is a consequence of 2.7. A proof is easy and will be omitted.

<u>2.8. Corollary</u>. Let W be a regular G-poset and let $d: W \longrightarrow \text{Sub-}G$ be an admissible map. Then there is a functor $\tau(d): W/G \longrightarrow Cat$ such that the categories W(d) and $(W/G)\int\tau(d)$ are isomorphic. For every element w of W, $\tau(d)([w])=(Gw)(d)$.■

2.9. Example. We shall use the notation of Section 1. Let $\mathcal{A}$ be an E-I-A-category and let $P=s_o(\mathcal{A})$. Define $G(\mathcal{A})=\prod_{x\in\tilde{\mathcal{A}}} G(x)$ and, for $a\in s_o(\mathcal{A})$,

$$G'(a)=\prod_{x\in(\tilde{\mathcal{A}}\setminus\{a\})} G(x) \quad \text{where } \{a\}=\{a_o,\dots,a_n\}.$$

The group $G(\mathcal{A})$ acts trivially on P. The $G(\mathcal{A})$-map $d':s_o(\mathcal{A})\longrightarrow \text{Sub-}G(\mathcal{A})$ such that $d'a=G'(a)$ is admissible and $s_o(\mathcal{A})(d')=s(\mathcal{A}_{Gr})$, because $G(\mathcal{A})/G'(a)=G(a)$.

For every element a of $s_o(\mathcal{A})$, the group $G(\mathcal{A})$ acts on the set $T(a)$ by the action of $G(a)$. Let $\tau:s_o(\mathcal{A})\longrightarrow G(\mathcal{A})\text{-Set}$ be the functor defined by this action, i.e. for every $a\in s_o(\mathcal{A})$, let $\tau(a)=T(a)$. Let $W_{\mathcal{A}}$ denote the $G(\mathcal{A})$-poset $s_o(\mathcal{A})\int\tau$. Then

$$\zeta(W_{\mathcal{A}})=s_o(\mathcal{A})\textstyle\int T_{\mathcal{A}}=s(\mathcal{A}_{Set}),$$

where $\mathcal{A}_{Set}$ is the subcategory of $\mathcal{A}$ with the same objects and such that

$$\mathcal{A}_{Set}(x,y)=\mathcal{A}(x,y),\ \mathcal{A}(x,x)=(id_x),$$

whenever x, y are objects of $\mathcal{A}$ and $x\neq y$. It is easy to check that

$$W_{\mathcal{A}}/G(\mathcal{A})=[s(\mathcal{A})].$$

There is an admissible map $d_{\mathcal{A}}:W_{\mathcal{A}}\longrightarrow \text{Sub-}G(\mathcal{A})$ such that for every element f of $T(a)$ $d_{\mathcal{A}}(a,f)=G'(a)$. It is easy to check that

$$W_{\mathcal{A}}(d_{\mathcal{A}})=s(\mathcal{A}).$$

Propositions 2.6, 2.7 and 2.8 imply that $W_{\mathcal{A}}$ is a regular $G(\mathcal{A})$-poset and that there is a functor $\tau_t:[s(\mathcal{A})]\longrightarrow Cat$ such that

$$s(\mathcal{A})=[s(\mathcal{A})]\textstyle\int\tau_t$$

and

$$\tau_t(a,[f])=(Gf)(d)=(G(a)/G(a)_f)(e)=\mathcal{G}(a)\textstyle\int G(a)f\ .$$

It is obvious that $\tau_t=\iota_{(Gr,Set)}\mathcal{M}_t$, where $\mathcal{M}_t$ is the functor defined in the end of the previous section. ∎

We shall now study categories associated to functors $\vartheta(d):W(d)\longrightarrow V(d')$ induced by G-posets maps $\vartheta:W\longrightarrow V$ satisfying condition $d\le d'\vartheta$. We need the following notation. Assume that H is a subgroup of G and that W_o is a H-poset. Then $G\times_H W_o$ is the G-poset such that $[(g,w)]\le[(g',w')]$, if and only if, $gH=g'H$ and $(g')^{-1}gw\le w'$. If W_o is a H-subposet of a G-poset W, then the map $\beta_H:G\times_H W_o\longrightarrow W$ such that $\beta_H[(g,w)]=gw$ is a G-poset map.

2.10. Proposition. Let $\zeta(\vartheta):\zeta(W) \longrightarrow \zeta(V)$ be the underlying poset map defined by ϑ and let v be an element of V .

(i) There exists an isomorphism of categories

$$t: (v\backslash\zeta(\vartheta))\int\sigma \longrightarrow v\backslash\vartheta(d),$$

where $\sigma: v\backslash\zeta(\vartheta) \longrightarrow Cat$ is the functor such that $\sigma(v\le\vartheta w)=(G/d'\vartheta w)(dw)$.

(ii) There exists an isomorphism of categories

$$\mu: (Gx_{d'v}(\zeta(\vartheta)/v))(d\beta) \longrightarrow \vartheta(d)/v,$$

where $\beta: Gx_{d'v}(\zeta(\vartheta)/v) \longrightarrow W$ is the G-poset map such that $\beta[g,(\vartheta w\le v)]=gw$.

Proof. We shall consider $v\backslash\zeta(\vartheta)$ $(\zeta(\vartheta)/v)$ as the subposet of $\zeta(W)$ consisting of all w such that $v\le\vartheta w$ $(\vartheta w\le v)$.

(i) If w and w' belong to $v\backslash\zeta(\vartheta)$ and $w\le w'$, then $\sigma(w\le w')$ is the functor induced by the G-map $\gamma: G/d'\vartheta w \longrightarrow G/d'\vartheta w'$ such that $\gamma(d'\vartheta w)=d'\vartheta w'$.

Let $t: (v\backslash\zeta(\vartheta))\int\sigma \longrightarrow v\backslash\vartheta(d)$ be a functor such that $t(w, gd'\vartheta w)$ is the element of $V(d')(v, g\vartheta w)$ determined by g. If $w\le w'$ and $m: gd'\vartheta w' \longrightarrow g'd'\vartheta w'$ is a morphism of $(G/d'\vartheta w')(dw')$ determined by an element g_o of G, then $t(w\le w', m): gw \longrightarrow g'w'$ is the morphism of W(d) determined also by g_o. A proof that t is an isomorphism of categories is easy and will be left to the reader.

(ii) Suppose that w' belongs to $\vartheta(d)/v$. Then we define $\mu[g',w']$ to be equal to the morphism $g'\vartheta w' \longrightarrow v$ of $V(d')$ determined by g'^{-1}. If $w\le w'$ and $m: [g,w] \longrightarrow [g',w']$ is the morphism of $(Gx_{d'v}(\zeta(\vartheta)/v))(d\beta)$ determined by $g'g^{-1}$, then $\mu(m): gw \longrightarrow g'w'$ is the morphism of W(d) determined also by $g'g^{-1}$.■

2.11. Corollary. (i) If $\zeta(\vartheta)$ is a right cofinal functor and $d=d'\vartheta$, then $\vartheta(d)$ is also a right cofinal functor.

(ii) If G acts trivially on V then there is an isomorphism of categories

$$(\zeta(\vartheta)/v)(d) \longrightarrow \vartheta(d)/v.$$

(iii) If W=V and ϑ is the identity map, then the natural inclusion

$$\varphi: (G/d'w)(dw) \longrightarrow \vartheta(d)/w$$

is a right cofinal functor.

Proof. The assertion (i) follows from 2.10.(i), because, in this case, the classifying space of the category $\sigma(w)$ is contractible for every w belonging to $v\backslash\zeta(\vartheta)$. Hence $B(v\backslash\zeta(\vartheta))$ is homotopy equivalent to $B(v/\vartheta(d))$. (See [12]1.2. and [1] or [5] 9.2.).

The statement (ii) is an immediate consequence of 2.10.(ii). The assertion (iii) follows from (i) and 2.10.(ii), because the natural inclusion

$$\zeta(G/d'w) \longrightarrow \zeta(G\times_{d'w}(W/w))$$

is a right cofinal functor.∎

<u>2.12. Corollary.</u> Let X be a functor from W(d) to Top. Then there exists a homotopy equivalence

$$\operatorname{hocolim}_{V(d')} \operatorname{hocolim}_{(G\times_{d'(-)}\zeta(\vartheta)/-)(d\beta)} X\beta \simeq \operatorname{hocolim}_{W(d)} X\ .∎$$

The result above follows immediately from 2.10.(ii) and from the general homotopy theory ([1],[5] 9.8,[4]). In particular, it yields the following facts.

<u>2.13. Corollary.</u> (i) Suppose that G acts trivially on V and that d'v=G whenever v belongs to V. Then V(d')=ζ(V) and there is a homotopy equivalence

$$\operatorname{hocolim}_{v\in\zeta(V)} \operatorname{hocolim}_{(\zeta(\vartheta)/v)(d)} X \simeq \operatorname{hocolim}_{W(d)} X.$$

(ii) The natural projection to the orbit set $\vartheta: W \longrightarrow W/G$ induces a homotopy equivalence

$$\operatorname{hocolim}_{[w]\in W/G} \operatorname{hocolim}_{(G(W/w))(d)} X \simeq \operatorname{hocolim}_{W(d)} X\ .$$

(iii) Suppose that W is a regular G-poset. Then there exist homotopy equivalences

$$\operatorname{hocolim}_{[w]\in W/G} E(G/dw)\times_{G_w/dw} X(w) \simeq \operatorname{hocolim}_{[w]\in W/G} \operatorname{hocolim}_{(Gw)(d)} X \simeq \operatorname{hocolim}_{W(d)} X\ .$$

<u>Proof.</u> The statement (i) is a consequence of 2.12. and 2.11.(ii). The assertion (ii) follows immediately from (i), because w" is an element of G(W/w) if and only if, there exists g∈G such that gw"≤w. The assertion (iii) can be obtained from (ii), because in this case, the inclusion $Gw \longrightarrow G(W/w)$ is a right cofinal functor. This fact is also a consequence of 2.8.∎

<u>2.14. Corollary.</u> Suppose that W=V and that ϑ is the identity map.

(i) There are homotopy equivalences

$$\operatorname{hocolim}_{w\in W(d')} E(G/d(w))\times_{d'(w)/d(w)} X(w) \simeq \operatorname{hocolim}_{w\in W(d')} \operatorname{hocolim}_{\vartheta(d)/w} X \simeq \operatorname{hocolim}_{W(d)} X$$

(ii) In particular, if $X:\mathcal{G}\int W \longrightarrow Top$, then, for any admissible map $d:W \longrightarrow Sub\text{-}G$, there are homotopy equivalences

$$\operatorname{hocolim}_{w\in W(d)} EG\times_{d(w)} X(w) \simeq \operatorname{hocolim}_{\mathcal{G}\int W} X \simeq EG\times_G \operatorname{hocolim}_{\zeta(W)} X .$$

<u>Proof.</u> The assertion (i) is a consequence of 2.11.(iii) and 2.3.(iii). The statement (ii) follows from (i), because $\mathcal{G}\int W=W(d_o)$, where $d_o w=(e)$ for every $w\in W$. ∎

<u>2.15. Example.</u> Consider the case where, for every $w\in W$, $X(w)$ is a one-point space. Then 2.14 implies that there are homotopy equivalences

$$EG\times_G BW \simeq \operatorname{hocolim}_{w\in W(d)} EG/d(w) \simeq EG\times_G \operatorname{hocolim}_{w\in W(d)} G/d(w)$$

where BW is the classifying space of the underlying poset of W with the G-action induced by the action of G on W.

In particular, if W is a normal G-poset, then

$$EG\times_G BW \simeq EG\times_G \operatorname{hocolim}_{w\in W(G)} EG/G_w ,$$

and if W is regular, then

$$EG\times_G BW \simeq EG\times_G \operatorname{hocolim}_{[w]\in W/G} EG/G_w . ∎$$

<u>2.16. Examples.</u> Let F be a G-subposet of Sub-G.

(i) Consider Example 2.4. Then 2.15 implies that there exists a homotopy equivalence

$$EG\times_G BF \simeq EG\times_G \operatorname{hocolim}_{O(F)} I$$

where $I:O(F)\longrightarrow Top$ is the natural inclusion such that $I(G/H)=G/H$.

(ii) Let $d:F^{op}\longrightarrow Sub\text{-}G$ be the admissible G-poset map such that, for every element H of F, $d(H)$ is equal to the centralizer $C_G(H)$ of H in G. The category $(F^{op}(d))^{op}$ will be denoted by $C_G(F)$. The morphisms of $C_G(F)$ can be regarded as group homomorphisms $\delta:H\longrightarrow H'$ of the form $\delta(-)=g(-)g^{-1}$, for some $g\in G$.

In the case where F is the G-poset of elementary abelian p-subgroups of G, we obtain the category $A(G,p)$ considered in [9].

Example 2.15 implies that there is a homotopy equivalence

$$EG\times_G BF^{op} \simeq \operatorname{hocolim}_{C_G(F)^{op}} EG\times_G(G/C_G(-)).$$

From the fact that the G-spaces BF and BF^{op} are equal, it follows that there is a homotopy equivalence

$$EG\times_G \operatorname{hocolim}_{O(F)} I \simeq EG\times_G \operatorname{hocolim}_{C_G(F)^{op}} G/C_G(-)\ .\blacksquare$$

The next result can be regarded as a specification of 1.12.

<u>2.17. Proposition.</u> Let W be a G-poset and let d: W——>Sub-G be an admissible map. For any functor X: W(d)——>Top, there are homotopy equivalences

$$\operatorname{hocolim}_{W(d)} X \simeq \operatorname{hocolim}_{s(W)(dq_W)} Xq_W \simeq \operatorname{hocolim}_{s(W)(G)} E(G/dq_W(-))\times_{G\_/dq_W(-)} Xq_W \simeq$$

$$\simeq \operatorname{hocolim}_{[w_o,\ldots,w_n]\in s(W)/G} E(G/dw_o)\times_{(G_{w_o}\cap\ldots\cap G_{w_n})/dw_o} X(w_o)$$

where $s(W)=sd(W)^{op}$ and q_W: S(W)——>W is the natural projection (from 2.5.(iii)).

<u>Proof.</u> The first homotopy equivalence is a consequence of 2.11.(i) and the fact that q_W is a right cofinal functor. The existence of the second one follows from 2.14.(i) and 2.5.(i). The third one is a consequence of 2.5.(iii) and 2.13.(iii) or 2.5.(ii).■

We end this paper with applying of the result above to the case of orbit categories.

<u>2.18. Corollary.</u> Let F be a G-subposet of Sub-G and let X: O(F)——>Top. There is a homotopy equivalence

$$\operatorname{hocolim}_{O(F)} X \simeq \operatorname{hocolim}_{[H_o,\ldots,H_n]\in S(F)/G} E(G/H_o)\times_{NH_o\cap\ldots NH_n} X(G/H_o).\blacksquare$$

For $X=EG\times_G I$ we obtain a homotopy equivalence

$$EG\times_G BF \simeq \operatorname{hocolim}_{[H_o,\ldots,H_n]\in S(F)/G} EG/(NH_o\cap\ldots\cap NH_n).$$

References.

[1] A.K.Bousfield, D.M.Kan "Homotopy Limits ,Completion and Localization" Lecture Notes in Math.304, 1972.

[2] G.E.Bredon "Equivariant cohomology theory" Lecture Notes in Math.34, 1967.

[3] K.H.Dovermann, M.Rothenberg "Equivariant surgery and classification of finite group action on manifolds" Mem.Am.Math.Soc.379, 1988.

[4] E.Dror Farjoun "Homotopy and homology of diagrams of spaces" in Lecture Notes in Math.1286,93-134.

[5] W.G.Dwyer, D.M.Kan "A classification theorem for diagrams of simplicial sets" Topology,Vol.23,No.2,139-155, 1984 .

[6] W.G.Dwyer, D.M.Kan "Function complexes for diagrams of simplicial sets" Indigationes Math.Vol.45,Fasc.2, 1983.

[7] P.J.Hilton, U.Stammbach "A course in Homological Algebra" Springer, 1971.

[8] S.Jackowski, J.E.McClure "Homotopy approximation for classifying spaces of compact Lie groups" in Lecture Notes in Math.1370, 1989.

[9] S.Jackowski, J.E.McClure "A homotopy decomposition for classifying spaces of compact Lie groups" preprint.

[10] W.Lück "Transformation groups and algebraic K-theory" to appear in Lecture Notes in Math.

[11] J.Słomińska "Equivariant Bredon cohomology of classifying spaces of families of subgroups" Bull.Ac.Sc.Pol.Sc.Math.Vol.XXIII,No.9-10, 1980, 503-505.

[12] R.W.Thomason "Homotopy colimits in the category of small categories" Proc.Camb.Phil.Soc.85, 1979, 91-109.

[13] W.G.Dwyer, D.M.Kan "Reducing equivariant homotopy theory to the theory of fibrations" in "Conference on Algebraic Topolology in honor of Peter Hilton", Contemporary Mathematics, vol.37, 1985.

On bordism rings with principal torsion ideal

Vladimir V. Vershinin
Institute of Mathematics
Siberian Branch of the Soviet Academy of Science
Novosibirsk 630090
USSR

1 Introduction

Bordism spectra have long been known [12] as a rich source of examples of homology comodules over the dual of the Steenrod algebra, and introducing singularities [2] allows us to vary these structures further. Such computations were pioneered for the unitary case MU in [3], and more recently for the symplectic case MSp in [13,14].

In this paper we introduce some new sequences of singularities into MSp, and study the homology and homotopy properties of the resulting spectra. We prove that they are multiplicative, characterise their homology comodules, and show that their homotopy rings have torsion ideals generated by a single element in dimensions of the form $2^i - 3$.

Our spectra are mainly of interest at the prime 2, where they lie in the poorly mapped territory between MSp and BP.

I owe many thanks to Nigel Ray, for much helpful advice on improving this paper.

2 Homotopy and homology computations

Nigel Ray's well known family of elements ϕ_i [11] lie in the symplectic bordism ring MSp_*. Each ϕ_i, $i \geq 1$, is indecomposable in MSp_{8i-3} and has $2\phi_i = 0$. It is also convenient to write ϕ_0 for the class θ_1 of the non-trivially framed circle in MSp_1.

Consider the following sequences of elements

$$\Delta_0 = \ (\phi_1, \phi_2, \ldots, \phi_{2^i}, \ldots),$$

$$\Delta_i = \ (\phi_0, \phi_1, \ldots, \phi_{2^{i-2}}, \phi_{2^i}, \phi_{2^{i+1}}, \ldots), \quad i \geq 1,$$

where Δ_i consists of all ϕ_j, j a power of 2, except for $\phi_{2^{i-1}}$.

Let Δ_i^n be the finite subsequence of Δ_i, $i \geq 0$, consisting of the first n elements. Thus

$$\Delta_i^n = (\phi_0, \phi_1, \ldots, \phi_{2^{n-2}}), \qquad i \geq n \geq 2$$

$$\Delta_i^1 = (\phi_0), \qquad i \geq 1$$

$$\Delta_i^n = (\phi_0, \phi_1, \ldots, \phi_{2^{i-2}}, \phi_{2^i}, \ldots, \phi_{2^{n-1}}), \quad i < n.$$

We propose to consider the theories $MSp_*^{\Delta_i^n}(\ \ )$, $i \geq 0$, i.e. symplectic bordism with singularities Δ_i^n, and their direct limit $MSp_*^{\Delta_i}(\ \ )$. We write the corresponding spectra as $MSp^{\Delta_i^n}$ and MSp^{Δ_i}. We studied $MSp^{\Delta_0^n}$ and MSp^{Δ_0} in [13], where we labelled them as MSp^{Σ_n} and MSp^{Σ} respectively.

Our main tools will be the Adams-Novikov spectral sequence [1,9] and the modified algebraic spectral sequence m.a.s.s. [13] which converges to the E_2 term of the Adams-Novikov spectral sequence. We recall [13,16] that the initial term $E_1^{q,s,t}$ of the m.a.s.s. for the spectrum X is isomorphic to

$$Ext_{\mathcal{A}'_p}(\mathbb{Z}/p, \overline{\overline{BP}}_* \otimes H_*(X; \mathbb{Z}/p)),$$

where $\mathcal{A}'_p$ is the dual to the factoralgebra $A_p/(Q_0)$ of the Steenrod algebra and Q_0 is the Bockstein operator. Also, $\overline{\overline{BP}}_*$ is the object associated to BP_* (the Brown-Peterson homology [1] of a point) by the filtration of the m.a.s.s.,

$$\overline{\overline{BP}}_* = \mathbb{Z}/p[h_0, h_1, \ldots, h_i, \ldots],$$

$$\deg h_i = \begin{cases} (2,0), & i = 0, \\ (1, 2(2^i - 1)), & i > 0. \end{cases}$$

In the case $X = MSp$ and $p = 2$ this initial term was computed in [13] and is isomorphic to

$$\mathbb{Z}/2[c_2, \ldots, c_k, \ldots, u_1, \ldots, u_j, \ldots, h_0, h_1, \ldots, h_m, \ldots],$$

$$k \neq 2^l - 1, \ \deg c_k = (0,0,4k), \ \deg u_j = (0,1,2(2^j - 1)),$$

$$\deg h_0 = (2,0,0), \ \deg h_m = (1,0,2(2^m - 1)), \ m \geq 1.$$

Our main result is the following.

Theorem 1. In the theory of symplectic bordism with singularities of type Δ_i^n there exists a multiplicative structure such that $\phi_{2i-1}^3 = 0$ in $MSp_*^{\Delta_i^n}$. The groups $H_k(MSp^{\Delta_i^n}; \mathbb{Z})$ are finitely generated and torsion free for all k, whilst $H_{2j-1}(MSp^{\Delta_i^n}; R) = 0$ for all j, where $R = \mathbb{Z}$ or $\mathbb{Z}/2$. The initial term of the m.a.s.s. for $MSp^{\Delta_i^n}$ is isomorphic to the ring

$$\mathbb{Z}/2[c_2, \ldots, c_k, \ldots, u_{i+1}, u_{n+2}, \ldots, h_0, \ldots, h_m, \ldots],$$

where the generators are the images of the corresponding elements in the initial term for MSp.

Proof. We will prove the theorem by induction by n. If $i = 0$, then all the statements of the theorem were proved in [13].

So we suppose that $i > 0$. If $n \leq i$, then we obtain the spectra MSp^{Σ_n} of [13] and the theorem is also proved there. So suppose it is true for some $n - 1$ such that $n - 1 > i$. Then $MSp^{\Delta_i^{n-1}}$ is a multiplicative spectrum. From the theorem 3.2 of [2] it follows that there exists an exact triangle.

$$\begin{array}{ccc} MSp_*^{\Delta_i^{n-1}}(X) & \xrightarrow{\beta} & MSp^{\Delta_i^{n-1}}(X) \\ \delta \nwarrow & & \swarrow \gamma \\ & MSp_*^{\Delta_i^n}(X), & \end{array} \qquad (1)$$

where X is any spectrum, the homomorphism β is multiplication by ϕ_{2^n-1} and δ is the Bockstein homomorphism; also γ is defined by considering manifolds with singularities Δ_i^{n-1} as manifolds with singularities Δ_i^n.

Let X be the spectrum defined by ordinary integral homology, or homology with coefficients in $\mathbb{Z}/2$. Then the above exact triangle can be rewritten as the exact sequence

$$\ldots \longrightarrow H_{k-2^{n+2}+3}(MSp^{\Delta_i^{n-1}};R) \xrightarrow{\beta} H_k(MSp^{\Delta_i^{n-1}};R) \xrightarrow{\gamma}$$

$$\longrightarrow H_k(MSp^{\Delta_i^n};R) \xrightarrow{\delta} H_{k-2^{n+2}+2}(MSp^{\Delta_i^{n-1}};R) \longrightarrow \ldots,$$

where $R = \mathbb{Z}$ or $\mathbb{Z}/2$. From the induction hypothesis that $H_{2j-1}(MSp^{\Delta_i^{n-1}};R) = 0$ it follows that $H_{2j-1}(MSp^{\Delta_i^n};R) = 0$ for all j. Hence, the long exact sequence splits into short exact sequences

$$0 \longrightarrow H_{2k}(MSp^{\Delta_i^{n-1}};R) \xrightarrow{\gamma} H_{2k}(MSp^{\Delta_i^n};R) \xrightarrow{\delta} H_{2k-2^{n+2}+2}(MSp^{\Delta_i^{n-1}};R) \longrightarrow 0, \quad (2)$$

from which it follows that $H_{2k}(MSp^{\Delta_i^n},\mathbb{Z})$ is finitely generated and torsion free. Consequently the Adams-Novikov spectral sequence, the algebraic spectral sequence and the m.a.s.s. for the spectrum $MSp^{\Delta_i^n}$ exist and converge.

Let $R = \mathbb{Z}/2$. Then from (2) we obtain the analogous sequence in cohomology

$$0 \longleftarrow H^{2k}(MSp^{\Delta_i^{n-1}};\mathbb{Z}/2) \xleftarrow{\gamma} H^{2k}(MSp^{\Delta_i^n};\mathbb{Z}/2) \xleftarrow{\delta} H^{2k-2^{n+2}+2}(MSp^{\Delta_i^{n-1}};\mathbb{Z}/2) \longleftarrow 0,$$

and also the long exact sequence

$$\begin{aligned}
0 \quad &\longrightarrow \operatorname{Hom}_{\mathcal{A}_2'}(\mathbb{Z}/2, \overline{\overline{BP}}_* \otimes H_{2k}(MSp^{\Delta_i^{n-1}};\mathbb{Z}/2)) \xrightarrow{\overline{\gamma}^0} \\
&\longrightarrow \operatorname{Hom}_{\mathcal{A}_2'}(\mathbb{Z}/2, \overline{\overline{BP}}_* \otimes H_{2k}(MSp^{\Delta_i^n};\mathbb{Z}/2)) \xrightarrow{\overline{\delta}^0} \\
&\longrightarrow \operatorname{Hom}_{\mathcal{A}_2'}(\mathbb{Z}/2, \overline{\overline{BP}}_* \otimes H_{2k-2^{n+2}+2}(MSp^{\Delta_i^{n-1}};\mathbb{Z}/2)) \xrightarrow{\overline{\beta}^0} \\
&\longrightarrow \operatorname{Ext}^1_{\mathcal{A}_2'}(\mathbb{Z}/2, \overline{\overline{BP}}_* \otimes H_{2k}(MSp^{\Delta_i^{n-1}};\mathbb{Z}/2)) \xrightarrow{\overline{\gamma}^1} \ldots \\
\ldots &\xrightarrow{\overline{\delta}^{s-1}} \operatorname{Ext}^{s-1}_{\mathcal{A}_2'}(\mathbb{Z}/2, \overline{\overline{BP}}_* \otimes H_{2k-2^{n+2}+2}(MSp^{\Delta_i^{n-1}};\mathbb{Z}/2)) \xrightarrow{\overline{\beta}^{s-1}} \\
&\longrightarrow \operatorname{Ext}^s_{\mathcal{A}_2'}(\mathbb{Z}/2, \overline{\overline{BP}}_* \otimes H_{2k}(MSp^{\Delta_i^{n-1}};\mathbb{Z}/2)) \xrightarrow{\overline{\gamma}^s} \\
&\longrightarrow \operatorname{Ext}^s_{\mathcal{A}_2'}(\mathbb{Z}/2, \overline{\overline{BP}}_* \otimes H_{2k}(MSp^{\Delta_i^n};\mathbb{Z}/2)) \xrightarrow{\overline{\delta}^s} \\
&\longrightarrow \operatorname{Ext}^s_{\mathcal{A}_2'}(\mathbb{Z}/2, \overline{\overline{BP}}_* \otimes H_{2k-2^{n+2}+2}(MSp^{\Delta_i^{n-1}};\mathbb{Z}/2)) \xrightarrow{\overline{\beta}^s} \ldots,
\end{aligned}$$

where the homomorphisms $\overline{\gamma}^s$ and $\overline{\delta}^s$ are induced by the morphisms γ and δ, respectively, and $\overline{\beta}^s$ is the connecting homomorphism.

We have the element u_{n+1} in $\mathrm{Ext}^1_{\mathcal{A}_2^i}(\mathbb{Z}/2, \overline{\overline{BP}}_* \otimes H_*(MSp^{\Delta_i^{n-1}}, \mathbb{Z}/2))$. The corresponding element ϕ_{2^n-1} in $MSp_*^{\Delta_i^{n-1}}$ goes to zero in $MSp_*^{\Delta_i^n}$. From consideration of the Adams-Novikov spectral squence it follows that under the mapping $\overline{\gamma}^1$ the element u_{n+1} is sent to 0. Using the fact that $\gamma_*: MSp_*^{\Delta_i^{n-1}}(\ ) \longrightarrow MSp_*^{\Delta_i^{n-1}}(\ )$ is a module map over MSp_*, we obtain that

$$u_{n+1} \cdot \mathrm{Ext}^{s-1}_{\mathcal{A}_2^i}(\mathbb{Z}/2, \overline{\overline{BP}}_* \otimes H_*(MSp; \mathbb{Z}/2))$$

is sent to 0 by $\overline{\gamma}^s$ and hence

$$u_{n+1} \cdot \mathrm{Ext}^{s-1}_{\mathcal{A}_2^i}(\mathbb{Z}/2, \overline{\overline{BP}}_* \otimes H_*(MSp^{\Delta_i^{n-1}}; \mathbb{Z}/2))$$

is sent to 0 by $\overline{\gamma}^s$. By our induction hypothesis, the dimension of $\mathrm{Ext}^{s-1}_{\mathcal{A}_2^i}(\mathbb{Z}/2, \overline{\overline{BP}}_* \otimes H_*(MSp^{\Delta_i^{n-1}}; \mathbb{Z}/2))$ for fixed q and t (i.e. $E_1^{q,s-1,t}$) as a vector space over $\mathbb{Z}/2$ coincides with the dimension of $u_{n+1} \cdot E_1^{q,s-1,t}$. Consequently, $\overline{\beta}^s$ is a monomorphism for $s = 0, 1, \ldots$. Thus, $\overline{\delta}^s$ is the zero homomorphism and $\overline{\gamma}^s$ is an epimorphism, $\overline{\gamma}^0$ is an isomorphism, and the kernel of $\overline{\gamma}^s$ for $s \geq 1$ is the module

$$u_{n+1} \cdot \mathrm{Ext}^{s-1}_{\mathcal{A}_2^i}(\mathbb{Z}/2, \overline{\overline{BP}}_* \otimes H_*(MSp^{\Delta_i^{n-1}}; \mathbb{Z}/2)).$$

This means that the initial term of the m.a.s.s. for the spectrum $MSp^{\Delta_i^n}$, as a module over the initial term of the m.a.s.s. for MSp, has the form indicated in the statement of our theorem.

We obtain the formulas for the first differential of the m.a.s.s. for $MSp^{\Delta_i^n}$ (for $t - s < 2^{n+3} - 3$) from the corresponding formulas in the m.a.s.s. for MSp [13,6]:

$$d_1(h_{i+1}) = h_0 u_{i+1}, \quad d_1(c_{2^i+2^k-1}) = h_{k+1} u_{i+1}, \quad k \neq i, \ k = 0, 1, \ldots.$$

The module structure of the E_1 term $E_1(MSp^{\Delta_i^n})$ of the m.a.s.s. for $MSp^{\Delta_i^n}$ generates the multiplicative structure, which in turn generates the ring structure in E_2 of the m.a.s.s. for $MSp^{\Delta_i^n}$. This is compatible with the module structure of E_2 of the m.a.s.s. for $MSp^{\Delta_i^n}$ over E_2 for MSp. The ring $E_2(MSp^{\Delta_i^n})$ has the following generators in dimensions $t - s < 2^{n+3} - 3$:

h_0; $\quad u_{i+1}$; $\quad c_m,\ m \neq 2^l - 1,\ m \neq 2^k + 2^i - 1,\ k, l = 0, 1, \ldots$;

$$\beta_N = \sum_{j=1}^{\nu} h_{n_j+1} c_{2^{n_\nu}+2^i-1} \cdots \hat{c}_{2^{n_j}+2^i-1} \cdots c_{2^{n_1}+2^i-1},$$

$$N = (n_\nu, \ldots, n_1),\ n_\nu > \ldots > n_1 \geq 0,\ n_j \neq i,\ \nu(N) = \nu \geq 1,\ \beta_{n_1} = h_{n_1+1};$$

h_{i+1}^2; $\quad c^2_{2^k+2^i-1}$; $\quad \xi_M = h_0 c_{2^{m_\nu}+2^i-1} \cdots c_{2^{m_1}+2^i-1} + h_{i+1}\beta_M$,

where M satisfies the same conditions as N.

There are some evident relations involving the above generators. In particular, we need

$$u_{i+1}\beta_N = 0$$

for all N. Evidently $E_2 = E_\infty$ if $t - s < 2^{n+3} - 3$. As a consequence, we obtain that in these dimensions, multiplication by u_{i+1} is monic on $E_\infty^{q,s,t}$ if $s \geq 1$, and hence this is true for E_2 of the Adams-Novikov spectral sequence for $MSp^{\Delta_i^n}$.

The obstruction to the existence of a multiplicative structure [8,4] in the theory $MSp_*^{\Delta_i^n}(\ )$ is an element $\gamma[P_n'] \in MSp^{\Delta_i^n}_{2^{n+3}-5}$ where $[P_n'] \in MSp^{\Delta_i^{n-1}}_{2^{n+3}-5}$. So the dimension $2^{n+3}-5$ of this element is odd. Consider its projection into the E_2 term of the Adams-Novikov spectral sequence. It must be of the form

$$\phi_{2^i-1}^{2j-1} Y \tag{3}$$

for some $j = 1, 2, \ldots$, and $Y \in \mathrm{Ext}_*^0(BP_*, BP_*(MSp^{\Delta_i^n}))$, and Y must be a cycle with respect to the differential d_3.

Now the relation $\phi_k^3 = \sum_{l=0}^{k-1} \phi_l M_l$ in MSp_* has been proved recently [7], for some $M_l \in MSp_*$. If we take $k = 2^{i-1}$, then we obtain that in $MSp_*^{\Delta_i^k}$, with $k \geq i$,

$$\phi_{2^i-1}^3 = 0.$$

Hence, we must have an element $x \in E_2^{0,3\cdot 2^{i+2}-8}$ of the Adams-Novikov spectral sequence for $MSp^{\Delta_i^n}$ such that $d_3(x) = \phi_{2^i-1}^3$. It follows that if $t - s < 2^{n+3} - 3$, then $d_r = 0$ for $r > 3$. Then in formula (3), Y must lie in $MSp_*^{\Delta_i^n}$, and if $j > 1$ then the obstruction is equal to zero (see the table below, which displays s against $t - s$):

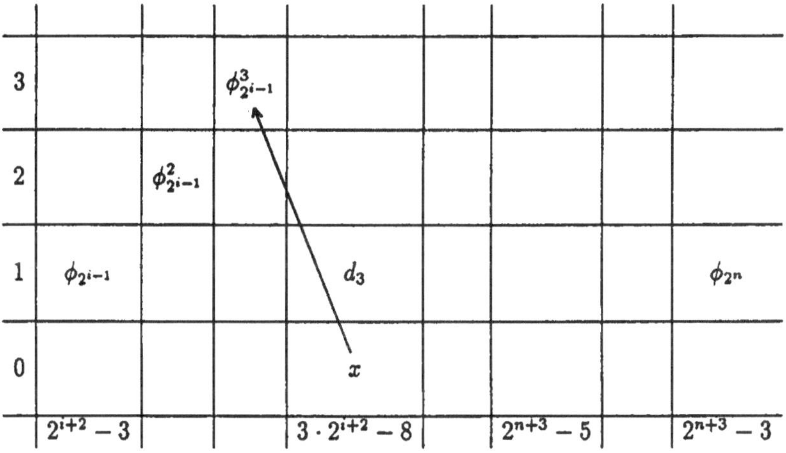

So the only possibility for the obstruction to be non-zero is that it has the form $\phi_{2^i-1}Y$, where $Y \in MSp^{\Delta_i^n}_{2^{n+3}-2^{i+2}-2}$. Thus the dimension of Y is equal to 2 mod 4. Hence, its projection to the E_∞ term of the m.a.s.s. must be of the form

$$\sum_j \beta_{N_j} Y_{N_j}$$

for some β_{N_j} and Y_{N_j}. But we have in the m.a.s.s. that $\beta_{N_j} u_{i+1} = 0$, so we obtain that $\phi_{2^i-1}Y = 0$. Thus the obstruction to multiplicativity is zero.

The obstruction to commutativity is in our case an element of order two and lies in dimension $2^{n+3}-4$ [8,4], and hence equals zero. For the obstruction $[\Gamma]$ to associativity, we have the condition

$$3[\Gamma] = \Delta \cdot \theta_1,$$

and $\theta_1 = 0$ in our theory $MSp_*^{\Delta_i^n}(\ )$ so long as $i > 0$.

If $i = 0$ then more detailed considerations in the m.a.s.s. for $MSp^{\Delta_0^n}$ show that the obstruction also vanishes.

Our induction is now complete.

Theorem 2. The multiplicative structure in the theory of symplectic bordism with singularities Δ_i^n may be chosen such that, for $p > 2$, its coefficient ring is isomorphic to the polynomial algebra

$$MSp_{(p)*}^{\Delta_i^n} = \mathbb{Z}_p[w_1, \dots, w_i, w_{i+2}, \dots, w_{n+1}, x_2, \dots, x_k, \dots],$$

where $k \neq 2^j - 1$ if $1 \leq j \leq n+1$, $j \neq i+1$, and $\deg w_l = 2(2^l - 1)$, $\deg x_k = 4k$.

Proof. We also prove this theorem by induction on n. As in the proof of theorem 1 we can suppose that $i > 0$ and $n - 1 > i$.

By the induction hypothesis, the multiplicative structure on $MSp^{\Delta_i^{n-1}}$ be chosen so that the ring $MSp_*^{\Delta_i^{n-1}}$ is polynomial and isomorphic to

$$\mathbb{Z}_p[w_1, \dots, w_i, w_{i+2}, \dots, w_n, x_2, \dots, x_k, \dots],$$

where $k \neq 2^j - 1$ if $1 \leq j \leq n$, $j \neq i+1$.

We have $\phi_{2^n-1} = 0$ in $MSp_{(p)*}^{\Delta_i^{n-1}}$ for $p > 2$. So from the Bockstein exact sequence we obtain short exact sequences

$$0 \longrightarrow MSp_{(p)*}^{\Delta_i^{n-1}} \longrightarrow MSp_{(p)*}^{\Delta_i^n} \longrightarrow MSp_{(p)*-2^{n+2}+2}^{\Delta_i^{n-1}} \longrightarrow 0.$$

Hence, the module $MSp_{(p)*}^{\Delta_i^n}$ is the direct sum

$$MSp_{(p)*}^{\Delta_i^n} \cong MSp_{(p)*}^{\Delta_i^{n-1}} \oplus MSp_{(p)*}^{\Delta_i^{n-1}} w_{n+1},$$

where $w_{n+1} \in MSp_*^{\Delta_i^n}$, and $\deg w_{n+1} = 2^{n+2} - 2$.

From the Adams-Novikov spectral sequence for $MSp^{\Delta_i^n}$ with $p = 2$, we deduce that

$$w_{n+2}^2 = m x_{2^{n+1}-1} + a_{n+1} + y_{n+1} w_{n+1},$$

where $a_{n+1} \in F^3 MSp_*^{\Delta_i^{n-1}}$ and $y_{n+1} \in F^2 MSp_*^{\Delta_i^{n-1}}$. Here $F^j MSp_*^{\Delta_i^{n-1}}$ means the module of filtration j in $MSp_*^{\Delta_i^{n-1}}/\mathrm{Tors} \subset \mathrm{Hom}_{A_2}(BP_*, BP_*(MSp^{\Delta_i^n}))$ corresponding to the m.a.s.s., m is odd and $x_{2^{n+1}-1}$ is defined so that h_{n+1} is associated to it.

It follows from theorem 4.6 in [8] that if in $MSp_*^{\Delta_i^n}(\ )$ there exists a multiplicative structure μ_n, then there also exists a multiplication $\tilde{\mu}_n$ such that

$$\tilde{\mu}_n(x, y) = \mu_n(x, y) + \mu_{n,n-1}(\mu_{n-1,n}(\delta(x), z), \delta(y)),$$

where z is an arbitrary element in the bordism group with singularities Δ_i^n in the corresponding dimension, and the pairings $\mu_{n,n-1}$ and $\mu_{n-1,n}$ are defined by the product of Δ_i^{n-1}-manifolds. In our case we obtain

$$\tilde{\mu}_n(w_{n+1}, w_{n+1}) = \mu_n(w_{n+1}, w_{n+1}) + 4z,$$

where z is an arbitrary element of $MSp^{\Delta^n_i}_{4(2^{n+1}-1)}$. Hence, we can vary the multiplicative structure in $MSp^{\Delta^n_i}_*(\ )$ so as to give

$$x_{2^{n+1}-1} = \pm(w^2_{n+1} - a_{n+1} - w_{n+1}y_{n+1}).$$

This means that, after suitable variation, the ring $MSp^{\Delta^n_i}_{(p)*}$ will become polynomial, and the theorem is proved.

We shall consider now the homology of the spectrum $MSp^{\Delta^n_i}$.

Theorem 3. There exists an isomorphism of comodules over $\mathcal{A}_*$, the dual of the Steenrod algebra, of the form

$$H_*(MSp^{\Delta^n_i}; \mathbb{Z}/2) \cong \mathbb{Z}/2[\xi^2_1, \ldots, \xi^2_{i-2}, \xi^4_{i-1}, \tilde{\xi}^2_i, \ldots, \tilde{\xi}^2_{n+1}, \xi^4_{n+2}, \ldots] \otimes \mathbb{Z}/2[c_2, \ldots, c_k, \ldots],$$

where the ξ_i are generators of $\mathcal{A}_*$, and $\tilde{\xi}_j \equiv \xi_j$ mod decomposable in $\mathcal{A}_*$. The elements c_k are primitive.

Proof. Suppose the theorem is true for $n-1$. The elements c_k in $H_*(MSp^{\Delta^n_i}; \mathbb{Z}/2)$ are defined as the images of $c_k \in H_*(MSp^{\Delta^{n-1}_i}; \mathbb{Z}/2)$ by the monomorphism γ of the exact sequence (2) for $R = \mathbb{Z}/2$. We can also regard the algebra

$$\mathbb{Z}/2[\xi^2_1, \ldots, \xi^2_{i-2}, \xi^4_{i-1}, \tilde{\xi}^2_i, \ldots, \tilde{\xi}^2_n, \xi^4_{n+1}, \ldots]$$

as the image of the morphism

$$\nu^{n-1}_*: H_*(MSp^{\Delta^{n-1}_i}; \mathbb{Z}/2) \longrightarrow H_*(H\mathbb{Z}/2; \mathbb{Z}/2),$$

where $\nu^{n-1}: MSp^{\Delta^{n-1}_i} \to H\mathbb{Z}/2$ corresponds to the generator of $H^0(MSp^{\Delta^{n-1}_i}; \mathbb{Z}/2)$.

Let us consider now the Atiyah-Hirzebruch spectral sequence

$$E^2 = H_*(BP; \pi_*(MSp^{\Delta^n_i})) \Longrightarrow MSp^{\Delta^n_i}_*(BP).$$

It is known that in the analogous Atiyah-Hirzebruch spectral sequence for $MSp^{\Sigma_n}_*(BP)$ [15]

$$d_{2(2^{n+1}-1)}(m_{n+1}) = \phi_{2^n-1}.$$

We have an evident morphism of spectra

$$\gamma: MSp^{\Delta^{n-1}_i} \longrightarrow MSp^{\Sigma_n},$$

inducing a morphism of spectral sequences, from which it follows that the element ϕ_{2^n-1} cannot be killed in E^r, for $r < 2(2^{n+1}-1)$. In $E^{2(2^{n+1}-1)}$ we have

$$d_{2(2^{n+1}-1)}(\overline{m}_{n+1}) = \phi_{2^n-1},$$

where $\overline{m}_{n+1} \equiv m_{n+1}$ modulo decomposable. This means that $\overline{m}_{n+1}$ is an infinite cycle in the spectral sequence for $MSp^{\Delta^n_i}(BP)$. Note that the canonical morphism $\pi: BP \longrightarrow H\mathbb{Z}/2$ also induces a map of Atiyah-Hirzebruch spectral sequences.

Let ω^n be the morphism

$$\omega^n : MSp^{\Delta^n_i} \longrightarrow H\mathbb{Z}$$

inducing the identity on $\pi_0(\ )$. Then the following compositions

$$BP_*(MSp^{\Delta^n_i}) \xrightarrow{\omega^n_*} BP_*(H\mathbb{Z}) \cong H\mathbb{Z}_*(BP),$$

$$HZ/2_*(MSp^{\Delta_*^n}) \xrightarrow{\omega_*^n} HZ/2_*(HZ) \overset{\chi}{\cong} HZ_*(HZ/2)$$

are the edge homomorphisms in the corresponding spectral sequences. Here χ interchanges HZ and $HZ/2$. We combine these morphisms in the commutative diagram

$$\begin{array}{ccccc} BP_*(MSp^{\Delta_*^n}) & \xrightarrow{\omega_*^n} & BP_*(HZ) & \cong & HZ_*(BP) \\ \pi_* \downarrow & & \pi_* \downarrow & & \pi_* \downarrow \\ HZ/2_*(MSp^{\Delta_*^n}) & \xrightarrow{\omega_*^n} & HZ/2_*(HZ) & \overset{\chi}{\cong} & HZ_*(HZ/2). \end{array} \tag{4}$$

The fact that $\overline{m}_{n+1}$ is an infinite cycle means that we have some element $\bar{\mu}_{n+1}$ in $BP_*(MSp^{\Delta_*^n})$ whose image under the edge homomorphism is $\overline{m}_{n+1}$. The image of $\overline{m}_{n+1}$ under the action of π_* is $\bar{\xi}_{n+1}^2$, where $\bar{\xi}_{n+1} \equiv \xi_{n+1}$ modulo decomposable elements. We add the following rectangle

$$\begin{array}{ccc} HZ/2_*(HZ) & \overset{\chi}{\cong} & HZ_*(HZ/2) \\ \eta_* \downarrow & & \eta_* \downarrow \\ HZ/2_*(HZ/2) & \overset{\chi}{\cong} & HZ/2_*(HZ/2) \end{array}$$

to the diagram (4) ($\eta: HZ \longrightarrow HZ/2$ denotes reduction mod 2) and obtain that the image of $\pi_*(\bar{\mu}_{n+1}) \in HZ/2_*(MSp^{\Delta_*^n})$ under $\nu_*^n = (\eta \circ \omega^n)_*$ in $HZ/2_*(HZ/2)$ is equal to $\tilde{\xi}_{n+1}^2 = \chi(\bar{\xi}_{n+1}^2)$, where $\nu^n: MSp^{\Delta_*^n} \to HZ/2$ corresponds to the generator of $H^0(MSp^{\Delta_*^n}; Z/2)$, and $\tilde{\xi}_{n+1} \equiv \xi_{n+1}$ modulo decomposable elements. We therefore also denote $\pi_*(\bar{\mu}_{n+1})$ by $\tilde{\xi}_{n+1}^2$.

We see from the exact sequence (2) that $H_*(MSp^{\Delta_*^n}; Z/2)$ is a module over $H_*(MSp^{\Delta_*^n}; Z/2)$ on two generators, in dimension 0 and $2(2^{n+1}-1)$ respectively. The element $\tilde{\xi}_{n+1}^2$ does not belong to the image of $H_*(MSp^{\Delta_*^{n-1}}; Z/2)$ under the action of γ_* and so may serve as the generator in dimension $2(2^{n+1}-1)$. Hence, every element of $H_*(MSp^{\Delta_*^n}; Z/2)$ can be written in the form

$$P_0 + P_1 \cdot \tilde{\xi}_{n+1}^2 \tag{5}$$

where P_0 and P_1 in $H_*(MSp^{\Delta_*^{n-1}}; Z/2)$ are polynomials in $\xi_1^2, \ldots, \xi_{i-2}^2, \xi_{i-1}^4, \tilde{\xi}_i^2, \ldots, \tilde{\xi}_n^2, \xi_{n+1}^4, \ldots, c_2, \ldots, c_k, \ldots$.

Writing the element $(\tilde{\xi}_{n+1}^2)^2$ in the form of (5) gives

$$(\tilde{\xi}_{n+1}^2)^2 = Q_0 + Q_1 \cdot \tilde{\xi}_{n+1}^2. \tag{6}$$

Applying ν_*^n to both sides of (6) yields that $Q_0 = \xi_{n+1}^4 + Q_0'$, where Q_0' does not contain ξ_{n+1}^4. The polynomial Q_1 also does not contain ξ_{n+1}^4, and the element ξ_{n+1}^4 becomes decomposable in $H_*(MSp^{\Delta_*^n}; Z/2)$.

Now we must prove that the ring $H_*(MSp^{\Delta_*^n}; Z/2)$ is polynomial. If this is false, then there must exist a polynomial P in the variables $c_2, \ldots, c_k, \ldots, \xi_1^2, \ldots, \xi_{i+1}^4, \tilde{\xi}_{i+2}^2, \ldots, \tilde{\xi}_{n+1}^2, \xi_{n+2}^4, \ldots$, which is identically zero. Writing P as a polynomial in $\tilde{\xi}_{n+1}^2$, we have

$$P = a_k(\tilde{\xi}_{n+1}^2)^k + \ldots + a_0,$$

where $a_k \neq 0$ and a_j does not contain $\tilde{\xi}_{n+1}^2$.

If $k=2l$, then we obtain from the relation (6):

$$a_{2l}\xi_{n+1}^{4l}+q_0+q_1\tilde{\xi}_{n+1}^2=0, \tag{7}$$

where $q_i\in H_*(MSp^{\Delta_i^{n-1}};\mathbf{Z}/2)$ and the polynomial q_i can involve ξ_{n+1}^4 only in powers less than l. The left hand side of (7) has the form (5), and so $a_{2l}\xi_{n+1}^{4l}+q_0=0$. Hence, $a_{2l}=0$, contradicting the definition of P. On the other hand, If $k=2l+1$, then using (6) we obtain that

$$(a_{2l+1}\xi_{n+1}^{4l}+q_1)\tilde{\xi}_{n+1}^2+q_0=0,$$

where $q_i\in H_*(MSp^{\Delta_i^{n-1}};\mathbf{Z}/2)$ and the polynomials q_i can involve ξ_{n+1}^4 only in powers less than l. Analogously, we obtain $a_{2l}+1=0$.

So our proof is complete. Remark that we have utilised the fact that γ and δ of (2) are morphisms of modules over $H_*(MSp^{\Delta_i^{n-1}};\mathbf{Z}/2)$.

Theorem 4. Symplectic bordism with singularities Δ_i admits a multiplicative structure, which can be chosen such that, for $p>2$,

$$MSp_{(p)*}^{\Delta_i}=\mathbf{Z}_p[w_1,\ldots,w_i,w_{i+2},\ldots,x_2,\ldots,x_k,\ldots],$$

where $k\neq 2^i-1$ if $j\neq i+1$, and $\deg w_l=2(2^l-1)$, $\deg x_k=4k$.

The torsion ideal of $MSp_*^{\Delta_i}$ is generated by the image of Nigel Ray's element ϕ_{2^i-1} and the relation

$$\phi_{2^i-1}^3=0$$

is fulfilled.

The groups $H_k(MSp^{\Delta_i};\mathbf{Z})$ are finitely generated and torsion free for all k, and are zero in odd dimensions; furthermore

$$H_*(MSp^{\Delta_i};\mathbf{Z}/2)\cong\mathbf{Z}/2[\xi_1^2,\ldots,\xi_{i-2}^2,\xi_{i-1}^4,\tilde{\xi}_i^2,\ldots\tilde{\xi}_j^2,\ldots]\otimes\mathbf{Z}/2[c_2,\ldots,c_k,\ldots].$$

Proof. This follows from the previous theorem by standard direct limit arguments.

Note that we may easily dualise our results so as to obtain the cohomology of $MSp^{\Delta_i^n}$ and MSp^{Δ_i} as modules over the Steenrod algebra. Specifically, let $\mathcal{D}_i^n$, $i\geq 0$, be the subalgebra of the Steenrod algebra for which a $\mathbf{Z}/2$-basis consists of the elements Sq^J (of the Milnor basis), where $J=(j_1,\ldots,j_k)$ and the j_l are such that

$$\begin{aligned} j_l\leq 1, &\quad \text{if} \quad l\neq i+1 \quad \text{and} \quad l\leq n+1,\\ j_l\leq 3, &\quad \text{if} \quad l=i+1 \quad \text{or} \quad l>n+1,\end{aligned}$$

and let $\mathcal{D}_i$ be the subalgebra of $\mathcal{A}_2$ with basis consisting of the elements Sq^J such that $j_l\leq 1$ if $l\neq i+1$, and $j_l\leq 3$ if $l=i+1$. Then there exist isomorphisms

$$H^*(MSp^{\Delta_i^n};\mathbf{Z}/2)\cong\bigoplus_w(\mathcal{A}_2/\mathcal{A}_2\overline{\mathcal{D}_i^n})\cdot S_w,$$

$$H^*(MSp^{\Delta_i};\mathbf{Z}/2)\cong\bigoplus_w(\mathcal{A}_2/\mathcal{A}_2\overline{\mathcal{D}_i})\cdot S_w,$$

as modules over the Steenrod algebra, where the S_w are generators with $w=(j_1,\ldots,j_k)$, $j_l\neq 2^m-1$, and $n(w)=\deg S_w=4\sum_{l=1}j_l$.

In fact we define the generators S_w in $H^*(MSp^{\Delta_i^n};\mathbb{Z}/2)$ as dual to the corresponding monomial on $c_2,\ldots,c_k,\ldots$. We then deduce that the operations Sq^{Δ_k}, $Sq^{2\Delta_{i+1}}$ and $Sq^{2\Delta_m}$ for $k=1,2,\ldots,m\geq n+2$ act trivially on S_w. Hence, we obtain an epimorphism

$$\bigoplus_w (\mathcal{A}_2/\mathcal{A}_2\overline{\mathcal{D}_i^n})\cdot S_w \longrightarrow H^*(MSp^{\Delta_i^n};\mathbb{Z}/2),$$

with w as before. Comparing the dimensions of the two vector spaces in this formula we see that it is an isomorphism.

We conclude this section with some results we shall require later.

Proposition 1. The Atiyah-Hirzebruch spectral sequence for $MSp_*^{\Delta_i^n}(MSp^{\Delta_i^n})$ collapses.

Proof. We again use induction. Let us suppose that the spectral sequence for $MSp_*^{\Delta_i^{n-1}}(MSp^{\Delta_i^{n-1}})$ collapses. Then maps of the corresponding spectral sequences are induced by $\gamma\colon MSp^{\Delta_i^{n-1}}\longrightarrow MSp^{\Delta_i^n}$ and $\delta\colon MSp^{\Delta_i^n}\longrightarrow S^{2^{n+2}-2}MSp^{\Delta_i^{n-1}}$, which on the level of the E^2 terms yield an exact sequence

$$\ldots\longrightarrow H_*(MSp^{\Delta_i^{n-1}};\mathbb{Z})\otimes MSp_*^{\Delta_i^{n-1}}\xrightarrow{\gamma_*}H_*(MSp^{\Delta_i^n};\mathbb{Z})\otimes MSp_*^{\Delta_i^n}\xrightarrow{\delta_*}$$

$$\longrightarrow H_*(MSp^{\Delta_i^{n-1}};\mathbb{Z})\otimes MSp_{*-2^{n+2}+2}^{\Delta_i^{n-1}}\longrightarrow\ldots.$$

First, we prove that the spectral sequence for $MSp_*^{\Delta_i^n}(MSp^{\Delta_i^{n-1}})$ collapses. For suppose it does not, and that the first nontrivial differential is d_r. This differential is a homomorphism of modules over $MSp_*^{\Delta_i^n}$, so there exists an element $x\in H^*(MSp^{\Delta_i^{n-1}};\mathbb{Z})$ such that $d_r(x\otimes 1)\neq 0$, where $x\otimes 1\in E^r_{q,0}$. The image of $E^r_{*,0}$ under the homomorphism δ_* is zero because the spectrum $S^{2^{n+2}-2}MSp^{\Delta_i^{n-1}}$ is $(2^{n+2}-3)$-connected. Hence $\delta_*(d_r(x\otimes 1))=0$ and so there exists an element $y\in E^r_{q-r+1,r}(n-1,n-1)$, the E^r term of the spectral sequence for $MSp_*^{\Delta_i^{n-1}}(MSp^{\Delta_i^{n-1}})$, such that $\gamma_*(y)=d_r(x\otimes 1)$.

Let us consider the element $x\otimes 1$, belonging to $E^r_{q,0}(n-1,n-1)$. In this spectral sequence we deduce that $d_r(x\otimes 1)=y+y'\neq 0$, (where y' is such that $\gamma_*(y')=0$), which contradicts the induction hypothesis.

Now we prove that the spectral sequence for $MSp_*^{\Delta_i^n}(MSp^{\Delta_i^n})$, whose E_2 term is isomorphic to $H_*(MSp^{\Delta_i^n};\mathbb{Z})\otimes MSp^{\Delta_i^n}$, collapses. Let σ be an element in $H_{2^{n+2}-2}(MSp^{\Delta_i^n};\mathbb{Z})$ whose image under δ_* is the unit of $H_*(MSp^{\Delta_i^{n-1}};\mathbb{Z})$. Every element of $H_*(MSp^{\Delta_i^n};\mathbb{Z})$ can be written in the form $a+b\sigma$, where $a,b\in\gamma_*(H_*(MSp^{\Delta_i^{n-1}};\mathbb{Z}))$. In $E^2_{p,q}(n,n-1)$ there are no elements of finite order if $q<2^{n+3}-3$, so $d_r(\sigma\otimes 1)=0$ for all r. The morphisms γ and δ are morphisms of module spectra over $MSp^{\Delta_i^{n-1}}$. We thus obtain the formula $d_r((a+b\sigma)\otimes 1)=0$ using the multiplicative properties of the Atiyah-Hirzebruch spectral sequence.

Corollary 1. The Atiyah-Hirzebruch spectral sequence for $(MSp^{\Delta_i^n})^*(MSp^{\Delta_i^n})$ collapses.

Proof. This follows from the duality between the homology and cohomology spectral sequences.

Corollary 2. The Atiyah-Hirzebruch spectral sequence for $(MSp^{\Delta_i})^*(MSp^{\Delta_i})$ collapses.

3 Splittings and the BiP spectra

Now we describe how to obtain more precise information about the ring $MSp_*^{\Delta_i}$. For this purpose we use the spectral sequence of the singularity ϕ_{2^i-1} (see [13,5]). If we take the theory $MSp_*^{\Delta_i}(\ )$ and add one more singularity ϕ_{2^i-1} we obtain the theory $MSp_*^{\Sigma}(\ )$, whose coefficient ring is isomorphic to the polynomial algebra

$$MSp_*^{\Sigma} \cong \mathbf{Z}[w_1,\dots,w_j,\dots,x_2,\dots,x_k,\dots],$$

where $j = 1,2,\dots$, $k \neq 2^l - 1$ and $\deg w_j = 2(2^j - 1)$, $\deg x_k = 4k$. The E_1 term of the spectral sequence is the polynomial algebra $MSp_*^{\Sigma}[U_{i+1}]$, where

$$\deg U_{i+1} = (1, 2^{i+2} - 3),\quad \deg w_j = (0, 2(2^j - 1)),\quad \deg x_k = (0, 4k).$$

If we filter this term properly we obtain the initial term of the m.a.s.s., and using the following formulas for the first differential

$$d_1(h_k) = \begin{cases} 0, & k \neq i+1, \\ h_0 U_k, & k = i+1, \end{cases}$$

$$d_1(c_l) = \begin{cases} 0, & l \neq 2^i + 2^{j-1} - 1, \\ h_j U_{i+1}, & l = 2^i + 2^{j-1} - 1,\ (j \neq i+1) \end{cases}$$

we obtain formulas for the first differential of the spectral sequence of the singularity:

$$d_1(U_{i+1}) = 0, \qquad d_1(w_k) = \begin{cases} 0, & k \neq i+1, \\ 2U_k, & k = i+1, \end{cases}$$

$$d_1(x_l) = \begin{cases} 0, & l \neq 2^i + 2^{j-1} - 1, \\ w_j U_{i+1}, & l = 2^i + 2^{j-1} - 1,\ (j \neq i+1). \end{cases}$$

Let P' be a manifold whose cobordism class $[P']$ in $MSp_*^{\Sigma_{i+1}}$, for $\Sigma_{i+1} = (\phi_0,\dots,\phi_{2^i-1})$, is the obstruction to the multiplicativity of the theory $MSp_*^{\Sigma_{i+1}}(\ )$. Since this theory is multiplicative [13], there exists a manifold Q with singularities of type Σ_{i+1} such that $\partial_0 Q = P'$. There are the following multiplicativity formulas for the first differential of the spectral sequence of the singularity:

$$d_1(ab) = d_1(a)b + (-1)^{(\deg a + \deg U_{i+1})\deg U_{i+1}} \frac{d_1(a)[\delta(Q)]d_1(b)}{U_{i+1}} + (-1)^{\deg a(\deg U_{i+1}+1)} a d_1(b),$$

where deg is the topological (second) degree of an element and δ is the Bockstein operator. In our case $\delta(Q)$ is the manifold with singularities of type $\Sigma_i = (\phi_0,\dots,\phi_{2^i-2})$ such that $\partial_0(\delta(Q)) = 2\phi_{2^i-1}$. So $[\delta(Q)]$ may be taken as w_{i+1}, because the only precondition is that

$$\partial W_{i+1} = 2\phi_{2^i-1},$$

where W_{i+1} represents the bordism class w_{i+1}. So we obtain the following formula

$$d_1(ab) = d_1(a)b + (-1)^{\deg a+1} \frac{d_1(a) w_{i+1} d_1(b)}{U_{i+1}} + a d_1(b).$$

It follows from this that w_{i+1}^2 and $y_{2^i+2^{j-1}-1} = x_{2^i+2^{j-1}-1}^2 - w_{i+1}w_jx_{2^i+2^{j-1}-1}$ are cycles of the differential d_1.

Let us denote by L the subring of E_1 generated by w_j, u_{i+1}, and $x_{2^i+2^{j-1}-1}$; then we have an isomorphism

$$E_2 \cong H(L,d_1)_*[x_k], \qquad k \neq 2^i + 2^{j-1} - 1, \quad j \neq i+1.$$

The spectrum MSp^Σ is a free free spectrum, hence from the E_2 term the spectral sequence of the singularity ϕ_{2^i-1} coincides with the Adams-Novikov spectral sequence, and $d_2 = 0$. From the fact that $\phi_{2^i-1}^3 = 0$ it follows that there exists an element $\xi \in E_3^{0,3\cdot2^{i+2}-8}$ such that

$$d_3(\xi) = \phi_{2^i-1}^3.$$

For $E_3^{0,*}$ in this dimension, we have $x_{3\cdot2^i-2}$ and $y_{2^i+2^{i-1}} = x_{2^i+2^{i-1}-1}^2 - w_iw_{i+1}x_{2^i+2^{i-1}-1}$ as the two generators.

We do not know the action of the differential d_3 exactly, so we denote by η the second generator (the first is ξ) such that $d_3(\eta) = 0$. We can choose other generators in E_3 so that they will be cycles for d_3. If we denote by L the subring of E_3 generated by ξ, and $H(L,d_1)$ except for η, then we obtain the isomorphism

$$E_\infty = E_4 \cong (H(\tilde{L},d_3))_*[x_k,\eta], \qquad k \neq 2^i + 2^{j-1} - 1, \quad j \neq i+1, \quad k \neq 3\cdot2^i - 2.$$

Obviously there are no extension problems, and if we eliminate the first grading s then we see that $E_\infty^* \cong MSp_*^{\Delta_i}$. It is not difficult to deduce that multiplication by $y_{2^i+2^{j-1}-1}$ for $j \neq i, i+1$, and by η or ξ^2, are monomorphisms $MSp_*^{\Delta_i} \longrightarrow MSp_*^{\Delta_i}$.

Let us denote the sequence of all these elements, and also x_k for $k \neq 2^i + 2^{j-1} - 1$, $k \neq 3\cdot2^i - 2$, $j \neq i+1$, by Ξ, and order it by dimension. We thus obtain a regular sequence in $MSp_*^{\Delta_i}$. In particular, multiplication by the first element of Ξ induces a map of spectra $\alpha\colon MSp^{\Delta_i} \longrightarrow MSp^{\Delta_i}$, whose mapping cone we denote by $MSp^{\Delta_i,1}$. After localisation we obtain the sequence

$$MSp_{(2)}^{\Delta_i} \xrightarrow{\alpha} MSp_{(2)}^{\Delta_i} \xrightarrow{\gamma^0} MSp_{(2)}^{\Delta_i,1} \tag{8}$$

which generates the commutative diagram

$$\begin{array}{ccccccc}
H_*(MSp_{(2)}^{\Delta_i};\mathbf{Z}_2) & \xrightarrow{\alpha_*} & H_*(MSp_{(2)}^{\Delta_i};\mathbf{Z}_2) & \xrightarrow{\gamma^0_*} & H_*(MSp_{(2)}^{\Delta_i,1};\mathbf{Z}_2) & \longrightarrow \\
\downarrow & & \downarrow & & \downarrow & \\
H_*(MSp_{(2)}^{\Delta_i};\mathbf{Q}) & \xrightarrow{\alpha_*} & H_*(MSp_{(2)}^{\Delta_i};\mathbf{Q}) & \xrightarrow{\gamma^0_*} & H_*(MSp_{(2)}^{\Delta_i,1};\mathbf{Q}) & \longrightarrow \\
\| & & \| & & \| & \\
\pi_*(MSp_{(2)}^{\Delta_i})\otimes\mathbf{Q} & \xrightarrow{\alpha_*} & \pi_*(MSp_{(2)}^{\Delta_i})\otimes\mathbf{Q} & \xrightarrow{\gamma^0_*} & \pi_*(MSp_{(2)}^{\Delta_i,1})\otimes\mathbf{Q} & \longrightarrow .
\end{array}$$

The induced map

$$\alpha_*\colon \pi_*(MSp_{(2)}^{\Delta_i})\otimes\mathbf{Q} \longrightarrow \pi_*(MSp_{(2)}^{\Delta_i})\otimes\mathbf{Q}$$

is a monomorphism, as is the canonical morphism $H_*(MSp_{(2)}^{\Delta_i};\mathbf{Z}_2) \to H_*(MSp_{(2)}^{\Delta_i};\mathbf{Q})$. Hence

$$\alpha_*\colon H_*(MSp_{(2)}^{\Delta_i};\mathbf{Z}_2) \longrightarrow H_*(MSp_{(2)}^{\Delta_i};\mathbf{Z}_2)$$

is also a monomorphism, and we have the short exact sequence

$$0 \longrightarrow H_*(MSp_{(2)}^{\Delta_i}; \mathbf{Z}_2) \xrightarrow{\alpha_*} H_*(MSp_{(2)}^{\Delta_i}; \mathbf{Z}_2) \xrightarrow{\gamma_*^0} H_*(MSp_{(2)}^{\Delta_i,1}; \mathbf{Z}_2) \longrightarrow 0.$$

The images of the elements of Ξ are generators of $H_*(MSp_{(2)}^{\Delta_i}; \mathbf{Z}_2)$ under the Hurewicz map, hence $H_*(MSp_{(2)}^{\Delta_i,1}; \mathbf{Z}_2)$ is free over $\mathbf{Z}_2$. It follows from this fact and the universal coefficient formula that α induces an epimorphism in cohomology

$$\alpha^*: H^*(MSp_{(2)}^{\Delta_i}; \mathbf{Z}_2) \longrightarrow H^*(MSp_{(2)}^{\Delta_i}; \mathbf{Z}_2).$$

Therefore α induces an epimorphism of E_2 terms of the the Atiyah-Hirzebruch spectral sequences for $(MSp_{(2)}^{\Delta_i})^*(MSp_{(2)}^{\Delta_i})$. Applying corollary 2 we see that $(MSp_{(2)}^{\Delta_i})^*(\alpha)$ is also an epimorphism, and so we have a short exact sequence

$$0 \longleftarrow (MSp_{(2)}^{\Delta_i})^*(MSp_{(2)}^{\Delta_i}) \longleftarrow (MSp_{(2)}^{\Delta_i})^*(MSp_{(2)}^{\Delta_i}) \longleftarrow (MSp_{(2)}^{\Delta_i})^*(MSp_{(2)}^{\Delta_i,1}) \longleftarrow 0.$$

Hence, there exists a map

$$\varphi: MSp_{(2)}^{\Delta_i} \longrightarrow MSp_{(2)}^{\Delta_i}$$

such that $\varphi\alpha = 1_{MSp_{(2)}^{\Delta_i}}$, and the sequence (8) shows that $MSp_{(2)}^{\Delta_i}$ is equivalent to a sum

$$MSp_{(2)}^{\Delta_i} \simeq MSp_{(2)}^{\Delta_i,1} \vee S^l MSp_{(2)}^{\Delta_i}.$$

Then there exists a map

$$\psi_1: MSp_{(2)}^{\Delta_i,1} \longrightarrow MSp_{(2)}^{\Delta_i}$$

such that $\gamma\psi_1 = 1_{MSp_{(2)}^{\Delta_i,1}}$, and a map

$$\mu^1: MSp_{(2)}^{\Delta_i} \wedge MSp_{(2)}^{\Delta_i,1} \longrightarrow MSp_{(2)}^{\Delta_i,1}$$

such that the diagram

$$\begin{array}{ccccc}
MSp_{(2)}^{\Delta_i} \wedge MSp_{(2)}^{\Delta_i} & \xrightarrow{1\wedge\alpha} & MSp_{(2)}^{\Delta_i} \wedge MSp_{(2)}^{\Delta_i} & \xrightarrow{1\wedge\gamma^0} & MSp_{(2)}^{\Delta_i} \wedge MSp_{(2)}^{\Delta_i,1} \\
\downarrow \mu & & \downarrow \mu & & \downarrow \mu \\
MSp_{(2)}^{\Delta_i} & \xrightarrow{\alpha} & MSp_{(2)}^{\Delta_i} & \xrightarrow{\gamma^0} & MSp_{(2)}^{\Delta_i,1}
\end{array}$$

is commutative. Therefore $\mu^1 = \gamma^0 \circ \mu \circ (1 \wedge \psi_1)$, and μ^1 is unique.

Analogously let

$$\mu^{1,1}: MSp_{(2)}^{\Delta_i,1} \wedge MSp_{(2)}^{\Delta_i,1} \longrightarrow MSp_{(2)}^{\Delta_i,1}$$

be such that $\mu^{1,1} = \mu^1 \circ (\psi_1 \wedge 1) = \gamma^0 \circ \mu \circ (\psi_1 \wedge \psi_1)$. Then $\mu^{1,1}$ is associative and commutative.

Taking the second element in Ξ and repeating the above procedure, we obtain a multiplicative spectrum $MSp_{(2)}^{\Delta_i,2}$, which is a direct summand in $MSp_{(2)}^{\Delta_i,1}$ (and so in $MSp_{(2)}^{\Delta_i}$), and a morphism of spectra

$$\gamma^1: MSp_{(2)}^{\Delta_i,1} \longrightarrow MSp_{(2)}^{\Delta_i,2}.$$

Using induction, we build a chain of morphisms

$$MSp_{(2)}^{\Delta_i} \longrightarrow MSp_{(2)}^{\Delta_i,1} \longrightarrow \ldots \longrightarrow MSp_{(2)}^{\Delta_i,k} \longrightarrow \ldots.$$

In this chain each spectrum is obtained from the previous one by attaching cells of increasing dimension. The direct limit of this diagram is a spectrum which we denote by BiP. Obviously it is multiplicative, and we have canonical maps

$$\gamma: MSp_{(2)}^{\Delta_i} \longrightarrow BiP, \qquad \psi: BiP \longrightarrow MSp_{(2)}^{\Delta_i},$$

where ψ is defined from the sequence of maps

$$\psi_1 \circ \ldots \circ \psi_n: MSp_{(2)}^{\Delta_i,n} \longrightarrow MSp_{(2)}^{\Delta_i}$$

for all n. Evidently, $\gamma\psi = 1_{BiP}$, and there exists a splitting

$$MSp_{(2)}^{\Delta_i} \simeq \bigvee_j S^{n_j} BiP,$$

where the n_j are the degrees of the elements from Ξ.

Using theorem 4 and the construction of the spectrum BiP we obtain an isomorphism of comodules over A_*:

$$H_*(BiP; \mathbb{Z}/2) \cong \mathbb{Z}/2[\xi_1^2, \ldots, \xi_{i-2}^2, \xi_{i-1}^4, \bar{\xi}_i^2, \ldots, \bar{\xi}_j^2, \ldots] \otimes E.$$

Here E is a $\mathbb{Z}/2$-module whose elements are primitive over A_* and whose generators we denote by e_n, of degree $4n$, where

$$n = 2^{m_1} + \ldots + 2^{m_\nu} - \nu \cdot 2^i - \nu, \qquad 0 \le m_1 < \ldots < m_\nu, \quad m_l \neq i, \quad \nu = 1, 2, \ldots,$$

or

$$n = 2^{i+1} + 2^i - 2 \qquad i > 0.$$

The summand with $n = 2^{i+1} + 2^i - 2$ arises from the nontriviality of the differential d_3 in dimension $4(2^{i+1} + 2^i - 2)$, which kills $\phi_{2^i-1}^3$.

If we turn to cohomology we obtain

$$H^*(BiP; \mathbb{Z}/2) \cong \bigoplus_n (A_2/A_2\overline{D}_i) \cdot S_n,$$

where the elements S_n are dual to e_n.

If $i = 0$ then our spectrum has the same homology as the spectrum BoP, constructed by David Pengelly in [10]. Using the methods of his paper (p.1116-1120) it is easy to prove that in this case BiP is equivalent to BoP.

We may now summarize all the above considerations in the following theorem.

Theorem 5. There exists an indecomposable multiplicative spectrum BiP, for each $i = 0, 1, \ldots$, such that

$$MSp_{(2)}^{\Delta_i} \simeq \bigvee_j S^{n_j} BiP$$

and

$$H_*(BiP; \mathbb{Z}/2) \cong \mathbb{Z}/2[\xi_1^2, \ldots, \xi_{i-2}^2, \xi_{i-1}^4, \bar{\xi}_i^2, \ldots, \bar{\xi}_j^2, \ldots] \otimes E.$$

as comodules over A_*.

4 References

1. J.F. Adams, Stable homotopy and generalized homology, University of Chicago Press, Chicago, 1974.
2. N.-A. Baas. On bordism theory of manifolds with singularities, Math. Scand. 33 (1973), 279-302.
3. N.-A. Baas and I. Madsen, On the realization of certain modules over the Steenrod algebra, Math. Scand. 31 (1972), 220-224.
4. B.I. Botvinnik, Multiplicativity in the cobordism theories of manifolds with singularities, Proc. of Inst. of Math. Novosibirsk 7 (1987), 44-61 (in Russian).
5. B.I. Botvinnik, V.V. Vershinin, Multiplicative properties of spectral sequence of singularities. Siberian Math. J. 28 (1987), No.4. 569-575.
6. B.I. Botvinnik, V.V. Vershinin and V.G. Gorbunov, Some applications of spectral sequences in cobordism theory, Preprint, Novosibirsk, 1986 (in Russian).
7. V.G. Gorbunov and N. Ray, Orientations of *Spin* bundles and symplectic cobordism, Preprint, Manchester University (1989).
8. O.K. Mironov, Existence of multiplicative structures in the theories of cobordism with singularities, Math. USSR Izvestiya 9 (1975).
9. S.P. Novikov, The methods of algebraic topology from the viewpoint of cobordism theory, Math. USSR Izvestiya 1 (1967).
10. D. Pengelly, The homotopy type of MSU, Amer. J. Math. 104 (1982), 1101-1123.
11. N. Ray, Indecomposable in Tors MSp_*, Topology, 10 (1971), 261-270.
12. R.E. Stong, Notes on cobordism theory, Princeton University Press, Princeton, N.J. 1968.
13. V.V. Vershinin, Symplectic cobordism with singularities, Math. USSR Izvestiya, 22 (1984), No.2., 211-226.
14. V.V. Vershinin, On the decomposition of certain spectra, Math. USSR Sbornik, 60 (1988), No.2., 283-290.
15. V.V. Vershinin and V.G. Gorbunov, Ray's elements as obstructions to orientability of symplectic cobordism, Sov. Math. Dokl. 32 (1985), 855-858.
16. V.V. Vershinin and V.G. Gorbunov, On a rectangle of spectral sequences, Proc. of Inst. of Math., Novosibirsk, 9 (1987), 41-59 (in Russian).

"Localization and the Sullivan Fixed Point Conjecture"

Amir H. Assadi[(1)]

University of Wisconsin (Madison, WI USA),
Max–Planck–Institut für Mathematik (Bonn, FRG)

Introduction:

Let K be a finite dimensional G–space, where G is a p–elementary abelian group, i.e. $G \cong (\mathbb{Z}/p\mathbb{Z})^n$. The Borel–Quillen–Hsiang localization theorem states that $H^*_G(K;\mathbb{F}_p) \longrightarrow H^*_G(K^G;\mathbb{F}_p)$ is an isomorphism modulo $H^*_G(\{\text{point}\};\mathbb{F}_p)$–torsion, where H^*_G is Borel's equivariant cohomology ([B] [Q] [Hw]). The above theorem is not true for infinite dimensional spaces in general. As we shall see below, the Sullivan conjecture implies that such a localization holds for infinite dimensional G–spaces $\text{Map}(E_G,K)$, where $\dim K < \infty$. Conversely, the main result of Section 2 proves that if the Borel–Quillen–Hsiang localization holds for $\text{Map}(E_G,X)$, then $E_G \times X$ is G–homotopy equivalent to $E_G \times K$ with $\dim K < \infty$. Here E_G is the usual universal contractible free G–space. This provides an answer to a problem posed in [A2]. This question and other problems of this nature arise naturally in the geometric and differential topological aspects of transformation groups of manifolds. In particular, at present most methods of constructing group actions on a given manifold yield only infinite dimensional free G–spaces. See [A1] [AB] [W] and their references.

While the localization theorem applies to p–elementary groups, and the Sullivan conjecture holds only for p–groups, we have formulated our results for all finite groups. The proof of the main topological results, (Theorem 2.4) is reduced to the case of cyclic groups of prime order using an inductive argument. The main tool which provides such a local–to–global passage is the

---

(1) Acknowledgement. Much stimulus for the question and the results of the present work was provided during my joint work [AB] and conversations with W. Browder, to whom I would like to express my gratitude.

It is a pleasure to thank G. Carlsson, H. Miller, and J. Lannes for conversations on their proofs of the Sullivan's fixed point conjecture.

The present version of this work was completed at the Max–Planck–Institut für Mathematik (Bonn). I would like to thank Professor F. Hirzebruch for his encouragement and support and the staff of the Institute for their hospitality during my stay at MPI.

Research partially supported by an NSF grant and the Max–Planck–Institut für Mathematik.

algebraic result (Theorem 1.1) of Section 1 which is a projectivity criterion for integral and modular representations occuring as the cohomology of certain G–spaces.

The proof of our converse of the localization theorem for $G = \mathbb{Z}/p\mathbb{Z}$ does not use the proof of the Sullivan conjecture, but merely a statement of this kind. Therefore, it seems appropriate to present the statement and proofs in a sufficiently flexible manner to accommodate the possible improvements. Since the Borel–Quillen Localization theorem is essentially of homological nature, so are the proofs of our theorems. Thus, "the quasicompletion functors" which are modeled homologically after Bousfield–Kan's completion functors will also work in the context of Section 2. This approach emphasizes those homological properties of these functors which are relevant for our purposes and how they are used in the course of the proof. To apply the converse to the localization theorem, one needs to develop computation tools. At present, Lannes' results in [L] are the best available for $G = (\mathbb{Z}/p\mathbb{Z})^n$. Such results in conjunction with our theorems yield more general results for finite groups which are not necessarily p–elementary abelian. In non–technical terms, let us mention one corollary:

Corollary: Let G be a finite group and let X be a free G–space. Then there exists a finite dimensional G–space K such that $E_G \times K$ is G–homotopy equivalent to X if and only if for each prime $p \mid |G|$, and a representative p–Sylow subgroup $G_p \subseteq G$, there exists a finite dimensional G_p–space $K(p)$ such that $E_{G_p} \times K(p)$ is G_p–homotopy equivalent to X.

An interesting feature of the localization theorem as pointed out by Quillen in [Q] is that it is valid for compact G–spaces even if they are infinite dimensional. This motivates the following.

Problem: Suppose $G = \mathbb{Z}/p\mathbb{Z}$ and X is a compact G–space. Does the Sullivan fixed point conjecture hold for X?

Section 1. Algebraic Preliminaries

Let G be a finite group, and let k be an algebraic closure of $\mathbb{F}_p$ = the field with p–elements. All modules are assumed to be finitely generated. A classical result of Rim [R] states that a $\mathbb{Z}G$–module M is $\mathbb{Z}G$–projective if and only if its restrictions $M \mid \mathbb{Z}P$ are $\mathbb{Z}P$–projective for all Sylow subgroups $P \subseteq G$. Chouinard has refined this result [Ch] by replacing the p–Sylow subgroups in Rim's theorem by (maximal) p–elementary abelian subgroups. Thus the projectivity of M is detected by all its restrictions to $M \mid \mathbb{Z}A$ for all p–elementary abelian $A \subseteq G$, i.e. $A \cong \mathbb{Z}_p \oplus \mathbb{Z}_p \oplus \dots \oplus \mathbb{Z}_p$. To decide the projectivity of $M \mid \mathbb{Z}A$, it suffices to consider the kA–module $M \otimes k$. Thus, let A be a p–elementary abelian group of rank n and with $\{e_1,\dots,e_n\}$ a set of generators, and let I be the augmentation ideal. It is possible to choose a k–subspace $L \subset I$ with $\dim_k L = r$ and such that $I \cong L \oplus I^2$ as k–vector spaces. Then L generates kA as a k–algebra and for each $\lambda \in L$, $(\lambda+1)^p = 1$. The elements $\sigma \in kA$ of the form $\sigma = \lambda+1$, $\lambda \in L$ (for such an L) are called "shifted units" and the cyclic subgroups $S \equiv \langle\sigma\rangle$ of order p are called "shifted cyclic subgroup". (See [Cj]). In [D] Dade has proved that a given kA–module M is kA–projective (hence kA–free since kA is local) if and only if $M | kS$ is kS–projective for all such shifted cyclic subgroups of kA. (Note that almost all shifted cyclic subgroups of kA do not come from cyclic subgroups of A.) We will fix L for the rest of the following discussion.

In [A2], the author proved the following projectivity criterion which will be used in Section 2.

<u>1.1 Theorem</u>. Suppose X is a connected G–space such that for each maximal p–elementary abelian subgroup $A \subseteq G$, the $H^*(-;k)$–spectral sequence $X \longrightarrow E_A \times_A X \longrightarrow BA$ collapses. Then $\bigoplus_{i>0} H^i(X;k)$ is a projective kG–module if and only if it is projective as a kC–module for every subgroup $C \subseteq G$ of order p. Similarly, $\bigoplus_{i>0} H^i(X;\mathbb{Z})$ is a projective $\mathbb{Z}G$–module if and only if it is $\mathbb{Z}C$–projective for all cyclic subgroups C of prime order.

Note that if X is a Moore space with G–action and $X^G \neq \emptyset$, then the conditions of Theorem 1.1 are satisfied, and we get a projectivity criterion for the cohomology of Moore spaces with G–action.

<u>Section 2. A Converse to the Localization Theorem</u>

By a "converse" we mean the following. Given a finite dimensional G–space X, the Borel–Quillen–Hsiang theorem tells us how to get information about the cohomology of the fixed point sets for G = p–elementary abelian. Now suppose instead of the finite dimensional G–space X, we are given only the Borel construction $\pi : Y \longrightarrow BG$ (or equivalently the corresponding infinite dimensional free G–space $\tilde{Y} = \pi^*(E_G)$) and the localized equivariant cohomology information of the type in the conclusion of Borel–Quillen–Hsiang theorem. Then we can recover a finite dimensional G–space X whose Borel construction $E_G \times_G X \longrightarrow BG$ is "the same" as $Y \longrightarrow BG$ (in the sense of fibre homotopy equivalence). The key to such a construction is a statement of the type of Sullivan's fixed point conjecture.

In this section, we will use "completion functors", homologically modeled after Bousfield–Kan's completion functors [BK]. For simplicity of exposition, we will assume that our functors are defined for all topological spaces; however, such functors may have smaller domains of definition in the course of applications, in which case, the appropriate modification of the following properties is necessary. Recall that the Bousfield–Kan $\mathbb{F}_p$–completion functor satisfies the following:

(C0) $\quad R$ is a functor from the category of topological spaces to itself.

(C1) $\quad R$ commutes with arbitrary disjoint unions and finite products.

(C2) There is a coefficient ring R associated to R such that if $f : X \longrightarrow Y$ induces an isomorphism $F_* : H_*(X;R) \longrightarrow H_*(Y;R)$, then $R(f)_* : H_*(R(X);R) \longrightarrow H_*(R(Y);R)$ is also an isomorphism.

(C3) There is a full subcategory of topological spaces $\mathrm{Top}(R)$ (associated to R) and there is a natural transformation τ : identity $\longrightarrow R$ which satisfy:

i) If $\pi_1(X) = 0$, then $X \in \mathrm{Top}(R)$.

ii) If $X \in \mathrm{Top}(R)$ then $R(X) \in \mathrm{Top}(R)$.

iii) For all $X \in \mathrm{Top}(R)$, the map $\tau(X) : X \longrightarrow R(X)$ induces an $H_*(-;R)$–isomorphism.

<u>Definition</u> (i) A functor R satisfying (C0)–(C3) above is called a "quasicompletion functor".

(ii) Let G be a category of groups and R be a quasicompletion functor. We say that <u>R is adapted to G</u> if the following is satisfied. Here E is a universal contractible G–space.

(C4) For all $G \in G$ and all finite dimensional G–spaces X such that $H_*(X;R)$ and $H_*(X^G;R)$ are finitely generated, the map of constants $X^G \longrightarrow \mathrm{Map}_G(E,X)$ induces an isomorphism $H_*(R(X^G);R) \longrightarrow H_*(\mathrm{Map}_G(E,R(X));R)$.

When $R = \mathbb{F}_p$ and R_p is the Bousfield–Kan [BK] $\mathbb{F}_p$–completion, then R_p is adapted to the category of all finite p–groups by the validity of the Sullivan's conjecture mentioned above. For the Bousfield–Kan $\mathrm{Top}(R)$ consists of $\mathbb{F}_p$–good spaces [BK].

<u>Remarks</u>. (a) The condition on $\pi_1(X)$ in (C3) may be weakened to $\bar{H}_*(\pi_1(X);R) = 0$, or even $H_1(X;R) = 0$ in applications.

(b) The condition (C4) is essentially the Sullivan fixed point conjecture which has been proved for p–group independently by G. Carlsson, J. Lannes, and H. Miller. The important special case where G acts trivially on X was done by H. Miller in [M]. See also [C] [L].

(c) Sullivan had stated his conjecture for p–groups. It is worth noticing that the Sullivan fixed point conjecture is not true for G–spaces where G is not a p–group. In [A3] the author has shown that for any finite group G which is not a p–group there exists a fixed–point free G–action on $\mathbb{R}^n$. These easily provide counterexamples. However, one may still ask the following:

Under which circumstances for a finite G–CW complex X the existence of an equivariant map $E_G \longrightarrow X$ implies that $X^G \neq \emptyset$?

We consider first the case $G = \mathbb{Z}/p\mathbb{Z}$. This case is sufficient for many applications.

<u>2.1. Theorem</u>. Let $G = \mathbb{Z}/p\mathbb{Z}$, and let X be a free G–space such that $\pi_1(X) = 0$ and $H_*(X)$ is finitely generated. Let R be the Bousfield–Kan $\mathbb{F}_p$–completion (or any quasicompletion functor adapted to $G = \{\mathbb{Z}/p\mathbb{Z}\}$ whose coefficient is $\mathbb{F}_p$). The conditions (A0)–(A2) together are necessary and sufficient for the existence of a finite dimensional G–complex Y such that: (i) $Y^G \in \mathrm{Top}(R)$, (ii) $H_*(Y^G)$ is finitely generated, and (iii) $E \times Y$ and X are G–homotopy

equivalent.

(A0) $\mathrm{Map}_G(E,R(X))$ belongs to the image of R up to $\mathbb{F}_p$–homology isomorphism.

(A1) There exists a finite dimensional complex $F \in \mathrm{Top}(R)$ with $H_*(F)$ finitely generated, and a map $\eta : F \longrightarrow \mathrm{Map}_G(E,X)$ such that "the induced map" $\hat{\eta} : R(F) \longrightarrow \mathrm{Map}_G(E,R(X))$ induces $H_*(-;\mathbb{F}_p)$–isomorphism. (See the remark below.)

(A2) The map $\lambda : \mathrm{Map}_G(E,R(X)) \longrightarrow \mathrm{Map}(E,R(X))$ induces a Borel–Quillen localized isomorphism in $H_G^*(-;\mathbb{F}_p)$–theory.

Remarks:

1. "The induced map" $\hat{\eta}$ is obtained as follows. The map η of (A1) has an adjoint map $\bar{\eta} : E \times F \longrightarrow X$. Then $\hat{\eta}$ is the adjoint map of the composition $E \times R(F) \longrightarrow R(E) \times R(F) \cong R(E \times F) \xrightarrow{R(\bar{\eta})} R(X)$.

2. Let us observe that for $G = \mathbb{Z}/p\mathbb{Z}$ and $t \in H^2(G;\mathbb{Z}/p\mathbb{Z})$ nilpotent for $p = \text{odd}$ or $t \in H^1(\mathbb{Z}/2\mathbb{Z};\mathbb{F}_2)$ for $p = 2$, the Tate cohomology $\hat{H}^*(G;\mathbb{F}_p)$ coincides with $H^*(G;\mathbb{F}_p)[\frac{1}{t}]$. When $X = \text{point}$, the localized equivariant cohomology reads: $H_G^*(\text{point};\mathbb{F}_p)[\frac{1}{t}] \cong \hat{H}^*(G;\mathbb{F}_p)$. Thus, we may denote the functor $H_G^*(-;\mathbb{F}_p)[\frac{1}{t}]$ by $\hat{H}_G^*(-;\mathbb{F}_p)$ for short and suggest the properties of Tate cohomology as well.

Proof. Suppose such a Y exists, and let $F = Y^G$. Consider "The map of constants" $F \longrightarrow \mathrm{Map}_G(E,Y)$ which becomes a homology equivalence upon applying R (by virtue of condition (C4)): $H_*(R(F);\mathbb{F}_p) \cong H_*(\mathrm{Map}_G(E,R(Y);\mathbb{F}_p))$. For simplicity of notation, H_* denotes homology with $\mathbb{F}_p$–coefficients throughout this proof.

Since $\mathrm{Map}_G(E,E)$ has the homotopy type of a point, one has $\mathrm{Map}_G(E,X) \simeq \mathrm{Map}_G(E,E \times Y) \simeq \mathrm{Map}_G(E,E) \times \mathrm{Map}_G(E,Y)$ on the level of path components. Thus one has the map $F \longrightarrow \mathrm{Map}_G(E,X)$ such that the composition below induces an H_*–equivalence:

$$R(F) \longrightarrow \mathrm{Map}_G(E,R(Y)) \longrightarrow \mathrm{Map}_G(E,E) \times \mathrm{Map}_G(E,R(Y)) \longrightarrow$$

$$\longrightarrow \mathrm{Map}_G(E,E \times R(Y) \longrightarrow \mathrm{Map}_G(E,R(E \times Y)) \longrightarrow \mathrm{Map}_G(E,R(X)) .$$

(Note that the homotopy fixed–point set is an invariant of G–maps which are non–equivariant homotopy equivalences.) Hence $R(F) \longrightarrow \mathrm{Map}_G(E,R(X))$ is also an H_*–equivalence and conditions (A0) and (A1) are seen to be necessary. To see the necessity of (A2), consider the diagram:

$$
\begin{array}{ccccccccc}
R(Y^G) & \xrightarrow{h_1} & Y^G & \longrightarrow & Y & \longrightarrow & \mathrm{Map}(E,E\times Y) & \longrightarrow & \mathrm{Map}(E,X) \\
\downarrow{\scriptstyle \sigma} & & & & & & & & \downarrow \\
\mathrm{Map}_G(E,R(Y)) & & \longrightarrow & & \mathrm{Map}_G(E,R(X)) & & \xrightarrow{\lambda} & & \mathrm{Map}(E,R(X)) \\
& \searrow & & \nearrow & & & & & \\
& & \mathrm{Map}_G(E,R(E\times Y)) & & & & & &
\end{array}
$$

(Diagram 1)

In the above, all maps which are not labeled induce H_*–equivalence, since various spaces involved belong to $\mathrm{Top}(R)$ and $R = \mathbb{F}_p$ by the hypothesis. σ is an H_*–equivalence since R satisfies (C4). h_1 induces $\hat{H}_G(-;\mathbb{F}_p)$–isomorphism by the Borel–Quillen localization theorem ([Q] [Hw]). It follows that λ also induces such an isomorphism, and condition (A2) is also necessary.

We sketch now the proof that the above conditions are also sufficient. Consider the composition $f_0 \equiv \iota \cdot \eta$ where $\eta : F \longrightarrow \mathrm{Map}_G(E,X)$, and $\iota : \mathrm{Map}_G(E,X) \longrightarrow \mathrm{Map}(E,X)$ are given by (A1) and the inclusion, respectively. The strategy is to add (finitely many) free G–cells to F so that the map f_0 is extended to a highly connected map. One obtains the diagram below:

$$
\begin{array}{ccc}
F & \xrightarrow{\eta} & \mathrm{Map}_G(E,X) \\
{\scriptstyle j}\downarrow & & \downarrow{\scriptstyle \iota} \\
Y_0 & \longrightarrow & \mathrm{Map}(E,X)
\end{array}
$$

(Diagram 2)

Here, $Y_0^G = F$ and we may assume that the cofibre of f is a Moore space with finitely generated homology, since $H_*(X)$ is assumed to be finitely generated. At this point, it may be helpful for the reader to consider the special case where $\pi_1(F) = 0$, where (C4) is true without any completion (according to the validity of the Sullivan conjecture in this case.) In this case the proof is much simpler technically since f is readily seen to satisfy the following claim.

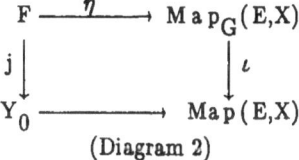

We claim that f induces an isomorphism for the functor $\hat{H}_G(-;\mathbb{F}_p)$ so that the reduced homology of the cofibre of f is cohomologically trivial as a G–module. From this claim, the proof of the theorem is completed as follows. Let C_f be the cofibre of f, and $\bar{H}^*(C_f)$ be its reduced homology. Then $\hat{H}(G;\bar{H}^*(C_f)) \cong \hat{H}_G(C_f,\{\text{point}\})$ vanishes ($\mathbb{F}_p$–coefficients), which is sufficient for the cohomological triviality of $H^*(C_f;\mathbb{F}_p)$ for $G = \mathbb{Z}/p\mathbb{Z}$. Since $H^*(C_f;\mathbb{Z})$ is finitely generated and we may assume it to be $\mathbb{Z}$–free as well, it follows that it is $\mathbb{Z}G$–projective ([R]). By standard arguments (e.g. [A3] Chapter I and II) we may add free G–cells to Y_0 and extend f to a homological equivalence, which we continue to call $f : Y \longrightarrow \mathrm{Map}(E,X)$. Since X and Y are 1–connected, this yields a homotopy equivalence. The evaluation map $\epsilon : E \times Map(E,X) \longrightarrow X$, $\epsilon(f,e) = f(e)$, $e \in E$, is equivariant and a homotopy equivalence. Hence it is a G–homotopy equivalence since both spaces are G–free. (Note that the action on $E \times Map(E,X)$ is the diagonal action and the action on $Map(E,X)$ is by conjugation, i.e. $f^g(\chi) \equiv gf(g^{-1}\chi)$.) This finishes the proof and it remains to establish the claim.

The proof of the claim is based on studying a number of commutative diagrams:

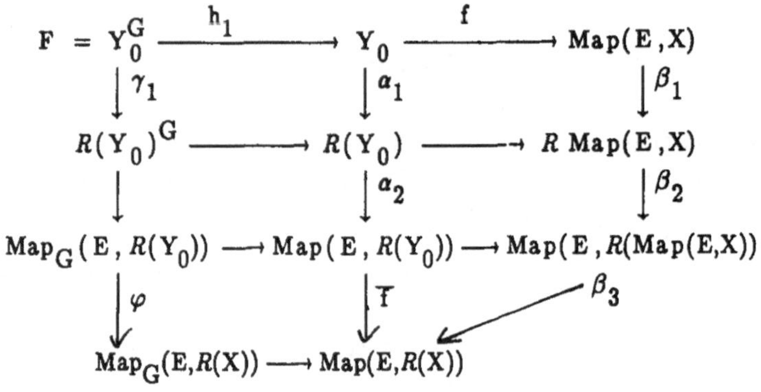

(Diagram 3)

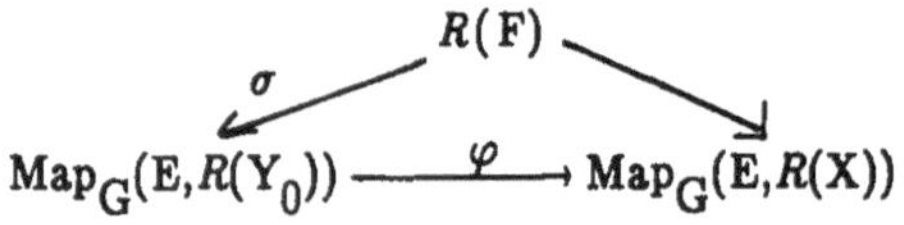

(Diagram 4)

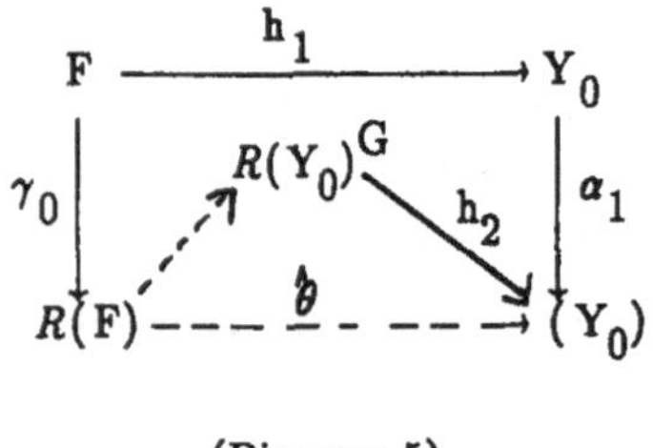

(Diagram 5)

In diagram 3 we have the following $\hat{H}_G$–isomorphisms: h_1, α_1, β_1, α_2, β_2 and β_3; and in diagram 4, we get σ and φ induce H_*–isomorphisms. In diagram 5, the dotted arrows exist by the functoriality of R and γ_0 induces an H_*–isomorphism. It follows from spectral sequence arguments that $\hat{\theta}$ induces a Borel–Quillen localized isomorphism. Combining these with a study of the diagram:

$$
\begin{array}{ccccc}
 & & \mathrm{Map}_G(E,\mathcal{R}(Y_0)) & \xrightarrow{\;h_3\;} & \mathrm{Map}(E,\mathcal{R}(Y_0)) \\
 & {}^{\alpha}\nearrow & \downarrow{\scriptstyle\varphi} & & \downarrow{\scriptstyle\hat{f}} \\
R(F) & & & & \\
 & {}_{\hat{\eta}}\searrow & & & \\
 & & \mathrm{Map}_G(E,\mathcal{R}(X)) & \xrightarrow{\;\lambda\;} & \mathrm{Map}(E,\mathcal{R}(X))
\end{array}
$$

we finally conclude that $\hat{f}$ induces a Borel–Quillen localized isomorphism. This is used in conjunction with a spectral sequence argument to show that in the diagram below f induces a Borel–Quillen localized isomorphism:

$$
\begin{array}{ccc}
Y_0 & \xrightarrow{\;f\;} & \mathrm{Map}(E,X) \\
{\scriptstyle\alpha_3}\downarrow & & \downarrow{\scriptstyle\beta_4} \\
\mathrm{Map}(E,\mathcal{R}(Y_0)) & \xrightarrow{\;\hat{f}\;} & \mathrm{Map}(E,\mathcal{R}(X))
\end{array}
$$

Here $\beta_4 = \beta_3 \cdot \beta_2 \cdot \beta_1$. Thus, $\hat{H}_G(C_f,\{\text{Point}\}) = 0$ and the claim is established.

□

For the case of finite complexes, we find obstructions in $\tilde{K}_0(\mathbb{Z}G)$ which are algebraic in nature and may be treated separately from the homotopy–theoretic side of such problems.

<u>2.2. Theorem</u>. Given G,X, and $\mathcal{R}$ as in Theorem 2.1, suppose that (A0)–(A2) are satisfied, and in (A1) F is a finite complex with similar properties. Then there is an obstruction $w(X) \in \tilde{K}_0'(\mathbb{Z}G)$ such that $w(X) = 0$ if and only if there exists a finite G–complex Y such that $E \times Y$ and X are G–homotopy equivalent and $Y^G = F$. The obstruction w(X) does not depend on F as long as F satisfies (A1). ($\tilde{K}_0'(\mathbb{Z}G)$ is a certain subquotient of $\tilde{K}_0(\mathbb{Z}G)$ in general).

<u>2.3. Remark</u>. It is not always true that w(X) is independent of F for any finite group G. For $G = \mathbb{Z}/p\mathbb{Z}$, this is a consequence of the triviality of the Swan homomorphism $\sigma_G : (\mathbb{Z}/p\mathbb{Z})^x \longrightarrow \tilde{K}_0(\mathbb{Z}G)$. Thus, in this case if such a finite Y exists, and if F′ is any finite complex which admits an $\mathbb{F}_p$–homotopy equivalence $F' \longrightarrow F$, then there exists a finite G–complex Y′ such that $(Y')^G = F'$ and $E \times Y'$ is G–homotopy equivalent to X as well (cf. [A3]).

<u>2.4. Example</u>. Let $\mathcal{R}$ be the Bousfield–Kan $\mathbb{F}_{23}$–completion functor, and let G be the cyclic

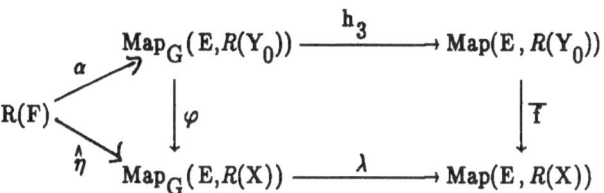

group of order 23 acting on $M \cong \mathbb{Z}/47\mathbb{Z}$ via the inclusion of $G \subset \mathrm{Aut}(\mathbb{Z}/47\mathbb{Z}) \cong \mathbb{Z}/46\mathbb{Z}$. The calculations of Swan shows that there is no finite G–complex X with $\bar{H}_*(X) \cong M$ as a $\mathbb{Z}G$–module. However, there are finite dimensional G–complexes Y such that $\bar{H}_*(Y) \cong M$ as $\mathbb{Z}G$–modules. For any such Y $\bar{H}(Y;\mathbb{F}_{23}) = 0$ and $Y^G \in \mathrm{Top}(R)$. However, it is not possible to find a finite G–complex K such that $E \times K$ and $E \times Y$ are G–homotopy equivalent.

Next, we briefly outline how we can generalize 2.1 from $\mathbb{Z}/p\mathbb{Z}$ to general finite groups. Let G be a finite group and p be any prime dividing order of G . We define the following sets of subgroups of G :

$$P_p(G) \equiv \{P \subseteq G \,\big|\, |P| \text{ is a p–power }\},\ P(G) \equiv \bigcup_{p\,|\,|G|} P_p(G)\,;$$

$$A_p(G) \equiv \{(P_1,P_2)\,|\,P_i \in P_p(G),\ i = 1,2;\ P_2 \lhd P_1 \text{ and}$$

$$P_1/P_2 \cong (\mathbb{Z}/p\mathbb{Z})^r \text{ for some } r \geq 0\},\ A(G) \equiv \bigcup_{p\,|\,|G|} A_p(G)\ .$$

The following proposition provides us with the necessary conditions for "finiteness" of G–spaces in the appropriate context. A similar result with appropriate modifications hold for finitely dominated G–spaces in the equivariant sense. As pointed out earlier, the recent proofs of the equivariant Sullivan conjecture show that the quasicompletion functors which are used in the following proposition form a nonempty set !

2.5. Proposition. Suppose that G is a finite group and p is any prime dividing $|G|$, and let R_p be the Bousfield–Kan completion or any quasi–completion functor whose associated coefficients is $\mathbb{F}_p$. Assume that Y is a finite dimensional G–space such that $H_*(Y^P;\mathbb{F}_p)$ is finitely generated for each $P \in P_p(G)$ and Y^P belong to $\mathrm{Top}(R_p)$. Let X be a free G–space such that $E_G \times Y$ and X are G–homotopy equivalent. Then for each $P \in P(G)$ and each $(P_1,P_2) \in A_p(G)$ the following hold:

(B0) All spaces $\mathrm{Map}_P(E,R_p(X))$ are $H_*(-;\mathbb{F}_p)$–equivalent to spaces in the image of R_p .

(B1) There exist finite dimensional complexes $F(P) \in \mathrm{Top}(R_p)$ with finitely generated $H_*(F(P);\mathbb{F}_p)$ and maps $\eta(P) : F(P) \longrightarrow \mathrm{Map}_P(E,X)$ such that $\hat{\eta} : R_p(F(P)) \longrightarrow \mathrm{Map}_P(E,R_p(X))$ is an $H_*(-;\mathbb{F}_p)$–equivalence.

(B2) The map $\lambda(P_1,P_2) : \mathrm{Map}_{P_1}(E,R_p(X)) \longrightarrow \mathrm{Map}_{P_2}(E,R_p(X))$ induces a Borel–Quillen localized isomorphism for the group $A \equiv P_1/P_2$.

Proof. Let $F(P) = Y^P$ and $\eta(P)$ as in Theorem 2.1 (where $F(P)$ and $\eta(P)$ are denoted by F and η respectively). Since the first two conditions are consequences of the properties of quasi–completion functors as in Theorem 2.1, we will justify the last condition only. Consider the following commutative diagram.

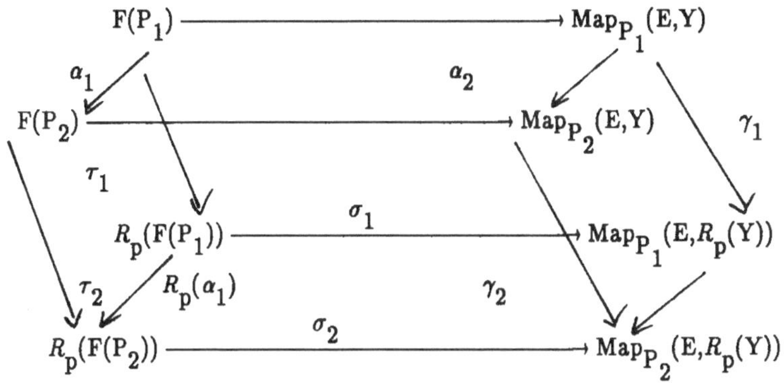

The maps $F(P_i) \longrightarrow Map_{P_i}(E,Y)$ are given by the maps of constants $Y^{P_i} \longrightarrow Map_{P_i}(E,Y)$, and the maps τ_i and σ_i induce H_*–isomorphisms, where H_* denotes homology with $\mathbb{F}_p$–coefficients as in 1.1. Moreover, since $\dim F(P_2) < \infty$, α_1 induces a Borel–Quillen localized isomorphism in $\hat{H}_A$–theory, where $A \equiv P_1/P_2$. Thus $\hat{H}_A(F(P_2),F(P_1)) = 0$. Comparison of the Serre spectal sequences of the Borel constructions of various spaces involved show that the map

$$Map_{P_1}(E,R_p(Y)) \longrightarrow Map_{P_2}(E,R_p(Y))$$

induces also a Borel–Quillen localized isomorphism as well. Since $E \times Y$ and X are G–homotopy equivalent, the map

$$\lambda : Map_{P_1}(E,R_p(X)) \longrightarrow Map_{P_2}(E,R_p(X))$$

induces a Borel–Quillen localized isomorphism, as in Theorem 2.1.

□

2.6. Theorem. Let G, p, and R_p be as in Proposition 2.1 above. Let X be a free G–space such that the conditions B(0)–B(2) of Proposition 2.1 are satisfied. Then there exists a finite dimensional G–space Y such that $H_*(Y^P;\mathbb{F}_p)$ are finitely generated, $Y^P \in Top(R_p)$ for each $P \in P_p(G)$, and $E \times Y$ and X are G–homotopy equivalent. If the complexes $F(P)$ are taken to be finite complexes, then there exists an obstruction $w(X) \in \tilde{K}_0'(\mathbb{Z}G)$ such that $w(X) = 0$ if and only if Y is G–homotopy equivalent to a finite G–complex.

Outline of proof: In order to prove that such a Y exists, we actually proceed to construct the p–singular set of Y, i.e. $S_p(Y) = \bigcup_{1 \neq P \in P_p(G)} Y^P$ for each $p \mid |G|$, in order to obtain maps $h_p : S_p(Y) \longrightarrow Map(E,X)$ which are equivariant and such that the induced maps $\hat{h}_p^P : R_p(S_p(Y)^P) \longrightarrow Map_P(E,R_p(X))$ induce H_*–isomorphisms for each $P \in P_p(G)$, where $H_* = H_*(-;\mathbb{F}_p)$ as before. By adding free G–cells to $\bigcup_p S_p(Y)$ we make the map $\bigcup_p h_p : \bigcup_p S_p(Y) \longrightarrow Map(E,X)$ highly connected and we obtain $f : Y_0 \longrightarrow Map(E,X)$ so that the cofibre C_f of f is a Moore space, and $S_p(Y_0) \equiv S_p(Y)$. Then we try to show that $\tilde{H}_*(C_f;\mathbb{Z})$ is $\mathbb{Z}G$–projective. In cases where we deal with finite complexes, the class $[\tilde{H}_*(C_f)] \in \tilde{K}_0(\mathbb{Z}G)$ will represent the finiteness obstruction w(X) which will be only well–defined up to ambiguity arising from different choices of $S_p(Y)$ in the course of this construction. This leads, then, to a well–defined obstruction, denoted again by w(X) (by abuse of notation) in a subquotient of $\tilde{K}_0(\mathbb{Z}G)$.

In order to show that $\tilde{H}_*(C_f;\mathbb{Z})$ is $\mathbb{Z}G$–projective, we use the projectivity criterion Theorem 1.1 to reduce the problem to showing that $H_*(C_f;\mathbb{Z}) \mid \mathbb{Z}C$ is $\mathbb{Z}C$–projective for each $C \subseteq G$, $|C| = p$. But in this case, we are in the situation of Theorem 2.1, since by construction $R_p(Y_0^C) \longrightarrow Map_C(E,R_p(X))$ induces a homology isomorphism, and other conditions are also satisfied, as one can check from the hypothesis. Hence the proof of Theorem 2.1 shows that $H_*(C_f) \mid \mathbb{Z}C$ is $\mathbb{Z}C$–projective for any such C.

Fix a $K \in P_p(G)$. It remains to show how to construct $S_p(Y)^K$. We proceed by induction on the lattice of p–subgroups $P_p(G)$. Suppose that $h^P : S_p(Y)^P \longrightarrow Map(E,X)$ is constructed for all subgroups P such that $K \subsetneq P$, $h_p^P : R_p(S_p(Y)^P) \longrightarrow Map_P(E,R_p(X))$ induces an H_*–isomorphism. Let L denote $S_p(Y)$ for short. We add free $W(K) \equiv N(K)/K$ cells to L^K and extend it to G–orbits (which are added to L in the usual fashion) so that the map $\alpha : L_0 \longrightarrow Map(E,X)$ in this way satisfies the following: the cofibre of $\alpha(K) : L_0^K \longrightarrow Map_K(E,K)$, call it $C(\alpha(K))$ has homology (i.e. $\tilde{H}_*(-;\mathbb{F}_p)$) only in one dimension, i.e. it is a $H_*(-;\mathbb{F}_p)$–Moore space. Now $\tilde{H}_*(C(\alpha(K)))$ is an $\mathbb{F}_p(W)$–module and we claim that it is $\mathbb{F}_p(W)$–free. Using the modular version of the projectivity criterion (Theorem 1.1), we need to check this for each cyclic subgroup of order p, say $C \subseteq W$, $|C| = p$. We have the exact sequence: $1 \longrightarrow K \longrightarrow K' \longrightarrow C \longrightarrow 1$ where $|K'| = p \cdot |K|$. Hence, by the induction hypothesis $R_p(L_0^{K'}) \longrightarrow Map_{K'}(E,R)_p(X))$ induces an H_*–isomorphism. Translating this into W–actions, we have $(L_0^K)^{C'} = L_0^{K'}$ and $R_p((L_0^K)^C \longrightarrow Map_C(E,R_p(X)^K)$ is a homology isomorphism. On the other hand, by studying the diagram

$$\begin{array}{ccc}
L_0^K & \xrightarrow{\alpha(K)} & \mathrm{Map}_K(E,X) \\
\downarrow & & \downarrow \\
(L_0^K)^C & \xrightarrow{\ell} & \mathrm{Map}_K(E,X)^C \\
\downarrow{=} & & \downarrow{=} \\
L_0^{K'} & \xrightarrow{\ell} & \mathrm{Map}_{K'}(E,X) \\
\downarrow & & \downarrow \\
R_p(L_0^{K'}) & \xrightarrow{\ell'} & \mathrm{Map}_{K'}(E,R_p(X))
\end{array}$$

as in Theorem 2.1, we conclude that $H_*(C(\alpha(K)))$ is cohomologically trivial, hence $\mathbb{F}_p G$–projective. This $\mathbb{F}_p G$–projective module can be killed and the map $\alpha(K)$ will be made more connected so we achieve the inductive step. □

We have the following interesting application:

2.7. Theorem. Let G, p, and R_p be as in Proposition 2.5. Let X be a G–space such that X and $\mathrm{Map}_P(E,X)$ belong to $\mathrm{Top}(R_p)$ for each $P \in P_p(G)$. Then there exists a finite dimensional G–complex K such that $E \times X$ and $E \times K$ are G–homotopy equivalent, if and only if for each cyclic subgroup C_i of order p_i there exists a finite dimensional C_i–complex K_i such that $E \times X$ and $E \times K_i$ are C_i–homotopy equivalent.

□

Section 3. Some Applications and Problems

To show that the theorems of Section 2 are useful, we need to verify the hypotheses in some geometrically interesting situations. This involves, in particular, cohomology computations of some equivariant function spaces, or equivalently, the space of sections of fibrations over $B(\mathbb{Z}/p\mathbb{Z})^n$ arising from Borel constructions. In this respect, J. Lannes'work [L] is quite relevant. Combined with some cohomology calculations of certain classifying spaces, Lannes' theorem leads to finiteness results, from which we derive the validity of the hypotheses of the main theorem 2.1 for $G = \mathbb{Z}/p\mathbb{Z}$. Then Theorem 2.7 allows us to derive the finiteness conclusions for a general finite group.

We recall below the following theorem of Lannes (conjectured by H. Miller in [Mm]). Let π be a p–elementary abelian group, and let K be the category of unstable algebras over the mod p Steenrod algebra. For any space X a homotopy class of maps $B\pi \longrightarrow X$ induces a homomorphism $H^*(X;\mathbb{F}_p) \longrightarrow H^*(B\pi;\mathbb{F}_p)$ in K.

3.1. Theorem (J. Lannes [L]). Let X be a simply–connected space such that $\dim H^i(X;\mathbb{F}_p) < \infty$ for all $i \geq 0$. Then the natural map

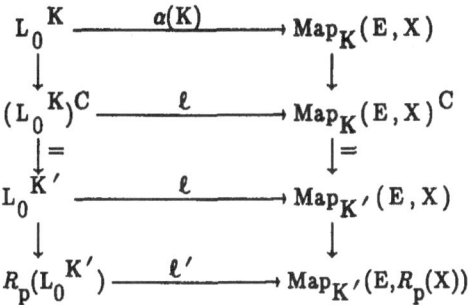

$$[B\pi,X] \longrightarrow \mathrm{Hom}_K(H^*(X;\mathbb{F}_p),\ H^*(B\pi;\mathbb{F}_p))$$

is bijective.

The first interesting case that we consider is a classical problem. Let X be a free G–space which is (non–equivariantly homotopy equivalent to the n–sphere S^n.

3.2. Problem. When does there exist a G–action on S^n such that $E_G \times S^n$ is G–homotopic to X ?

In homotopy theory, this is a problem about spherical fibrations. Let $\mathscr{H}_+(S^n)$ be the monoid of self–maps of degree one of S^n. Then the spherical fibration $X \longrightarrow X/G \longrightarrow BG$ is classified by a map $\lambda : BG \longrightarrow B\,\mathscr{H}_+(S^n)$ provided that G acts on X by degree one homeomorphisms. Problem 3.2 now translates into a lifting problem for the fibration $B\,\mathrm{Top}_+(S^n) \longrightarrow B\,\mathscr{H}_+(S^n)$ for the map λ. A more refined question is the following:

3.3 Problem. When is a spherical fibration over BG fibre homotopy equivalent to an ortthogonal fibration ?

This problem involves a similar lifting problem for the fibration $B0(n+1) \longrightarrow B\,\mathscr{H}_+(S^n)$ for λ.

According to Theorem 3.1 this is reduced to a lifting problem on the level of cohomology over the Steenrod algebra (which is not an easy problem in general either !). Now let us recall that according to Theorem 2.7, it suffices to solve the lifting problem of 3.2 for $\mathbb{Z}/p\mathbb{Z}$. (Note that Bousfield–Kan's completion [BK] suffices in this case). The lifting problem of 3.3 for $G = \mathbb{Z}/p\mathbb{Z}$ in fact can be solved on the level of cohomology due to deep calculations of the structure of $H^*(B\,\mathscr{H}_+(S^n);\mathbb{F}_p)$ over the Steenrod algebra due to F. Cohen [CLM] and related computations of J. Milgram and Madsen–Milgarm (Cf. [MJ], [Mj] and [MM] for example).

Positive solutions to Problem 3.3 for $G = \mathbb{Z}/p\mathbb{Z}$ and Theorem 2.7 give a partial answer to Problem 3.2. Namely, let X be a free G–space such that $X \simeq S^n$. Then there exists a finite dimensional G–complex K such that $E_G \times K$ is G–homotopy equivalent to X. In fact K may be taken equivariantly finitely dominated in the appropriate context. This result is the first step towards a complete solution of Problems 3.2 and 3.3 via methods of equivariant surgery, and it suggests that there are interesting relationships between Problem 3.3 and Atiyah's theorem on the K–theory of BG (to the effect that K(BG) is the I–adic compeltion of the representation ring R(G) CF. [At]).

Another interesting case is to consider G–actions on simply–connected Moore spaces. Let X be a Moore space on which a finite group of square–free order acts freely. Suppose that $\bar{H}_*(X)$ has the following property with respect to the induced $\mathbb{Z}G$–module structure: For each prime order subgroup $C \subset G$, $\bar{H}_*(X)|\mathbb{Z}C \cong P \oplus Q$, where P is $\mathbb{Z}C$–projective and Q is indecomposable. (P and Q depend on C). Then there exists a finite dimensional G–space K such that $E_G \times K$ is G–homotopy equivalent to X. The proof of the existence of the G–space K is reduced to the special case $G = \mathbb{Z}/p\mathbb{Z}$, thanks to Theorem 2.7 above. In this case, $\hat{H}^*(C;Q)$ is isomorphic to either $\hat{H}^*(C;\mathbb{Z})$ or $\hat{H}^*(C;I)$, where I is the augmentation ideal. This allows one to modify the arguments (involving the Sullivan fixed point conjecture and Lannes' Theorem 3.1) for the above special case $X = S^n$ in order to construct the desired K.

Finally, the above discussion leads us to the following conjecture which has intersting implications for the topological realizability of homotopy actions and the Steenrod problem, cf. [A2] for related discussions.

3.4. Conjecture. Let M be a finitely generated $\mathbb{Z}$–torsion free $\mathbb{Z}G$–module, where G is a finite group. Suppose that there exists a Moore space X with G–action such that $\bar{H}_*(X)$ is isomorphic to M as $\mathbb{Z}G$–modules. Then there exists a finite dimensional Moore G–space K with the same property.

References

[A1] *Assadi, A.*: "Extensions libres des actions des groupes finis", Proc. Aarhus Top. Conf. 1982, Springer LNM 1051 (1984).

[A2] *Assadi, A.*: "Homotopy Actions and Cohomology of Finite Groups", Proc. Conf. Transf. Groups, Poznan, July 1985, Springer–Verlag LNM 1217 (1986) 26–57.

[A3] *Assadi, A.*: "Finite Group Actions on Simply–connected Manifolds and CW complexes", Memoirs AMS 257 (1982).

[AB] *Assadi, A. – Browder, W.*: "Construction of finite group actions on simply–connected manifolds" (to appear).

[At] *Atiyah, M.F.*: "Characters and Cohomology of Finite Groups", Publ. I.H.E.S.

[B] *Borel, A. et al*: "Seminar on Transformation Groups", Annals of Math. Studies, Princeton University Press, Princeton, N.J.

[BK] *Bousfield–Kan*: "Homotopy Limits, Localization, and Completion", Springer–Verlag LNM no. 304 (1972).

[Br] *Brown, K.*: "Cohomology of Groups", Springer–Verlag GTM, no. 87 (1984).

[Cj] *Carlson, J.*: "The varieties and the cohomology ring of a module", J. Algebra 85 (1983), 104–143.

[C] *Carlsson, G.*: "The Homotopy Limit Problem", (Preprint 1986).

[Ch] *Chouinard, L.*: "Projectivity and relative projectivity for group rings", J. Pure Appl. Alg. 7 (1976), 287–302.

[CE] *Cartan, H. – Eilenberg, S.*: "Homological Algebra", Princeton University Press, Princeton, N.J.

[CLM] *Cohen, F.R. – Lada, T.J. – May, P.J.*: "The Homology of Iterated Loop Spaces", Springer LNM 533, (1976).

[D] *Dade, E.*: "Endo–permutation modules over p–groups II", Ann. of Math. 108 (1978), 317–346.

[Hw] *Hsiang, W.Y.*: "Cohomology Theory of Topological Transformation Groups", Springer, Berlin (1975).

[L] *Lannes, J.*: "Sur la Cohomologie Modulo p des p–Groupes Abeliens Elementaires", (Preprint 1986).

[M] *Miller, H.R.*: "The Sullivan Conjecture on Maps from Classifying Spaces", Annals of Math. 120, (1984), 39–87.

[Mm] *Miller, H.R.*: "Massey–Peterson Towers and Maps from Classifying Spaces", Proc. Alg. Top. Aarhus 1982, Springer LNM 1051 (1984).

[Mj] *Milgram, J.*: "A Survey of the Classifying Spaces Associated to Spherical Fiberings and Surgery", Proc. Symp. Pure Math. 32 AMS (1978) 79–90.

[MJ] *Milgram, J.*: "The mod–2 Spherical Characteristic Classes", Ann. Math. 92 (1970) 238–261.

[MM] *Madsen, I. – Milgram, J.*: "The Classifying Spaces for Surgery and Cobordism of Manifolds", Ann. Math. Studies, Princeton University Press (1979), Princeton, N.J.

[Q] *Quillen, D.*: "The spectrum of an equivariant cohomology ring I", and "II" Ann. of Math. 94 (1971), 549–573 and 573–602.

[R] *Rim, D.S.*: "Modules over finite groups", Ann. Math. 69 (1959), 700–712.

[Su] *Sullivan, D.*: "Genetics of Homotopy Theory and the Adams Conjecture", Ann. Math. 100, (1974) 1–79.

[W] *Weinberger, S.*: "Constructions of group actions: A survey of recent developments", Contemporary Math. Vol. 36 A.M.S. (1985).

Characteristic numbers and group actions

SUNG SOOK KIM

Let G denote the finite cyclic group of order n. The problem of determining necessary and sufficient conditions for F to be the fixed point set of a smooth cyclic group action on some sphere has been solved when n is a prime power. P. A. Smith proved that F must be a $\mathbb{Z}_n$-homology sphere. If n is odd it is also known that F is unitary. L. Jones has shown that these conditions are also sufficient to realize F as the fixed point set of smooth cyclic group action on some sphere when n is a prime power [J]. In the general case it is known that F is a union of smooth manifolds, unitary if n is odd. And if $\mathbb{Z}_n$ acts on some even dimensional sphere, then the Euler characteristic number of F is 2 if action is orientation-preserving and 0 if action is orientation-reversing. We may ask about possible restrictions on Pontryagin numbers of components of the fixed set F. If n is a prime power, the Pontryagin numbers of the fixed point set all vanish by the P. A. Smith theorem and the Hirzebruch signature theorem.

It is natural to ask whether such restrictions hold for other types of smooth cyclic groups acting on spheres. In case where n is not a prime power, by work of R. Oliver [Ol] and its extention of A. Assadi [Asd] and K. Pawałowski [Pa 1-4], we can construct exotic actions on spheres such that Pontryagin classes of fixed point sets do not vanish. But in these examples the Pontryagin numbers all vanish because these actions bound group actions on disks. R. Schultz has shown in [S2] that if G is a cyclic group whose order is not a prime power, then there are smooth actions of G on spheres such that the fixed point sets have nonzero Pontryagin numbers provided the dimension of the fixed point set is greater than 16. Testing to see if the lower bound in dimensions is necessary. There are two possibilities. First, more sophisticated computations might make it possible to remove the restriction on dimensions. Second, there might be some unusual things happening in low dimensions (compare Ewing's result for $\mathbb{Z}_p$ actions on spheres). Main Theorem is evidence for the first one.

Main Theorem. *Let G be a cyclic group of order p, where p is an odd prime, and let $q \neq p$ be another odd prime. For each $r > 0$ there is a smooth G-action on some $\mathbb{Z}_q$-homology sphere such that the fixed point set is a closed connected $4r$-dimensional manifold with nonzero Pontryagin numbers. In fact, there are subgroups $Fix_{4r}^{p,q}$ of the oriented bordism group Ω_{4r} such that*

(i) *every element of $Fix_{4r}^{p,q}$ contains a representative that is the fixed point set of some smooth G-action on some $\mathbb{Z}_q$-homology sphere,*
(ii) $Fix_{4r}^{p,q} \otimes \mathbb{Q} = \Omega_{4r}^{SO} \otimes \mathbb{Q}$, *for all $r \geqslant 1$ and $p \neq q$.*

Remark: If $p = q$, then the fixed point set of some smooth G-action on some $\mathbb{Z}_q$-homology sphere is a rational homology sphere by the P. A. Smith Theorem. It follows that the fixed point set maps to zero in $\Omega_*^{SO} \otimes \mathbb{Q}$.

Acknowledgments: I wish to express my gratitude to my advisor, professor Reinhard Schultz for his invaluable guidance, encouragement, and for his generous support during my research. I am indebeted to professor Mikiya Masuda for his encouragement and fruitful discussions.

1. PRELIMINARIES

In [S2], R. Schultz showed the existence of closed smooth manifolds F that are fixed point sets of smooth $\mathbb{Z}_{pq}$-actions on homotopy spheres and have nontrivial Pontryagin numbers. In fact, sufficient conditions for such F were obtained and one can see that the argument works in our setting. We shall give a brief explanation in this section.

We assume F is connected and unitary; *i.e.*, the stable normal bundle ν_F of F has a prescribed complex structure. Let G denote the cyclic group of order p, where p is an odd prime, and let ξ be a complex G-vector bundle over F such that

$$\xi^G = F \tag{1.1}$$

and $\xi = \nu_F \in \widetilde{K}(F)$ if we forget the action on ξ.
Decompose ξ into eigenbundles of the G-action as follows:

$$\xi = \bigoplus_{k=1}^{p-1} \xi_k \otimes t^k, \tag{1.2}$$

where ξ_k is a complex vector bundle and the generator $g \in G$ acts on $t^k (= \mathbb{C})$ as multiplication by ζ^k $(\zeta = e^{2\pi i/p})$.

We define

$$S_\xi(g) = \text{Constant}\langle \mathcal{L}(F) \prod_{k=1}^{p-1} \prod_j \frac{\zeta^k e^{x_{kj}} + 1}{\zeta^k e^{x_{kj}} - 1}, [F] \rangle,$$

where $\mathcal{L}(F)$ denotes the Atiyah-Singer $\mathcal{L}$-class [AS] of the bundle tangent to F, $[F]$ denotes the fundamental class of F, $\langle . , . \rangle$ is the natural Kronecker pairing between cohomology and homology, and the symbols x_{kj} have the usual interpretation (roots of the Chern classes of ξ_k) as formal two-dimensional cohomology classes such that the total Chern class of ξ_k is $c(\xi_k) = \prod_j (1 + x_{kj})$. We note that if ξ is the equivariant normal bundle of F in some $\mathbb{Z}_q$-homology sphere Σ, then $S_\xi(g)$ is the right hand side of the G-signature formula for the G-manifold Σ. So $S_\xi(g)$ vanishes. Moreover the Euler characteristic $\chi(F)$ of F is 2 by the Lefschetz Fixed Point Theorem. The following theorem says that the G-signature and Euler charateristic conditions are almost sufficient to realize a unitary manifold F as the fixed set of a smooth G action on a $\mathbb{Z}_q$-homology sphere.

Theorem 1.1. *Let F^{4r} be a closed unitary manifold and let ξ be a complex G-vector bundle over F satisfying the condition (1.1). Suppose*

(1) $\chi(F) = 2$,
(2) $S_\xi(g) = 0$.

Then for some positive integer m the m-fold connected sum $F \#.... \#F$ is the fixed point set of a smooth G action on a $\mathbb{Z}_q$-homology sphere.

Proof: We shall give an outline of the proof. See [S2] for the details.

Step 1: Let $F_0 = F - int(D^{4r})$. Since $\chi(F_0) = 1$ by (1), there is a finite G-CW complex K such that

(i) $K^G = F_0$,
(ii) K is a $\mathbb{Z}_q$-homology G-disk.

One can imitate this construction in the smooth case, starting with the closed disk bundle of $\xi|F_0$ and adding G-handles corresponding to the G-cells that one adds to form K. Denote the resulting G-manifold by L_1.

Step 2: As in [S2] Theorem 2.1. an equivariant surgery problem with domain ∂L_1 is defined. Since ∂L_1^G is already a standard sphere, the surgery obstruction to making ∂L_1 into a semilinear sphere lies in a Wall group $L^h_{dim\ \partial L_1}(\mathbb{Z}[G], 1)$. However this group is trivial because dim ∂L_1 and $|G| = p$ are both odd [B]. Hence we can perform equivariant surgery of ∂L_1 to get a semilinear sphere. Let W be the G-manifold obtained by attaching the surgery cobordism to L_1. It may not be a $\mathbb{Z}_q$-homology G-disk. The obstruction to making it into a $\mathbb{Z}_q$-homology G-disk lies in $L^h_{dim\ L_1}(\mathbb{Z}_{(q)}[G], 1)$. However by construction it is only well defined modulo the images of $L^h_{dim\ L_1}(\mathbb{Z}[G], 1)$. The corresponding cokernel turns out to have a finite exponent. This means that if we take a boundary connected sum of finite copies, say m_1 copies, of W, then we can perform surgery on $m_1 W$ to make it into a $\mathbb{Z}_q$-homology G-disk without touching the boundary. Denote the resulting G-manifold by L_2.

Step 3: We show when a connected sum of finite copies, say m_2 copies, of ∂L_2, bounds a semilinear disk. Under this condition, we can attach the semilinear disk to a boundary connected sum of m_2 copies of L_2 along the boundary to get a desired $\mathbb{Z}_q$ - homology G-sphere having $m_2 m_1 F$ as the fixed point set.

If V denotes the tangential representation at the fixed points of the G-action on ∂L_2, then the set Θ^V_G of G h-cobordism classes of semilinear spheres having V as the tangential representation forms an abelian group. A semilinear sphere bounds a semilinear disk if and only if it determines the zero class in Θ^V_G. The results of Browder and Petrie [BP] imply that two types of invariants on Θ^V_G detect elements of Θ^V_G up to finite ambiguity; one is a modified Atiyah-Singer invariant and the other is an invariant detecting the knot type of the fixed point set [MS]. As noted in [MS], the second invariant is a generalization of the classical signature invariant of a knot. In our case ∂L_2 bounds a G-disk, so that the latter invariant vanishes. Condition (2) means that the modified Atiyah-Singer invariant vanishes. Thus if we take a connected sum of finitely many copies of ∂L_2, then it bounds a semilinear disk.

2. THE 4-DIMENSIONAL CASE

In this section we treat the 4-dimensional case of Main Theorem. In order to prove the 4-dimensional case we need to find a pair (F^4, ξ) as in (1.1) which satisfies the conditions (1) and (2) in Theorem 1.1. If we find such a pair (F^4, ξ) with Sign $F \neq 0$, then Main Theorem follows from Theorem 1.1 because Ω_4 is infinite cyclic and $[F] \neq 0 \in \Omega_4$ if Sign $F \neq 0$. We produce such a pair (F^4, ξ) in the following way. Let μ be a fiber homotopically trivial vector bundle over $S^2 \times S^2$ and take a fiber homotopical trivialization ω.

$$\begin{array}{ccc} E(\mu) & \xrightarrow{\omega} & (S^2 \times S^2) \times \mathbb{R}^n \\ \downarrow & & \downarrow \\ S^2 \times S^2 & \xrightarrow{\mathrm{Id}} & S^2 \times S^2 \end{array}$$

Make ω transverse to the zero section via a proper homotopy, and take $\bar{\omega}$ to be transverse to the zero section. Let $F = \bar{\omega}^{-1}$ (the zero section), and let $f = \bar{\omega}|F : F \longrightarrow S^2 \times S^2$. This is a degree one map, and it follows that

$$\begin{aligned} TF &\underset{\text{stably}}{\cong} f^*(T(S^2 \times S^2) + \mu - (S^2 \times S^2) \times \mathbb{R}^n) \\ &\underset{\text{stably}}{\cong} f^*\mu. \end{aligned}$$

Therefore

(2.1) $\qquad \nu_F \underset{\text{stably}}{\cong} f^*(-\mu)$, where ν_F is a stable normal bundle of F.

The following lemma is well known.

Lemma 2.1.

(1) *Let $\eta_{\mathbb{C}}$ be the complex Hopf line bundle over S^2. Then $\eta_{\mathbb{C}}^{2l} \in$ kernel of J for every integer l, where $J : \widetilde{KO}(S^2) \to J(S^2)$.*

(2) *Let $\eta_{\mathbb{H}}$ be the quaternionic Hopf line bundle over S^4. Then $24\eta_{\mathbb{H}} \in$ kernel of J, where $J : \widetilde{KO}(S^4) \to J(S^4)$.*

Let $\eta_i (i = 1, 2)$ be the pullback of $\eta_{\mathbb{C}}$ by the projection map from $S^2 \times S^2$ to the i -th factor and let $\bar{\eta}$ be the pullback of $\eta_{\mathbb{H}}$ by a degree one map from $S^2 \times S^2$ to S^4. Set

$$-\mu_k = u_k(\eta_1^{2a_k} \oplus \eta_2^{2b_k}) \oplus 24v_k\bar{\eta}, \quad \text{where} \quad u_k, v_k \in \mathbb{Z}$$

and

$$\mu = \sum_{k=1}^{p-1} \mu_k.$$

Then μ is fiber homotopically trivial since each μ_k is fiber homotopically trivial by Lemma 2.1. As observed above we have a degree one map $f : F \longrightarrow S^2 \times S^2$ such that $\nu_F = f^*(-\mu)$. We note that if each μ_k is a complex vector bundle, then so is $f^*(-\mu)$. We give ν_F a complex structure by $f^*(-\mu)$ so that F is a unitary manifold. Set $\xi_k = f^*(-\mu_k)$ and $\xi = \bigoplus_{k=1}^{p-1} \xi_k \otimes t^k$. Then the pair (F^4, ξ) satisfies (1.1). Now we need another lemma.

Lemma 2.2. *Let (F^4,ξ) be a pair as in (1.1) and let $\xi = \bigoplus_{k=1}^{p-1}\xi_k\otimes t^k$ be the decomposition as in (1.2). Then condition (2) in Theorem 1.1 reduces to the following equation.*

(2.2)

$$\frac{1}{12}\sum_{k=1}^{p-1}p_1(\xi_k)[F]=\sum_{k=1}^{p-1}\frac{\zeta^k+\zeta^{-k}}{(\zeta^k-\zeta^{-k})^2}p_1(\xi_k)[F]+2(\sum_{k=1}^{p-1}\frac{1}{\zeta^k-\zeta^{-k}}c_1(\xi_k))^2[F].$$

Proof:

$$\begin{aligned}S_\xi(g)&=\text{Constant}\times\mathcal{L}(F)\prod_{k=1}^{p-1}[(\frac{\zeta^k e^{x_{k1}}+1}{\zeta^k e^{x_{k1}}-1})(\frac{\zeta^k e^{x_{k2}}+1}{\zeta^k e^{x_{k2}}-1})][F]\\&=\text{Constant}\times\mathcal{L}(F)\prod_{k=1}^{p-1}[1-\frac{2\zeta^k}{\zeta^{2k}-1}c_1(\xi_k)\\&\quad+\frac{\zeta^k}{(\zeta^k-1)^2}(c_1(\xi_k)^2-2c_2(\xi_k))+\frac{(2\zeta^k)^2}{(\zeta^{2k}-1)^2}c_2(\xi_k)][F],\end{aligned}$$

where $\mathcal{L}(F)=1+\dfrac{p_1(F)}{12}$.

Since $S_\xi(g)=0$, we have

$$\begin{aligned}0=&\mathcal{L}(F)\prod_{k=1}^{p-1}[1-\frac{2\zeta^k}{\zeta^{2k}-1}c_1(\xi_k)\\&+\frac{\zeta^k}{(\zeta^k-1)^2}(c_1(\xi_k)^2-2c_2(\xi_k))+\frac{(2\zeta^k)^2}{(\zeta^{2k}-1)^2}c_2(\xi_k)][F],\end{aligned}$$

which can be rewritten as

$$\begin{aligned}0=&\frac{1}{4}\,\text{Sign}(F)+\prod_{k=1}^{p-1}[1-\frac{2\zeta^k}{\zeta^{2k}-1}c_1(\xi_k)\\&+\frac{\zeta^k}{(\zeta^k-1)^2}(c_1(\xi_k)^2-2c_2(\xi_k))+\frac{(2\zeta^k)^2}{(\zeta^{2k}-1)^2}c_2(\xi_k)][F]\\=&\frac{1}{4}\,\text{Sign}(F)+[\sum_{k=1}^{p-1}\frac{\zeta^k}{(\zeta^k-1)^2}p_1(\xi_k)+\sum_{k=1}^{p-1}\frac{(2\zeta^k)^2}{(\zeta^{2k}-1)^2}c_2(\xi_k)\\&+\sum_{\substack{k=1\\k<j}}^{p-1}(\frac{2\zeta^k}{\zeta^{2k}-1})(\frac{2\zeta^j}{\zeta^{2j}-1})c_1(\xi_k)c_1(\xi_j)][F]\end{aligned}$$

Since $p_1(\xi_k)=c_1(\xi_k)^2-2c_2(\xi_k)$ and $\text{Sign}\,F=-\frac{1}{3}\Sigma_{k=1}^{p-1}p_1(\xi_k)[F]$, by an elementary calculation we have the formula (2.2) in Lemma 2.2. Q.E.D.

We need to find examples of manifolds F and vector bundles ξ_k for which equation (2.2) is satisfied.

By elementary calculations we have

$$c_1(-\mu_k) = 2u_k(a_k x + b_k y), \text{ where } x, y \in H^2(S^2 \times S^2; \mathbb{Z})$$
$$c_1(-\mu_k)^2 = 8a_k b_k {u_k}^2 xy$$
$$p_1(-\mu_k) = -48 v_k xy.$$

Since f is a degree one map, we have

$$c_1(\xi_k)^2[F] = 8a_k b_k {u_k}^2$$
$$p_1(\xi_k)[F] = -48 v_k.$$

To find a solution of (2.2), we may restrict attention to the special case in which

$$u_k = v_k = 0 \qquad for\ \frac{p+1}{2} \leqslant k \leqslant p-1.$$

$$a_k = a,\ b_k = b,\ u_k = u \qquad for\ 1 \leqslant k \leqslant \frac{p-1}{2}.$$

The latter implies that $c_1(\xi_k) = c_1(\xi_j)$ for $1 \leqslant k,\ j \leqslant \frac{p-1}{2}$. Then (2.2) becomes

$$\text{(2.3)} \qquad \sum_{k=1}^{\frac{p-1}{2}} v_k = 12 \sum_{k=1}^{\frac{p-1}{2}} \frac{\zeta^k + \zeta^{-k}}{(\zeta^k - \zeta^{-k})^2} v_k - N\Big(\sum_{k=1}^{\frac{p-1}{2}} \frac{1}{\zeta^k - \zeta^{-k}}\Big)^2,$$

where $N = 4abu^2$. In order to find $\{v_k\}$ and N, multiply both sides of (2.3) by

$$\prod_{k=1}^{\frac{p-1}{2}} (\zeta^k - \zeta^{-k})^2 = (-1)^{\frac{p-1}{2}} p \ \text{(compare page 72 of [ST])}.$$

Then

(2.4)

$$(-1)^{\frac{p-1}{2}} p \sum_{k=1}^{\frac{p-1}{2}} v_k = 12 \sum_{k=1}^{\frac{p-1}{2}} v_k(\zeta^k + \zeta^{-k}) \prod_{\substack{j=1 \\ j \neq k}}^{\frac{p-1}{2}} (\zeta^j - \zeta^{-j})^2 - N \sum_{k=1}^{\frac{p-1}{2}} \prod_{\substack{j=1 \\ j \neq k}}^{\frac{p-1}{2}} (\zeta^j - \zeta^{-j})^2.$$

The right hand side is invariant under the complex conjugation $\zeta \to \zeta^{-1}$, so it can be expressed by

$$\sum_{j=1}^{\frac{p-1}{2}} \Big[\Big(\sum_{k=1}^{\frac{p-1}{2}} a_{jk} v_k + \alpha_j N\Big)(\zeta^j + \zeta^{-j})\Big],$$

where a_{jk} and α_j are integers which are independent of v_k and N. On the other hand, since $1 = -\sum_{j=1}^{\frac{p-1}{2}}(\zeta^j + \zeta^{-j})$ the left hand side of (2.4) can be expressed as a linear combination of the numbers $\zeta^j + \zeta^{-j}$ $(1 \leqslant j \leqslant \frac{p-1}{2})$. However, since the elements $\zeta^j + \zeta^{-j}$ $(1 \leqslant j \leqslant \frac{p-1}{2})$ are linearly independent over $\mathbb{Z}$, the coefficients of $\zeta^j + \zeta^{-j}$ on both sides of (2.4) must coincide with each other for each j. Therefore we obtain a system of $\frac{p-1}{2}$ linear equations in $\{v_k\}$and N. This system can be written

$$(2.5) \qquad A\begin{pmatrix} v_1 \\ v_2 \\ \cdots \\ v_{\frac{p-1}{2}} \end{pmatrix} = \begin{pmatrix} \alpha_1 \\ \alpha_2 \\ \cdots \\ \alpha_{\frac{p-1}{2}} \end{pmatrix} N,$$

where A is a $\frac{p-1}{2} \times \frac{p-1}{2}$ matrix and α_j are constants given by the previous formula. Note that A and the α_j are independent of $\{v_k\}$ and N. If A is nonsingular, then we can solve the equation (2.5). Moreover, if $v_1 + v_2 + \dots\dots + v_{\frac{p-1}{2}} \neq 0$, then $p_1(\xi) \neq 0$, so that $\operatorname{Sign} F \neq 0$.

Lemma 2.3. *A is a nonsingular matrix.*

Proof: Consider

$$(2.6) \qquad \sum_{k=1}^{\frac{p-1}{2}} v_k = 12 \sum_{k=1}^{\frac{p-1}{2}} \frac{\zeta^k + \zeta^{-k}}{(\zeta^k - \zeta^{-k})^2} v_k.$$

The only solution of this equation is the trivial one ($v_k = 0$ for all k) if and only if A is a nonsingular matrix (compare (2.3)).

Step 1: Since

$$\operatorname{trace}\left(\frac{\zeta^k + \zeta^{-k}}{(\zeta^k - \zeta^{-k})^2}\right) = 2 \sum_{k=1}^{\frac{p-1}{2}} \frac{\zeta^k + \zeta^{-k}}{(\zeta^k - \zeta^{-k})^2},$$

by taking the trace of (2.6) we obtain

$$(2.7) \qquad (p-1) \sum_{k=1}^{\frac{p-1}{2}} v_k = 24 \sum_{k=1}^{\frac{p-1}{2}} v_k \sum_{k=1}^{\frac{p-1}{2}} \frac{\zeta^k + \zeta^{-k}}{(\zeta^k - \zeta^{-k})^2}.$$

Note that $\frac{\zeta^k + \zeta^{-k}}{(\zeta^k - \zeta^{-k})^2} = \frac{1}{2}\Phi_2(\zeta^k)$ (see [E], page 447). Since the trace of $\Phi_2(\zeta^k)$ is $-\frac{p^2-1}{24}$ (see page 28 Of [K]), equation (2.7) implies that $\sum_{k=1}^{\frac{p-1}{2}} v_k = 0$.

Step 2: By Step 1 we obtain $0 = \sum_{k=1}^{\frac{p-1}{2}} \frac{\zeta^k + \zeta^{-k}}{(\zeta^k - \zeta^{-k})^2} v_k$. Since the algebraic numbers $\Phi_2(\zeta^k)$, for $1 \leqslant k \leqslant \frac{p-1}{2}$, are linearly independent over the rationals (see page 448 of [E]), it follows that $v_k = 0$ for all k.

Lemma 2.4. *Let $\{v_k\}$ be the solution of the system of equations (2.5). If $N \neq 0$, then $\sum_{k=1}^{\frac{p-1}{2}} v_k \neq 0$.*

Proof: Suppose $\sum_{k=1}^{\frac{p-1}{2}} v_k = 0$. Then (2.3) becomes

$$12\sum_{k=1}^{\frac{p-1}{2}} \frac{\zeta^k + \zeta^{-k}}{(\zeta^k - \zeta^{-k})^2} v_k = N\Big(\sum_{k=1}^{\frac{p-1}{2}} \frac{1}{\zeta^k - \zeta^{-k}}\Big)^2. \tag{2.8}$$

The trace of the left hand side of (2.8) is $24\sum_{j=1}^{\frac{p-1}{2}} v_j \sum_{k=1}^{\frac{p-1}{2}} \frac{\zeta^k+\zeta^{-k}}{(\zeta^k-\zeta^{-k})^2} = 0$. The trace of the right hand side of (2.8) is $N\sum_{j=1}^{p-1}(\sum_{k=1}^{\frac{p-1}{2}} \frac{1}{\zeta^{kj}-\zeta^{-kj}})^2$. Since $\zeta^{kj} - \zeta^{-kj} = 2i \sin\frac{2\pi kj}{p}$, we have

$$\left(\sum_{k=1}^{\frac{p-1}{2}} \frac{1}{\zeta^{kj} - \zeta^{-kj}}\right)^2 = -\frac{1}{4}\left(\sum_{k=1}^{\frac{p-1}{2}} \frac{1}{\sin\frac{2\pi kj}{p}}\right)^2 \leqslant 0,$$

and when $j = 1$,

$$-\frac{1}{4}\Big(\sum_{k=1}^{\frac{p-1}{2}} \frac{1}{\sin\frac{2\pi k}{p}}\Big)^2 < 0.$$

Therefore the right hand side of (2.8) is not equal to zero if $N \neq 0$, and consequently $\sum v_k = 0$ implies $N = 0$. Q.E.D.

Now we have to consider condition (1) in Theorem 1.1. For the condition (1), we take the connected sum of F with finitely many copies of $S^1 \times S^3$ (or $S^2 \times S^2$), if necesary, to achieve it. Then we can pull back each ξ_k to $F' = F\#(S^1 \times S^3)\#......\#(S^1 \times S^3)$ (or $F' = F\#(S^2 \times S^2)\#.....\#(S^2 \times S^2)$) by the degree one map $F' \to F$ obtained by collapsing $(S^1 \times S^3)$'s (or $(S^2 \times S^2)$'s) to a point. We note that the condition (2) is preserved under this procedure. Finally we note that $[F'] = [F] \in \Omega_4$. Since $\mathrm{Sign}\, F' = \mathrm{Sign}\, F \neq 0$ if $N \neq 0$, we have proved the first statement in Main Theorem. To obtain the subgroup $Fix_4^{p,q}$ described in Main Theorem, take the additive subgroup of Ω_4^{SO} generated by F and F'.

3. THE GENERAL CASE

The purpose of this section is to give an outline of the proof of the general case of Main Thorem. I am grateful to Professor Schultz for suggesting the approach used here. The reader is referred to [S2] for more details.

Outline of the proof of Main Theorem:

Step 1: Let p^A be a $4r$-dimensional monomial in the rational Pontryagin classes of a vector bundle. The results of [S2] show that for each A there is a closed oriented $4r$-manifold M_A, a $G(= \mathbb{Z}_p)$ vector bundle ξ over M_A and a $\mathbb{Z}_q$-homology G-disk Δ_A such that

(1) $p^B(M_A)[M_A] = 0$ if and only if $B \neq A$,
(2) ξ is invariant under Aut G,

(3) $\partial\Delta_A$ is a semilinear sphere and $\partial\Delta_A^G = S^{4r-1}$,
(4) $\Delta_A^G \cong M_A - int(D^{4r})$ and $\nu(\Delta_A^G) \cong \xi|(M_A - int(D^{4r}))$, where $\nu(\Delta_A^G)$ denotes the G normal bundle of Δ_A^G.
(5) the local representations V_A at the fixed points are equivalent for all A.

Note: : If $\alpha \in \operatorname{Aut} G$ and Y is a G-space with action map $\Phi : G \times Y \to Y$, then $\alpha_* Y$ is the G-space with action map $\alpha_*\Phi(g, y) = \Phi(\alpha(g), y)$. An object is said to be invariant under Aut G if $\alpha_* Y$ is G-equivariantly isomorphic to Y for all $\alpha \in \operatorname{Aut} G$.

The proof is similar to the proof of Theorem 1.1. If $\partial\Delta_A$ is G-diffeomorphic to a linear G-sphere, then we can glue a linear G-disk to Δ_A along the boundary to get a $\mathbb{Z}_q$-homology G-sphere Σ_A having a manifold M'_A orientation preservingly homeomorphic to M_A as the fixed point set. If this is true for each A, then every connected sum $\#_A(\#^{n_A} M'_A)$ can also be realized as a fixed point set. But the subgroup of Ω_{4r}^{SO} generated by the manifolds M'_A has finite index in the oriented bordism group (since Pontryagin numbers detect Ω_{4r}^{SO} up to finite ambiguity), it follows that the subgroup $Fix_{4r}^{p,q}$ generated by the M'_A has properties (i) and (ii) in the conclusion of Theorem B. However, $\partial\Delta_A$ is not necessarily G-diffeomorphic to a linear G-sphere. The following steps are necessary to overcome this difficulty.

Step2: Let V be the tangential G representation at the fixed point of $\partial\Delta_A^G$. Then $\partial\Delta_A$ determines an element in Θ_V^G. Let $F\Theta_V^G$ be the group of G framed semilinear spheres as defined in [MS] for example. We can see that the natural homomorphism $F\Theta_V^G \to \Theta_V^G$ has a finite cokernel. In particular, if we replace Δ_A by the connected sum of Δ_A with itself enough times, then we may assume that $\partial\Delta_A$ comes from an element in $F\Theta_V^G$. Choose a G framing F on $\partial\Delta_A$. Then signature defects $\delta(\partial\Delta_A : F)$, $\delta(\partial\Delta_A^G : F)$ and an equivariant signature defect $\delta^G(\partial\Delta_A : F)(g)$ $(g \neq 1)$ are defined for $\partial\Delta_A$ with the framing F, and the group $F\Theta_V^G \otimes \mathbb{Q}$ is detected by these invariants (see [DR1-2], [MS]).

Step 3: Using the methods of [MS], we can prove that for a suitable choice of F we have $\delta(\partial\Delta_A^G : F) = 0$. Furthermore, we also have $\delta^G(\partial\Delta_A : F)(g) = S_\xi(g)$ for $g \neq 1$, where ξ is the G vector bundle over M_A described previously.

Our assumption that ξ is invariant under automorphisms of $\mathbb{Z}_p$ implies that $\delta^G(\partial\Delta_A, F) = S_\xi$ is constant and rational valued on $G - \{1\}$ (in general the values lie in $\mathbb{Q}[e^{2\pi i/p}]$). It follows that the subspace of $\Theta_V^G \otimes \mathbb{Q}$ spanned by the images of the manifolds $\partial\Delta_A$ has dimension at most 2. It follows from the results of [DR1-2], [MS] and from elementary results in equivariant stable homotopy theory that connected sum of finitely many, say k, copies of $\partial\Delta_A$ with itself bounds a G framed manifold U such that $U^G \cong D^{4r}$ and the framing on U extends that on $\partial\Delta_A$. We paste together $k\Delta_A$ and U along the boundary to get a closed G manifold $N_0(A)$. By construction we can see $N_0(A)^G \cong kM_A\#\Sigma_0$ for some homotopy $4r$-sphere Σ_0 and $\nu(N_0(A)^G) = f_0^*(\#^k\xi)$, where f_0 is a map from $kM_A\#\Sigma_0$ to kM_A obtained by collapsing Σ_0 to a point.

By the preceding paragraph, the subspace of $F\Theta_V^G \otimes \mathbb{Q}$ generated by the classes $(\partial\Delta_A : F)$ is detected by $\delta(\partial\Delta_A : F)$ and $\delta^G(\partial\Delta_A : F)(g_0)$, where g_0 is a fixed generator of G. The next step is to show that we can choose an equivariant framing F' on some multiple connected sum $k'\partial\Delta_A$ such that $\delta(k'\partial\Delta_A : F') = 0$. Since $\delta(\partial\Delta_A : F)$ is given by the signature of $N_0(A)$, it will suffice to find a well-behaved modification of the previous construction such that Sign $N_0(A) = 0$; more precisely, we wish to modify an equivariant connected sum of several copies of $N_0(A)$ with itself so that the signature vanishes. Suppose Sign $N_0(A) \neq 0$ so that dim $N_0(A)$ is divisible by 4. Set dim $N_0(A) = 4s$. According to [KM], every multiple of some non-zero integer b_s (depending only on s) can be represented as the signature of a closed almost parallelizable manifold of dimension $4s$. Hence we can choose an integer l_0 and a closed smooth almost parallelizable manifold W of dimension $4s$ such that $l_0 \operatorname{Sign} N_0(A) + p \operatorname{Sign} W = 0$. Take the connected sum of $\#^{l_0} N_0(A)$ and $G \times W$ along a free G orbit to get a closed G manifold, say $N_1(A)$. Then we have $\operatorname{Sign} N_1(A) = 0$. In case $\operatorname{Sign} N_0(A) = 0$, we set $N_1(A) = N_0(A)$ and we let $l_0 = 1$ and $W = \emptyset$. Thus $\operatorname{Sign} N_1(A) = 0$ in all cases. Let n_0 be the order of Σ_0 in the Kervaire-Milnor group, so that $n_0 \Sigma_0 = S^{4r}$. Consider $N_2(A) = n_0 N_1(A)$. It has the following properties:

$$N_2(A)^G \cong \#^l M_A, \text{ where } l = n_0 l_0 k, \tag{3.1}$$

$$\operatorname{Sign} N_2(A) = 0, \tag{3.2}$$

$$\operatorname{Sign}(g, N_2(A)) \text{ is constant if } g \neq 1. \tag{3.3}$$

The last of these is true because $\nu(N_2(A)^G) \cong \#^l \xi$, and the latter is invariant under Aut G.

Step 4: The calculations of Ewing [E] yield

$$\operatorname{Sign}(g, N_2(A)) \neq 0 \text{ if } p^A = p_1. \tag{3.4}$$

Therefore if we set $N = N(p_1)$, it follows from (3.2), (3.3), and (3.4) that

$$a \operatorname{Sign}(G, N^r) + b \operatorname{Sign}(G, N_2(A)) = 0, \tag{3.5}$$

for some integers a and b. Take $aN^r \# bN_2(A)$ and adjust the Euler characteristic number of the fixed point set to equal 2 by taking the connected sum with finitely many copies of products of two linear G-spheres. We denote the resulting G manifold by $N_3(A)$. This gives a G-surgery problem in the sense of [DP2]. By (3.5) $\operatorname{Sign}(G, N_3(A)) = 0$. Hence the G-surgery obstruction has finite order, say t, because $L^h_{4s}(R[G]) = \mathbb{Z}^{\frac{p+1}{2}} \oplus$ torsion , where R is a subring of $\mathbb{Q}$ which contains $1/p$ and $1/2$ ($R = \mathbb{Z}_{(q)}$ in our case), and the torsion free part is detected by the G-signature. The surgery obstruction is additive with respect to connected sum, so t times the connected sum of $N_3(A)$ with itself can be converted into a $\mathbb{Z}_q$-homology G-sphere, say $N^*(A)$, by surgery without touching the fixed point set. We note that

$$[N^*(A)^G] = t(a[N^G]^r + b[N_2(A)^G]) = ta[N^G]^r + tbl[M_A] \tag{3.6}$$

by (3.1), where [ ] denotes the appropriate cobordism class.

Step 5: By the 4-dimensional case there is a $\mathbb{Z}_q$-homology G-sphere Q such that $\dim Q^G = 4$ and $[Q^G] \neq 0 \in \Omega_4$. Take the r-fold product Q^r and adjust the Euler characteristic number of the fixed point set to be 2 as before. Then we again get G-surgery problem. Since $\mathrm{Sign}(G, Q^r) = 0$, we conclude by the same reasoning as in Step 4 that $u(Q^G)^r$ will be the fixed point set of some $\mathbb{Z}_q$-homology G-sphere, say $\widetilde{Q}$, for some integer u; *i.e.*, we have

$$[\widetilde{Q}^G] = u[Q^G]^r \tag{3.7}$$

Since $[N^G]$ and $[Q^G]$ are elements of Ω_4^{SO} and Ω_4^{SO} is infinite cyclic, we have

$$v[N^G]^r + wu[Q^G]^r = 0 \tag{3.8}$$

for some integers v and w. Now consider $vN^*(A)\#wta\widetilde{Q}$. This is a $\mathbb{Z}_q$-homology G-sphere and the oriented cobordism class of the fixed point set is

$$\begin{aligned} v[N^*(A)^G] + wta[\widetilde{Q}^G] &= v(ta[N^G]^r + tbl[M_A]) + wtau[Q^G]^r \quad \text{by (3.6) and (3.7)} \\ &= ta(v[N^G]^r + wu[Q^G]^r) + vtbl[M_A] \\ &= vtbl[M_A] \quad \text{by (3.8).} \end{aligned}$$

Of course, the integers v, t, b, l all depend on A. For each A, let $y_A \in \Omega_{4r}^{SO}$ be the class of $v_A t_A b_A l_A [M_A]$. It follows that every class in the subgroup $Fix_{4r}^{p,q}$ generated by the set of all y_A is represented by the fixed point set of a smooth $\mathbb{Z}_p$-action on some $\mathbb{Z}_q$-homology sphere. Furthermore, by property (1) in Step 1 we know that the classes y_A generate a subgroup of finite index in Ω_{4r}^{SO}. Therefore the subgroup $Fix_{4r}^{p,q}$ has all the properties described in Main Theorem. Q.E.D.

REFERENCES

[Asd] A. Assadi, *Finite Group Actions on Simply Connected Manifolds and CW Complexes*, Memoirs Amer. Math. Soc. 257 (1982).

[AS] M. F. Atiyah and I. M. Singer, *The index of elliptic operators*, III,Ann. of Math 87 (1968), 546-604.

[AW] J. F. Adams and G. Walker, *On complex Stiefel manifolds*, Proc.Camb. Phil.Soc. 61 (1965), 81-103.

[B] A. Bak, *Odd dimensional surgery groups of odd torsion groups vanish*, Topology 14 (1975), 367-374.

[BP] W. Browder and T.Petrie, *Diffeomorphisms of manifolds and semifree actions on homotopy spheres*, Bull. Amer. Math. Soc. 77 (1971), 160-163.

[tD] T. tom Dieck, *Bordism of G-manifolds and integrality theorems*, Topology 9 (1970), 345–358.

[DP1] K. H. Dovermann and T. Petrie, *G-Surgery II*, Memoirs Amer. Math. Soc. 260 (1982).

[DP2] ———, *An induction theorem for equivariant surgery (G-Surgery III)*, Amer.J.Math. 105 (1983), 1369-1403.

[DP3] ———, *Smith equivalence for representations of odd order groups*, Topology 24 (1985), 283–305.

[DR1] K. H. Dovermann and M. Rothenberg, *An equivariant surgery sequence and equivariant homeomorphism and diffeomorphism classification*, Topology Symposium (Siegen, 1979), Lecture Notes in Math. Vol. 788, Springer, Berlin-HeidelbergNew York-Toyko, 1980, pp. 257–280.
[DR2] ———, *Equivariant Surgery and Classification of Finite Group Actions on Manifolds*, Memoirs Amer. Math. Soc. 379 (1988).
[E] J. Ewing, *Spheres as fixed point sets*, Quart.J.Math.Oxford **27** (1976), 445-455.
[H1] F. Hirzebruch, *Topological Methods in Algebraic Geometry*, Spinger-Verlag, new York, 1966.
[H2] ———, *Hilbert modular surfaces*, L'enseignement mathematique **19** (1973), 183-281.
[J] L. Jones, *The converse to the fixed point theorem of P. A. Smith II*, Indiana Univ. Math. J. **22** (1972), 309–325*correction* **24** (1975), 1001–1003.
[Kaw] K. Kawakubo, *The index and the Todd genus of $\mathbb{Z}_p$-actions*, Amer. J. Math. **97** (1975), 182–204.

[K] S.S. Kim, Ph.D. Thesis, Purdue University,1988.
[KM] M. Kervaire and J. Milnor, *Groups of homotopy spheres*, Ann. of Math. **77** (1963), 504–537.
[MS] M. Masuda and R. Schultz, *Equivariant inertia groups and rational invariants for nonfree actions* (to appear).
[O] R. Oliver, *Fixed point sets of finite group actions on acyclic complexes*, Comment. Math. Helv. **50** (1975), 155–177.
[Pa1] K. Pawałowski, *Fixed points of cyclic group actions on disks*, Bull. Acad. Polon. Sci. Math. Astr. Phys. **26** (1978), 1011-1015.
[Pa2] ———, *Group actions with inequivalent representations at fixed points*, Math.Z. **187** (1984), 29-47.
[Pa3] ———, *Equivariant thickening for compact Lie group actions*, Mathematica Gottingensis **71** (1986).
[Pa4] ———, *Fixed points of smooth group actions on disks and euclidean spaces*, preprint, 1986.
[S] R. Stong, *Notes on Corbordism Theory*, Mathematical Notes, Princeton University Press, 1968.
[ST] I. N. Stewart and D. O. Tall, *Algebraic Number Theory (second edition)*, Champman and Hall, New York, 1987.
[S1] R. Schultz, *Nonlinear analogs of linear group actions on spheres*, Bull. Amer.Math. Soc **11** (1984), 263-285.
[S2] ———, *Pontryagin numbers and periodic diffeomorphisms of spheres, to appear in the Proceedings of the Conference on Group Actions, Osaka, 1987 (in the Springer Lecture Notes Series).*

Sung Sook Kim
Department of Mathematics
Paichai University
Suh-Ku, Doma-Dong
Taejon, 302-735, KOREA
Currently visiting:
Korea Institute of Technology

Remarks on one fixed point A_5-actions on homology spheres

[(1)] Masaharu Morimoto and [(2)] Katsuhiro Uno

[(1)] Department of Mathematics, College of Liberal Arts & Sciences
Okayama University, Tsushimanaka, Okayama, 700 Japan
[(2)] Department of Mathematics, Faculty of Science
Osaka University, Toyonaka, Osaka, 560 Japan

0. Introduction

Let G be a group. If G acts on a set X (from the left), we denote by X^G the set of all the G-fixed elements in X, that is,

$$X^G = \{x \in X \mid gx = x \text{ for all } g \in G\}.$$

If X^G consists of a single element, then the action is called a one fixed point action. In this paper, we consider one fixed point actions of the alternating group A_5 on five letters, which is isomorphic to the icosahedral group, on homology spheres. A typical example is the Poincaré sphere Σ with a standard A_5-action. (A precise definition of a standard A_5-action of the Poincaré sphere Σ will be found in Section 1.) One of our main results is the following theorem.

THEOREM 0.1. *For any integer k with $k \geq 2$, the standard sphere S^{3k} has a smooth one fixed point A_5-action which is A_5-cobordant to $\Sigma(k)$, the k-fold cartesian product of Σ with the diagonal A_5-action.*

It may be worthwhile to mention that K. H. Dovermann, M. Masuda and T. Petrie investigated in Section 2 of [8] a standard A_5-action on the Poincaré sphere. One of their main results is as follows.

THEOREM ([8] Proposition 2.2). *There exists a nonsingular real algebraic A_5-variety which is A_5-diffeomorphic to the Poincaré sphere Σ with standard A_5-action.*

On the other hand, in the same paper, they showed that ;

THEOREM ([8] Theorem 1.3). *Suppose that a compact Lie group G acts smoothly on a closed smooth manifold M. If M is G-cobordant to a nonsingular G-variety, then M is G-diffeomorphic to a nonsingular real algebraic G-variety.*

Thus our theorem and the two theorems above yield that ;

The first author is partially supported by Grant-in-Aid for Scientific Research 01740048 and the second author by 01740041.

COROLLARY 0.2. *For any integer k with $k \geq 2$, the standard sphere S^{3k} has a one fixed point A_5-action, with which S^{3k} is A_5-diffeomorphic to a nonsingular real algebraic A_5-variety.*

This generalizes [9] Theorem A and [8] Theorem 1.1, in which they show that if $n = 6$ or $12k$ for some k with $k \geq 2$, then there exists a real algebraic action of A_5 with exactly one fixed point on a nonsingular variety which is diffeomorphic to a homotopy sphere of dimension n.

Perhaps a word about the history of one fixed point actions on spheres is in order. The first example was given by E. Stein [29], who showed that S^7 has a one fixed point smooth action of $SL(2,5)$. Immediately after, T. Petrie found various one fixed point smooth actions on homotopy spheres ([23] and [24]), and in 1986, E. Laitinen and P. Traczyk [15] turned our attention to the question of whether S^6 has a one fixed point smooth action of A_5 or not. At present, we know that all S^6 and S^n $(n \geq 9)$ have one fixed point smooth actions of A_5 ([16] – [21]). Moreover, N. P. Buchdahl, S. Kwasik and R. Schultz [6] showed that all S^n $(n \geq 6)$ have one fixed point locally linear A_5-actions. However, they also proved that none of S^n $(n \leq 5)$ has a one fixed point locally linear action of any finite group. (Compare with M. Furuta [11], and [17]).

We now mention how the present paper is organized. Our first observation will be made on the singular sets. In general, if a group G acts on X, then for any subgroup H of G the *H-singular set* $X_{s(H)}$ is defined by

$$X_{s(H)} = \bigcup_{g \in H \setminus \{1\}} \{x \in X \mid gx = x\}.$$

In particular, if H is the entire group G, then we write X_s instead of $X_{s(G)}$ for convenience and call it the singular set of X. For three dimensional homology spheres with one fixed point smooth action of A_5, there are at most four A_5-homeomorphism classes of singular sets, which we call types. (Proposition 1.9.) These types give a device for studying cobordisms or surgeries, which will be considered in this paper. However, determining the A_5-homeomorphism classes itself may be of interest in its own right. As an example, we shall determine the type of the Poincaré sphere Σ with a standard A_5-action. (Theorem 1.13.) We next consider equivariant cobordisms in Section 2, where the main result will give a sufficient condition for two A_5-manifolds to be cobordant (Theorem 2.1.), and, in Section 3, G-normal maps and G-normal cobordisms will be studied following T. Petrie [23]. It seems that the notions and terminologies introduced in Sections 2 and 3 may be useful in the other situations. Thus, we state the definitions and the assertions more generally than is necessary for the proof of our theorem. The proof of our theorem above will be given in Section 4 assuming a key lemma, whose proof will be found in Section 5.

Throughout this paper, a G-action on a smooth manifold is understood to be a *smooth* G-action and a G-map is assumed to be continuous unless otherwise stated. We denote by $\mathbf{Z}$, $\mathbf{R}$ and $\mathbf{C}$ the ring of integers, the real number field and the complex number field, respectively, on which an action of any group is understood to be trivial. For a set X, we denote by $|X|$ the cardinality of X.

1. Types of the singular sets

In this section we first investigate the singular sets of three dimensional homology spheres with one fixed point A_5-actions. We denote the cyclic group of order m by C_m and the dihedral group of order $2m$ by D_{2m}. Also, A_4 means the alternating group on four letters, which is isomorphic to the tetrahedral group. For elements $g_1, g_2, \ldots, g_n$ of A_5, the subgroup of A_5 generated by $g_1, g_2, \ldots, g_n$ is denoted by $< g_1, g_2, \ldots, g_n >$. In the first two lemmas, we summarize some data on A_5. Perhaps, it may be supposed that the reader is familiar with them. (cf. [12] Chapter 2, Section 2.4.)

LEMMA 1.1.

(1) *The isomorphism classes of nontrivial subgroups of A_5 are C_2, C_3, C_5, D_4, D_6, D_{10} and A_4.*

(2) *Any two subgroups of A_5 are isomorphic if and only if they are conjugate.*

We now put $x = (1,2)(3,4)$, $y = (3,5,4)$, $z = (1,2,3,5,4)$ and $u = (1,3)(2,4)$ in A_5.

LEMMA 1.2.

(1) *We have $x^2 = y^3 = z^5 = u^2 = (uz^2)^3 = 1, xyx = y^{-1}, xzx = z^{-1}, ux = xu \neq 1, y = z^{-1}uz^{-1}$ and $uzu = zuxz$.*

(2) *The subgroup $< x >$ is properly contained in the following seven subgroups.*

$$< x, u > (\cong D_4), \quad < x, y > (\cong D_6), \quad < x, uyu > (\cong D_6),$$
$$< x, z > (\cong D_{10}), \quad < x, uzu > (\cong D_{10}), \quad < x, z^2uz > (\cong A_4) \text{ and } A_5$$

(3) *The above $< x, z^2uz >$ contains $< x, u >$.*

Throughout this paper, unless otherwise stated, the above elements x, y, z and u are fixed and we write the subgroups of A_5 as follows.

$$C_2 =< x >, C_3 =< y >, C_5 =< z >$$

$$D_4 =< x, u >, D_6 =< x, y >, D_{10} =< x, z > \quad \text{and} \quad A_4 =< x, z^2uz >$$

It might be helpful to keep the following figure in mind.

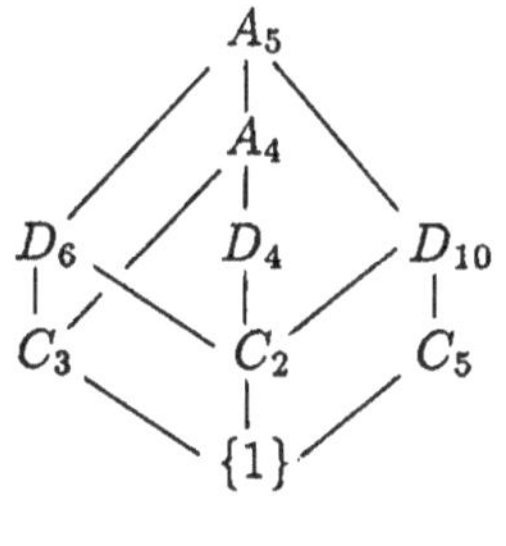

Figure 1.3.

In this section, A_5-actions in the following family will be considered.

Definition 1.4. We denote by $\mathcal{S}$ the family of topological A_5-spaces X satisfying the following conditions (1) – (5):

(1) $X = X_s$.
(2) $|X^{A_5}| = 1$.
(3) $X^H = X^K$ whenever $H \subset K \subset A_5, H \cong D_4$ and $K \cong A_4$.
(4) $|X^H| = 2$ whenever $H \subset A_5$ and $H \cong D_{2m}$ for some $m = 2, 3$ or 5.
(5) X^H is homeomorphic to S^1 whenever $H \subset A_5$ and $H \cong C_m$ for some $m = 2, 3$ or 5.

Moreover, we let $\mathcal{MS}$ be the family of all closed, oriented, three dimensional smooth A_5-manifolds X whose singular sets X_s lie in $\mathcal{S}$.

PROPOSITION 1.5. *Let X be a three dimensional homology sphere having a smooth A_5-action with exactly one fixed point. Then X lies in $\mathcal{MS}$, that is, the singular set X_s of X belongs to $\mathcal{S}$.*

Remark. It is well known that the Poincaré sphere Σ with standard A_5-action is a homology sphere with one fixed point action, (cf. the paragraph following Proposition 1.9). Thus, Σ belongs to $\mathcal{MS}$.

PROOF: Denote by $p(A_5)$ the fixed point of X. Let V be the tangential representation of X at $p(A_5)$. Clearly $\dim V = 3$ and $V^{A_5} = 0$. Since dimensions of nontrivial irreducible real A_5-representations are at least 3, V is irreducible. Thus, from the character table of A_5, we can conclude that if H is a noncyclic (resp. nontrivial cyclic) subgroup of A_5, then $\dim V^H = 0$ (resp. 1). Also, $\dim X^H$ is 0 or 1 accordingly (if all components have the same dimension). The condition (5) of $\mathcal{S}$ follows immediately from Smith's theorem. Also, the conditions (4) and (3) follow subsequently to $X^{C_m} \cong S^1$ and $X^{D_4} \cong S^0$ by Smith's theorem since $D_{2m}/C_m \cong C_2$ and $A_4/D_4 \cong C_3$, respectively. Thus, X lies in $\mathcal{MS}$. This completes the proof.

Let $\mathcal{H}$ be the set of all subgroups of A_5 isomorphic to A_4, D_{10} or D_6. Once we fix a space X in $\mathcal{MS}$, for a subgroup H in $\mathcal{H}$ or $H = A_5$, we denote by $p(H)$ the point in X_s with isotropy subgroup H. Now note that the numbers of subgroups of A_5 isomorphic to C_2, C_3 and C_5 are 15, 10 and 6, respectively. Since $X_s^g = X_s^{<g>}$ for all $g \in A_5$, X_s is a union of at most 31 circles. Furthermore, if H_1 and H_2 are distinct cyclic subgroups ($\neq \{1\}$) of A_5, then they generate a noncyclic subgroup. Hence $X_s^{H_1} \cap X_s^{H_2}$ is either one point or S^0. This means that X_s is a union of exactly 31 circles. These circles intersect at the points $p(H)$ for some H. For example, in order to find the points at which the circle $X_s^{C_2}$ intersects with the other circles, it suffices to look for the subgroups that properly contains C_2. By observations like this, we get the following.

PROPOSITION 1.6. *Let X be in $\mathcal{MS}$. Then ;*

(1) *The circle $X_s^{C_2}$ intersects with the other circles at exactly six points $p(A_5)$, $p(A_4)$, $p(D_6)$, $p(uD_6u) = up(D_6)$, $p(D_{10})$ and $p(uD_{10}u) = up(D_{10})$.*
(2) *The circle $X_s^{C_3}$ intersects with the other circles at exactly four points $p(A_5)$, $p(D_6)$, $p(z^2A_4z^3)$ and $p(z^3A_4z^2)$.*
(3) *The circle $X_s^{C_5}$ intersects with the other circles at exactly two points $p(A_5)$ and $p(D_{10})$.*

Next, we conversely see how many circles in X_s intersect at the points in the above proposition. For instance, to see the intersection at $p(D_6)$, we must look for nontrivial cyclic subgroup of A_5 properly contained in D_6. In this case, $< x >$, $< y >$, $< xy >$ and $< xy^2 >$ satisfy this condition. Therefore, the four circles X_s^x, X_s^y, X_s^{xy} and $X_s^{xy^2}$ intersect at $p(D_6)$. Similarly, counting the number of cyclic subgroups with the desired property, we get the following.

PROPOSITION 1.7. *Let X be in $\mathcal{MS}$. Then the following shows how many circles in X_s intersect at the points in Proposition 1.6.*

(1) *4 circles intersect at each $p(D_6)$ and $p(uD_6u) = up(D_6)$.*
(2) *6 circles intersect at each $p(D_{10})$ and $p(uD_{10}u) = up(D_{10})$.*
(3) *7 circles intersect at each $p(A_4)$, $p(z^2A_4z^3)$ and $p(z^3A_4z^2)$.*
(4) *31 circles intersect at $p(A_5)$.*

Now imagine that we walk on the circle $X_s^{C_2}$ starting from and ending at $p(A_5)$. Since $ux = xu$, the action of u gives a diffeomorphism of $X_s^{C_2}$ fixing $p(A_5)$ and $p(A_4)$ and interchanging $p(D_{2m})$ and $p(uD_{2m}u)$ for $m = 3$ and 5. (See also Proposition 1.6 (1).) Hence, on $X_s^{C_2}$, we must meet the intersection points $p(H)$'s in one of the following order. (Note : In each case, we do not specify a direction.)

(1) $p(A_5) - p(D_6) - p(D_{10}) - p(A_4) - p(uD_{10}u) - p(uD_6u) - p(A_5)$
(2) $p(A_5) - p(D_6) - p(uD_{10}u) - p(A_4) - p(D_{10}) - p(uD_6u) - p(A_5)$
(3) $p(A_5) - p(D_{10}) - p(D_6) - p(A_4) - p(uD_6u) - p(uD_{10}u) - p(A_5)$
(4) $p(A_5) - p(D_{10}) - p(uD_6u) - p(A_4) - p(D_6) - p(uD_{10}u) - p(A_5)$

Definition 1.8. According as the above (1) - (4), we say that $X_s \in \mathcal{S}$ (or $X \in \mathcal{MS}$) is of *type* $(A_5 - D_6 - D_{10} - A_4)$, $(A_5 - D_6 - uD_{10}u - A_4)$, $(A_5 - D_{10} - D_6 - A_4)$ or $(A_5 - D_{10} - uD_6u - A_4)$, respectively.

Remark. In the above definition of types, the subgroups C_2, A_4, D_6, D_{10}, uD_6u and $uD_{10}u$ are fixed (e.g. $D_6 =< x, y >$), and Lemma 1.2 (2) implies that they are those subgroups that contain $C_2 =< x >$. Hence by Lemma 1.1 (2), if we choose another involution in A_5 instead of $x = (1,2)(3,4)$, then we get conjugate ordered sets of subgroups, which means that, in some sense, the definition of type does not depend on the choice of an involution. However, notice that, fixing $C_2 =< x >$, the subgroups D_6 and D_{10} are choice-free. For example, there is no reason why $< x, uyu >$ is called uD_6u instead of D_6. So, once we fix an involution, the subgroups in the description of types should be considered just as themselves not as those isomorphism classes. Also, since there is an automorphism μ of A_5 such that $\mu(C_2) = C_2$, $\mu(D_6) = D_6$ and $\mu(D_{10}) = uD_{10}u$, equivalent manifolds may have different types. Here we say that two A_5-manifolds X and Y are equivalent if there are an automorphism μ of A_5 and a diffeomorphism σ from X onto Y such that the following diagram commutes. (Here the horizontal arrows mean actions of A_5.)

$$\begin{array}{ccc} A_5 \times X & \longrightarrow & X \\ {\scriptstyle \mu\times\sigma}\downarrow & & {\scriptstyle \sigma}\downarrow \\ A_5 \times Y & \longrightarrow & Y \end{array}$$

Now Lemma 1.1 and Propositions 1.5, 1.6 and 1.7 yield that the equivariant homeomorphism classes of the singular sets in S are determined by the above types. Namely;

PROPOSITION 1.9. *Let X and Y be in $\mathcal{MS}$. Then X_s and Y_s in S are A_5-homeomorphic if and only if they have the same type in the sense of Definition 1.8.*

Now we concentrate on a standard A_5-action on the Poincaré sphere Σ and figure out the type of its singular set. As is well known, for a nontrivial representation $\rho: A_5 \to SO(3)$, $\Sigma = \Sigma(\rho)$ is defined to be a (left) coset space of $SO(3)$ by a subgroup $\rho(A_5)$. Here $SO(3)$ is the special orthogonal group of degree three over the real number field. The *standard A_5-action* on $\Sigma(\rho)$ is the action $A_5 \times \Sigma(\rho) \to \Sigma(\rho)$; $(g, a\rho(A_5)) \mapsto \rho(g)a\rho(A_5)$. Notice that, by the definition, $\Sigma(\rho)$ is a three dimensional homology sphere with one fixed point A_5-action. Thus Proposition 1.5 implies that Σ lies in $\mathcal{MS}$. Moreover, it is easy to see that the tangential A_5-representation $T_p(\Sigma(\rho))$ at the fixed point p of $\Sigma(\rho)$ is A_5-isomorphic to the A_5-module $V(\rho)$ associated with ρ. Using this fact, we can see that $\Sigma(\rho)$ is A_5-diffeomorphic to $\Sigma(\rho')$ if and only if the associated characters χ_ρ and $\chi_{\rho'}$ coincide with each other. Note that A_5 has two inequivalent 3-dimensional irreducible real representations. Thus, there are two A_5-diffeomorphism types of the Poincaré spheres. (Furthermore, there are two A_5-homotopy types of the Poincaré spheres.)

Let us begin our computation. We use the special unitary group $SU_2(\mathbf{C})$ of degree 2 over the complex number field,

$$SU_2(\mathbf{C}) = \left\{ \begin{pmatrix} a & b \\ -\bar{b} & \bar{a} \end{pmatrix} \mid a, b \in \mathbf{C},\ |a|^2 + |b|^2 = 1 \right\},$$

for computational convenience. It is a double cover of $SO(3)$, i.e., $SU_2(\mathbf{C})$ has the center $\{\pm 1\}$ of order 2 and $SU_2(\mathbf{C})/\{\pm 1\}$ is isomorphic to $SO(3)$. Moreover, there is an injective homomorphism $\tilde{\rho}$ from the binary icosahedral group $SL(2,5)$ to $SU_2(\mathbf{C})$ such that the image of $\tilde{\rho}$ contains the center of $SU_2(\mathbf{C})$. Also, the factor group $\tilde{\rho}(SL(2,5))/\{\pm 1\}$ is isomorphic to A_5. Hence, not only as topological spaces, but also as A_5-spaces, Σ is diffeomorphic to $SU_2(\mathbf{C})/\tilde{\rho}(SL(2,5))$. (For these and related facts, we refer the reader to [28] §4.4.) We now construct $\tilde{\rho}$ concretely. Put

$$\begin{aligned}
\alpha &= \frac{1}{2}(\zeta + \zeta^{-1})^2 = 2\cos^2\frac{2\pi}{5}, \\
\gamma &= -\zeta^3 = -(\cos\frac{6\pi}{5} + \sqrt{-1}\sin\frac{6\pi}{5}), \\
\delta &= \frac{\zeta + \zeta^{-1}}{\zeta^2 - \zeta^{-2}} = -\frac{\sqrt{-1}\cos\frac{2\pi}{5}}{\sin\frac{4\pi}{5}} \quad \text{and} \\
\delta' &= \frac{1}{\zeta^2 - \zeta^{-2}} = -\frac{\sqrt{-1}}{2\sin\frac{4\pi}{5}},
\end{aligned}$$

where $\zeta = \cos\frac{2\pi}{5} + \sqrt{-1}\sin\frac{2\pi}{5}$. Note that $\overline{\alpha} = \alpha$, $\overline{\delta} = -\delta$ and $\overline{\delta'} = -\delta'$. Also, we define six matrices as follows:

$$A = \begin{pmatrix} 0 & 1 \\ -1 & 0 \end{pmatrix}, C = \begin{pmatrix} \gamma & 0 \\ 0 & \overline{\gamma} \end{pmatrix}, D = \begin{pmatrix} \delta & \delta' \\ \delta' & -\delta \end{pmatrix},$$

$$L = \frac{\sqrt{2}}{2}\begin{pmatrix} 1 & 1 \\ -1 & 1 \end{pmatrix}, M = \sqrt{\frac{-1}{1+\alpha^2}}\begin{pmatrix} -1 & \alpha \\ \alpha & 1 \end{pmatrix}, N = \sqrt{-1}\begin{pmatrix} \delta & \delta' \\ -\delta' & \delta \end{pmatrix}$$

It is easy to show that the above six matrices lie in $SU_2(\mathbf{C})$. Also, write E and B to mean the identity matrix in $SU_2(\mathbf{C})$ and the matrix $C^{-1}DC^{-1}$, respectively. Then, we have the following.

LEMMA 1.10.
(1) $A^2 = D^2 = C^5 = B^3 = (DC^2)^3 = -E$, $ACA^{-1} = C^{-1}$, $ADA^{-1} = -D$, $ABA^{-1} = B^{-1}$ and $DCD = CDAC$.
(2) *A, C and D generate a subgroup of $SU_2(\mathbf{C})$ isomorphic to the binary icosahedral group $SL(2,5)$.*
(3) $N^{-1}CN = D^{-1}CD, M^{-1}BM = B^{-1}$ and $L^{-1}DL = DA$.

PROOF: (1). For the equations not involving B, see p.93 of [28]. Also, noticing $D^{-1} = -D$, we have

$$B^3 = (C^{-1}DC^{-1})^3 = -C(DC^2)^{-3}C^{-1} = -CEC^{-1} = -E \text{ and}$$

$$ABA^{-1} = (AC^{-1}A^{-1})(ADA^{-1})(AC^{-1}A^{-1}) = -CDC = B^{-1}.$$

Also, (2) is found loc.cit., and (3) follows by easy computations.

By Lemma 1.10 (2), we henceforth identify $SL(2,5)$ with the subgroup of $SU_2(\mathbf{C})$ generated by the above A, C and D, that is, we regard $SL(2,5) =< A, C, D >$ and define $\tilde{\rho}$ to be the inclusion map from $SL(2,5)$ to $SU_2(\mathbf{C})$. We now write A', B', C' and D' to denote the elements of $SL(2,5)/\{\pm 1\}(\cong A_5)$ obtained as the images of A, B, C and D, respectively, under the natural epimorphism from $SU_2(\mathbf{C})$ onto $SO(3)$. Then, since Lemma 1.2(1) gives a generating set and relations of A_5, considering $\tilde{\rho}$ modulo the center $\{\pm 1\}$, Lemma 1.10 (1) yields the following. (See also p.93 of [28].)

LEMMA 1.11. *There exists an injective homomorphism ρ from A_5 to $SU_2(\mathbf{C})/\{\pm 1\}(\cong SO(3))$ sending x, y, z and u into A', B', C' and D', respectively.*

Remark. By direct calculation, we have $\chi_\rho(z) = \zeta + \zeta^{-1} + 1 = (1+\sqrt{5})/2$, where χ_ρ is the character associated with ρ. If we put $\zeta = \cos\frac{4\pi}{5} + \sqrt{-1}\sin\frac{4\pi}{5}$, then we similarly have an injective homomorphism ρ' from A_5 to $SO(3)$. However, this gives the Poincaré sphere whose A_5-diffeomorphism type is different from that of $\Sigma(\rho)$ since $\chi_{\rho'}(z) = (1-\sqrt{5})/2$.

In our computation the Poincaré sphere $\Sigma = \Sigma(\rho)$ given by the above homomorphism ρ will be used. For any matrix Q in $SU_2(\mathbf{C})$, we denote by $\overline{Q}$ the left coset of $\tilde{\rho}(SL(2,5))$ in $SU_2(\mathbf{C})$ containing Q. Thus, we may consider $\overline{Q}$ as a point in Σ. Now the points on the intersections of the circles in the singular set Σ_s are obtained as follows.

LEMMA 1.12. *It follows that* $p(A_4) = \overline{L}$, $p(D_6) = \overline{M}$, $p(D_{10}) = \overline{N}$, $p(uD_6u) = \overline{DM}$ *and* $p(uD_{10}u) = \overline{DN}$.

PROOF: Let Q be a matrix in $SU_2(\mathbf{C})$. Then $\overline{Q}$ lies in $\Sigma^{C_2} = \Sigma^x$ if and only if AQ and Q lie in the same coset of $SL(2,5)$, that is,

$$Q^{-1}AQ \text{ lies in } SL(2,5).$$

Since any two elements in $SL(2,5)$ of order 4 are $SL(2,5)$-conjugate, the above is equivalent to

$$Q^{-1}AQ = Q'AQ'^{-1}$$

for some Q' in $SL(2,5)$. Since $\overline{Q} = \overline{QQ'}$, we can conclude that if $\overline{Q}$ lies in Σ^x, then we can take Q from the centralizer $C_{SU_2(\mathbf{C})}(A)$ of A in $SU_2(\mathbf{C})$:

$$C_{SU_2(\mathbf{C})}(A) = \{Q \in SU_2(\mathbf{C}) \mid QA = AQ\}.$$

Conversely, for all Q in $C_{SU_2(\mathbf{C})}(A)$ the point $\overline{Q}$ clearly lies in Σ^x. On the other hand, by an easy computation, it follows that Q lies in $C_{SU_2(\mathbf{C})}(A)$ if and only if it has real entries. Thus, we get

$$C_{SU_2(\mathbf{C})}(A) = \left\{ \begin{pmatrix} \cos\theta & \sin\theta \\ -\sin\theta & \cos\theta \end{pmatrix} \mid \theta \in \mathbf{R} \right\}.$$

However, since $C_{SL(2,5)}(A) = \{\pm E, \pm A\}$, we may write

$$\Sigma^x = \left\{ \overline{Q} \mid Q = \begin{pmatrix} \cos\theta & \sin\theta \\ -\sin\theta & \cos\theta \end{pmatrix}, 0 \le \theta \le \pi/2 \right\}.$$

Namely, the above gives the circle Σ^x, on which the points are parameterized by θ in $0 \le \theta \le \pi/2$ with $\theta = 0$ and $\theta = \pi/2$ being identified. Now since L, MD and N have real entries, it follows that $\overline{L}$, $\overline{MD}(= \overline{M})$ and $\overline{N}$ lie in Σ^x. Next, notice that Lemma 1.10 (3) implies that

$$M^{-1}BM,\ N^{-1}CN \text{ and } L^{-1}DL$$

lie in $SL(2,5)$. Thus, for example, BM and M lie in the same coset of $SL(2,5)$, which implies that $\overline{M}$ lies in Σ^y. Likewise, $\overline{N}$ and $\overline{L}$ lie in Σ^z and Σ^u, respectively. Thus, $\overline{M} \in \Sigma^x \cap \Sigma^y (= \Sigma^{D_6})$, $\overline{N} \in \Sigma^x \cap \Sigma^z (= \Sigma^{D_{10}})$, and $\overline{L} \in \Sigma^x \cap \Sigma^u (= \Sigma^{D_4} = \Sigma^{A_4})$, and we obtain the first three equalities. Finally, since $\rho(u) = D'$, the last two equalities follow clearly.

The type of standard actions can be determined as follows.

THEOREM 1.13. *Let* $\rho\colon A_5 \to SO(3)$ *be a nontrivial representation. Then the singular set* $\Sigma(\rho)_s$ *of the Poincaré sphere* $\Sigma(\rho)$ *with standard* A_5*-action is of type* $(A_5 - D_6 -$

$uD_{10}u - A_4$) (resp. $(A_5 - D_6 - D_{10} - A_4)$) if $\chi_\rho(z) = (1+\sqrt{5})/2$ (resp. $(1-\sqrt{5})/2$), where χ_ρ is the character associated with ρ.

PROOF: First let $\rho\colon A_5 \to SO(3)$ be the homomorphism in Lemma 1.11. Then, we have $\chi_\rho(z) = (1+\sqrt{5})/2$. Recall the six matrices in Lemma 1.10. We look at the points $\overline{DM}$, $\overline{N}$ and $\overline{L}$. For a matrix Q in $SU_2(\mathbf{C})$, let us denote its (i,j) entry by $(Q)_{i,j}$. Then we have;

$$(DM)_{1,1} = \sqrt{\frac{-1}{1+\alpha^2}}(\alpha\delta' - \delta) = \frac{\cos\frac{2\pi}{5} - \cos^2\frac{2\pi}{5}}{\sin\frac{4\pi}{5}\sqrt{1+4\cos^4\frac{2\pi}{5}}} = 0.3568\ldots,$$

$$(DM)_{1,2} = \sqrt{\frac{-1}{1+\alpha^2}}(\alpha\delta + \delta') = \frac{1+4\cos^3\frac{2\pi}{5}}{2\sin\frac{4\pi}{5}\sqrt{1+4\cos^4\frac{2\pi}{5}}} = 0.9342\ldots,$$

$$(N)_{1,1} = \sqrt{-1}\delta = \frac{\cos\frac{2\pi}{5}}{\sin\frac{4\pi}{5}} = 0.5257\ldots,$$

$$(N)_{1,2} = \sqrt{-1}\delta' = \frac{1}{2\sin\frac{4\pi}{5}} = 0.8507\ldots \text{ and}$$

$$(L)_{1,1} = (L)_{1,2} = 0.7071\ldots.$$

Note that they are all real and positive. Also, we have $0 < (DM)_{1,1} < (N)_{1,1} < (L)_{1,1}$. Hence, by the argument in the proof of Lemma 1.12, we can determine the order of the points on $\Sigma(\rho)^x$ as in the statement of the theorem, which shows that the type of $\Sigma(\rho)_s$ is $(A_5 - D_6 - uD_{10}u - A_4)$.

Let μ be an automorphism of A_5 which is not an inner automorphism. (Note : Those μ are actually given by conjugation by some elements that lie in the symmetric group on five letters but not in A_5.) Then, we have $\chi_{\rho\mu}(z) = (1-\sqrt{5})/2$. And there is such an automorphism μ that satisfies $\mu(C_2) = C_2$, $\mu(D_6) = D_6$, $\mu(D_{10}) = uD_{10}u$ and $\mu(A_4) = A_4$. (e.g. The conjugation by the transposition $(1,2)$.) Hence it follows that the type of $\Sigma(\rho\mu)_s$ is $(A_5 - D_6 - D_{10} - A_4)$. Since A_5-diffeomorphism type of a Poincaré sphere is determined by the character, this completes the proof of Theorem 1.13.

Remark. For any nontrivial real A_5-representation $\rho\colon A_5 \to SO(3)$ and any type γ of the singular set, there exists a three dimensional homology sphere with one fixed point A_5-action whose tangential representation at the unique fixed point is isomorphic to $V(\rho)$, whose type is γ and which is A_5-cobordant to $\Sigma(\rho)$, where $V(\rho)$ is the A_5-module associated with ρ. This will be proved in [4].

2. Existence of equivariant cobordisms

Let G be a finite group. In this section, we suppose that a G-manifold, a real G-module and a real G-vector bundle possess a G-invariant riemannian metric, a G-invariant inner product, and a G-invariant metric, respectively, and that all are oriented.

We explain notations and terminologies which will be used in the rest of this paper. Let X be a topological G-space. For a real G-module M, let $\varepsilon_X(M)$ be a real G-vector bundle whose total space is $X \times M$ with diagonal G-action, base space is X and fiber is M. If the base space is clear from the context, we write $\varepsilon(M)$ instead of $\varepsilon_X(M)$. Let ξ and η be real G-vector bundles over X. If there is a G-vector bundle isomorphism α from $\xi \oplus \varepsilon_X(M)$ to $\eta \oplus \varepsilon_X(M)$ for some real G-module M, we say that ξ and η are *stably G-isomorphic*. We usually write this isomorphism simply by $\alpha\colon \xi \to \eta$ instead of the precise description such as $\alpha\colon \xi \oplus \varepsilon_X(M) \to \eta \oplus \varepsilon_X(M)$ and call it a *stable* isomorphism from ξ to η. This notation will be used even if α is actually an isomorphism from ξ to η (not from $\xi \oplus \varepsilon_X(M)$ to $\eta \oplus \varepsilon_X(M)$). However, if this is the case, we call α an *unstable* isomorphism from ξ to η. Stable or unstable isomorphisms of a particular type will have the following special names. A *stable* (resp. an *unstable*) *G-trivialization* is a stable (resp. an unstable) G-vector bundle isomorphism $\alpha\colon \xi \to \varepsilon_X(V)$ for some real G-module V. Note that a stable G-trivialization α is actually an isomorphism from $\xi \oplus \varepsilon_X(M)$ to $\varepsilon_X(V \oplus M)$ for some real G-module M.

For a real G-vector bundle ξ over X and $H \subseteqq G$, let ξ^H denote the H-fixed bundle of ξ over X^H, and let ξ_H be its orthogonal complement in $\xi\,|_{X^H}$. So, in particular, we may write $\varepsilon_X(M)^H = \varepsilon_{X^H}(M^H)$ and $\varepsilon_{X^H}(M)_H = \varepsilon_{X^H}(M_H)$, where M^H is the submodule of H-fixed elements in M and M_H is its orthogonal complement in M. Suppose that we have an unstable G-isomorphism $\alpha_X\colon \xi \oplus \varepsilon_X(M) \to \xi' \oplus \varepsilon_X(M)$. Then, for each subgroup H of G we have unstable $N_G(H)$-isomorphisms

$$\alpha_X^H\colon \xi^H \oplus \varepsilon_{X^H}(M^H) \to \xi'^H \oplus \varepsilon_{X^H}(M^H) \text{ and}$$
$$\alpha_{X^H}\colon \xi_H \oplus \varepsilon_{X^H}(M_H) \to \xi'_H \oplus \varepsilon_{X^H}(M_H).$$

Finally, we give the following remark, which is used freely thereafter. Let $\alpha_X\colon \xi \to \varepsilon_X(V)$ and $\alpha_Y\colon \xi' \to \varepsilon_Y(V)$ be stable G-trivializations. Then they are precisely unstable isomorphisms

$$\alpha_X\colon \xi \oplus \varepsilon_X(M_X) \to \varepsilon_X(V \oplus M_X) \quad \text{and}$$
$$\alpha_Y\colon \xi' \oplus \varepsilon_Y(M_Y) \to \varepsilon_Y(V \oplus M_Y),$$

for some real G-modules M_X and M_Y, which may be different. However taking stabilizations of them if necessary, we can regard them as

$$\alpha_X\colon \xi \oplus \varepsilon_X(M) \to \varepsilon_X(V \oplus M) \quad \text{and}$$
$$\alpha_Y\colon \xi' \oplus \varepsilon_Y(M) \to \varepsilon_Y(V \oplus M),$$

for the same real G-module M (e.g. $M = M_X \oplus M_Y$).

In the rest of this section, we assume that our group G is A_5. As in the previous section, denote by $\mathcal{H}$ the family of all subgroups of A_5 isomorphic to A_4, D_{10} or D_6. And for a manifold X in $\mathcal{MS}$ and H in $\mathcal{H}$, we let $p(H) = p_X(H)$ be the point in X with isotropy subgroup H. Also, let $T(X)$ be the tangent bundle of X, and let $T_p(X)$ be the fiber of $T(X)$ over p in X. Now we can state our main theorem in this section.

THEOREM 2.1. *Let X and Y be manifolds in $\mathcal{MS}$. Suppose that they have stable A_5-trivializations $\alpha_X\colon T(X) \to \varepsilon_X(V)$ and $\alpha_Y\colon T(Y) \to \varepsilon_Y(V)$ for the same real A_5-module V. Then X and Y are A_5-cobordant.*

Remark. This V must be an irreducible real A_5-module of dimension three.

Let $\alpha_X: T(X) \oplus \varepsilon_X(M) \to \varepsilon_X(V \oplus M)$ be an unstable A_5-trivialization such that the map between the fibers over the point $p(A_5)$,

$$\alpha_X^{D_4} |_{p(A_5)}: T_{p(A_5)}(X)^{D_4} \oplus M^{D_4} \to V^{D_4} \oplus M^{D_4} \quad \text{(i.e.,} \quad M^{D_4} \to M^{D_4}),$$

is orientation preserving. The set of those trivializations can be written as a disjoint union of two subsets. The elements in one subset are called type plus and those in the other subset are called type minus. We now explain this fact. First, notice that we can adopt the orientation of V so that the restriction

$$\alpha_X |_{p(A_5)}: T_{p(A_5)}(X) \oplus M \to V \oplus M$$

preserves the orientation. This orientation on V is regarded as the orientation of V_{D_4}, since $V_{D_4} = V$ as sets. Consequently, the restrictions

$$\alpha_X^{D_4} |_{p(A_5)}: T_{p(A_5)}(X)^{D_4} \oplus M^{D_4} \to V^{D_4} \oplus M^{D_4} \quad \text{and}$$
$$\alpha_{X D_4} |_{p(A_5)}: T_{p(A_5)}(X)_{D_4} \oplus M_{D_4} \to V_{D_4} \oplus M_{D_4}$$

preserve the orientation. This implies that the restrictions

$$\alpha_X^{D_4} |_{p(A_4)}: T_{p(A_4)}(X)^{D_4} \oplus M^{D_4} \to V^{D_4} \oplus M^{D_4} \quad \text{and}$$
$$\alpha_{X D_4} |_{p(A_4)}: T_{p(A_4)}(X)_{D_4} \oplus M_{D_4} \to V_{D_4} \oplus M_{D_4}$$

both preserve the orientation or both reverse the orientation. In the former case, we say that α_X is of type *plus* (or +), while in the latter case we say that α_X is of type *minus* (or −). Here we give an example.

Example 2.2 (cf. [8] or [9] Section 2). Let V be an irreducible real A_5-module of dimension three, and let ρ be the homomorphism from A_5 to $SO(3)$ associated with V. Then, ρ determines the Poincaré sphere Σ with standard action. The tangential representation of Σ at the unique fixed point is isomorphic to V. Moreover there exists an unstable A_5-trivialization $\alpha_{\Sigma+}: T\Sigma \to \varepsilon_\Sigma(V)$, which is of type plus.

Remark. Let $X \in \mathcal{MS}$, and let $\alpha_X: T(X) \to \varepsilon_X(V)$ be a stable A_5-trivialization. We write it as an unstable isomorphism $\alpha_X: T(X) \oplus \varepsilon(M) \to \varepsilon(V \oplus M)$. In the case where $\dim M^{A_5} \geq 1$ (i.e. $M \supseteqq \mathbf{R}$ as A_5-modules), we can find an unstable A_5-isomorphism $\alpha'_X: T(X) \oplus \varepsilon(M) \to \varepsilon(V \oplus M)$ such that $\alpha'^{D_4}_X |_{p(A_5)}: M^{D_4} \to M^{D_4}$ is orientation preserving, as follows. If $\alpha_X^{D_4} |_{p(A_5)}$ is orientation preserving, then obviously we can set $\alpha'_X = \alpha_X$. Clearly, the map $-1: \mathbf{R} \to \mathbf{R}$ is orientation reversing. If $\alpha_X^{D_4} |_{p(A_5)}$ is orientation reversing, then we can take α'_X as the composition of α_X and the stabilization of $\varepsilon_X(-1): \varepsilon_X(\mathbf{R}) \to \varepsilon_X(\mathbf{R})$. Thus, for the proof of Theorem 2.1, we may restrict stable A_5-trivializations α_X to ones such that $\alpha_X^{D_4} |_{p(A_5)}: M^{D_4} \to M^{D_4}$ are orientation preserving (i.e. ones having a type plus or minus).

The rest of this section is devoted to proving the above theorem. The proof consists of several lemmas (Lemmas 2.3 – 2.9), in which we state the assertions in general context. In the first lemma, we shall show that, given a stable A_5-trivialization of any type, we can construct another stable A_5-trivialization of the other type. So, to prove the theorem, we may assume that X and Y both have the trivializations of type minus.

LEMMA 2.3. *Let X lie in $\mathcal{MS}$. Suppose that there is a stable A_5-trivialization α_{X_ϱ}: $T(X) \to \varepsilon_X(V)$ of type $\varrho = +$ or $-$, then there also exists a stable A_5-trivialization $\alpha_{X_{-\varrho}}: T(X) \oplus \varepsilon_X(\mathbf{R}) \to \varepsilon_X(V \oplus \mathbf{R})$ of type $-\varrho$.*

PROOF: Note that V is an irreducible A_5-module of dimension three. So, we have a homomorphism ρ from A_5 to $SO(3)$ as in Example 2.2. Then, we can take a covering $\tilde{\rho}: SL(2,5) \to SU_2(\mathbf{C})$ of ρ, and identify $SU_2(\mathbf{C})$ with $S(\mathbf{H})$, the unit sphere of the quaternion field $\mathbf{H}$. Thus we obtain a four dimensional real $SL(2,5)$-module $\mathbf{H}_{adj}$ by sending $(g, a) \in SL(2,5) \times \mathbf{H}$ to $\tilde{\rho}(g)a\tilde{\rho}(g)^{-1} \in \mathbf{H}$. This $\mathbf{H}_{adj}$ can be regarded as an A_5-module, and moreover is isomorphic to $V \oplus \mathbf{R}$. Hence, we identify them, namely, $\mathbf{H}_{adj} = V \oplus \mathbf{R}$. Let $f: X \to S(\mathbf{H}_{adj})$ be the A_5-map obtained by pinching the outside of the open A_5-disk neighborhood of the fixed point $p(A_5)$ of X. We may assume that $f(p(A_5)) = 1 \in \mathbf{H}_{adj}$. Then, we get $f(p(H)) = -1$ for H in $\mathcal{H}$. For an integer k, define an A_5-trivialization $\mathrm{Twist}_k(\alpha_{X_\varrho}): T(X) \oplus \varepsilon_X(\mathbf{R}) \to \varepsilon_X(V \oplus \mathbf{R})$ by

$$\mathrm{Twist}_k(\alpha_{X_\varrho})(a) = \alpha_{X_\varrho}(a) f(p(a))^k \quad \text{for} \quad a \in T(X) \oplus \varepsilon_X(\mathbf{R}),$$

where $p: T(X) \oplus \varepsilon_X(\mathbf{R}) \to X$ is the bundle projection, and the multiplication is taken in the fiber $\mathbf{H}_{adj} = V \oplus \mathbf{R}$. Since $\mathrm{Twist}_k(\alpha_{X_\varrho})$ is of type $(-1)^k \varrho$, taking $\mathrm{Twist}_1(\alpha_{X_\varrho})$, the lemma is proved.

Later "Twist" is used again. Note also that if k is even, then the restriction of $\mathrm{Twist}_k(\alpha_{X_\varrho})$ to the fiber over $p(H)$ ($H \in \mathcal{H}$ or $H = G$) is the same as that of α_{X_ϱ}. Concerning the type of the singular sets, we have the following, by which, in the proof of the theorem, we may assume that X_s and Y_s have the same type. (Note : A_5-surgery does not change cobordism classes.)

LEMMA 2.4. *Let X be a manifold in $\mathcal{MS}$, and let $\alpha_X: T(X) \to \varepsilon_X(V)$ be a stable A_5-trivialization of type ϱ. Choose an arbitrary type γ of the singular set. Then, one can perform A_5-surgery on X of isotropy type (C_2) to obtain a manifold X' in $\mathcal{MS}$ of type γ and a stable A_5-trivialization $\alpha_{X'}: T(X') \to \varepsilon_{X'}(V)$ of type ϱ.*

PROOF: Let X_s be of type $(A_5 - H_1 - H_2 - A_4)$. Take a point p of X^{C_2} between the points $p(H_1)$ and $p(H_2)$ which are points in X with isotropy subgroups H_1 and H_2, respectively. Further take embeddings $\phi_i: S^0 \to X^{C_2}$, $i = 1, 2$, such that

(1) $\phi_1(1)$ lies between $p(A_5)$ and $p(H_1)$,
(2) $\phi_1(-1)$ lies between $p(H_1)$ and p,
(3) $\phi_2(1)$ lies between p and $p(H_2)$, and
(4) $\phi_2(-1)$ lies between $p(H_2)$ and $p(A_4)$.

Consider D_4-surgery on X^{C_2} along $\mathrm{ind}_{C_2}^{D_4} \phi_i: D_4/C_2 \times S^0 \to X^{C_2}$, $i = 1, 2$. Then X^{C_2} is changed to a D_4-space $Y(X)$ consisting of five circles as in Figure 2.5. We note that the D_4-diffeomorphism type of $Y(X)$ is independent of any initially given X in $\mathcal{MS}$. From this observation it holds that by A_5-surgery of isotropy type (C_2) on X, we can obtain a manifold X' in $\mathcal{MS}$ of type γ. Since A_5-surgery employed here is of isotropy type (C_2) and of dimension 0, there are no obstructions to obtaining a stable A_5-trivialization of type ϱ after the surgery. This proves the lemma.

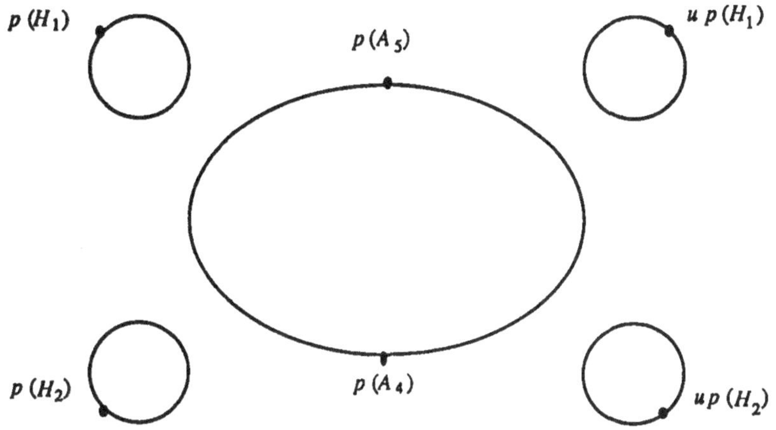

Figure 2.5.

We denote the points in X and Y with isotropy subgroup H in $\mathcal{H}$ by $p(H) = p_X(H)$ and $q(H) = p_Y(H)$, respectively, for notational convenience. Since we may assume that $T_{p(A_5)}(X)$ and $T_{q(A_5)}(Y)$ are isomorphic in the proof of the theorem, the next lemma shows that we can choose closed A_5-regular neighborhoods $RN(A_5, X_s)$ of X_s in X and $RN(A_5, Y_s)$ of Y_s, between which there is an orientation preserving A_5-diffeomorphism.

LEMMA 2.6. *Let X and Y be manifolds in $\mathcal{MS}$. Then, we can choose closed A_5-regular neighborhoods $RN(A_5, X_s)$ and $RN(A_5, Y_s)$ such that there exists an orientation preserving A_5-diffeomorphism $\beta: RN(A_5, X_s) \to RN(A_5, Y_s)$ if and only if X and Y have singular sets of the same type and the tangential A_5-representations $T_{p(A_5)}(X)$ and $T_{q(A_5)}(Y)$ are isomorphic to each other.*

PROOF: Since the 'only if' part is obvious, we prove the 'if' part. Note that the two A_5-modules $T_{p(A_5)}(X)$ and $T_{q(A_5)}(Y)$ have orientations since X and Y are oriented. Multiplying the real number -1 if necessary, we may assume that the isomorphism from $T_{p(A_5)}(X)$ to $T_{q(A_5)}(Y)$ preserves the orientation. By the equivariant tubular neighborhood theorem, $T_{p(A_5)}(X)$ (resp. $T_{q(A_5)}(Y)$) can be regarded as an open A_5-disk neighborhood of $p(A_5)$ (resp. $q(A_5)$). Taking the restriction to the unit disk, we obtain an orientation preserving A_5-diffeomorphism from a closed A_5-disk neighborhood $RN(A_5, p(A_5))$ of $p(A_5)$ to $RN(A_5, q(A_5))$ of $q(A_5)$. Let H be a subgroup in $\mathcal{H}$. If C is a nontrivial cyclic subgroup of H, then X^C and Y^C are connected, and hence

$$\operatorname{res}^H_C T_{p(H)}(X) \cong \operatorname{res}^{A_5}_C T_{p(A_5)}(X) \cong \operatorname{res}^{A_5}_C T_{q(A_5)}(Y) \cong \operatorname{res}^H_C T_{q(H)}(Y).$$

This implies that $T_{p(H)}(X)$ is H-isomorphic to $T_{q(H)}(Y)$. Similarly to the above, we can obtain an orientation preserving H-diffeomorphism from a closed H-disk neighborhood $RN(H,p(H))$ of $p(H)$ to $RN(H,q(H))$ of $q(H)$. We put

$$X_{\mathcal{H}} = \bigcup_{H\in\mathcal{H}} X^H.$$

A closed A_5-tubular neighborhood $RN(A_5, X_{\mathcal{H}})$ of $X_{\mathcal{H}}$ is obtained as the disjoint union of $RN(A_5,p(A_5))$ and $RN(H,p(H))$'s, where H runs over $\mathcal{H}$. From the above argument, there exists an orientation preserving A_5-diffeomorphism from $RN(A_5, X_{\mathcal{H}})$ to $RN(A_5, Y_{\mathcal{H}})$. Let $\mathcal{C}$ be the set of all nontrivial cyclic subgroups of A_5. Since Proposition 1.9 implies that X_s is A_5-homeomorphic to Y_s, for $C \in \mathcal{C}$ we can choose a closed, thin, $N_{A_5}(C)$-tubular neighborhood $RN(N_{A_5}(C), X^C)$ of X^C. Then a closed equivariant regular neighborhood $RN(A_5, X_s)$ can be obtained as follows.

$$RN(A_5, X_s) = RN(A_5, X_{\mathcal{H}}) \cup \bigcup_{C\in\mathcal{C}} RN(N_{A_5}(C), X^C).$$

Since $\dim X_s = 1$ and $N_{A_5}(C)/C \cong C_2$ for all $C \in \mathcal{C}$, the above orientation preserving A_5-diffeomorphism from $RN(A_5, X_{\mathcal{H}})$ to $RN(A_5, Y_{\mathcal{H}})$ can be easily extended to one from $RN(A_5, X_s)$ to $RN(A_5, Y_s)$. This proves Lemma 2.6.

Now suppose that X and Y lie in $\mathcal{MS}$ and that there are stable A_5-trivializations $\alpha_X: T(X) \to \varepsilon_X(V)$ and $\alpha_Y: T(Y) \to \varepsilon_Y(V)$ of type minus. Moreover, assume that there is an orientation preserving A_5-diffeomorphism β from a closed A_5-regular neighborhood $RN(A_5, X_s)$ of X_s in X to $RN(A_5, Y_s)$ of Y_s. Since the general 'fiber' of $RN(A_5, X_s)$ over X^{C_2} is a two dimensional disk, for each integer k we obtain an A_5-selfdiffeomorphism of $RN(A_5, X_s)$ by equivariantly twisting the 'fiber' k-times along X^{C_2}. Denote this selfdiffeomorphism by γ_k, and set $\beta_k = \beta\gamma_k$. For any subgroup H of A_5, we can choose a closed equivariant regular neighborhood $RN(N_{A_5}(H), X_{s(H)})$ of $X_{s(H)}$ in X so that

$$RN(N_{A_5}(L), X_{s(L)}) \subset RN(N_{A_5}(K), X_{s(K)}) \quad \text{whenever} \quad \{1\} \neq L \subset K \subset A_5,$$

and we can regard $RN(N_{A_5}(H), Y_{s(H)}) = \beta(RN(N_{A_5}(H), X_{s(H)}))$.

Now consider the stable, real A_4-vector bundle map

$$\phi_{i,j} = (\alpha_Y \mid)(d\beta_j \mid)(\mathrm{Twist}_{2i}(\alpha_X) \mid)^{-1}: RN(A_4, X_{s(D_4)}) \times V \to RN(A_4, Y_{s(D_4)}) \times V.$$

For "Twist", see the proof of Lemma 2.3. Notice that $\phi = \phi_{i,j}$ is actually a map from $RN(A_4, X_{s(D_4)}) \times (V \oplus M)$ to $RN(A_4, Y_{s(D_4)}) \times (V \oplus M)$ for some real A_5-module M. Moreover, $\phi\mid_{p(A_5)}: V \oplus M \to V \oplus M$ (the restriction of ϕ to the fiber over the A_5-fixed point) is an A_5-isomorphism and the map $\phi\mid_{p(A_4)}: V \oplus M \to V \oplus M$ is an A_4-isomorphism, which are independent of i and j. Let $\mathbf{R}$, U and W be irreducible real A_4-modules of dimension $1, 2$ and 3, respectively. Then, $\mathrm{res}^{A_5}_{A_4} V \cong W$, and for adequate integers ℓ, m and n, we have

$$\mathrm{res}^{A_5}_{A_4}(V \oplus M) \cong \ell\mathbf{R} \oplus mU \oplus nW.$$

In particular, $\dim(V \oplus M) = \ell + 2m + 3n$. In this situation we have the following.

LEMMA 2.7. *Suppose that the map $\phi_{0,0} |_{p(A_5)}$ is regularly A_5-homotopic to the identity map. (Note: By the remark above, this becomes true for all $\phi_{i,j}$.) If $\ell \geq 3$ and $n \geq 3$, then there exist integers i and j such that the A_4-vector bundle map $\phi = \phi_{i,j}$ is regularly A_4-homotopic to the product map of the base map with the identity map on the fiber $V \oplus M$.*

PROOF: We use the notation: $G = A_5$, $H = A_4$, $D = D_4$ and $C = C_2$. First consider $\phi = \phi_{i,j}$ for arbitrarily fixed i and j. It may be assumed without loss of generality that

$$\phi |_{RN(G,p(G))}: RN(G, X^G) \times (V \oplus M) \to RN(G, Y^G) \times (V \oplus M)$$

is the product map of the base map with the identity map on the fiber.

We note that the space $Aut(H, V \oplus M)$ of H-automorphisms of $V \oplus M$ is homeomorphic to

$$GL_\ell(\mathbf{R}) \times GL_m(\mathbf{C}) \times GL_n(\mathbf{R})$$

by Schur's lemma. And $\phi |_{p(H)}: \ell\mathbf{R} \oplus mU \oplus nW \to \ell\mathbf{R} \oplus mU \oplus nW$ is a direct sum of isomorphisms $\phi_L: L \to L$, where $L = \ell\mathbf{R}, mU$ or nW. Since $Aut(H, mU) \cong GL_m(\mathbf{C})$ is connected, ϕ_{mU} is regularly H-homotopic to the identity map on mU. Without loss of generality, we may assume that ϕ_{mU} is the identity map. Since $\mathrm{Twist}_{2i}(\alpha_X)$ and α_Y are of type minus, $\phi_{\ell\mathbf{R}}$ and ϕ_{nW} are orientation preserving. This fact implies that $\phi_{\ell\mathbf{R}}$ (resp. ϕ_{nW}) is regularly H-homotopic to the identity map on $\ell\mathbf{R}$ (resp. nW). Hence we may assume that $\phi_{\ell\mathbf{R}}$ and ϕ_{nW} are the identity maps, and consequently that $\phi |_{p(H)}$ is the identity map. Thus we can assume that

$$\phi |_{RN(H,X^H)}: RN(H, X^H) \times (V \oplus M) \to RN(H, Y^H) \times (V \oplus M)$$

is the product map of the base map with the identity map on the fiber.

The H-space $X_{s(D)} \setminus \mathrm{Int}RN(H, X^D)$ consists of six line segments, among which there is no H-fixed one. Let $[a,b]$ be one of the two line segments lying in X^C. If $\phi |_{[a,b]}$ is regularly C-homotopic to the product map ι of the base map with the identity map on the fiber relatively to the boundary $\{a,b\}$, then we can conclude that the map ϕ is regularly H-homotopic to the product map of the base map with the identity map on the fiber. The obstruction σ to constructing a regular C-homotopy between $\phi |_{[a,b]}$ and ι (relatively to the boundary) lies in $\pi_1(Aut(C, V \oplus M))$. We note that $Aut(C, V \oplus M) = Aut(C, (V \oplus M)^C) \times Aut(C, (V \oplus M)_C)$. Thus the obstruction σ can be written as (σ_1, σ_2), where $\sigma_1 \in \pi_1(GL_{\ell+2m+n}(\mathbf{R})) \cong \mathbf{Z}/2$ and $\sigma_2 \in \pi_1(GL_{2n}(\mathbf{R})) \cong \mathbf{Z}/2$.

Now consider the effect of changing the choice of i and j. If we replace $\phi_{i,j}$ by $\phi_{i+k,j}$, then σ_1 does (resp. does not) change if k is odd (resp. even). The change of j has similar effect on σ_2. However, $\phi_{i,j}$ and $\phi_{i,j+k}$ give the same σ_1. Thus, we can find integers i and j such that the obstruction σ vanishes. Therefore, we have proved Lemma 2.7.

Return to the proof of Theorem 2.1. By the argument given so far, we may assume that we are in the situation in the paragraph preceding Lemma 2.7. Now take M sufficiently large so that M includes at least three isomorphic copies of each irreducible real A_5-modules. Then, the conditions $\ell \geq 3$ and $n \geq 3$ are satisfied. If $\phi |_{p(A_5)}$ (for α_X

and α_Y) is not regularly A_5-homotopic to the identity map, then, in the following way, we can construct $\alpha'_Y: T(Y) \oplus \varepsilon_Y(M') \to \varepsilon_Y(V \oplus M')$ such that $\phi' |_{p(A_5)}$ for α_X and α'_Y is regularly A_5-homotopic to the identity map. Let α''_Y be the map

$$\alpha_Y \oplus \varepsilon_Y(\phi |_{p(A_5)}): (T(Y) \oplus \varepsilon_Y(M)) \oplus \varepsilon_Y(V \oplus M) \to \varepsilon_Y((V \oplus M) \oplus (V \oplus M)).$$

Then we can obtain a stable A_5-trivialization α'_Y of type minus by the method used in the proof of Lemma 2.3 (for $M' = M \oplus V \oplus M \oplus \mathbf{R}$) and it is easy to see that $\phi' |_{p(A_5)}$ for α_X and α'_Y is regularly A_5-homotopic to the identity map. Thus we may assume that the conclusion of Lemma 2.7 holds. Hence, changing α_X and β suitably, we may assume that ϕ is regularly A_4-homotopic to the product map of the base map with the identity map on the fiber.

For a subgroup H of A_5, we define an $N_{A_5}(H)$-manifold $Z(H, X)$ by

$$Z(H, X) = X \setminus \mathrm{Int} RN(N_{A_5}(H), X_{s(H)}).$$

We define $Z(H, Y)$ similarly. Glue $Z(H, Y)$ and $Z(H, X)$ along the boundary by the restriction of the A_5-diffeomorphism β from $RN(A_5, X_s)$ to $RN(A_5, Y_s)$ and get a closed $N_G(H)$-manifold $Z(H)$ with free H-action, that is,

$$Z(H) = Z(H, Y) \cup_{\beta|} Z(H, X).$$

Then, by Lemma 2.7 we may assume that the A_4-manifold $Z(D_4)$ has the tangent bundle stably A_4-isomorphic to $\varepsilon_{Z(D_4)}(V)$. Since the D_4-action on $Z(D_4)$ is free, we have the classifying map $f_{D_4}: Z(D_4)/D_4 \to BD_4$ of the principal D_4-bundle $Z(D_4) \to Z(D_4)/D_4$, where BD_4 is the classifying space of principal D_4-bundles.

LEMMA 2.8. *The map $(Z(D_4)/D_4, f_{D_4})$ is null cobordant.*

PROOF: We abbreviate $Z(D_4)$ to Z. Let $w_j = w_j(Z/D_4) \in H^j(Z/D_4; \mathbf{Z}/2)$ be the j-th Stiefel-Whitney class of the manifold Z/D_4. For each partition $k + k_1 + ... + k_r = 3$ and each element $c \in H^k(BD_4; \mathbf{Z}/2)$, the element

$$< w_{k_1} ... w_{k_r} f^*_{D_4}(c), [Z/D_4] >$$

in $\mathbf{Z}/2$ is called a bordism Stiefel-Whitney number of $(Z/D_4, f_{D_4})$, where $[Z/D_4]$ is the orientation class in $H_3(Z/D_4; \mathbf{Z}/2)$. From bordism theory, it follows that the bordism class $[Z/D_4, f_{D_4}]$ is null if all the bordism Stiefel-Whitney numbers are zero, (see P. E. Conner and E. E. Floyd [7] Chapter II Theorem 17.2 or F. Uchida [30] Theorems 2.15 and 2.18). Since Z/D_4 is orientable, closed, three dimensional manifold, its tangent bundle is stably trivial. Thus, all the Stiefel-Whitney classes vanish. It follows that $[Z/D_4, f_{D_4}]$ is null if $f^*_{D_4}(c) = 0$ for all elements $c \in H^3(BD_4; \mathbf{Z}/2)$. Let $D_4 = C_2 \times C'_2$, where $C'_2 \subset D_4$. Then $BD_4 = BC_2 \times BC'_2$, and we can regard $H^*(BC_2; \mathbf{Z}/2)$ as the polynomial ring $\mathbf{Z}/2[\ a\ ]$ of indeterminate a and $H^*(BC'_2; \mathbf{Z}/2)$ as $\mathbf{Z}/2[\ b\ ]$ of indeterminate b. Let π be the projection from BD_4 to BC_2. We observe the cohomology element $(\pi f_{D_4})^*(a^2)$ in $H^2(Z/D_4; \mathbf{Z}/2)$. The restriction of the homomorphism $\rho: A_5 \to SO(3)$ associated with V to the subgroup C_2 is conjugate to the homomorphism given by

$$x \mapsto \begin{pmatrix} -1 & 0 & 0 \\ 0 & -1 & 0 \\ 0 & 0 & 1 \end{pmatrix},$$

where x is the generator of C_2. Thus, $(\pi f_{D_4})^*(a^2)$ coincides with the second Stiefel-Whitney class $w_2(\varepsilon_Z(V)/D_4)$ of the vector bundle $\varepsilon_Z(V)/D_4$ over Z/D_4. Since the tangent bundle $T(Z/D_4)$ (which is stably isomorphic to $\varepsilon_Z(V)/D_4$) is stably trivial, $w_2(T(Z/D_4)) = 0 = w_2(\varepsilon_Z(V)/D_4)$. Thus, $(\pi f_{D_4})^*(a^2) = 0$ and $f_{D_4}^*(a^2) = 0$. This implies $f_{D_4}^*(a^3) = f_{D_4}^*(a^2)f_{D_4}^*(a) = 0$. Similarly we obtain $f_{D_4}^*(b^2) = 0$ and $f_{D_4}^*(b^3) = 0$. Since $H^3(BD_4; \mathbf{Z}/2)$ has a basis $\{a^3, a^2b, ab^2, b^3\}$ over $\mathbf{Z}/2$, we see that $f_{D_4}^*(c) = 0$ for all elements $c \in H^3(BD_4; \mathbf{Z}/2)$. Consequently, $(Z/D_4, f_{D_4})$ is null cobordant. This completes the proof.

Let $f_{A_5}: Z(A_5)/A_5 \to BA_5$ be the classifying map of the principal A_5-bundle $Z(A_5) \to Z(A_5)/A_5$. The following lemma completes the proof of Theorem 2.1.

LEMMA 2.9. *The map $(Z(A_5)/A_5, f_{A_5})$ is null cobordant. Consequently, X and Y are A_5-cobordant (relatively to the singular set) to each other.*

PROOF: If all the cobordism Stiefel-Whitney numbers of $(Z(A_5)/A_5, f_{A_5})$ are zero, then $(Z(A_5)/A_5, f_{A_5})$ is null cobordant. First note that all the Stiefel-Whitney classes of $Z(A_5)/A_5$ vanish. Thus, $[Z(A_5)/A_5, f_{A_5}]$ is null if $f_{A_5}^*(a) = 0$ for all elements $a \in H^3(BA_5; \mathbf{Z}/2)$. Since D_4 is a Sylow 2-subgroup of A_5,

$$\pi_{A_5/D_4}^*: H^3(Z(A_5)/A_5; \mathbf{Z}/2) \to H^3(Z(A_5)/D_4; \mathbf{Z}/2)$$

is injective (see G. E. Bredon [5] p.121). It follows that $(Z(A_5)/A_5, f_{A_5})$ is null cobordant if $(Z(A_5)/D_4, f'_{D_4})$ is null cobordant, where $f'_{D_4}: Z(A_5)/D_4 \to BD_4$ is the classifying map. It is easy to see that $(Z(A_5)/D_4, f'_{D_4})$ is cobordant to $(Z(D_4)/D_4, f_{D_4})$ which is null cobordant by Lemma 2.8. Thus, $(Z(A_5), f_{A_5})$ is null cobordant. This implies that $Z(A_5)$ is null A_5-cobordant, and also that X and Y are A_5-cobordant (relatively to the singular set) to each other. This completes the proof.

3. G-normal maps

Let G be a finite group. In this section, we introduce the notion of G-normal maps and G-normal cobordisms defined by T. Petrie [23] and prove Proposition 3.5 below, in which we construct a G-normal map and a G-normal cobordism from a real G-module. These will be used in the next section for the proof of Theorem 0.1.

Given a finite G-CW-complex X, the G-poset $\Pi(X)$ associated with X is defined by

$$\Pi(X) = \coprod_{H \subseteqq G} \pi_0(X^H);$$

see [19] and [23]. For $\alpha \in \Pi(X)$, we set $G_\alpha = \{g \in G \mid g\alpha = \alpha\}$. A G-vector bundle ξ (with G-invariant metric) over X gives a G_α-vector bundle $\pi_\alpha\xi$ over X_α (the underlying space of α) by

$$\xi|X_\alpha = \xi^H|X_\alpha \oplus \pi_\alpha\xi,$$

where H is the subgroup such that $\alpha \in \pi_0(X^H)$. The collection $\pi\xi = \{\pi_\alpha \xi \mid \alpha \in \Pi(X)\}$ is called a $\Pi(X)$-vector bundle over X. If ξ happens to be the tangent bundle $T(X)$ of a compact G-manifold X (with G-invariant riemannian metric), then $\pi T(X)$ is simply denoted by νX. Let $\varrho\xi$ denote either ξ or $\pi\xi$. The *stabilization* $s(\varrho\xi)$ of $\varrho\xi$ is defined by

$$s(\varrho\xi) = \varrho(\xi \oplus \varepsilon(M)),$$

where M is a real G-module with G-invariant inner product. The *stabilization* of a G-(or $\Pi(X)$-)vector bundle isomorphism $b: \varrho\xi \to \varrho\xi'$ is defined by

$$s(b) = b \oplus \varrho(id_{\varepsilon(M)}): s(\varrho\xi) \to s(\varrho\xi').$$

Let X and Y be compact, oriented G-manifolds, and let ξ be a G-vector bundle over Y with fiber-dim $\xi = \dim Y$. By a term G-normal map we mean a triple of the following maps. First, f is a G-map from $(X, \partial X)$ to $(Y, \partial Y)$ of degree one. Secondly, b is a stable G-vector bundle isomorphism from $T(X)$ to $f^*\xi$. And finally, c is a $\Pi(X)$-vector bundle isomorphism from νX to $\pi f^*\xi$ such that $\pi(b) = s(c)$ (cf. [**19**] and [**23**]), where $s(c)$ is a stabilization of c. Note that in the current paper, we use the term 'a G-normal map' in the sense of [**19**] and [**23**] not in the sense of [**18**] nor [**20**]. A G-normal map is denoted by, for example, $\mathbf{w} = (f; b; c): (X, \partial X; TX; \nu X) \to (Y, \partial Y; f^*\xi; \pi f^*\xi)$. However, if the boundaries ∂X and ∂Y of X and Y, respectively, are empty, then we write it by $\mathbf{w} = (f; b; c): (X; TX; \nu X) \to (Y; f^*\xi; \pi f^*\xi)$.

Given two G-normal maps $\mathbf{w} = (f; b; c): (X, \partial X; TX; \nu X) \to (Y, \partial Y; f^*\xi; \pi f^*\xi)$ and $\mathbf{w}' = (f'; b'; c'): (X', \partial X'; TX'; \nu X') \to (Y, \partial Y; f'^*\xi; \pi f'^*\xi)$, the notion of a G-normal cobordism

$$\mathbf{W} = (F; B; C): (W, \partial W; TW; \nu W) \to (I \times Y, \partial(I \times Y); F^*(\varepsilon_I(\mathbf{R}) \times \xi); \pi F^*(\varepsilon_I(\mathbf{R}) \times \xi))$$

between them can be given generalizing naturally the corresponding concept in ordinary surgery theory. Here $I = [0, 1]$.

Let $f: X \to Y$ be a G-map. For a prime p, f is called a *$\{p\}$-equivalence* if $f^P: X^P \to Y^P$ is a mod p homology equivalence for every nontrivial p-subgroup P of G. If f is a $\{p\}$-equivalence for any prime p, then we call f a $\mathcal{P}$-equivalence. A G-map f is called a *singularity equivalence* if it satisfies one of the conditions (1) and (2) below :

(1) The restriction $f_s: X_s \to Y_s$ of f to the singular set gives an equivalence of homology with integral coefficients.

(2) The reduced projective class group $\widetilde{K}_0(\mathbf{Z}[G])$ of the integral group ring $\mathbf{Z}[G]$ is trivial and f is a $\mathcal{P}$-equivalence.

Let $f: (X, \partial X) \to (Y, \partial Y)$ be a G-map. It is called a *boundary equivalence* if its restriction $\partial f: \partial X \to \partial Y$ to the boundaries is a homology equivalence.

Let w be the orientation homomorphism $w: G \to \{1, -1\}$ given by $w(g) = 1$ (resp. -1) if g in G preserves (resp. reverses) the orientation of Y. Using this, we can define an involutive anti-automorphism $-$ of the integral group ring $\mathbf{Z}[G]$ by $\overline{g} = w(g)g^{-1}$ for all g in G. Let $G(X)$ be the subset of G consisting of all elements g of order two such that $\dim X^g = [(n-1)/2]$, where $n = \dim X$. The form parameter $\Gamma G(Y)$ on $\mathbf{Z}[G]$ for $\lambda = (-1)^{[n/2]}$ is defined to be the smallest form parameter containing all elements of $G(Y)$. The following two results may be fundamental in G-surgery theory.

LEMMA 3.1 ([18] Theorem A). *Let Y be a compact, connected, simply connected and oriented G-manifold of dimension $n \geq 5$, and let $\mathbf{w} = (f; b; c): (X, \partial X; TX; \nu X) \to (Y, \partial Y; f^*\xi; \pi f^*\xi)$ be a G-normal map. Suppose that the following conditions hold.*

(1) $2 \dim Y_s < n$.
(2) $\dim Y^L < [(n-1)/2]$ *whenever* $L \supsetneqq \{1, g\}$ *for some* $g \in G(Y)$.
(3) f *is a boundary and singularity equivalence.*

Then $\mathbf{w}$ determines an element $\sigma(\mathbf{w})$ in the Bak group $W_n(\mathbf{Z}[G], \Gamma G(Y); w)$, and if $\sigma(\mathbf{w}) = 0$, then one can perform G-surgery keeping the boundary and the singular set fixed to convert $\mathbf{w}$ so that $f: X \to Y$ is a homotopy equivalcence.

Remark. In [20] Theorems A and B, it is proved that, under the same conditions as in Lemma 3.1, one can perform G-surgery as above if and only if $\sigma(\mathbf{w}) = 0$. Namely, $\sigma(\mathbf{w})$ gives the G-surgery obstruction.

LEMMA 3.2. *Let Y satisfy the same assumptions as in Lemma 3.1, and further, suppose that it is without boundary and satisfies the strong gap hypothesis, i.e., $2(\dim Y_s + 1) < \dim Y$. Then the following hold.*

(1) *Let $\mathbf{w} = (f; b; c): (X; TX \nu X) \to (Y; f^*\xi; \pi f^*\xi)$ be a G-normal map and assume that f is a singularity equivalence. Then $\sigma(\mathbf{w})$ lies in the Wall group $L_n^h(G)$ of homotopy equivalence and it gives the G-surgery obstruction.*
(2) *If $\mathbf{w}' = (f'; b'; c'): (X'; TX'; \nu X') \to (Y; f'^*\xi; \pi f'^*\xi)$ is another G-normal map and if there exists a G-normal cobordism*

$$\mathbf{W} = (F; B; C): (W, \partial W; TW; \nu W) \to (I \times Y, \partial(I \times Y); F^*(\varepsilon_I(\mathbf{R}) \times \xi); \pi F^*(\varepsilon_I(\mathbf{R}) \times \xi))$$

between $\mathbf{w}$ and $\mathbf{w}'$ such that $F: W \to I \times Y$ is a $\mathcal{P}$-equivalence, then one has $\sigma(\mathbf{w}) = \sigma(\mathbf{w}')$.

PROOF: This lemma may be well known. We refer the reader to [20] Theorem D for the details.

In the rest of this section, we fix a real G-module V with G-invariant inner product, and construct several G-manifolds. We denote by $S(V)$ (resp. $D(V)$) the unit sphere (resp. closed unit disk) of V. Consider $\mathbf{R}$ as the real 1-dimensional trivial G-module with standard inner product. The tangent bundle $T(V)$ of V is identified with $\varepsilon_V(V)$. The G-vector bundle $\varepsilon_{S(V)}(\mathbf{R}) \oplus T(S(V))$ can be regarded as the restriction of $T(V)$ to $S(V)$ by the standard G-isomorphism. Here $\varepsilon_{S(V)}(\mathbf{R})$ should be understood to be the normal bundle $\nu(S(V), V)$ of $S(V)$ in V by the above identification.

For an integer k with $k \geq 1$, let $\mathbf{R}^k$ be the k-fold direct sum of $\mathbf{R}$. The G-vector bundle $\varepsilon_{S(V)}(\mathbf{R}^k) \oplus T(S(V))$ can be identified with $\varepsilon_{S(V)}(\mathbf{R}^{k-1} \oplus V)$ by the standard isomorphism

$$\varepsilon_{S(V)}(\mathbf{R}^k) \oplus T(S(V)) = \varepsilon_{S(V)}(\mathbf{R}^{k-1}) \oplus \varepsilon_{S(V)}(\mathbf{R}) \oplus T(S(V)) = \varepsilon_{S(V)}(\mathbf{R}^{k-1}) \oplus \varepsilon_{S(V)}(V).$$

Here the restriction of the isomorphism to $\varepsilon_{S(V)}(\mathbf{R}^{k-1})$ should be understood as the identity map.

For a positive integer j, we define $V(j)$ here to be the j-fold direct sum of V, and we put $Y(j) = S(\mathbf{R} \oplus V(j))$ and $X(j+1)' = Y(j) \times Y(1)$. Denote by $p(j)_+$ and $p(j)_-$ the points $(1,0)$ and $(-1,0)$, respectively, of $Y(j)$, where $1 \in \mathbf{R}$ and $0 \in V$. The tangential representation at $(p(j)_+, p(1)_+)$ in $X(j+1)'$ is isomorphic to $V(j+1)$. By pinching the outside of the equivariant open disk neighborhood of $(p(j)_+, p(1)_+)$ in $X(j+1)'$, we get a degree one G-map $h(j+1)' : X(j+1)' \to Y(j+1)$. We note that

$$\varepsilon(\mathbf{R}) \oplus T(Y(j+1)) = \varepsilon(\mathbf{R} \oplus V(j+1)), \quad \text{and}$$

$$\varepsilon(\mathbf{R}^2) \oplus h(j+1)'^* T(Y(j+1)) = \varepsilon(\mathbf{R}^2 \oplus V(j+1)).$$

An unstable G-vector bundle isomorphism

$$b(j+1)' : \varepsilon(\mathbf{R}^2) \oplus T(X(j+1)') \to \varepsilon(\mathbf{R}^2 \oplus V(j+1))$$

is defined to be the standard isomorphism

$$\{\varepsilon(\mathbf{R}) \oplus T(S(\mathbf{R} \oplus V(j)))\} \times \{\varepsilon(\mathbf{R}) \oplus T(S(\mathbf{R} \oplus V(1)))\} = \{\varepsilon(\mathbf{R} \oplus V(j))\} \times \{\varepsilon(\mathbf{R} \oplus V(1))\}.$$

Thus we obtain a G-normal map $\mathbf{v}(j+1)' = (h(j+1)'; b(j+1)'; \pi b(j+1)')$. On the other hand, let $\mathbf{1}_G(j+1) = (id; id; \pi id)$ be the identity G-normal map on $Y(j+1)$. Here the second id is the identity map on $T(Y(j+1))$. By our specified identification, we obtain $s(id) : \varepsilon(\mathbf{R}^2) \oplus T(Y(j+1)) \to \varepsilon(\mathbf{R}^2 \oplus V(j+1))$.

Lemma 3.3. *The above G-normal map $\mathbf{v}(j+1)'$ is G-normally cobordant to $\mathbf{1}_G(j+1)$.*

Proof: Let S_1 be the sphere of radius 3 with center being the origin in $\mathbf{R} \oplus V(j)$, S_2 the sphere of radius 3 with center being the origin in $\mathbf{R} \oplus V(1)$. We identify $S_1 \times S_2$ with $X(j+1)$. Define S_3 by

$$S_3 = \{(x, u, y, v) \in \mathbf{R} \oplus V(j) \oplus \mathbf{R} \oplus V(1) \mid x = 5, <u,u> + <y,y> + <v,v> = 1\}.$$

We identify S_3 as $Y(j+1)$. It is easy to find a compact, orientable, codimension one submanifold W of $\mathbf{R} \oplus V(j) \oplus \mathbf{R} \oplus V(1)$ and a G-map $F : W \to I \times Y(j+1)$, where $I = [0,1]$, such that

(1) the bounday ∂W of W is $S_1 \times S_2 \cup S_3$,
(2) the G-collar neighborhood of ∂W in W is $CN_{12} \cup CN_3$, where

$$CN_{12} = \{(x,u,y,v) \in \mathbf{R} \oplus V(j) \oplus \mathbf{R} \oplus V(1) \mid <x,x> + <u,u> = 9, \text{ and } 4 \le <y,y> + <v,v> \le 9\} \text{ and}$$
$$CN_3 = \{(x,u,y,v) \in \mathbf{R} \oplus V(j) \oplus \mathbf{R} \oplus V(1) \mid 4 \le x \le 5, <u,u> + <y,y> + <v,v> = 1\}, \text{and}$$

(3) F is an extension of $h(j+1)' : S_1 \times S_2 \to \{1\} \times Y(j+1)$ and $id : S_3 \to \{0\} \times Y(j+1)$.

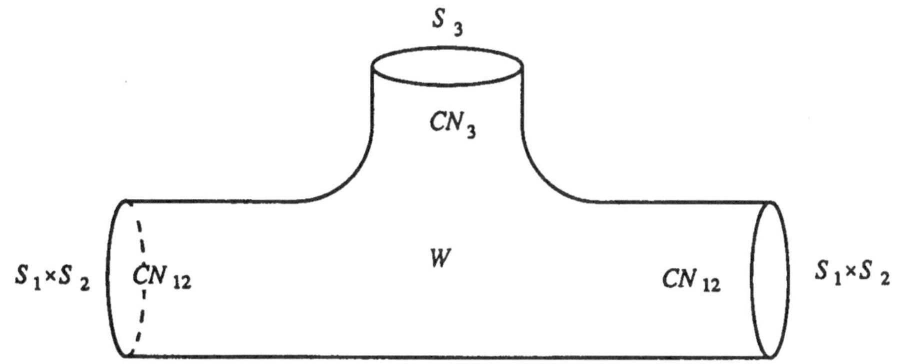

Figure 3.4.

We identify $T(\mathbf{R} \oplus V(j) \oplus \mathbf{R} \oplus V(1))$ with $\varepsilon(\mathbf{R} \oplus V(j) \oplus \mathbf{R} \oplus V(1))$ by the standard identification. There is a standard unstable G-vector bundle isomorphism $B': \varepsilon(\mathbf{R}) \oplus T(W) \to \varepsilon(\mathbf{R} \oplus V(j) \oplus \mathbf{R} \oplus V(1))$ mapping the first $\varepsilon(\mathbf{R})$ to the normal bundle of W in $\mathbf{R} \oplus V(j) \oplus \mathbf{R} \oplus V(1)$. The restriction of B' to $S_1 \times S_2$ coincides with $b(j+1)'$, however, the restriction of B' to S_3 does not coincide with $s(id)$. Take a smooth function $\lambda: W \to \mathbf{R}$ such that $\lambda(x) = 0$ for all $x \in W \setminus CN_3$, $0 \leq \lambda(x) \leq \pi/2$ for all $x \in CN_3$, and $\lambda(x) = \pi/2$ for all $x \in S_3$. Let $p: TW \to W$ be the projection, and write $B'(x,y)$ for $x \in \varepsilon(\mathbf{R})$ and $y \in TW$ in the form

$$B'(x,y) = (B'_1(x,y), B'_2(x,y), B'_3(x,y), B'_4(x,y)) \in \varepsilon(\mathbf{R}) \oplus \varepsilon(V(j)) \oplus \varepsilon(\mathbf{R}) \oplus \varepsilon(V(1)).$$

Define an unstable isomorphism $B: \varepsilon(\mathbf{R}) \oplus T(W) \to \varepsilon(\mathbf{R} \oplus V(j) \oplus \mathbf{R} \oplus V(1))$ by

$$\begin{aligned} B(x,y) =&(\{\cos \lambda(p(y))\} B'_1(x,y) + \{\sin \lambda(p(y))\} B'_3(x,y), B'_2(x,y), \\ &- \{\sin(p(y))\} B'_1(x,y) + \{\cos \lambda(p(y))\} B'_3(x,y), B'_4(x,y)). \end{aligned}$$

Then the restriction of B to $S_1 \times S_2$ coincides with $b(j+1)'$, and also the restriction to S_3 coincides with $s(id)$. Thus $\mathbf{W} = (F; B; \pi B)$ is a G-normal cobordism between $\mathbf{v}(j+1)'$ and $\mathbf{1}_G(j+1)$. This completes the proof.

Let k be an integer greater than j. Define $\mathbf{1}_G(k-j-1)'$ to be the identity G-normal map on the $(k-j-1)$-fold cartesian product $Y(1)^{k-j-1}$ of $Y(1)$, and $\mathbf{v}(k, j+1)$ to be the G-normal map $\mathbf{v}(j+1)' \times \mathbf{1}_G(k-j-1)'$. Composing $\mathbf{v}(k, j+1)$, where $j = 1, 2, \ldots, k-1$, we obtain a G-normal map $\mathbf{v}(k) = (h(k); b(k); \pi b(k))$, where $h(k): Y(1)^k \to Y(k)$ and $b(k): \varepsilon(\mathbf{R}^k) \oplus T(Y(1)^k) \to \varepsilon(\mathbf{R}^k \oplus V(k))$ (an unstable isomorphism). (Note : $Y(1)^k$ is the k-hold cartesian product of $Y(1)$.) Then Lemma 3.3 yields ;

PROPOSITION 3.5. *The G-normal map $\mathrm{v}(k) = (h(k); \mathfrak{b}(k); \pi\mathfrak{b}(k))$ is G-normally cobordant to $\mathbf{1}_G(k)$.*

4. A proof of Theorem 0.1

In the current section G denotes the alternating group A_5 on five letters, and we let Σ be the Poincaré sphere with a standard G-action. Here we prove our main theorem ;

THEOREM 0.1. *For any integer k with $k \geq 2$, the standard sphere S^{3k} has a smooth one fixed point A_5-action which is A_5-cobordant to $\Sigma(k)$, the k-fold cartesian product of Σ with the diagonal A_5-action.*

Let V be the tangential representation at the unique G-fixed point of Σ. Then as a real G-module, V is irreducible, and for a subgroup H, $\dim V^H = 0, 1,$ or 3 if H is noncyclic, nontrivial cyclic, or trivial, respectively, (cf. Definition 1.4). Using this V we can define several manifolds as in the previous section. (e.g. $Y(1) = S(\mathbf{R} \oplus V)$) So, we keep the same notation as there, but remember that our V here is a special one given above. We first obtain the following by Petrie's transverality construction (see [**19**] Section 3, or [**23**]).

LEMMA 4.1. *There are a G-manifold $X(1)''$ with exactly one fixed point and a G-normal map $\mathrm{u}(1) = (g(1); \mathfrak{b}(1)''; \mathfrak{c}(1)'')$, where $g(1){:}\, X(1)'' \to Y(1)$ and $\mathfrak{b}(1)''{:}\, T(X(1)'') \to g(1)^*T(Y(1))$, such that the following hold :*

(1) *$X(1)''^G = g(1)^{-1}(p(1)_+)^G$ (and of course, $| X(1)''^G | = 1$). (The G-fixed point in $X(1)''$ is also denoted by $p(1)_+$.)*
(2) *If H is a proper, noncyclic subgroup of G, then*

$$X(1)''^H = \{p(1)_+\} \coprod g(1)^{-1}(p(1)_-)^H \quad \text{and} \quad | g(1)^{-1}(p(1)_-)^H | = 1.$$

(3) *For every maximal subgroup H of G, there exists an H-normal cobordism $\mathbf{U}_H(1) = (G_H(1);\ B_H(1)'';\ C_H(1)'')$ between $\mathrm{res}^G_H \mathrm{u}(1)$ and $\mathrm{res}^G_H\ \mathbf{1}_G(1)$, where $\mathbf{1}_G(1)$ is the identity G-normal map on $Y(1)$.*

For an integer k with $k \geq 2$, let $X(k)$ (resp. $\Sigma(k)$) be the k-fold cartesian product of $X(1)''$ (resp. Σ). Then we have ;

LEMMA 4.2. *The G-manifold $X(k)$ has exactly one G-fixed point and is G-cobordant to $\Sigma(k)$.*

PROOF: It suffices to prove that $X(1)''$ has one G-fixed point and is G-cobordant to Σ. By Lemma 4.1 (1), the former is clear. On the other hand, from Lemma 4.1 (2), the restriction $g(1)^H{:}\, X(1)''^H \to Y(1)^H$ of $g(1)$ to $X(1)''^H$ is a homotopy equivalence if H is isomorphic to A_4 or D_n, where $n = 4, 6$ or 10. Now let C be an arbitrary nontrivial cyclic subgroup of G. Then $X(1)''^C$ is a disjoint union of circles. Among the circles, those which do not contain the G-fixed point $p(1)_+$ consist of points with isotropy subgroup C. Kill the circles not containing the G-fixed point $p(1)_+$ by G-surgery on $X(1)''$ of isotropy types $(C_2), (C_3)$ and (C_5) simultaneously. Then we can modify $\mathrm{u}(1)$ so that $g(1)^C{:}\, X(1)''^C \to Y(1)^C$ is a homotopy equivalence. In particular, the singular set

$X(1)''_s$ of $X(1)''$ belongs to $\mathcal{S}$. Furthermore, the stable isomorphism $b(1)''$ gives a stable A_5-trivialization (of type minus) for the module V. Since Σ has an A_5-trivialization for V (cf. Example 2.2), Theorem 2.1 implies that $X(1)''$ is G-cobordant to Σ, which completes the proof.

Now we are going to perform G-surgery on $X(k)$, keeping the G-fixed point set fixed, to obtain a homotopy sphere of dimension $3k$. Let $\mathbf{w}(k) = (f(k); b(k); c(k))$ be the G-normal map obtained by composing the k-fold cartesian product $\mathbf{u}(1) \times ... \times \mathbf{u}(1)$ of $\mathbf{u}(1) = (g(1); b(1)''; c(1)'')$ with $\mathbf{v}(k) = (h(k); b(k); \pi b(k))$, which are defined in Section 3. Here notice that the target manifold of $f(k)$ is the $3k$-dimensional sphere $Y(k) = S(\mathbf{R} \oplus V \oplus ... \oplus V)$. Then by Proposition 3.5 and Lemma 4.1 (3), we have ;

LEMMA 4.3. *For every maximal subgroup H of G, there exists an H-normal cobordism $\mathbf{W}_H(k) = (F_H(k); B_H(k), C_H(k))$ between $\mathrm{res}^G_H \mathbf{w}(k)$ and $\mathrm{res}^G_H \mathbf{1}_G(k)$, where $\mathbf{1}_G(k)$ the identity G-normal map on $Y(k)$.*

Hereafter we abbreviate the notation by omitting (k) if there seems to be no confusions. For example, we use $\mathbf{w}$ instead of $\mathbf{w}(k)$. Then we have ;

LEMMA 4.4. *One can perform G-surgery of the above G-normal map $\mathbf{w} = (f; b; c)$ keeping the G-fixed point set fixed and one can perform H-surgery of the above H-normal cobordisms $\mathbf{W}_H = (F_H; B_H, C_H)$ between $\mathrm{res}^G_H \mathbf{w}$ and $\mathrm{res}^G_H \mathbf{1}_G$ so that*

(1) *the new $f: X \to Y$ is a $\mathcal{P}$-equivalence, and*
(2) *if $k \geq 4$, then all $F_H: W_H \to I \times \mathrm{res}^G_H Y$, for maximal subgroups H of G, are $\mathcal{P}$-equivalences, where $I = [0, 1]$.*

The above will be proved in the next Section by using Lemma 4.3 and a case by case arguments. Here we assume it to go on. Note that Lemma 4.4 (1) implies that f is a singularity equivalence because it is shown in I. Reiner and S. Ullom [26] or [25] that, for all subgroups H of A_5, the reduced projective class groups $\widetilde{K}_0(\mathbf{Z}[H])$ of the integral group rings $\mathbf{Z}[H]$ are trivial. Since f is trivially a boundary equivalence, we may apply Lemmas 3.1 and 3.2. In the case where $k = 2$ (resp. ≥ 3), by Lemma 3.1 (resp. Lemma 3.2), f gives $\sigma(\mathbf{w})$ which lies in the Bak group $W_2(\mathbf{Z}[G], \Gamma; triv.)$ (resp. the Wall group $L^h_{3k}(G)$), where Γ is the smallest form parameter on $\mathbf{Z}[G]$ containing all elements of G of order two.

LEMMA 4.5. *The element $\sigma(\mathbf{w})$ is zero.*

PROOF: In the case where $k = 2$, the result follows from $W_2(\mathbf{Z}[G], \Gamma; triv.) = 0$ ([21] Proposition 1.1), and when $k = 3$, it follows from $L^h_1(G) = 0$ (A. Bak and M. Kolster [3] Corollary 4.4). We now show $\sigma(\mathbf{w}) = 0$ in the case where $k \geq 4$. It suffices to show that $\mathrm{res}^G_H \sigma(\mathbf{w}) = 0$ for $H = D_4, D_6$ and D_{10} by the Dress induction theorem [2] Section 12, [10] or [29] since every maximal 2-hyperelementary subgroup of G is conjugate to one of D_4, D_6 and D_{10}. But for $H \cong D_6$, D_{10} or $A_4 (> D_4)$, we have H-normal cobordisms $\mathbf{W}_H$ between $\mathrm{res}^G_H \mathbf{w}$ and $\mathrm{res}^G_H \mathbf{1}_G$ satisfying the condition (2) of Lemma 4.4. Thus Lemma 3.2 (2) implies that $\mathrm{res}^G_H \sigma(\mathbf{w}) = \sigma(\mathbf{1}_H) = 0$ for all $H = D_4, D_6$ and D_{10}. This proves the lemma.

Hence by the above lemma, we can perform G-surgery, keeping the singular set fixed, to converting $\mathbf{w} = (f; b; c)$ so that $f: X \to Y$ is a homotopy equivalence. In particular, since Y is a homotopy sphere, so is X. Also, recall that this resulting X is G-cobordant to $\Sigma(k) = \Sigma \times ... \times \Sigma$ (the k-fold cartesian product) by Lemma 4.2. Summarizing the above, we get the following.

PROPOSITION 4.6. *Let k be an integer with $k \geq 2$. Then, there exists a homotopy sphere X of dimension $3k$ with G-action such that*

(1) *X has exactly one G-fixed point,*
(2) *X is G-cobordant to $\Sigma(k) = \Sigma \times ... \times \Sigma$ and*
(3) *for all maximal subgroups H of G, $\mathrm{res}^G_H X$ is H-cobordant to $\mathrm{res}^G_H Y(k)$, where $Y(k) = S(\mathbf{R} \oplus V \oplus ... \oplus V)$.*

COMPLETION OF PROOF OF THEOREM 0.1: It suffices to modify (the underlying manifold of) the above X in Proposition 4.6 to the standard sphere of dimension $3k$ with the same properties. Consider the equivariant connected sum X' of X with $G \times_H \mathrm{res}^G_H X$ at the points of isotropy type (H) for $H = A_4$ and D_{10}. (For details, see [19] Section 3, p. 248.) Then X' also satisfies all the properties (1) – (3) with X being replaced by X'. In fact, the property (2) of X' follows from the properties (2) and (3) of X, and the property (3) of X' follows from the property (3) of X. On the other hand, let $\Theta(3k)$ denote the group of $3k$-dimensional homotopy spheres, which is known to be a finite group by [17] Theorem 1.2. For a $3k$-dimensional homotopy sphere Z with G-action, we denote by $[Z]$ the element in $\Theta(3k)$ which corresponds to the underlying space of Z. Then for our X and X' we have

$$[X'] = (1 + \ell)[X] \quad \text{in} \quad \Theta(3k),$$

where if $H = A_4$ then $\ell = 5$ and if $H = D_{10}$ then $\ell = 6$. Now since 5 and 6 are relatively prime, we can obtain the standard sphere, which corresponds to the identity element of $\Theta(3k)$, as the underlying space of an equivariant connected sum X'' of X with several $(G \times_{A_4} \mathrm{res}^G_{A_4} X)$'s and $(G \times_{D_{10}} \mathrm{res}^G_{D_{10}} X)$'s. Then the resulting X'' also satisfies all the properties (1) – (3) with X being replaced by X''. This proves the theorem.

5. A proof of Lemma 4.4

In this section G still means A_5. We restate Lemma 4.4.

LEMMA 4.4. *One can perform G-surgery of the G-normal map $\mathbf{w} = (f; b; c)$ keeping the G-fixed point set fixed and one can perform H-surgery of the H-normal cobordisms $\mathbf{W}_H = (F_H; B_H, C_H)$ between $\mathrm{res}^G_H \mathbf{w}$ and $\mathrm{res}^G_H 1_G$ so that*

(1) *the new $f: X \to Y$ is a $\mathcal{P}$-equivalence, and*
(2) *if $k \geq 4$, then all $F_H: W_H \to I \times \mathrm{res}^G_H Y$, for maximal subgroups H of G, are $\mathcal{P}$-equivalences, where $I = [0, 1]$.*

This is proved essentially in [19] Section 5. However, we give a proof also here for reader's convenience. For two subgroups H and K of G with $N_G(K) \subseteqq H \neq G$, we use the following notations.

$$X^{>K} = \{x \in X \mid G_x \supsetneqq K\} \text{ and } W_H^{>K} = \{w \in W_H \mid H_w \supsetneqq K\}$$

We first give the following lemma.

Lemma 5.1. *Let H and K be subgroups of G with $N_G(K) \subseteqq H \neq G$. Suppose that there exist a closed H-regular neighborhood $U_{>K}$ of $HW_H^{>K}$ in W_H and an H-diffeomorphism $\psi_{>K}: U_{>K} \to I \times (X \cap U_{>K})$ relative to $X \cap U_{>K}$. Suppose further that there exist a closed H-neighborhood $N_{>K}$ of $HX^{>K} \cup HW_H^{>K}$ in W_H and an H-diffeomorphism $\Psi_{>K}: N_{>K} \to I \times (X \cap N_{>K})$ relative to $X \cap N_{>K}$ such that $N_{>K} \supseteqq U_{>K}$, $\Psi_{>K} \mid U_{>K} = \psi_{>K}$, $N_{>K}$ is orthogonal to both X and Y, and $\Psi_{>K}^{-1}(\{1\} \times (X \cap N_{>K})) = Y \cap N_{>K}$.*
Then one can perform equivariant surgery of isotropy type (K) to modify $\mathrm{w} = (f; b; c)$ and $\mathrm{W}_H = (F_H; B_H; C_H)$ so that

1. *$f^K: X^K \to Y^K$ is a homotopy equivalence,*
2. *$F_H^K: W_H^K \to I \times Y^K$ is also a homotopy equivalence, and*
3. *there exists a closed H-regular neighborhood $U_K \supseteqq U_{>K}$ of HW_H^K in W_H and an H-diffeomorphism $\psi_K: U_K \to I \times (X \cap U_K)$ relative to $X \cap U_K$ such that $\psi_K \mid U_{>K} = \psi_{>K}$.*

Proof: See [19] Theorem 4.2.

Let H and L be subgroups of $G = A_5$ such that $L \subseteqq H$. We say that w and W_H are *good* for (H, L) if the following conditions (1) and (2) are satisfied for all subgroups K such that $L \subseteqq K \subseteqq H$ and $K \neq \{1\}$.

1. $f^K: X^K \to Y^K$ are homotopy equivalences.
2. $F_H^K: W_H^K \to I \times Y^K$ are also homotopy equivalences.

In the following, we modify w and W_H step by step using Lemma 5.1 in order to obtain the goodness of them for many pairs (H, L) of subgroups of A_5 satisfying $N_G(L) \subseteqq H \neq G$

Proposition 5.2.

1. *If $n = 3k \geq 6$, then one can modify w and W_H so that they are good for (A_4, A_4), (D_6, D_6), (D_{10}, D_{10}), (A_4, D_4), (D_{10}, C_5), (D_6, C_3) and (A_4, C_2). In particular, for any nontrivial subgroup K, the resulting $f^K: X^K \to Y^K$ is a homotopy equivalence.*
2. *If $n = 3k \geq 12$, then one can further modify them so that they are good also for $(A_4, \{1\})$, $(D_6, \{1\})$ and $(D_{10}, \{1\})$.*

Proposition 5.2 (1) guarantees the property (1) in Lemma 4.4 and (2) guarantees the property (2) there. Thus Proposition 5.2 proves Lemma 4.4.

PROOF: We shall modify $\mathbf{w}$ and $\mathbf{W}_H$ step by step to obtain good ones.

Step 1: $(H,L) = (A_4, A_4)$. In this step, $H = K = L = A_4 = N_G(A_4)$. Thus $HW_H{}^{>K} = \emptyset$ and $HX^{>K} = X^G =$ one point. (X^H consists of $2k$ points.) Consider $U_{>K}$ as the empty set. It is easy to see that $W_H{}^H$ includes a path connecting X^G with a point p_+ in Y^G. Take a closed H-tubular neighborhood $N_{>H}$ of this path in $W_H{}^H$. Then $N_{>K}$ is clearly H-diffeomorphic to $I \times (X \cap N_{>H})$. Then, by Lemma 5.1 for $(H,K) = (A_4, A_4)$, we can perform equivariant surgery of isotropy type (A_4) to obtain a new G-normal map $\mathbf{w} = (f; b; c)$ and an H-normal cobordism $\mathbf{W}_H = (F_H; B_H, C_H)$ between $\mathrm{res}^G_H \mathbf{w}$ and $\mathrm{res}^G_H \mathbf{1}_G$ which are good for $(H,L) = (A_4, A_4)$.

Step 2: $(H,L) = (D_6, D_6)$ and (D_{10}, D_{10}). The argument is quite similar to one in Step 1 above, and we omit it.

Step 3: $(H,L) = (A_4, D_4)$. Since we have done already for $K = A_4$ in Step 1, it suffices to modify $\mathbf{w}$ and $\mathbf{W}_H$ for $K = D_4$. So we put $K = D_4$. From the construction, $X^K = X^H$ and $f^K: X^K \to Y^K$ is already a homotopy equivalence. Thus, it suffices to modify $F_H{}^K: W_H{}^K \to I \times Y^K$ to be a homotopy equivalence. Since $\dim W_H{}^K = 1$, $W_H{}^K$ is a union of $W_H{}^H$ and several circles liing in $\mathrm{Int} W_H{}^K$. Perform H-surgery on W_H of isotropy type (D_4) to kill these circles. Then the resulting $F_H{}^K$ is a homotopy equivalence as is required, and thus $\mathbf{w}$ and $\mathbf{W}_H$ are good for $(H,L) = (A_4, D_4)$.

Step 4: $(H,L) = (D_{10}, C_5)$. We have already done for $K = D_{10}$ and are required to modify $\mathbf{w}$ and $\mathbf{W}_H$ for $K = C_5$. So, we put $K = C_5$. Note that $HX^K = X^K$, $HX^{>K} = X^H$, $HW_H{}^K = W_H{}^K$ and

$$HW_H{}^{>K} = W_H{}^H \cong I \times X^H.$$

In particular, $HX^{>K} \subsetneqq HW_H{}^{>K}$. Take a closed H-tubular neighborhood U of $W_H{}^H$ in W_H. Regard $U_{>K}$ and $N_{>K}$ in Lemma 5.1 as this U, and apply Lemma 5.1. Then, by equivariant surgery of isotropy type (C_5) we can obtain good $\mathbf{w}$ and $\mathbf{W}_H$ for $(H,L) = (D_{10}, C_5)$.

Step 5: $(H,L) = (D_6, C_3)$. We have already done for $K = D_6$ and are required to modify $\mathbf{w}$ and $\mathbf{W}_H$ for $K = C_3$. So, we put $K = C_3$. The proper subgroups B of G which properly include C_3 are isomorphic to D_6 or A_4. Further, such a subgroup B isomorphic to D_6 is unique, i.e. $B = D_6$, and the number of such B's isomorphic to A_4 is two which we denote by $A(1)$ and $A(2)$. Then, $HX^{>K} = X^H \cup X^{A(1)} \cup X^{A(2)}$ and $HW_H{}^{>K} = W_H{}^H = I \times X^H$. Then $W_H{}^H$ has a closed H-tubular neighborhood $U_{>K}$ in W_H which is H-diffeomorphic to $I \times (X \cap U_{>K})$. Recall that $\dim X^{>K} = 0$ and $\dim W_H{}^K = k \geq 2$. We can embedd $I \times (HX^{>K} \setminus HW_H{}^{>K})$ to $HW_H{}^K \setminus U_{>K}$ H-equivariantly, and obtain a closed H-neighborhood $N_{>K}$ of $HX^{>K} \cup HW_H{}^{>K}$, which is H-diffeomorphic to $I \times (X \cap N_{>K})$ and contains $U_{>K}$. Now apply Lemma 5.1 and obtain good $\mathbf{w}$ and $\mathbf{W}_H$ for $(H,L) = (D_6, C_3)$.

Step 6: $(H,L) = (A_4, C_2)$. We have already done for $K = A_4$ and D_4. Thus, we are required to do for $K = C_2$. So put $K = C_2$. We list the subgroups of G which properly include C_2: Those isomorphic to A_4 and D_4 are respectively unique. The number of those isomorphic to D_6 is two and we denote them by $D(3)$ and $D(4)$. The number of those isomorphic to D_{10} is two and we denote them by $D(5)$ and $D(6)$. Then, we have

$$X^{>K} = X^H \cup X^{D(3)} \cup X^{D(4)} \cup X^{D(5)} \cup X^{D(6)}$$

and $HW_H^{>K} = W_H^H$. First take a closed H-tubular neighborhood $U_{>K}$ of W_H^H in W_H. Then, $U_{>K}$ is H-diffeomorphic to $I \times (X \cap U_{>K})$. Next note that $\dim X^{>K} = 0$. Thus we can embedd $I \times (HX^{>K} \setminus HW_H^{>K})$ to $HW_H^K \setminus U_{>K}$ H-equivariantly and obtain a closed H-neighborhood $N_{>K}$ of $HX^{>K} \cup HW_H^{>K}$, which is H-diffeomorphic to $I \times (X \cap N_{>K})$ and contains $U_{>K}$. Now apply Lemma 5.1 and obtain good $\mathbf{w}$ and $\mathbf{W}_H$ for $(H, L) = (A_4, C_2)$.

Step 7: $(H, L) = (A_4, \{1\})$. Here we suppose that $n = 3k \geq 12$. We have already done for $K = A_4, D_4$ amd C_2. Furthermore, w satisfies the condition (1) of goodness for all K , (even for $K \cong C_3$). So put $K = C_3$, which is remaining, and we will modify $\mathbf{W}_H$ to satisfy the condition (2) of goodness. First note that $HW_H^{>K} = W_H^H \cong I \times X^H$. The maps $F_H^{>K} = F^H: W_H^H \to I \times Y^H$ and $f^K: X^K \to Y^K$ are homotopy equivalences. Let $U(H, H)$ be a closed H-tubular neighborhood of W_H^H. However, we can not use Lemma 5.1 since $N_G(K) \nsubseteq H$. On the other hand, since $k \geq 4$, we have

$$\dim W_H^K = k + 1 \geq 5 \text{ and } 2(\dim W_H^{>K} + 1) = 4 < k + 1 = \dim W_H^K.$$

Now try to modify $\mathbf{W}_H$ by H-surgery keeping $U(H, H) \cup \partial W_H$ fixed so that F_H^K is a homotopy equivalence. The obstruction $\sigma(F_H^K)$ lies in the Wall group $L_{k+1}^h(N)$, where $N = N_H(K)/K$, which is of course trivial. If $k + 1$ is odd, then $\sigma(F_H^K) = 0$ because $L_{k+1}^h(N) = 0$, and thus we can perform required H-surgery. In the general case, we need an add hoc argument. Let $\mathbf{1}(G)$ be the identity G-normal map on the closed G-disk $D(\mathbf{R} \oplus V(k))$, where $V(k)$ is the k-fold direct sum of V. Let $\mathbf{1}(G)_-$ and $F_{D_6 -}$ be the reversed copies of $\mathbf{1}(G)$ and F_{D_6}, respectively. Now glue these

$$\mathrm{res}_K^G \mathbf{1}(G)_-,\ \mathrm{res}_K^{D_6} F_{D_6 -}, \mathrm{res}_K^H F_H \text{ and } \mathrm{res}_K^G \mathbf{1}(G).$$

We denote the resulting K-normal map by $\mathbf{W} = (F; B; C)$, where $F: W \to \mathrm{res}_K^G S(\mathbf{R}^2 \oplus V(k))$. Then it is clear that $\sigma(F^K) = \sigma(F_H^K)$ in $L_{k+1}^h(N)$. Let $\mathbf{W}_- = (F_-; B_-; C_-)$ be the reversed copy of $\mathbf{W}$. Then, $\sigma(F_-^K) = -\sigma(F_H^K)$. Take an H-connected sum of $\mathbf{W}_H$ with $\mathrm{ind}_K^H \mathbf{W}_-$ at the points in $\mathrm{Int} W_H$ with isotropy type (K), and denote by $\mathbf{W}'_H = (F'_H, B'_H, C'_H)$, $F'_H: W'_H \to I \times Y$ the resulting H-normal cobordism. Then we get $\sigma(F_H'^K) = 0$. Hence we can perform H-surgery to modify $\mathbf{W}'_H$ so that F'_H is a homotopy equivalence satisfying $F_H'^{>K} = F_H^{>K}$. Replacing the initial $\mathbf{W}_H$ by the obtained $\mathbf{W}'_H$, the condition (2) of goodness is satisfied.

Step 8: $(H, L) = (D_6, \{1\})$ and $(D_{10}, \{1\})$. The argument is quite similar to one in Step 7, and we omit it.

This completes the proof of Proposition 5.2 and thus of Lemma 4.4.

References

1. Bak, A., *The computation of surgery groups of finite groups with abelian 2-hyperelementary subgroups*, Lecture Notes in Math. (ed. by M. R. Stein) **551** (1976), 384–407, Springer, Berlin - Heidelberg - New York - Tokyo.
2. Bak, A., "*K*-Theory of Forms," Princeton University Press, Princeton, 1981.
3. Bak, A. and Kolster, M., *The computation of odd dimensional projective surgery groups of finite groups*, Topology **21** (1982), 35–63.
4. Bak. A. and Morimoto M., *Equivariant surgery theory and its applications*, in preparation.
5. G. E. Bredon, "Introduction to Compact Transformation Groups," Academic Press, New York - London, 1972.
6. Buchdahl, N. P., Kwasik, S. and Schultz, R., *One fixed point actions on low-dimensional spheres*, preprint, Tulane University and Purdue University.
7. Conner P. E. and Floyd E. E., "Differentiable Periodic Maps," Ergebniss der Mathematik und Ihere Grenzgebiete Neue Folge Band **33**, Springer, Berlin - Heidelbrg - New York - Tokyo, 1964.
8. Dovermann, K. H., Masuda, M. and Petrie, T., *Fixed point free algebraic actions on varieties diffeomorphic to* $\mathbf{R}^n$, preprint.
9. Dovermann, K. H. and Masuda, M., *Fixed point free low dimensional real algebraic actions of* A_5 *on contractible varieties*, preprint.
10. Dress, A., *Induction and structure theorems for orthogonal representations of finite groups*, Ann. Math. **102** (1975), 291–325.
11. Furuta, F., *A remark on a fixed point of finite group action on* S^4, Topology **28** (1989), 35–38.
12. Grove, L.C. and Benson, C. T., "Finite Reflection Groups," (Second Edition) Graduate Texts in Mathematics **99**, Springer, Berlin - Heidelberg - New York - Tokyo, 1985.
13. Kervaire, M. A. and Milnor, J. W., *Groups of homotopy spheres I*, Ann. Math. **77** (1963), 504–537.
14. Kolster, M., *Even dimensional projective surgery groups of finite groups*, Lecture Notes in Math (ed. by R. K. Dennis) **967** (1982), 239–279, Springer, Berlin - Heidelberg - New York - Tokyo.
15. Laitinen, E.and Traczyk, P., *Pseudofree representations and 2–pseudofree actions on spheres*, Proc. Amer. Math. Soc. **97** (1986), 151–157.
16. Morimoto, M., *On one fixed point actions on spheres*, Proc. Japan Acad. **63** Ser. A (1987), 95–97.
17. Morimoto, M., S^4 *does not have one fixed point actions*, Osaka J. Math. **25** (1988), 575–580.
18. Morimoto, M., *Bak groups and equivariant surgery*, *K*-Theory **2** no. 4 (1989), 465–483.
19. Morimoto, M., *Most of the standard spheres have one fixed point actions of* A_5, Lecture Notes in Math (ed. by K. Kawakubo) **1375** (1989), 240-258, Springer, Berlin - Heidelberg - New York - Tokyo.
20. Morimoto, M., *Bak groups and equivariant surgery II*, to appear in *K*-Theory.
21. Morimoto, M., *Most of the standard spheres have one fixed point actions of* A_5. *II*, preprint.
22. Oliver, R., "Whitehead Groups of Finite Groups," London Math. Soc. Lecture Note Series **132**, Cambridge University Press, Cambridge, 1988.
23. Petrie, T., *One fixed point actions on spheres, I*, Adv. Math. **46** (1982), 3–14.
24. Petrie, T., *One fixed point actions on spheres, II*, Adv. Math. **46** (1982), 15–70.
25. Reiner, I., *Class groups and Picard groups of group rings and order*, Regional Conf. Ser. in Math. **26** (1976), A.M.S.
26. Reiner, I. and Ullom, S., *Remarks on class groups of integral group rings*, Symposia Mathematica (ed. by A. Dress) **13** (1974), 501–516, Academic Press, London.
27. Shaneson, J., *Wall's surgery obstruction groups for* $G \times \mathbf{Z}$, Ann. Math. **90** (1969), 296–334.
28. Springer, T.A., "Invariant Theory," Lecture Notes in Mathematics **585**, Springer, Berlin - Heidelberg - New York - Tokyo, 1977.
29. Stein, E., *Surgery on products with finite fundamental group*, Topology **16** (1977), 473–493.
30. Uchida, F., "Transformation Groups and Cobordism Theory," (Japanese) Kinokuniya Suugakusousho **2**, Kinokuniya, Tokyo, 1974.
31. Wall, C. T. C., *Classification of hermitian forms. VI Group rings*, Ann. Math. **103** (1976), 1–80.

A NOTE ON THE MOD 2 COHOMOLOGY OF $SL(\mathbb{Z})$

Dominique Arlettaz

Institut de mathématiques
Université de Lausanne
CH-1015 Lausanne, Switzerland

1. Introduction

Let $SL(\mathbb{Z})$ be the infinite special linear group of the ring of integers $\mathbb{Z}$, and for $n \geq 2$, $w_n \in H^n(SL(\mathbb{Z}); \mathbb{Z}/2)$ the n-th Stiefel-Whitney class of the inclusion $SL(\mathbb{Z}) \hookrightarrow GL(\mathbb{R})$ $(w_1 = 0)$. The mod 2 cohomology of $SL(\mathbb{Z})$ satisfies [3, Proposition 1.2]

$$H^*(SL(\mathbb{Z}); \mathbb{Z}/2) \cong \mathbb{Z}/2[w_2, w_3, w_4, \ldots] \otimes A ,$$

where A is an (unknown) commutative graded algebra. The corresponding result for the infinite general linear group $GL(\mathbb{Z})$ is

$$H^*(GL(\mathbb{Z}); \mathbb{Z}/2) \cong \mathbb{Z}/2[w_1, w_2, w_3, \ldots] \otimes A .$$

It is not hard to deduce from the knowledge of the integral homology of $SL(\mathbb{Z})$ in dimensions 1, 2 and 3 [1] that A contains no element of degree 1 and 2, and exactly one element of degree 3.

Dwyer and Friedlander have formulated in [6] a version of the Quillen-Lichtenbaum conjecture concerning the map between the algebraic K-theory spectrum and the étale K-theory spectrum, and they have shown that if their version of the 2-adic Quillen-Lichtenbaum conjecture is true for the ring $\mathbb{Z}$, i.e., if the map $BGL(\mathbb{Z}[\frac{1}{2}])^+ \to \widetilde{K}^{\text{ét}}(\mathbb{Z}[\frac{1}{2}])$ induces an isomorphism on mod 2 cohomology, then the following conjecture holds [6, Corollary 4.3].

Conjecture. $A = \Lambda[u_3, u_5, \ldots, u_{2n+1}, \ldots]$, *where* $\deg u_{2n+1} = 2n+1$.

A part of this work was done during a stay at McMaster University : it is a pleasure to express my gratitude for its hospitality.

Notice that if this is true, then the 2-torsion subgroup of $H_4(SL(\mathbf{Z});\mathbf{Z})$ is cyclic and the isomorphism $H_4(SL(\mathbf{Z});\mathbf{Z}) \cong K_4\mathbf{Z} \oplus \mathbf{Z}/2$ [2, Theorems 1.1 and 1.3] implies that the 2-torsion subgroup of $K_4\mathbf{Z}$ is trivial.

Let us mention the following simple fact. The rational cohomology of $SL(\mathbf{Z})$ is known [5] : $H^*(SL(\mathbf{Z});\mathbf{Q}) = \Lambda[x_5, x_9, \ldots, x_{4n+1}, \ldots]$, where $\deg x_{4n+1} = 4n+1$. This produces elements $u_5, u_9, \ldots, u_{4n+1}$ in $H^*(SL(\mathbf{Z});\mathbf{Z}/2)$ and it is possible to check that they are actually in A, but it is not clear whether or not they are exterior.

The purpose of this paper is to make the above conjecture more plausible by detecting exterior classes of odd degree in $H^*(SL(\mathbf{Z});\mathbf{Z}/2)$. Consider the finite field $\mathbf{F}_p$ (p an odd prime), and denote by f_p the reduction mod p : $SL(\mathbf{Z}) \to SL(\mathbf{F}_p)$ and by f_p^* the induced homomorphism $H^*(SL(\mathbf{F}_p);\mathbf{Z}/2) \to H^*(SL(\mathbf{Z});\mathbf{Z}/2)$. By [7], the ring $H^*(SL(\mathbf{F}_p);\mathbf{Z}/2)$ is generated by the Chern classes $c_k \in H^{2k}(SL(\mathbf{F}_p);\mathbf{Z}/2)$ and by the classes $e_k \in H^{2k-1}(SL(\mathbf{F}_p);\mathbf{Z}/2)$, $k \geq 2$. For any odd prime p and any integer $k \geq 2$, we have [3, Lemma 1.4]

$$f_p^*(c_k) = w_k^2 ,$$

and we introduce the notation

$$\varepsilon_k := f_p^*(e_k) .$$

If $p \equiv 1 \pmod 4$, then $e_k^2 = 0$ and consequently,

$$\varepsilon_k^2 = 0 .$$

This is wrong if $p \equiv 3 \pmod 4$, but we have proved [3, Lemma 1.5] that in this case $\varepsilon_k = W_k + \gamma_k$, where $W_k = w_{2k-1} + \sum_{2 \leq j < k} w_j w_{2k-1-j}$ and

$$\gamma_k^2 = 0 .$$

Now, let us formulate our main result.

Theorem. *Let k be any integer ≥ 2.*
(a) If $p \equiv 5 \pmod 8$, then $\varepsilon_k \neq 0$ in $H^{2k-1}(SL(\mathbf{Z});\mathbf{Z}/2)$.
(b) If $p \equiv 3 \pmod 8$, then $\gamma_k \neq 0$ in $H^{2k-1}(SL(\mathbf{Z});\mathbf{Z}/2)$.

Both cases provide the following direct consequence.

Corollary. *$H^*(SL(\mathbf{Z});\mathbf{Z}/2)$ contains an exterior element of degree $2n+1$ for all $n \geq 1$.*

The corresponding results hold also for $GL(\mathbf{Z})$ ($k \geq 1$ in this case, but $\varepsilon_1 = 0$, respectively $\gamma_1 = 0$). Observe finally that the theorem is wrong if $p \equiv 1$ or $7 \pmod 8$.

2. Proof of the theorem

Proposition 1. *The theorem is true for any positive integer* $k \equiv 2 \pmod 4$.

Proof. According to Quillen's terminology for the computation of $H^*(SL(\mathbb{F}_p);\mathbb{Z}/2)$ for an odd prime number p [7], the class $e_k \in H^{2k-1}(SL(\mathbb{F}_p);\mathbb{Z}/2)$ is defined for any integer $k \geq 2$ as the image, under the reduction mod 2 : $H^{2k-1}(SL(\mathbb{F}_p);\mathbb{Z}/(p^k-1)) \to H^{2k-1}(SL(\mathbb{F}_p);\mathbb{Z}/2)$, of the cohomology class $\widetilde{e}_k \in H^{2k-1}(SL(\mathbb{F}_p);\mathbb{Z}/(p^k-1))$ introduced in [7, p.559]; this element $\widetilde{e}_k$ has the property that the Bockstein homomorphism associated with the short exact sequence $\mathbb{Z} \rightarrowtail \mathbb{Z} \twoheadrightarrow \mathbb{Z}/(p^k-1)$ maps $\widetilde{e}_k$ onto $\widehat{c}_k \in H^{2k}(SL(\mathbb{F}_p);\mathbb{Z})$, where $\widehat{c}_k$ is the k-th integral Chern class of $SL(\mathbb{F}_p)$ [7, Lemma 5]. Therefore, if β denotes the Bockstein homomorphism associated with the short exact sequence $\mathbb{Z} \rightarrowtail \mathbb{Z} \twoheadrightarrow \mathbb{Z}/2$, then $\beta(e_k) = \frac{1}{2}(p^k-1)\widehat{c}_k$ for any odd prime p and any integer $k \geq 2$. Now, look at the commutative diagram

$$\begin{array}{ccc} H^{2k-1}(SL(\mathbb{F}_p);\mathbb{Z}/2) & \xrightarrow{f_p^*} & H^{2k-1}(SL(\mathbb{Z});\mathbb{Z}/2) \\ \downarrow{\scriptstyle\beta} & & \downarrow{\scriptstyle\beta} \\ H^{2k}(SL(\mathbb{F}_p);\mathbb{Z}) & \xrightarrow{f_p^*} & H^{2k}(SL(\mathbb{Z});\mathbb{Z}) . \end{array}$$

It follows from [4, p.36] that the composition $H^*(BU;\mathbb{Z}) \to H^*(SL(\mathbb{F}_p);\mathbb{Z}) \to H^*(SL(\mathbb{Z});\mathbb{Z})$ (the first homomorphism being induced by the Brauer lifting, the second by f_p) coincides, away from the p-torsion, with the homomorphism $H^*(BU;\mathbb{Z}) \to H^*(SL(\mathbb{Z});\mathbb{Z})$ induced by the inclusion $SL(\mathbb{Z}) \hookrightarrow GL(\mathbb{C})$, and consequently that $f_p^*(\widehat{c}_k)$ is, up to p-torsion, the k-th integral Chern class $c_k(SL(\mathbb{Z}))$ of the inclusion $SL(\mathbb{Z}) \hookrightarrow GL(\mathbb{C})$. By commutativity of the diagram, $\beta(\varepsilon_k) = \beta(f_p^*(e_k)) = f_p^*(\beta(e_k))$ is then an odd multiple of $\frac{1}{2}(p^k-1)\,c_k(SL(\mathbb{Z}))$.

The order of $c_k(SL(\mathbb{Z}))$ in $H^{2k}(SL(\mathbb{Z});\mathbb{Z})$ is known for $k \equiv 2 \pmod 4$ [1] and it turns out that its 2-primary part is exactly the 2-primary part of (p^k-1) , if $p \equiv 3$ or $5 \pmod 8$. It is then easy to conclude that $\beta(\varepsilon_k)$ does not vanish in this case and to deduce assertion (a) of the theorem. If $p \equiv 3 \pmod 8$, remember that $\varepsilon_k = W_k + \gamma_k$ and observe that

$$W_k = w_{2k-1} + \sum_{2 \leq j < k} w_j w_{2k-1-j} = Sq^1\Big(w_{2k-2} + \sum_{1 \leq j < \frac{k}{2}} w_{2j} w_{2k-2-2j}\Big)$$

by Wu's formula. Thus, $\beta(W_k) = 0$ and $\beta(\gamma_k) = \beta(\varepsilon_k) \neq 0$, which implies (b').

Remark 2. In particular, ε_2 (if $p \equiv 5 \pmod 8$), respectively γ_2 (if $p \equiv 3 \pmod 8$), is the unique non-trivial (exterior) class of degree 3 of A. This will be one of the basic points in the proof of the theorem for the integers $k \not\equiv 2 \pmod 4$.

Notice that the proof of Proposition 1 does not work if $k \not\equiv 2 \pmod 4$: if $k \equiv 0 \pmod 4$, the exact order of $c_k(SL(\mathbf{Z}))$ is not understood, and if k is odd, the order of $c_k(SL(\mathbf{Z}))$ is 2 but the 2-primary part of $(p^k - 1)$ is greater than 2 for $p \equiv 5 \pmod 8$.

For the next step of our argument, we need to determine the action of the Steenrod algebra on $H^*(SL(\mathbf{F}_p); \mathbf{Z}/2)$; for simplicity we formulate actually the results for $GL(\mathbf{F}_p)$, i.e. for c_k and $e_k \in H^*(GL(\mathbf{F}_p); \mathbf{Z}/2)$ where $k \geq 1$ (set $c_1 = e_1 = 0$ for $SL(\mathbf{F}_p)$).

Lemma 3. *For any odd prime p and for $1 \leq i \leq k$, $Sq^{2i-1}c_k = 0$ and*

$$Sq^{2i}c_k = \sum_{0 \leq j \leq i} \binom{k-j-1}{i-j} c_j c_{k+i-j} .$$

Proof. This follows from the Wu's formula for the Chern classes in $H^*(BU; \mathbf{Z}/2)$.

Lemma 4. *(a) For any odd prime p and for $1 \leq i < k$,*

$$Sq^{2i}e_k = \binom{k-1}{i} e_{k+i} + \sum_{1 \leq j \leq i} \binom{k-j-1}{i-j}(c_j\, e_{k+i-j} + c_{k+i-j}\, e_j) .$$

(b) For any prime $p \equiv 1 \pmod 4$ and for $1 \leq i \leq k$, $Sq^{2i-1}e_k = 0$.
(c) For any prime $p \equiv 3 \pmod 4$ and for $1 \leq i \leq k$,

$$Sq^{2i-1}e_k = \begin{cases} 0 & , \quad \text{if } k-i \text{ is odd,} \\ \displaystyle\sum_{0 \leq j \leq i-1} \binom{k-j-1}{i-j-1}\, c_j c_{k+i-j-1} & , \quad \text{if } k-i \text{ is even.} \end{cases}$$

Proof. Let us call ρ^* the homomorphism $H^*(F\Psi^p; \mathbf{Z}/2) \to H^*(BC^m; \mathbf{Z}/2)$ introduced in [7, p. 563]; here $F\Psi^p$ is a space with the same cohomology as $GL(\mathbf{F}_p)$, C the cyclic group of order $p-1$, and we choose the integer $m \geq 2k-1$. The kernel of ρ^* is the ideal generated by the elements c_j and e_j for $j > m$: thus, ρ^* is injective in dimensions $\leq 4k-2$.

If $p \equiv 1 \pmod 4$, then $H^*(BC^m; \mathbf{Z}/2) \cong \mathbf{Z}/2[x_1, x_2, \ldots, x_m] \otimes \Lambda[y_1, y_2, \ldots, y_m]$ with $\deg x_j = 2$ and $\deg y_j = 1$ for $1 \leq j \leq m$. Consider the differential d of degree -1 defined by $d(x_j) = y_j$ and $d(y_j) = 0$ for $1 \leq j \leq m$: d commutes with the Steenrod squares. According to [7, p. 564], $\rho^*(c_k) = \sigma_k$ and $\rho^*(e_k) = d(\sigma_k)$, where σ_k denotes the k-th elementary symmetric function of $x_1, x_2, \ldots, x_m$. Therefore, we obtain

$$\rho^*(Sq^{2i}e_k) = Sq^{2i}d(\sigma_k) = d(Sq^{2i}\sigma_k) = d\Big(\binom{k-1}{i}\sigma_{k+i} + \sum_{1 \leq j \leq i} \binom{k-j-1}{i-j}\sigma_j\sigma_{k+i-j}\Big)$$

by Wu's formula, and deduce assertion (a) from $d(\sigma_{k+i}) = \rho^*(e_{k+i})$ and $d(\sigma_j\sigma_{k+i-j}) = \sigma_j d(\sigma_{k+i-j}) + d(\sigma_j)\sigma_{k+i-j} = \rho^*(c_j e_{k+i-j} + c_{k+i-j}e_j)$. Similarly, (b) follows from

$$\rho^*(Sq^{2i-1}e_k) = Sq^{2i-1}d(\sigma_k) = d(Sq^{2i-1}\sigma_k) = 0\,.$$

If $p \equiv 3 \pmod 4$, then $H^*(BC^m;\mathbf{Z}/2) \cong \mathbf{Z}/2[y_1, y_2, \ldots, y_m]$ and $\rho^*(c_k) = \tau_k^2$, $\rho^*(e_k) = Sq^{k-1}\tau_k$, where τ_k is the k-th elementary symmetric function of $y_1, y_2, \ldots, y_m$. We must investigate $\rho^*(Sq^{2i}e_k) = Sq^{2i}Sq^{k-1}\tau_k$ and $\rho^*(Sq^{2i-1}e_k) = Sq^{2i-1}Sq^{k-1}\tau_k$. The Adem relations and the Wu's formula produce the equalities

$$\rho^*(Sq^{2i}e_k) = Sq^{k+i-1}Sq^i\tau_k = Sq^{k+i-1}\Big(\tbinom{k-1}{i}\tau_{k+i} + \sum_{1\le j\le i}\tbinom{k-j-1}{i-j}\tau_j\tau_{k+i-j}\Big)\,,$$

$$\begin{aligned}\rho^*(Sq^{2i-1}e_k) &= (k-i-1)Sq^{k+i-1}Sq^{i-1}\tau_k = (k-i-1)(Sq^{i-1}\tau_k)^2\\ &= (k-i-1)\sum_{0\le j\le i-1}\tbinom{k-j-1}{i-j-1}\tau_j^2\tau_{k+i-j-1}^2\,.\end{aligned}$$

Thus, $Sq^{k+i-1}\tau_{k+i} = \rho^*(e_{k+i})$ and $Sq^{k+i-1}\tau_j\tau_{k+i-j} = Sq^j\tau_j Sq^{k+i-j-1}\tau_{k+i-j} + Sq^{j-1}\tau_j Sq^{k+i-j}\tau_{k+i-j} = \rho^*(c_j e_{k+i-j} + c_{k+i-j}e_j)$ provide (a), and $\tau_j^2\tau_{k+i-j-1}^2 = \rho^*(c_j c_{k+i-j-1})$ implies (c).

This enables us to prove the theorem for the integers $k \not\equiv 2 \pmod 4$.

Proposition 5. *The theorem is true for any positive integer* $k \equiv 0 \pmod 4$.

Proof. Consider first the case $p \equiv 5 \pmod 8$. By Lemma 4, $Sq^4 e_k = e_{k+2} + c_2 e_k + c_k e_2 \in H^{2k+3}(SL(\mathsf{F}_p);\mathbf{Z}/2)$ for $k \equiv 0 \pmod 4$, and consequently $Sq^4\varepsilon_k = f_p^*(e_{k+2} + c_2 e_k + c_k e_2) = \varepsilon_{k+2} + w_2^2\varepsilon_k + w_k^2\varepsilon_2$. If $\varepsilon_k = 0$, then $\varepsilon_{k+2} + w_k^2\varepsilon_2 = 0$. Apply again Sq^4 : $Sq^4\varepsilon_{k+2} = f_p^*(Sq^4 e_{k+2}) = f_p^*(c_2 e_{k+2} + c_{k+2}e_2) = w_2^2\varepsilon_{k+2} + w_{k+2}^2\varepsilon_2$, and $Sq^4(w_k^2\varepsilon_2) = w_{k+1}^2 Sq^2\varepsilon_2 + (w_{k+2}^2 + w_2^2 w_k^2)\varepsilon_2$. Therefore $w_2^2(\varepsilon_{k+2} + w_k^2\varepsilon_2) + w_{k+1}^2 Sq^2\varepsilon_2 = 0$, but this implies $w_{k+1}^2 Sq^2\varepsilon_2 = 0$ because of our assumption. This is impossible since ε_2 is the unique non-trivial element of degree 3 in A and $Sq^2\varepsilon_2$ does not vanish [2, Corollary 3.11].

If $p \equiv 3 \pmod 8$ and $\gamma_k = 0$, then ε_k is a polynomial in Stiefel-Whitney classes. Look again at $Sq^4\varepsilon_k = \varepsilon_{k+2} + w_2^2\varepsilon_k + w_k^2\varepsilon_2$: because we know from Proposition 1 that $\varepsilon_2 = W_2 + \gamma_2$ and $\varepsilon_{k+2} = W_{k+2} + \gamma_{k+2}$, where γ_2 and γ_{k+2} are non-trivial and do not belong to $\mathbf{Z}/2[w_2, w_3, w_4 \ldots]$, we may conclude that $\gamma_{k+2} + w_k^2\gamma_2 = 0$. As above, we then apply Sq^4 and obtain $w_{k+1}^2 Sq^2\gamma_2 = 0$: this is also a contradiction since γ_2 is the unique non-trivial element of degree 3 in A.

Proposition 6. *The theorem is true for any integer* $k \equiv 1 \pmod 4$, $k \ge 5$.

Proof. If $p \equiv 5 \pmod 8$, consider $Sq^4\varepsilon_k = f_p^*(Sq^4e_k) = f_p^*(c_2e_k + c_ke_2) = w_2^2\varepsilon_k + w_k^2\varepsilon_2$ for $k \equiv 1 \pmod 4$, according to Lemma 4 : the vanishing of ε_k would imply $w_k^2\varepsilon_2 = 0$, but this is not the case. If $p \equiv 3 \pmod 8$ and $\gamma_k = 0$, then ε_k is a polynomial in Stiefel-Whitney classes and the calculation of $Sq^4\varepsilon_k$ produces the contradiction $w_k^2\gamma_2 = 0$.

Remark 7. It is actually possible to prove Proposition 1 for $k \geq 6$ by using the same argument.

Proposition 8. *The theorem is true for any positive integer* $k \equiv 3 \pmod 4$.

Proof. Let p be a prime $\equiv 5 \pmod 8$ and compute again $Sq^4\varepsilon_k = \varepsilon_{k+2} + w_2^2\varepsilon_k + w_k^2\varepsilon_2$. If $\varepsilon_k = 0$, then $\varepsilon_{k+2} + w_k^2\varepsilon_2 = 0$. Apply now Sq^2 : $Sq^2\varepsilon_{k+2} = f_p^*(Sq^2e_{k+2}) = 0$ by Lemma 4 and consequently $Sq^2(w_k^2\varepsilon_2) = w_k^2Sq^2\varepsilon_2 = 0$; this is impossible (cf. proof of Proposition 5). Similarly, if $p \equiv 3 \pmod 8$ and $\gamma_k = 0$, then $\gamma_{k+2} + w_k^2\gamma_2 = 0$ and $Sq^2\gamma_{k+2} + Sq^2(w_k^2\gamma_2) = 0$; but this is a contradiction since $Sq^2\gamma_{k+2} = Sq^2\varepsilon_{k+2} + Sq^2W_{k+2} = Sq^2W_{k+2} \in \mathbb{Z}/2[w_2, w_3, w_4, \ldots]$ and $Sq^2(w_k^2\gamma_2) = w_k^2Sq^2\gamma_2 \notin \mathbb{Z}/2[w_2, w_3, w_4, \ldots]$.

References

[1] D. Arlettaz : *Chern-Klassen von ganzzahligen und rationalen Darstellungen diskreter Gruppen*, Math. Z. **187** (1984), 49–60.

[2] D. Arlettaz : *On the algebraic K-theory of* $\mathbb{Z}$, J. Pure Appl. Algebra **51** (1988), 53–64.

[3] D. Arlettaz : *Torsion classes in the cohomology of congruence subgroups*, Math. Proc. Cambridge Philos. Soc. **105** (1989), 241–248.

[4] M. Bökstedt : *The rational homotopy type of* $\Omega Wh^{Diff}(*)$, in Algebraic Topology Aarhus 1982, Lecture Notes in Math. **1051** (Springer 1984), 25–37.

[5] A. Borel : *Cohomologie réelle stable de groupes* S*-arithmétiques classiques*, C.R. Acad. Sci. Paris Sér. A **274** (1972), 1700–1702.

[6] W.G. Dwyer and E.M. Friedlander : *Conjectural calculations of general linear group homology*, in Applications of Algebraic K-theory to Algebraic Geometry and Number Theory, Contemp. Math. **55** Part I (1986), 135–147.

[7] D. Quillen : *On the cohomology and* K*-theory of the general linear groups over a finite field*, Ann. of Math. **96** (1972), 552–586.

Characteristic classes and 2–modular representations for some sporadic simple groups – II

Ch. B. Thomas
Max–Planck–Institut für Mathematik
Gottfried–Claren–Straße 26
D–5300 Bonn 3

Dedicated to the memory of J. Frank Adams, teacher, colleague and friend.

0. Introduction

As in an earlier paper [Th] we are concerned with calculating the cohomology ring $H^*(G,\mathbb{Z})$ of a sporadic simple group G away form the prime 2. This is easiest when the prime ℓ concerned divides $|G|$ to the first power, for $H^*(G,\mathbb{Z})_{(\ell)}$ is then periodic and all one has to do is identify a maximal generator. We complete this part of our programme is section two below. However our main purpose is at least to begin the determination of $H^*(G,\mathbb{Z})_{(\ell)}$ when an ℓ–Sylow subgroup G_ℓ is elementary abelian, and the ℓ–torsion is detected by the subgroup of $H^*(G_\ell,\mathbb{Z})$ left invariant by the action of the normaliser $N(G_\ell)$ of G_ℓ in G. We do this for several of the Mathieu groups M_k and for Janko's group J_1, postponing possible consideration of the general case to a future paper. As elsewhere in the theory of simple groups M_{24} provides an excellent test for the general method, since $M_{24,3}$ is an elementary non–abelian group of order 27, and the complete description of the stable elements in its cohomology is not easy.

A further motive for writing this paper is the wish to understand the relation between $H^*(G,K_*\mathbb{F}_{2^t})$ and the modular representations of G over the finite field $\mathbb{F}_{2^t}$. In most of the cases we consider the Chern subring in ordinary cohomology localised away from 2 is generated by the classes of one or two representations of low degree. This suggests a simple structure for $R\mathbb{F}_{2^t}(G)$ as a λ–ring with conjugation, particularly when $t = 1$ and one exploits the prime factorisation of $K_{2k-1}(\mathbb{F}_2) \simeq \mathbb{Z}/2^k-1$. However with the exception of J_1, which behaves much like a group with periodic cohomology, our results only suggest ways of studying modular representations, since we are faced with the familiar convergence problems of the Atiyah–Hirzebruch spectral sequence. Indeed the generic situation for groups of composite order seems to be that there are universal cycles, which cannot be detected by Chern classes in either the characteristic zero or the modular case. However cohomology does at least make plain which representations are important for the λ–ring structure: as an elementary example consider M_{11}, which has irreducible 2–modular representations of degrees 1, 10, 44 and 16. Using eigenvalues it is easy to see that $\rho_{44} = \lambda^2(\rho_{10}) - (1)$, but because of their characters when restricted to $M_{11,11}$ ρ_{16} and its conjugate cannot be obtained in this way. However $\rho_{16} + \overline{\rho_{16}} \cong \lambda^2(\rho_{10}) - \rho_{10} - (3)$, showing that this situation is simpler over the prime field $\mathbb{F}_2$. This is reflected in cohomology by the fact that

$$H^9(M_{11}, \mathbb{Z}/_4 5_{-1})_{(11)} \cong \mathbb{Z}/_{11}, \text{ but } H^9(M_{11}, \mathbb{Z}/_2 5_{-1})_{(11)} \text{ is trivial.}$$

As a harder example the reader may like to consider M_{23} in the same way.

The final section of this paper is devoted to 2–torsion in the cohomology of J_1. We include it as a supplement to the partial calculations already in the literature, see [Ch], and also because it represents one of the last contributions to mathematics by J. Frank Adams.

1. Mathieu groups

We recall that the five simple Mathieu groups were originally constructed as examples of multiply transitive groups; the two quintuply transitive groups $M_{12} \hookrightarrow S_{12}$ and $M_{24} \hookrightarrow S_{24}$ contain the other three examples as stabilising subgroups. For a description of the various ways in which the groups M_k have been described we refer the reader to the "Atlas" – we shall be mainly concerned with the second series:

$$PSL(3, \mathbb{F}_4) \hookrightarrow M_{22} \hookrightarrow M_{23} \hookrightarrow M_{24} .$$

Since the projective special linear group arises as a stabilising subgroup in this series, we denote it by M_{21}. The importance of the projective special linear group M_{21} is that it carries much of the structure of $H^*(M_k, \mathbb{Z})_{(3)}$, indeed for the first three groups $M_{k,3}$ is an elementary abelian group of rank 2. Furthermore, with N and Z as usual denoting normaliser and centraliser, we have

(i)	$Z(M_{k,3}) = M_{k,3}$	$(k = 21, 22 \text{ and } 23)$, and
(ii)	$N(M_{k,3})/Z(M_{k,3}) \cong Q_8$	(quaternion group, $k = 21, 22$) and
	$\cong SD_{16}$	(semi–dihedral group, $k = 23$) .

The group SD_{16} has presentation $\{s,t : s^8 = t^2 = 1 ,\ t^{-1}st = s^3\}$. The isomorphisms are not immediately apparent from the tables in the Atlas, but an alternative source is the paper of Z. Janko, [J]. Since the centraliser is as small as possible the action of the quotient group on $M_{k,3}$ is faithful. When $k = 21$ we write G for the normaliser, it is a split extension of the form

$$C_3^a \times C_3^b \rightarrowtail G \twoheadrightarrow Q_8^{s,t} .$$

We shall pick a convenient basis for the normal subgroup as a vector space over an extension field of $\mathbb{F}_3$ below. From now on we use the following notation:
Let K be a finite abelian group generated as a direct product by a,b,... The one-dimensional representation $\hat{\alpha}$ of K is faithful on the subgroup <a> , maps the remaining generators b,... to 1, and $\alpha = c_1(\hat{\alpha}) \in H^2(K,\mathbb{Z})$. The group M_{24} has a representation τ, the Todd representation, in $GL(11,\mathbb{F}_2)$ described in [Td], which when lifted to characteristic zero has the partial character:

class	1	3^6	5^4	7_1^3	7_2^3	11	23_1	23_2	3^8
χ_τ	11	2	1	$\frac{1-\sqrt{-7}}{2}$	$\frac{1-\sqrt{-7}}{2}$	0	$\frac{-1+\sqrt{-23}}{2}$	$\frac{-1-\sqrt{-23}}{2}$	-1

Here 7^3 denotes a conjugacy class consisting of three disjoint 7–cycles with three 1–cycles omitted from the notation, etc. We shall also denote by τ its restriction to any of the smaller Mathieu groups contained in M_{24} .

Away from the primes 2 and 3 we have

THEOREM 1 (i) $H^*(M_{24},\mathbb{Z}[\frac{1}{6}])$ is generated by the classes c_3, c_4, c_{10} and c_{11} of the 11–dimensional representation τ (suitably restricted to a representative family of Sylow subgroups).

(ii) If $k = 11, 12, 22$ or 23 $H^*(M_k,\mathbb{Z}[\frac{1}{6}])$ has the same generators away from the prime 11. In all four cases

$$c_{10}(\tau \mid M_{k,11}) = c_5^2(\rho_k \mid M_{k,11}) ,$$

where ρ_k can be identified from the table

k	11	12	22	23
$\deg(\rho_k)$	16	16	280	896

Remark. The anomalous behaviour at 11 is explained by the splitting of a single conjugacy class of permutations on passing from a symmetric to a Mathieu group.

For a proof see [Th].

Let $Ch(\ )_{(\ell)}$ denote the Chern subring of the even-dimensional cohomology localised at the prime ℓ .

THEOREM 2 <u>The subring</u> $Ch(M_k)_{(3)}$ <u>of</u> $H^*(M_k,\mathbb{Z})_{(3)}$ <u>is generated by</u> $c_i(\tau|M_{k,3})$, $i = 3,4$. <u>At least when</u> $k = 22$ <u>or</u> 23 $Ch(M_k)_{(3)}$ <u>is properly contained in</u> $H^*(M_k,\mathbb{Z})_{(3)}$.

Proof. We calculate the 3–primary part of $H^*(G,\mathbb{Z})$, where G is the normaliser of a representative 3–Sylow subgroup in $PSL(3,\mathbb{F}_4)$. The spectral sequence for the defining short exact sequence is trivial, so $H^*(G,\mathbb{Z})_{(3)} = H^*(C_3 \times C_3,\mathbb{Z})^{Q_8} = E_2^{*,0}$. The odd dimensional contribution is an exterior algebra on a 3–dimensional generator, compare [Le]. In even dimensions proceed as follows: Let V be a 2–dimensional vector space over $\mathbb{F}_3$ and consider the induced action of Q_8 on the symmetric algebra $S(V^*)$. Take coefficients in $\mathbb{F}_9$ rather than $\mathbb{F}_3$, so as to diagonalise the action of an element s of order 4 in Q_8 . Here we use the usual presentation of Q_8 as $\{s,t : s^4 = 1, s^2 = t^2, t^{-1}st = s^{-1}\}$, and represent Q_8 in $SL(2,\mathbb{F}_9)$ by

$$s \longmapsto \begin{bmatrix} i & 0 \\ 0 & -i \end{bmatrix}, \quad t \longmapsto \begin{bmatrix} 0 & -1 \\ 1 & 0 \end{bmatrix} .$$

Having extended the scalars choose a basis of eigenvectors $\{A,B\}$ for $\mathbb{F}_9 \otimes_{\mathbb{F}_3} S(V^*) = \mathbb{F}_q[\alpha,\beta]$ with $sA = iA$ and $sB = -iB$. Formally one first chooses A and then takes B to be the image of A under the Frobenius map ψ . As an automorphism ψ fixes α and β , and on the coefficients $\psi(\lambda) = \lambda^3$. We may further suppose that over the extension field $\mathbb{F}_q$ the bases $\{A,B\}$ and $\{\alpha,\beta\}$ are related by the equations

$$A = i\alpha + \beta , \quad B = \alpha + i\beta .$$

The remark in the preamble about the choice of basis is now clear – G_3 is to be generated by a and b dual to the classes α and β . Now $\mathbb{F}_9[A,B]^{<s>}$ has an $\mathbb{F}_9$–basis consisting of all monomials A^jB^k with $j + 3k \equiv 0 \bmod 4$, which is equivalent to $(k-j) \equiv 0 \bmod 4$. Since t induces the automorphism $A \longmapsto -B$, $B \longmapsto A$, one type of invariant polynomial is "evenly symmetric" in A and B , i.e. one considers symmetric polynomials of the form

$$A^jB^k + A^kB^j = \sigma^+_{jk} , \text{ where}$$

j and k are both even, and $(k-j) \equiv 0 \bmod 4$. The second type must satisfy $A^jB^k - A^kB^j = \sigma^-_{jk}$, where j and k are both odd and $(k-j) \equiv 0(4)$. The first few invariant polynomials are $A^2B^2 = -(\alpha^2 + \beta^2)^2$, $A^4 + B^4 = -(\alpha^4 + \beta^4)$, $A^5B - AB^5 = (\alpha^2 + \beta^2)(\alpha^3\beta + \alpha\beta^3)$, ... One sees immediately that $S(V^*)^{Q_8}$ has two generators of degree 4, one of degree 6, ... On the other hand by counting dimensions we see that all but one of the irreducible representations of G factor through the quotient group Q_8, and the exception, obtained by induction form the trivial representation, restricts to the regular representation minus a trivial summand on $C_3 \times C_3$. An easy calculation now shows that $Ch(G)_{(3)}$ is generated by c_6 and c_8 of this restriction, and hence is properly contained in $H^{even}(G,\mathbb{Z})_{(3)}$.

This argument applies immediately to the Mathieu groups M_{21} and M_{22} since the stable elements in the cohomology of $M_{k,3}$ coincide with those invariant under the normaliser, see [Sw]. Inspection of the character table again shows that $Ch(M_{22})_{(3)}$ is generated by the Chern classes of the regular representation of $C_3 \times C_3$. For M_{23} the argument follows the same pattern, except that one replaces Q_8 by SD_{16}, represented over $\mathbb{F}_q$ by

$$s \longmapsto \begin{bmatrix} \zeta & 0 \\ 0 & \zeta^3 \end{bmatrix}, \quad t \longmapsto \begin{bmatrix} 0 & 1 \\ 1 & 0 \end{bmatrix}, \quad \text{where } \zeta = 1-i$$

is a primitive 8th–root of unity. A basis of eigenvectors is given by $\{A,B\}$, where $sA = \zeta A$, $sB = \zeta^3 B$, and because t has order 2 rather than 4 the invariant polynomials are $A^jB^k + A^kB^j$ with $j + 3k \equiv 0 \bmod 8$. As one would except this subalgebra is smaller than for M_{22}, but A^2B^2 still provides a generator in degree 4, which is not describable as a Chern class.

The situation for the largest Mathieu group M_{24} is more complicated, since $M_{24,3}$ is a non–abelian group of order 27 and exponent 3. This cohomology of this group has been worked out by G. Lewis, see [Le], and using this multiplicative relations one can give a surprisingly simple description of the Chern subring. However the determination of the 3–primary part of $H^*(M_{24},\mathbb{Z})$ is harder, since we can no longer apply Swan's normaliser theorem.

2. Other sporadic simple groups

In this section we consider the twelve sporadic simple groups omitted from our previous paper [Th]. Loosely speaking these fall into three classes – the Fischer groups, those closely related to the Monster, and the oddments J_3, Ru, O'N, Ly and J_4. Because the last five groups are best described by means of faithful modular representations, our method works particularly well for them. However we start by sumarising the information for primes $\ell \geq 5$ dividing the order to the first power only, i.e. for which $H^*(G,\mathbb{Z})_{(\ell)}$ is periodic. A blank space means that the prime concerned does not divide the order; a space containing a dash (–) means that the Sylow subgroup G_ℓ is not cyclic. An asterisk (*) against an entry means that a generator for the periodic cohomology may be taken to be the appropriate Chern class of a non–trivial irreducible representation of smallest degree

G \ ℓ	5	7	11	13	17	19	23	29	31	37	43	47	67
J_3	4				16	18*							
Ru	–	12*		26*				28					
O'N	8*	–	22*			12			30				
Ly	–	12*	10*						12	36			44
J_4	8*	6*	–				44*	56*	20	24	28		
HN	–	12*	20*			18							
Th	–	–		24*		36*			30				
B	–	–	20*	26*	32*	36*	22		30			46	
He	–	–			16								
Fi_{22}	–	12*	10	12									
Fi_{23}	–	12*	20*	12	32*		22						
F'_{24}	–	–	20*	26*	32*		22	28					

For the first groups we can summarise the information from our table in the following result:

THEOREM 3 Let the pair (G,q) be as shown

J_3	Ru	O'N	Ly	J_4
2	5	7	5	11

,

and let R be the coefficient ring $\mathbb{Z}[\frac{1}{6}]$ if $G = J_3$ and $\mathbb{Z}[\frac{1}{6q}]$ otherwise. Then $\hat{H}^*(G,R)$ is a sum of polynomial algebras, each of which is generated by a Chern class of a restricted irreducible representation.

Proof. This follows the lines of the argument in [Th], and depends on an examination of (a) the character tables and (b) the listed maximal subgroups of G in the Atlas.

Remarks on the individual groups.

Ru: Perhaps the most revealing representation is that of the related group 2.G in the orthogonal group $O_{28}(\mathbb{Z}[i])$ reduced modulo 5. So far as odd torsion in cohomology is concerned 2.G behaves like G, and the Chern classes of this $\mathbb{F}_5$–representation pick up maximal generators for 7 and 13, and the square of a maximal generator for 29.

Ly: This is perhaps the most interesting group among the oddments, since the period for the prime 31 (equal to $\frac{2}{5}(31-1)$) is so low. This is explained by Ly containing $G_2(\mathbb{F}_5)$ as a maximal subgroup (this group of Lie type has periodic cohomology for the primes 7 and 31, the period for both the latter being 12). The remaining maximal subgroups of interest to us are the cyclic by cyclic extensions 67 : 22 and 37 : 18, and the semi–direct product $3^5 : (2 \times M_{11})$, which detect 67–, 37– and 11–torsion respectively. However in order to realise a maximal generator for 31 as a Chern class one must go in the Atlas to χ_{39} taking the value 43 110 144 at the identity.

J_4: This is usually thought of as a subgroup of $GL_{112}(\mathbb{F}_2)$. However comparison with other groups in this class suggests that one look for a more geometrically motivated representation over the Galois field $\mathbb{F}_{11}$.

Further calculations along the lines of those carried out for the Mathieu groups in the previous section seem possible, although not very rewarding. With the exceptions of HN, Ly and B, the orders of which are divisible by 5^6, all the groups on our list have the property that, if $\ell \geq 5$, then ℓ divides the order to at most the third power. Thus, if ℓ^2 is the highest power occuring, calculation of both $Ch(G)_{(\ell)}$ and $H^*(G,\mathbb{Z})_{(\ell)}$ as in Theorem 2 seems to be straightforward. A Sylow subgroup G_ℓ is necessarily abelian, and the image of the restriction map coincides with the subgroup invariant under the action of the normaliser $N(G_\ell)$. For ℓ^3 dividing the order one is again forced to use Lewis' calculations for the non–abelian group of order ℓ^3 and exponent ℓ. The situation is straightforward enough in principle, although certainly numerically complicated. The groups most accessible to this attack would seem to be Fi_{22} and Fi_{23}.

3. Janko's first group J_1 (revisited)

In our previous paper [Th] we exploited the fact that away from the prime two J_1 behaves like a group with periodic cohomology to calculate $H^*(J_1,\mathbb{Z}[\frac{1}{2}])$. With the exception of $\ell = 11$ the ℓ–periods all divide 12, which points to the importance of the $\mathbb{F}_{11}$–representation φ used originally by Janko to define the group. Indeed the dimension of φ equals 7 and is minimal for a positive non–trivial representation over any field. The values of φ on the different conjugacy classes are given by:

class	1	2	3	$5^{(1)}$	$5^{(2)}$	7	$10^{(1)}$	$10^{(2)}$	$15^{(1)}$	$15^{(2)}$	$19^{(i)}$
φ	7	-1	1	$\frac{-(1+\sqrt{5})}{2}$	$\frac{-(1-\sqrt{5})}{2}$	-1	$\frac{3-\sqrt{5}}{2}$	$\frac{3+\sqrt{5}}{2}$	$1+\sqrt{5}$	$1-\sqrt{5}$	$1+\lambda(a_i)$

Here λ is one of 3 irreducible characters of degree 6 for the normaliser $N(J_{1,19})$, and a_1, a_2, a_3 represent three conjugacy classes of elements of order 19.

All elements of order two are conjugate, a 2–Sylow subgroup is elementary abelian of order 8, and any positive representation of J_1 must restrict to a direct sum of copies of the trivial and regular representations. For example $\varphi|J_{1,2}$ equals $\rho_{reg} - (1)$. The calculations are completed in even dimensions by

THEOREM 4 (J.F. Adams) $H^{even}(J_1,\mathbb{Z})_{(2)}$ *may be presented by* 5 *generators* x,y,z,u,v *of dimensions* 6,8,10,12,14 *respectively, and two relations* $r_{20} = 0$, $r_{24} = 0$, *where*

$$r_{20} = x^2y + xv + yu + z^2,$$
$$r_{24} = x^4 + x^2u + xyz + y^3 + zv + u^2.$$

Proof. This is a more complicated version of that of Theorem 2, and we again use the symmetric algebra $S(V^*)$ associated with the 3–dimensional vector space V over $\mathbb{F}_2$. Write $S(V^*)$ as a polynomial algebra $\mathbb{F}_2[\alpha,\beta,\gamma]$, and let K be a subgroup of order 21 in $GL(V) \cong GL(3,\mathbb{F}_2)$ acting in the obvious way on $S(V^*)$. This is an accurate model for the cohomology of J_1, since the normaliser of $J_{1,2}$ is a cyclic–by–cyclic extension of the form 7 : 3.

In order to find generators for $S(V^*)^K$ we embed $\mathbb{F}_2[\alpha,\beta,\gamma]$ in $\mathbb{F}_8[\alpha,\beta,\gamma]$, and let ψ be the Frobenius automorphism as in section 1. Over $\mathbb{F}_8$ we can find a new basis $\{A,B,C\}$ of $S(V^*)$ consisting of linearly independent eigenvectors corresponding to the eigenvalues η, η^2 and η^4 for an element $k \in K$ of order 7. Here η is a primitive 7th root of unity.

Step 1 $S(V^*)^K$ has an $\mathbb{F}_2$–base consisting of the symmetric sums

$$\sigma_{ijk} = A^iB^jC^k + B^iC^jA^k + C^iA^jB^k ,$$

where $i + 2j + 4k \equiv 0 \bmod 7$.

This is proved by showing that monomials of the form $A^iB^jC^k$ are k–invariant, and then taking the sum in order to allow for the group extension.

Step 2 Write

$$\begin{aligned} x &= \sigma_{111} = ABC \\ y &= \sigma_{130} = AB^3 + BC^3 + CA^3 \\ z &= \sigma_{320} = A^3B^2 + B^3C^2 + C^3A^2 \\ u &= \sigma_{510} = A^5B + B^5C + C^5A \\ v &= \sigma_{700} = A^7 + B^7 + C^7 . \end{aligned}$$

Step 3 Use induction on the degree of the symmetric sums σ_{ijk} to show that the five polynomials above actually do generate the invariant elements. Direct calculation shows that they also belong to $\mathbb{F}_2[\alpha,\beta,\gamma]$, rather than to the polynomial ring over $\mathbb{F}_8$. Furthermore the two relations r_{20} and r_{24} are satisfied. (This can be proved more slickly using Steenrod operations.)

Step 4 The relations are exhaustive. We have to show that the ring epimorphism

$$R = \mathbb{F}_2[x,y,z,u,v]/(r_{20},r_{24}) \xrightarrow{f} S(V^*)^K$$

is a monomorphism. We begin by localising so as to invert $x = ABC$.

LEMMA 5 The map $R \longrightarrow R(x^{-1})$ is mono.

Proof. One first shows by successive formation of quotients that the sequence x,y,v,r_{20},r_{24} in

$\mathbb{F}_2[x,y,z,u,v]$ is regular. From this it follows that multiplication by x is (1–1) on the quotient ring R.

Now extend the scalars in the localised ring from $\mathbb{F}_2$ to $\mathbb{F}_8$, noting that we have one generator and one relation less. Replace r_{24} by

$$r'_{24} = x^4 + x^2u + y^3 + \frac{yzu + z^3}{x} + u^2 \text{, and write}$$

$U = A^2/_B$, $V = B^2/_C$, $W = C^2/_X$. Then U,V,W are fixed by k and permuted by an element h of order 3 in K. Write $y/_x = U + V + W = g_1$, $z/_x = UV + VW + WU = g_2$ and $x = UVW = g_3$. Then $u/_x = U^2V + V^2W + W^2U = g_4$, say, and

$$R(x^{-1}) = \frac{\mathbb{F}_2\,[g_1,g_2,g_3,g_4]}{(r_{12})}\,(x^{-1})\text{, where}$$

$r_{12} = g_4^2 + (g_1g_2 + g_3)g_4 + (g_1^3g_3 + g_2^3 + g_3^2)$ and r'_{24} and r_{20} can be expressed in terms of it.

Given the algebraic independence of U, V and W it is now clear that f is a monomorphism after inversion of x and extension of scalars. Given Lemma 5 the same is true for the original map.

COROLLARY 5 <u>The 2–primary part of the Chern subring</u> $Ch(J_1)_{(2)}$ <u>is properly contained in</u> $H^{even}(J_1,\mathbb{Z})_{(2)}$.

Proof. This is a matter of evaluating the total Chern class of the regular representation of an elementary abelian group of rank 3. It turns out that the only non–vanishing classes are

$$\begin{aligned} c_4 &= \alpha^4 + \ldots + \alpha^2\beta^2 + \ldots + \alpha\beta\gamma(\alpha + \beta + \gamma)\ , \\ c_6 &= \alpha^2\beta^4 + \ldots + \alpha\beta\gamma(\alpha^3 + \beta^3 + \gamma^3 + \alpha\beta\gamma) \text{ and} \\ c_7 &= \alpha^4(\beta^2\gamma + \gamma^2\beta) + \ldots\ . \end{aligned}$$

This calculation serves as a useful check on that in Theorem 4, and the existence of the classes x and z of degrees 6 and 10 shows that there are invariant elements other than Chern classes. Furthermore, and the same argument applies to the Mathieu groups, comparison of spectral sequences shows that the class x (for example) is a universal cycle in the Atiyah–Hirzebruch spectral sequence converging to the completed representation ring $\widehat{R(J_1)}$. Here no localisation of coefficients is involved, and we have yet further examples for which the Grothendieck filtration of $R(G)$ is definitely finer than the topological.

References

Ad	A. Adem et. al.	The geometry of cohomology of the Mathieu group M_{12} (these proceedings)
Atlas	J.H. Conway et al.	Atlas of finite groups, Clarendon Press (Oxford) 1985.
Ch	G.R. Chapman	Generators and relations for the cohomology ring of Janko's first group in the first twenty.one dimensions, in "Groups–St. Andrews 1981", Cambridge University Press, 1982.
J	Z. Janko	A characterisation of the Mathieu simple groups, I & II, J. Algebra 9 (1968) 1–19 and 20–41.
Le	G. Lewis	Integral cohomology rings of groups of order p^3, Trans. Amer. Math. Soc. 132 (1968) 501–29
Sw	R.G. Swan	The p–period of a finite group, Ill. J. Math. 4 (1960) 341–6
Td	J.A. Todd	On representations of the Mathieu groups as collineation groups, J. London Math. Soc. 34 (1959) 406–416
Th	C.B. Thomas	Characteristic classes and 2–modular representations for some sporadic simple groups, to appear in Contemporary Mathematics (Proceedings of the Northwestern Homotopy Theory Conference 1988)

Bonn, April 1989

The abelianization of the theta group in low genus

by

Steven H. Weintraub

Department of Mathematics
Louisiana State University
Baton Rouge, LA 70803-4918

Let S_g denote Siegel space of degree g, i.e., the space of g-by-g symmetric complex matrices with positive-definite imaginary part.

The classical theta function with characteristic m is the function $\theta_m : S_g \times \mathbb{C}^g \to \mathbb{C}$ ($\mathbb{C}^g$ consisting of row vectors) defined by the equation

$$\theta_m(\tau, z) = \sum_{p \in \mathbb{Z}^g} exp(\pi i[(p + m'/2)\tau\ {}^t(p + m'/2) + 2(p + m'/2)\ {}^t(z + m''/2)]).$$

Here the characteristic m is a row vector $m = (m_1', \dots, m_g', m_1'', \dots, m_g'')$ with each entry equal to zero or one. The parity $e(m)$ of a characteristic m is $e(m) = m_1'm_1'' + \dots + m_g'm_g'' \in \mathbb{Z}/2\mathbb{Z}$ and m is called even (odd) as $e(m) = 0 (= 1)$.

Theta functions satisfy the following transformation law [I]:

Theorem. *Let $T \in Sp_{2g}(\mathbb{Z})$ and write T as a block matrix*

$$T = \begin{pmatrix} A & B \\ C & D \end{pmatrix}.$$

Then

$$\theta_{T\cdot m}(z(C\tau + D)^{-1}, (A\tau + B)(C\tau + D)^{-1})$$
$$= \gamma_m(T) det(C\tau + D)^{\frac{1}{2}} exp[(\pi i z)(C\tau + D)^{-1}\ {}^tz]\theta_m(\tau, z)$$

where

$$T \cdot m = mT^{-1} + ((D\ {}^tC)_0\ (B\ {}^tA)_0)\ (mod\ 2)$$

where $(\ )_0$ denotes the row vector obtained by taking the diagonal elements of the matrix, and $\gamma_m(T)$ is an eighth root of unity, known as the theta multiplier of T for the characteristic m.

The action of the symplectic group $Sp_{2g}(\mathbb{Z})$ on characteristics (by $m \mapsto T \cdot m$) preserves parity, and is transitive on characteristics of a given parity. We let $\Gamma(m)$ be the stabilizer of m, i.e.

$$\Gamma(m) = \{T \in Sp_{2g}(\mathbb{Z}) \mid T \cdot m = m\}.$$

We observe that if Γ_2 is the principal congruence subgroup of level 2 in $Sp_{2g}(\mathbb{Z})$, i.e.

$$\Gamma_2 = \{T \in Sp_{2g}(\mathbb{Z}) \mid T = I (mod\, 2)\}$$

then $\Gamma_2 \subset \Gamma(m)$ for every m. (Indeed, $\Gamma_2 = \bigcap_m \Gamma(m)$.)

The "standard" theta function is the theta function with characteristic

$$m_0 = (0, \ldots, 0),$$

and the stabilizer $\Gamma(m_0)$ of this characteristic is classically known as the theta subgroup of $Sp_{2g}(\mathbb{Z})$. This subgroup is usually denoted $\Gamma(1,2)$, and consists of the matrices T as above with the g-by-g matrices A^tB and C^tD having only even entries along their diagonals. Since we are interested in what happens for different values of the genus g, we shall often denote this group by $\Gamma_g(1,2)$.

Let us now explain the connection between these groups and the topology of Riemann surfaces.

Let M^2 be a Riemann surface of genus g with intersection pairing I, and choose a symplectic basis $\{e_i, f_i\}$ for $H_1(M^2 : \mathbb{Z})$. Then I induces the $\mathbb{Z}/2\mathbb{Z}$ intersection pairing $I(mod\, 2)$ on $H_1(M^2 : \mathbb{Z}/2\mathbb{Z})$. A quadratic refinement of this $\mathbb{Z}/2\mathbb{Z}$ intersection pairing is a map $q : H_1(M^2 : \mathbb{Z}) \to \mathbb{Z}/2\mathbb{Z}$ such that

$$q(x+y) - q(x) - q(y) = I[x,y]\,(mod\, 2)$$

for all x, y in $H_1(M^2 : \mathbb{Z})$. In this algebraic situation we may explicitly describe all quadratic refinements of the bilinear form $I(mod\, 2)$. It is easy to check that they arise as follows:

The quadratic refinements q_m of $I(mod\, 2)$ are in one-to-one correspondence to the characteristics m (as defined above), where $q_m : H_1(M^2 : \mathbb{Z}) \to \mathbb{Z}/2\mathbb{Z}$ is given by

$$q_m(\sum x_i e_i + y_i f_i) = \sum x_i y_i + \sum x_i m_i' + \sum y_i m_i''\,(mod\, 2).$$

A lengthy but routine computation also shows: Let $f : M^2 \to M^2$ with

$$T = f_* : H_1(M^2 : \mathbb{Z}) \to H_1(M^2 : \mathbb{Z})$$

the induced map on homology. Then

$$q_{T\cdot m}(T(x)) = q_m(x) \quad for\ every \quad x \in H_1(M^2 : \mathbb{Z})$$

where $T \cdot m$ is defined as in 3.1.

Thus, in particular, we see that the stabilizer of the form q_m is the subgroup $\Gamma(m)$ of $Sp_{2g}(\mathbb{Z})$.

(The quadratic refinements of a given non-singular bilinear form are classified up to isomorphism by their Arf invariants (in $\mathbb{Z}/2\mathbb{Z}$), and the Arf invariant of q_m is just the parity $e(m)$.)

The quadratic refinements have a geometric interpretation which follows from the work of E. Brown [B], or, alternatively, from that of D. Johnson [J]. Namely, every

Riemann surface admits a spin structure, and indeed, exactly 2^{2g} spin structures, and these spin structures are in $1-1$ correspondence with the 2^{2g} quadratic refinements q_m of $I(mod\, 2)$. Thus a self-diffeomorphism f of the Riemann surface M preserves a spin structure w_m of M if and only if its induced map f_* on $H_1(M:\mathbf{Z})$ satisfies $f_* \in \Gamma(m)$. (The above discussion is taken from [LMW, sections 2 and 3].)

The eighth root of unity $\gamma_m(T)$ which appears above depends on the choice of square root of $det(C\tau + D)$ in the above transformation law. (This choice can be made independently of τ and z.) Because of this choice, the function

$$\gamma_m : \Gamma(m) \to \{eighth\ roots\ of\ unity\}$$

is *not* a homomorphism, but its square

$$(\gamma_m)^2 : \Gamma(m) \to \{fourth\ roots\ of\ unity\} \qquad (*)$$

is a homomorphism.

In [I, theorem 3] Igusa gave a method for computing $(\gamma_m)^2(T)$ for $T \in \Gamma_2$ and m an even characteristic. In case $m = m_0$ the result is particularly simple. Letting $\gamma_0 = \gamma_{m_0}$,

$$(\gamma_0)^2(T) = (-1)^{\frac{1}{2}\, tr(D-I)}$$

for any $T \in \Gamma_2$ where T is written as a block matrix as above and I denotes the g-by$g$ identity matrix. (Note that for any even m

$$(\gamma_m)^2 : \Gamma_2 \to \{square\ roots\ of\ unity\}.)$$

The work of [JM] contains two main results. The first is an algorithm for computing $(\gamma_0)^2(T)$ for any $T \in \Gamma(1,2)$. The second, which is our main point of interest, is the following. From $(*)$ we have an epimorphism from $\Gamma(1,2)$ onto $\mathbf{Z}/4\mathbf{Z}$, which of course factors through the abelianization of $\Gamma(1,2)$. They show

Theorem. [JM, theorem 1.1 (i)] *For $g \geq 3$, the map $T \mapsto (\gamma_0)^2(T)$ gives an isomorphism from the abelianization of the genus g theta group $\Gamma_g(1,2)$ to $\mathbf{Z}/4\mathbf{Z}$.*

Our main result here, which is considerably easier than theirs, is that this theorem is false for $g = 1,2$. To be precise:

Theorem 1. i) *The abelianization of $\Gamma_1(1,2)$ is isomorphic to $\mathbf{Z}/4\mathbf{Z} \oplus \mathbf{Z}$.*
ii) *There is an epimorphism from the abelianization of $\Gamma_2(1,2)$ onto $\mathbf{Z}/4\mathbf{Z} \oplus \mathbf{Z}/2\mathbf{Z}$.*

In the course of our proof we shall exhibit elements of $\Gamma_g(1,2)$, $g = 1,2$, whose images generate the given groups. (As the reader will see, the argument for (i) uses nothing that has not been known been known for decades, and it is certainly possible that this fact has been noticed before. Part (ii) is essentially new.)

Proof. We have mentioned that $\Gamma_2 \subset \Gamma(1,2)$, and that $Sp_{2g}(\mathbf{Z}) = \Gamma_1$ acts transitively on the even characteristics. For $g = 1$, there are 3 even characteristics, and $[\Gamma_1 : \Gamma_2] = 6$, so $[\Gamma_1(1,2) : \Gamma_2] = 2$. For $g = 2$, there are 10 even characteristics, and

$[\Gamma_1 : \Gamma_2] = 720$, so $[\Gamma_1(1,2) : \Gamma_2] = 72$. For convenience we let $\Gamma = \Gamma_g(1,2)$, the value of g being clear from the context.

i) The group Γ contains the element $S = \left(\begin{smallmatrix} 0 & 1 \\ -1 & 0 \end{smallmatrix}\right)$, and $S \notin \Gamma_2$, so Γ is generated by Γ_2 and S.

We now pass to the projective group $PSp_2(\mathbb{Z}) = Sp_2(\mathbb{Z})/\pm 1$. Then we have a commutative diagram, with the two vertical arrows being $2-1$ maps,

$$\begin{array}{ccccccccc} 1 & \to & \Gamma_2 & \to & \Gamma & \to & \mathbb{Z}/2\mathbb{Z} & \to & 1 \\ & & \downarrow & & \downarrow & & \| & & \\ 1 & \to & P\Gamma_2 & \to & P\Gamma & \to & \mathbb{Z}/2\mathbb{Z} & \to & 1 \end{array} .$$

Recall that $PSp_2(\mathbb{Z}) = PSL_2(\mathbb{Z})$ acts on $\mathcal{S}_1$, the ordinary upper-half plane, by fractional linear transformations. The subgroup $P\Gamma_2$ acts *freely* on $\mathcal{S}_1$, so $P\Gamma_2$ is isomorphic to $\pi_1(\mathcal{S}_1/P\Gamma_2)$. However $(\mathcal{S}_1/P\Gamma_2) = P^1(\mathbb{C}) - \{0,1,\infty\}$ (as is classically known) and so $\pi_1(\mathcal{S}_1/P\Gamma_2)$ is free on two generators, which we choose to be the loops around ∞ and 0. These loops are represented by the matrices $\bar{R} = \left(\begin{smallmatrix} 1 & 2 \\ 0 & 1 \end{smallmatrix}\right)$ and $\bar{U} = \left(\begin{smallmatrix} 1 & 0 \\ -2 & 1 \end{smallmatrix}\right)$ respectively.

If $\bar{S}$ is the image of S in $P\Gamma$, then matrix multiplication shows $\bar{S}^2 = 1$ and $\bar{S}\bar{R}\bar{S}^{-1} = \bar{U}$ so in terms of generators and relations we find

$$\begin{aligned} P\Gamma_2 &= \langle \bar{R}, \bar{U} \rangle \\ P\Gamma &= \langle \bar{R}, \bar{U}, \bar{S} \mid \bar{S}^2 = 1,\ \bar{S}\bar{R}\bar{S}^1 = \bar{U} \rangle \cdot \\ &= \langle \bar{R}, \bar{S} \mid \bar{S}^2 = 1 \rangle \end{aligned}$$

Lifting back to Γ_1, the elements $\bar{R}$, $\bar{U}$, and $\bar{S}$ lift to elements R, U and S given by the matrices as written. Now however, S is of order 4, so we find

$$\begin{aligned} \Gamma &= \langle R, U, S \mid S^4 = 1,\ SRS^{-1} = U \rangle \\ &= \langle R, S \mid S^4 = 1,\ RS^2 = S^2 R \rangle \end{aligned}$$

(reflecting the fact that $P\Gamma$ is the quotient of Γ by its center $\{1, S^2\}$), and the abelianization is as claimed.

ii) It will be convenient for us to regard $Sp_2(\mathbb{Z}) \times Sp_2(\mathbb{Z})$ as the subgroup of $Sp_4(\mathbb{Z})$ consisting of matrices of the form

$$\left\{ \begin{pmatrix} a_1 & & b_1 & \\ & a_2 & & b_2 \\ c_1 & & d_1 & \\ & c_1 & & d_2 \end{pmatrix} \middle| \; W_i = \begin{pmatrix} a_i & b_i \\ c_i & d_i \end{pmatrix} \in Sp_2(\mathbb{Z}), \quad i = 1,2 \right\}.$$

We write such a matrix as a pair (W_1, W_2).

As we have observed, the subgroups $\Gamma(m)$ are all conjugate for m even. Thus, instead of investigating Γ we may investigate $\Gamma' = \Gamma((1,1,1,1))$, the stabilizer of the even characteristic $m = (1,1,1,1)$. Then by (3.3) and (4.3) of [LW2] (see also section 2.2 of [LW1]), Γ' is generated by Γ_2, $Sp_2(\mathbb{Z}) \times Sp_2(\mathbb{Z})$, and the element

$$F = \left(\begin{array}{cc|cc} & -1 & & \\ 1 & & & \\ \hline & & & -1 \\ & & 1 & \end{array}\right).$$

(From this description it is also easy to see that $[\Gamma' : \Gamma_2] = 72$, as stated above.)

Let us consider the subgroup Λ of Γ' generated by $Sp_2(\mathbb{Z}) \times Sp_2(\mathbb{Z})$ and F. Then $[\Lambda : Sp_2(\mathbb{Z}) \times Sp_2(\mathbb{Z})] = 2$, but F is an element of order 4 (though $F^2 \in Sp_2(\mathbb{Z}) \times Sp_2(\mathbb{Z})$) and conjugation by F gives an automorphism of $Sp_2(\mathbb{Z}) \times Sp_2(\mathbb{Z})$ of order two given by the formula $F(W_1, W_2)F^{-1} = (-W_2, -W_1)$, so Λ is a non-split extension of $Sp_2(\mathbb{Z}) \times Sp_2(\mathbb{Z})$ by $\mathbb{Z}/2\mathbb{Z}$.

To avoid confusion we shall denote the principal congruence subgroup of level two of $Sp_2(\mathbb{Z})$ by G_2. Now the abelianization of Γ' certainly maps onto the abelianization of $\Gamma'/\Gamma_2 = \Lambda/(\Lambda \cap \Gamma_2)$, and

$$\Lambda \cap \Gamma_2 = \{(W_1, W_2) \mid W_i \in G_2, \quad i = 1, 2\}.$$

Note that $F^2 \in \Lambda \cap \Gamma_2$ so we have a split extension (with F projecting non-trivially to $\mathbb{Z}/2\mathbb{Z}$)

$$1 \to (Sp_2(\mathbb{Z})/G_2) \times (Sp_2(\mathbb{Z})/G_2) \to \Lambda/\Lambda \cap \Gamma_2 \to \mathbb{Z}/2\mathbb{Z} \to 1.$$

It is well-known that $Sp_2(\mathbb{Z})/G_2$ is isomorphic to the symmetric group on three symbols, whose abelianization is $\mathbb{Z}/2\mathbb{Z}$, and, indeed, the element $S = \left(\begin{smallmatrix} 0 & 1 \\ -1 & 0 \end{smallmatrix}\right)$ of $Sp_2(\mathbb{Z})$ has non-trivial image under the abelianization map.

If we define elements V_1, V_2 of Λ by $V_1 = (-S, I)$ and $V_2 = (I, -S)$, where I is the 2×2 identity matrix, it then follows easily that the abelianization of Γ'/Γ_2 is the abelianization of $\Sigma/\Sigma \cap \Gamma_2$, where Σ is the group generated by V_1, V_2, and F. But the quotient group is

$$\langle V_1, V_2, F \mid V_1^2 = V_2^2 = F^2 = 1, \; V_1V_2 = V_2V_1, \; FV_1F^{-1} = V_2 \rangle$$

so has abelianization $(\mathbb{Z}/2\mathbb{Z}) \oplus (\mathbb{Z}/2\mathbb{Z})$. Thus we have so far that the the abelianization of Γ' maps onto $(\mathbb{Z}/2\mathbb{Z}) \oplus (\mathbb{Z}/2\mathbb{Z})$, with the element V_1 (or equivalently V_2) mapping to $(1,0)$ and F to $(0,1)$. Note, however, that $V_1^2 = (-I, I)$ (resp. $V_2^2 = (I, -I)$) and by [I, theorem 3] $(\gamma_{(1,1,1,1)})^2(V_1^2) = (\gamma_{(1,1,1,1)})^2(V_2^2) = -1$, so V_1 (resp. V_2) is an element of order 4, and the theorem follows.

Let us write $\mathbb{Z}$ additively but identify $\mathbb{Z}/2\mathbb{Z}$ with $\{1, -1\}$ and $\mathbb{Z}/4\mathbb{Z}$ with $\{1, i, -1, -i\}$. Let $\alpha = (\alpha_1, \alpha_2) : \Gamma_1(1,2) \to \mathbb{Z}/4\mathbb{Z} \oplus \mathbb{Z}$ and $\beta' = (\beta_1', \beta_2') : \Gamma' \to \mathbb{Z}/4\mathbb{Z} \oplus \mathbb{Z}/2\mathbb{Z}$ be the maps constructed in the proof of the theorem.

We have actually stated the theorem for Γ, not Γ'. To obtain an explicit map for Γ we must conjugate our elements above by a suitable element of $Sp_4(\mathbb{Z})$. Such an element is

$$H = \begin{pmatrix} 1 & 1 & 1 & 0 \\ 1 & 1 & 0 & 1 \\ -1 & 0 & 1 & -1 \\ 0 & -1 & -1 & 1 \end{pmatrix}$$

Letting $\widetilde{V}_i = HV_iH^{-1}$ and $\widetilde{F} = HFH^{-1}$ we obtain the (fearsome looking) elements

$$\widetilde{V}_1 = \begin{pmatrix} -1 & 0 & -2 & -2 \\ -2 & 2 & 0 & -1 \\ 2 & -2 & -1 & 0 \\ 0 & 1 & 2 & 2 \end{pmatrix} \qquad \widetilde{V}_2 = \begin{pmatrix} 2 & -2 & -1 & 0 \\ 0 & -1 & -2 & -2 \\ 1 & 0 & 2 & 2 \\ -2 & 2 & 0 & 1 \end{pmatrix}$$

$$\widetilde{F} = \begin{pmatrix} 2 & -3 & -2 & 0 \\ 3 & -2 & 0 & 2 \\ -2 & 0 & -2 & -3 \\ 0 & 2 & 3 & 2 \end{pmatrix}$$

We define $\beta = (\beta_1, \beta_2) : \Gamma_2(1,2) \to \mathbb{Z}/4\mathbb{Z} \oplus \mathbb{Z}/2\mathbb{Z}$ by $\beta(T) = \beta'(H^{-1}TH)$.

Corollary 2.

i) *The map* $\alpha_1 : \Gamma_1(1,2) \to \mathbb{Z}/4\mathbb{Z}$ *satisfies* $\alpha_1(R) = 1$, $\alpha_1(S) = i$. *Indeed, the map* α_1 *agrees with* $(\gamma_0)^2$. *The map* $\alpha_2 : \Gamma_2(1,2) \to \mathbb{Z}$ *satisfies* $\alpha_2(R) = 1$, $\alpha_2(S) = 0$.

ii) *The map* $\beta_1 : \Gamma_2(1,2) \to \mathbb{Z}/4\mathbb{Z}$ *satisfies* $\beta_1(\widetilde{V}_1) = \beta_1(\widetilde{V}_2) = i$, $\beta_1(\widetilde{F}) = 1$. *Indeed,* β_1 *agrees with* $(\gamma_0)^2$. *The map* $\beta_2 : \Gamma_2(1,2) \to \mathbb{Z}/2\mathbb{Z}$ *satisfies* $\beta_2(\widetilde{V}_1) = \beta_2(\widetilde{V}_2) = 1$, $\beta_2(\widetilde{F}) = -1$. *Indeed,* β_2 *agrees with the map* $T \to (-1)^{d(H^{-1}TH)}$, *where*

$$d(M) = det\begin{pmatrix} m_{12} & m_{14} \\ m_{32} & m_{34} \end{pmatrix} = det\begin{pmatrix} m_{21} & m_{23} \\ m_{41} & m_{43} \end{pmatrix} \ (mod\, 2) \text{ for } M = (m_{ij}) \in \Gamma'.$$

Proof. All of part i) with the exception of the agreement with $(\gamma_0)^2$ follows directly from the proof of the theorem. The agreement with $(\gamma_0)^2$ follows as one may calculate that $(\gamma_0)^2(R) = 1$, $(\gamma_0)^2(S) = i$, and these two elements are generators.

As for part ii), note that in the proof of the theorem β_1' was defined through $(\gamma_{(1,1,1,1)})^2$, so the agreement of β_1 with $(\gamma_0)^2$ is a tautology. Then the given values for $(\gamma_0)^2$ follow by computation. The value of $\beta_2(\widetilde{V}_i)$ and $\beta_2(\widetilde{F})$ is given by the proof, and one can easily check that the given map is a homomorphism taking the prescribed values.

(The calculation of $(\gamma_0)^2$ in i) is quite classical, but the easiest way to calculate $(\gamma_0)^2$ in ii) is to use the formula of [JM].)

Acknowledgement. The author is partially supported by NSF grant DMS-8803552.

References

[B] Brown, E. The Kervaire invariant of a manifold, Proc. Symp. Pure Math. (AMS) **22**(1970), 65-71.

[I] Igusa, J.-I. On the graded ring of theta-constants, Am. J. Math. **86**(1964), 219-246.

[J] Johnson, D. Spin structures and quadratic forms on surfaces, J. London Math. Soc. (2) **22**(1980), 365-373.

[JM] Johnson, D. and Millson, J.J. Modular Lagrangians and the theta multiplier, to appear.

[LMW] Lee, R., Miller, E. Y. and Weintraub, S.H. Rochlin invariants, theta functions, and the holonomy of some determinant line bundles, J. reine angew. Math. **392**(1988), 187-218.

[LW1] Lee, R. and Weintraub, S.H. Cohomology of $Sp_4(\mathbb{Z})$ and related groups and spaces, Topology **24**(1985), 391-410.

[LW2] Lee, R. and Weintraub, S.H. On the transformation law for theta-constants, J. Pure Appl. Algebra **44**(1987), 273-285.

LIST OF TALKS

PLENARY TALKS

Kunio Murasugi
Invariants of Graphs with Applications to Knot Theory.

F. Thomas Farrell
Topological Rigidity.

Alexander B. Goncharov
Projective Geometry and Algebraic K-theory.

Hans-Werner Henn
Some Finiteness Results in the Category of Unstable Modules over the Steenrod Algebra.

Lowell Jones
The Space of Stable Pseudo-isotopies on a Non-positively Curved Manifolds.

Zbigniew Marciniak
Geometric Approach to Units in Group Rings of Infinite Groups.

Bob Oliver
Self-maps of Classifying Spaces of Compact Lie Groups.

Elmer Rees
The Fundamental Groups of Algebraic Varieties.

Melvin Rothenberg
Equivariant rational homotopy and classification of G-manifolds.

Nobuaki Yagita
BP-cohomology of BG for a Compact Lie Group.

SECTIONAL TALKS

Alexandro Adem
Cohomology of Sporadic Simple Groups.

Boris Apanasov
Conformal Structures on Hyperbolic Manifolds and Varieties of Representations.

Dominique Arlettaz
On the Cohomology of Congruence Subgroups.

Stanisław Betley
Homology Groups of GL(R) with Twisted Coefficients.

Boris Botwinnik
The Geometrical Point of View on the Adams-Novikov Spectral Sequence.

William Browder
Smooth Exotic Actions on Products of Spheres.

Frank Connolly
On the Rigidity of Certain Groups.

Steven R. Costenoble
Application of Equivariant Orientation Theory.

Jim Davis
Alexander Polynomials of Periodic Knots.

Ryszard Doman
Rational Moore G-spaces and co-Hopf G-spaces.

Karl Heinz Dovermann
Topological Invariants of Real Algebraic Group Actions.

Giora Dula
Relative Attaching Map in Thom Spaces.

Thomas Fiedler
Knots and the Topology of Complex Curves on Complex Surfaces.

Alexander Harshiladze
The Browder-Livsay Groups for Abelian 2-Groups.

Jean-Claude Hausmann
Topological Spaces Associated to Robot Arms.

Johannes Huebschmann
Perturbation theory and cohomology of groups.

Francis E. A. Johnson
Flat Complex Algebraic Manifolds and Flat Kähler manifolds.

Klaus Heinz Kamps
Aspects of Abstract Homotopy Theory.

Sung Sook Kim
Characteristic Numbers and Group Actions.

John Klippenstein
Applications of a Relationship between K-theory Operations and Cohomology Operations.

Andrzej Kozlowski
Characteristic Classes of Transfers of Vector Bundles.

Errki J. Laitinen
A Splitting Principle for Fixed Point Functor.

Wolfgang Lück
Analytic and Topological Torsion for Manifolds with Boundary and Symmetries.

Mikiya Masuda
Semifree SU(2)-actions on Homology Spheres and the Rochlin Invariant.

Sergiej Matveev
Theory of Complexity of 3-manifolds.

James McClure
Topological Hochschild Homology of the bu-Spectrum.

Aleksandr S. Mishchenko
Fredholm Structures on Infinite-dimensional Manifolds and Their Homological Description.

Masaharu Morimoto
One Fixed Point Actions on Spheres.

Hans Jorgen Munkholm
On the Boundedly Controlled K-theory over an Open Cone.

Roin Nadiradze
Realization of Elements in the Sp- and Sc-cobordism Theories.

Nguyen Viet Dong
On the Cohomology of the Unipotent Subgroup of the General Linear Group $GL(3, F(q))$.

Nguyen Huynh Phan
On the Topology of the Space of Reachable Symmetric Linear Systems.

Dietrich Notbohm
Maps Between Classifying Spaces.

Andrei Pazhitnov
On the Exactness of Novikov Inequalities for the Manifold with Free Abelian Fundamental Group.

Eric Pedersen
Controlled Surgery and Applications to Group Actions.

Stewart Priddy
The Stable Type of BG.

Pham Anh Mingh
Transfer Map and the Hochschild-Serre Spectral Sequence.

Dieter Puppe
Critical Point Theory with Symmetries.

Jonathan Rosenberg
The KO Assembly Map and Positive Scalar Curvature.

Julius Rudiak
Orientability of Bundles and Fibrations.

Michał Sadowski
Equivariant Splittings Induced by Some Toral Actions.

Reinhard Schultz
Positive scalar curvature and spherical space forms.

Jolanta Słomińska
Homotopy Colimits over EI-categories.

Larry Smith
Fake Lie Groups and Maximal Tori.

Paweł Traczyk
New Criteria for Periodic Knots.

Evgenii Troitsky
Some Aspects of the C^*-index Theorem.

Vladimir Vershinin
On Spectra Realizing Some Modules over the Steenrod Algebra.

Peter Webb
The Structure of Mackey Functors.

Andrzej Weber
A Filtration in the Intersection Homology Groups.

Steve Weintraub
Cohomology of certain Siegel Modular Varieties.

CURRENT ADDRESSES OF PARTICIPANTS AND AUTHORS

Alexandro Adem
Department of Mathematics
University of Wisconsin
MADISON, WI 53706
U.S.A.

Piotr Akhmetev
Steklov Institute
Soviet Academy of Sciences
MOSCOW 117333
Soviet Union

Christopher Allday
Department of Mathematics
University of Hawaii at Manoa
HONOLULU, HI 96822
U.S.A.

Paweł Andrzejewski
Instytut Matematyki
Uniwersytet Szczeciński
PL-70-451 SZCZECIN
Poland

Boris Apanasov
Institute of Mathematics
Soviet Academy of Sciences
NOVOSIBIRSK 630090
Soviet Union

Dominique Arlettaz
Department of Mathematics
Universite de Lausanne
CH-1015 LAUSANNE
Switzerland

Amir Assadi
Department of Mathematics
University of Wisconsin
MADISON, WI 53706
U.S.A.

Hans Joachim Baues
Max-Planck-Institut
für Mathematik
D-5300 BONN 3
Germany

Stanisław Betley
Instytut Matematyki
Uniwersytet Warszawski
PL-00-913 WARSZAWA 59
Poland

Agnieszka Bojanowska
Instytut Matematyki
Uniwersytet Warszawski
PL-00-913 WARSZAWA 59
Poland

Boris Botwinnik
Computer Center
Soviet Academy of Sciences
KHABAROVSK 680063
Soviet Union

Cezary Bowszyc
Instytut Matematyki
Uniwersytet Warszawski
PL-00-913 WARSZAWA 59
Poland

William Browder
Department of Mathematics
Princeton University
PRINCETON, NJ 08544
U.S.A.

Frank Connolly
Department of Mathematics
University of Notre Dame
NOTRE DAME, IN 46556
U.S.A.

R. Costenoble
Department of Mathematics
Hofstra University
HEMPSTEAD, NY 11550
U.S.A.

James F. Davis
Department of Mathematics
Indiana University
BLOOMINGTON, IN 47405
U.S.A.

Andrzej Dawidowicz
Zakład Matematyki
Wyższa Szkoła Pedagogiczna
PL-10-561 OLSZTYN
Poland

Ryszard Doman
Instytut Matematyki
Uniwersytet im. A. Mickiewicza
PL-60-769 POZNAŃ
Poland

Wojciech Dorabiala
Instytut Matematyki
Uniwersytet Szczeciński
PL-70-451 SZCZECIN
Poland

Ronald M. Dotzel
Department of Mathematics
University of Missouri-St.Louis
ST.LOUIS, MO 63121
U.S.A.

Karl-Heinz Dovermann
Department of Mathematics
University of Hawaii at Manoa
HONOLULU, HI 96822
U.S.A.

Emmanuel Dror-Farjoun
Department of Mathematics
The Hebrew University
91904 JERUSALEM
Israel

Giora Dula
Department of Mathematics
Purdue University
WEST LAFAYETTE, IN 47907
U.S.A.

Grzegorz Dylawerski
Instytut Matematyki
Uniwersytet Gdański
PL-80-308 GDAŃSK
Poland

Zdzisław Dzedzej
Instytut Matematyki
Uniwersytet Gdański
PL-80-308 GDAŃSK
Poland

John Ewing
Department of Mathematics
Indiana University
BLOOMINGTON, IN 47405
U.S.A.

F. Thomas Farrell
Department of Mathematics
Columbia University
NEW YORK, NY 10027
U.S.A.

Thomas Fiedler
Institute of Mathematics
Akademie der Wissenschaften
D-1086 BERLIN
Germany

Paweł Gajer
Instytut Matematyki
Uniwersytet Wrocławski
PL-50-384 WROCLAW
Poland

Andrzej Gaszak
Instytut Matematyki
Uniwersytet im. A. Mickiewicza
PL-60-769 POZNAŃ
Poland

Charles H. Giffen
Department of Mathematics
University of Virginia
CHARLOTTESVILLE, VA22903
U.S.A.

Jacek Gocłowski
Zakład Matematyki
Wyższa Szkoła Pedagogiczna
PL-10-561 OLSZTYN
Poland

Marek Golasiński
Instytut Matematyki
Uniwersytet im. M. Kopernika
PL-87-100 TORUŃ
Poland

Alexander B. Goncharov
Steklov Institute
Soviet Academy of Sciences
MOSCOW 117133
Soviet Union

Andrzej Granas
Department of Mathematics
Universite de Montreal
MONTREAL, Quebec H3G 3J7
Canada

Bogusław Hajduk
Instytut Matematyki
Uniwersytet Wrocławski
PL-50-384 WROCLAW
Poland

Alexander Harshiladze
I.Z.M.I.R.
Soviet Academy of Sciences
TROITSK 142092
Soviet Union

Akiro Hattori
Department of Mathematics
University of Tokyo
TOKYO 113
Japan

Jean-Claude Hausmann
Department of Mathematics
Universite de Geneve
CH-1211 GENEVE 24
Switzerland

Hans-Werner Henn
Department of Mathematics
Universität Heidelberg
D-6900 HEIDELBERG
Germany

Johannes Hübschmann
Department of Mathematics
Universität Heidelberg
D-6900 HEIDELBERG
Germany

Soren Illman
Department of Mathematics
University of Helsinki
SF-00100 HELSINKI 10
Finland

Paul Iqodt
Department of Mathematics
K.U.L.
B-8500 KORTRIJK
Belgium

Marek Izydorek
Instytut Matematyki
Politechnika Gdańska
PL-80-952 GDAŃSK
Poland

Stefan Jackowski
Instytut Matematyki
Uniwersytet Warszawski
PL-00-913 WARSZAWA 59
Poland

Jan Jaworowski
Department of Mathematics
Indiana University
BLOOMINGTON, IN 47405
U.S.A.

Jerzy Jezierski
Katedra Zastosowań Matematyki
S.G.G.W.
PL-02-766 WARSZAWA
Poland

Jerzy Jodel
Instytut Matematyki
Uniwersytet Gdański
PL-80-308 GDAŃSK
Poland

Francis E. A. Johnson
Department of Mathematics
University College
LONDON WC1E 6BT
Great Britain

Lowell Jones
Department of Mathematics
State University of New York
STONY BROOK, NY 11790
U.S.A.

Yoshinobu Kamishima
Department of Mathematics
Hokkaido University
SAPPORO 060
Japan

Klaus Heiner Kamps
Department of Mathematics
Fernuniversität
D-5800 HAGEN
Germany

Cherry Kearton
Department of Mathematics
University of Durham
DURHAM DH1 3LE
Great Britain

Sung Sook Kim
Department of Mathematics
Korea Institute of Technology
TAEJON, 305-701
South Korea

John Klippenstein
Department of Mathematics
University of British Columbia
VANCOUVER, B.C. V6T 1Y4
Canada

Julius Korbas
Institute of Mathematics
Slovak Academy of Sciences
CS-81473 BRATISLAVA
Czechoslovakia

Ulrich Koschorke
Department of Mathematics
Universität Siegen
D-5900 SIEGEN 21
Germany

Andrzej Kozlowski
Department of Mathematics
Wayne State University
DETROIT,MI 48202
U.S.A.

Tadeusz Koźniewski
Instytut Matematyki
Uniwersytet Warszawski
PL-00-913 WARSZAWA 59
Poland

Józef Krasinkiewicz
Instytut Matematyczny
Polska Akademia Nauk
PL 00-950 WARSZAWA
Poland

Jan Kubarski
Instytut Matematyki
Politechnika Łódzka
PL-93-590 ŁÓDŹ
Poland

Errki J. Laitinen
Department of Mathematics
University of Helsinki
SF-00100 HELSINKI
Finland

L. Gaunce Lewis
Department of Mathematics
Syracuse University
SYRACUSE, NY 13244
U.S.A.

Marek Lewkowicz
Instytut Matematyki
Uniwersytet Wrocławski
PL-50-384 WROCLAW
Poland

Wladyslaw Lorek
Instytut Matematyki
Uniwersytet Wrocławski
PL-50-384 WROCLAW
Poland

Wolfgang Lück
Department of Mathematics
University of Kentucky
LEXINGTON, KY40506
U.S.A.

Oleg W. Manturov
Department of Mathematics
Moscow State University
MOSCOW 129344
Soviet Union

Ewa Marchow
Instytut Matematyki
Uniwersytet im. A. Mickiewicza
PL-60-769 POZNAŃ
Poland

Zbigniew Marciniak
Instytut Matematyki
Uniwersytet Warszawski
PL-00-913 WARSZAWA 59
Poland

Tadeusz Marx
Katedra Ekonometrii i Inform.
S.G.G.W.
PL-02-528 WARSZAWA
Poland

Mikiya Masuda
Department of Mathematics
Osaka City University
OSAKA 558
Japan

Sergiej Matveev
Department of Mathematics
Chelabinsk University
CHELABINSK 454014
Soviet Union

James McClure
Department of Mathematics
University of Kentucky
LEXINGTON, KY 40506
U.S.A.

Piotr Mikrut
Instytut Matematyki
Uniwersytet Wrocławski
PL-50-384 WROCLAW
Poland

Aleksandr S. Mishchenko
Department of Mathematics
Moscow State University
MOSCOW 129344
Soviet Union

Masaharu Morimoto
Department of Mathematics
Okayama University
OKAYAMA 700
Japan

Hans Jorgen Munkholm
Department of Mathematics
Odense Universitet
DK-5320 ODENSE
Denmark

Kunio Murasugi
Department of Mathematics
University of Toronto
TORONTO, Ontario M5S 1A1
Canada

Roin Nadiradze
Institute of Mathematics
Georgian Academy of Sciences
TBILISI 380093
Soviet Union

Ikumitsu Nagasaki
Department of Mathematics
Osaka University
OSAKA 560
Japan

Adam Neugebauer
Instytut Matematyki
Uniwersytet im. A. Mickiewicza
PL-60-769 POZNAŃ
Poland

Nguyen Viet Dong
Department of Mathematics
University of Hanoi
HANOI
Vietnam

Nguyen Huynh Phan
Department of Mathematics
Pedagogical Univ. of Vinh
NGHE TINH
Vietnam

Dietrich Notbohm
SFB 170
Georg August Universität
D-3400 GÖTTINGEN
Germany

Krzysztof Nowiński
Instytut Matematyki Stosowanej
Uniwersytet Warszawski
PL-00-913 WARSZAWA 59
Poland

Robert Oliver
Department of Mathematics
Aarhus Universitet
DK-8000 AARHUS C
Denmark

Krzysztof Pawałowski
Instytut Matematyki
Uniwersytet im. A. Mickiewicza
PL-60-769 POZNAŃ
Poland

Andriej Pazhitnov
Institute of Chemical Physics
Soviet Academy of Sciences
MOSCOW 117977
Soviet Union

Eric Pedersen
Department of Mathematics
State University of New York
BINGHAMTON, NY 13901
U.S.A.

Charya Peterson
SFB 170
Georg-August Universitaet
D-3400 GÖTTINGEN
Germany

Franklin Peterson
Department of Mathematics
Massachusets Institute of Techn.
CAMBRIDGE, MA 02139
U.S.A.

Pham Anh Mingh
Department of Mathematics
University of Hanoi
HANOI
Vietnam

Stewart Priddy
Department of Mathematics
Northwestern University
EVANSTON, IL 60208
U.S.A.

Dieter Puppe
Department of Mathematics
Universität Heidelberg
D-6900 HEIDELBERG
Germany

Volker Puppe
Department of Mathematics
Universität Konstanz
D-7750 KONSTANZ
Germany

Elmer Rees
Department of Mathematics
Edinburgh University
EDINBURGH EH9 3JZ
Great Britain

Jonathan Rosenberg
Department of Mathematics
University of Maryland
COLLEGE PARK, MD 20742
U.S.A.

Shmuel Rosset
Department of Mathematics
Tel Aviv University
69978 RAMAT AVIV
Israel

Melvin Rothenberg
Department of Mathematics
University of Chicago
CHICAGO, IL 60637
U.S.A.

Julius Rudiak
M.I.S.I.
MOSCOW 129337
Soviet Union

Sławomir Rybicki
Instytut Matematyki
Politechnika Gdańska
PL-80-952 GDAŃSK
Poland

Michał Sadowski
Instytut Matematyki
Uniwersytet Gdański
PL-80-308 GDAŃSK
Poland

Jan Samsonowicz
Instytut Matematyki
Politechnika Warszawska
PL-00-661 WARSZAWA
Poland

Reinhard Schultz
Department of Mathematics
Purdue University
WEST LAFAYETTE, IN 47907
U.S.A.

Roland Schwänzl
Department of Mathematics
Universität Osnabrück
D-4500 OSNABRÜCK
Germany

Jolanta Słomińska
Instytut Matematyki
Uniwersytet im. M. Kopernika
PL-87-100 TORUŃ
Poland

Larry Smith
Department of Mathematics
Georg August Universität
D-3400 GÖTTINGEN
Germany

Stanisław Spież
Instytut Matematyczny
Polska Akademia Nauk
PL-00-950 WARSZAWA
Poland

Mihail Stanko
Steklov Institute
Soviet Academy of Sciences
MOSCOW 117133
Soviet Union

Boris Sternin
M.I.M.S.
MOSCOW 109028
Soviet Union

Andrzej Szczepański
Instytut Matematyki
Uniwersytet Gdański
PL-80-308 GDAŃSK
Poland

Laurence Taylor
Department of Mathematics
University of Notre Dame
NOTRE DAME, IN 46556
U.S.A.

Paweł Traczyk
Instytut Matematyki
Uniwersytet Warszawski
PL-00-913 WARSZAWA 59
Poland

Krzysztof Trautman
Instytut Matematyki
Uniwersytet Warszawski
PL-00-913 WARSZAWA 59
Poland

Evgenii Troitsky
Department of Mathematics
Moscow State University
MOSCOW 129344
Soviet Union

Katsuhiro Uno
Department of Mathematics
Osaka University
OSAKA, 560
Japan

Vladimir Vershinin
Institute of Mathematics
Soviet Academy of Sciences
NOVOSIBIRSK 630090
Soviet Union

Rainer Vogt
Department of Mathematics
Universität Osnabrück
D-4500 OSNABRÜCK
Germany

Peter Webb
Department of Mathematics
University of Oregon
EUGENE, OR 97403
U.S.A.

Andrzej Weber
Instytut Matematyki
Uniwersytet Warszawski
PL-00-913 WARSZAWA 59
Poland

Steven Weintraub
Department of Mathematics
Louisiana State University
BATON ROUGE, LA 70803
U.S.A.

Michael Weiss
Department of Mathematics
Aarhus Universitet
DK-8000 AARHUS
Denmark

Urs Würgler
Institute of Mathematics
Universität Bern
CH-3012 BERN
Switzerland

Nobuaki Yagita
Department of Mathematics
Musashi Institute of Technology
TOKYO 158
Japan

Vol. 1394: T.L. Gill, W.W. Zachary (Eds.), Nonlinear Semigroups, Partial Differential Equations and Attractors. Proceedings, 1987. IX, 233 pages. 1989.

Vol. 1395: K. Alladi (Ed.), Number Theory, Madras 1987. Proceedings. VII, 234 pages. 1989.

Vol. 1396: L. Accardi, W. von Waldenfels (Eds.), Quantum Probability and Applications IV. Proceedings, 1987. VI, 355 pages. 1989.

Vol. 1397: P.R. Turner (Ed.), Numerical Analysis and Parallel Processing. Seminar, 1987. VI, 264 pages. 1989.

Vol. 1398: A.C. Kim, B.H. Neumann (Eds.), Groups – Korea 1988. Proceedings. V, 189 pages. 1989.

Vol. 1399: W.-P. Barth, H. Lange (Eds.), Arithmetic of Complex Manifolds. Proceedings, 1988. V, 171 pages. 1989.

Vol. 1400: U. Jannsen. Mixed Motives and Algebraic K-Theory. XIII, 246 pages. 1990.

Vol. 1401: J. Steprans, S. Watson (Eds.), Set Theory and its Applications. Proceedings, 1987. V, 227 pages. 1989.

Vol. 1402: C. Carasso, P. Charrier, B. Hanouzet, J.-L. Joly (Eds.), Nonlinear Hyperbolic Problems. Proceedings, 1988. V, 249 pages. 1989.

Vol. 1403: B. Simeone (Ed.), Combinatorial Optimization. Seminar, 1986. V, 314 pages. 1989.

Vol. 1404: M.-P. Malliavin (Ed.), Séminaire d´Algèbre Paul Dubreil et Marie-Paul Malliavin. Proceedings, 1987–1988. IV, 410 pages. 1989.

Vol. 1405: S. Dolecki (Ed.), Optimization. Proceedings, 1988. V, 223 pages. 1989. Vol. 1406: L. Jacobsen (Ed.), Analytic Theory of Continued Fractions III. Proceedings, 1988. VI, 142 pages. 1989.

Vol. 1407: W. Pohlers, Proof Theory. VI, 213 pages. 1989.

Vol. 1408: W. Lück, Transformation Groups and Algebraic K-Theory. XII, 443 pages. 1989.

Vol. 1409: E. Hairer, Ch. Lubich, M. Roche. The Numerical Solution of Differential-Algebraic Systems by Runge-Kutta Methods. VII, 139 pages. 1989.

Vol. 1410: F.J. Carreras, O. Gil-Medrano, A.M. Naveira (Eds.), Differential Geometry. Proceedings, 1988. V, 308 pages. 1989.

Vol. 1411: B. Jiang (Ed.), Topological Fixed Point Theory and Applications. Proceedings. 1988. VI, 203 pages. 1989.

Vol. 1412: V.V. Kalashnikov, V.M. Zolotarev (Eds.), Stability Problems for Stochastic Models. Proceedings, 1987. X, 380 pages. 1989.

Vol. 1413: S. Wright, Uniqueness of the Injective III$_1$ Factor. III, 108 pages. 1989.

Vol. 1414: E. Ramirez de Arellano (Ed.), Algebraic Geometry and Complex Analysis. Proceedings, 1987. VI, 180 pages. 1989.

Vol. 1415: M. Langevin, M. Waldschmidt (Eds.), Cinquante Ans de Polynômes. Fifty Years of Polynomials. Proceedings, 1988. IX, 235 pages. 1990.

Vol. 1416: C. Albert (Ed.), Géométrie Symplectique et Mécanique. Proceedings, 1988. V, 289 pages. 1990.

Vol. 1417: A.J. Sommese, A. Biancofiore, E.L. Livorni (Eds.), Algebraic Geometry. Proceedings, 1988. V, 320 pages. 1990.

Vol. 1418: M. Mimura (Ed.), Homotopy Theory and Related Topics. Proceedings, 1988. V, 241 pages. 1990.

Vol. 1419: P.S. Bullen, P.Y. Lee, J.L. Mawhin, P. Muldowney, W.F. Pfeffer (Eds.), New Integrals. Proceedings, 1988. V, 202 pages. 1990.

Vol. 1420: M. Galbiati, A. Tognoli (Eds.), Real Analytic Geometry. Proceedings, 1988. IV, 366 pages. 1990.

Vol. 1421: H.A. Biagioni, A Nonlinear Theory of Generalized Functions, XII, 214 pages. 1990.

Vol. 1422: V. Villani (Ed.), Complex Geometry and Analysis. Proceedings, 1988. V, 109 pages. 1990.

Vol. 1423: S.O. Kochman, Stable Homotopy Groups of Spheres: A Computer-Assisted Approach. VIII, 330 pages. 1990.

Vol. 1424: F.E. Burstall, J.H. Rawnsley, Twistor Theory for Riemannian Symmetric Spaces. III, 112 pages. 1990.

Vol. 1425: R.A. Piccinini (Ed.), Groups of Self-Equivalences and Related Topics. Proceedings, 1988. V, 214 pages. 1990.

Vol. 1426: J. Azéma, P.A. Meyer, M. Yor (Eds.), Séminaire de Probabilités XXIV, 1988/89. V, 490 pages. 1990.

Vol. 1427: A. Ancona, D. Geman, N. Ikeda, École d'Eté de Probabilités de Saint Flour

XVIII, 1988. Ed.: P.L. Hennequin. VII, 330 pages. 1990.

Vol. 1428: K. Erdmann, Blocks of Tame Representation Type and Related Algebras. XV. 312 pages. 1990.

Vol. 1429: S. Homer, A. Nerode, R.A. Platek, G.E. Sacks, A. Scedrov, Logic and Computer Science. Seminar, 1988. Editor: P. Odifreddi. V, 162 pages. 1990.

Vol. 1430: W. Bruns, A. Simis (Eds.), Commutative Algebra. Proceedings. 1988. V, 160 pages. 1990.

Vol. 1431: J.G. Heywood, K. Masuda, R. Rautmann, V.A. Solonnikov (Eds.), The Navier-Stokes Equations – Theory and Numerical Methods. Proceedings, 1988. VII, 238 pages. 1990.

Vol. 1432: K. Ambos-Spies, G.H. Müller, G.E. Sacks (Eds.), Recursion Theory Week. Proceedings, 1989. VI, 393 pages. 1990.

Vol. 1433: S. Lang, W. Cherry, Topics in Nevanlinna Theory. II, 174 pages. 1990.

Vol. 1434: K. Nagasaka, E. Fouvry (Eds.), Analytic Number Theory. Proceedings, 1988. VI, 218 pages. 1990.

Vol. 1435: St. Ruscheweyh, E.B. Saff, L.C. Salinas, R.S. Varga (Eds.), Computational Methods and Function Theory. Proceedings, 1989. VI, 211 pages. 1990.

Vol. 1436: S. Xambó-Descamps (Ed.), Enumerative Geometry. Proceedings, 1987. V, 303 pages. 1990.

Vol. 1437: H. Inassaridze (Ed.), K-theory and Homological Algebra. Seminar, 1987–88. V, 313 pages. 1990.

Vol. 1438: P.G. Lemarié (Ed.) Les Ondelettes en 1989. Seminar. IV, 212 pages. 1990.

Vol. 1439: E. Bujalance, J.J. Etayo, J.M. Gamboa, G. Gromadzki. Automorphism Groups of Compact Bordered Klein Surfaces: A Combinatorial Approach. XIII, 201 pages. 1990.

Vol. 1440: P. Latiolais (Ed.), Topology and Combinatorial Groups Theory. Seminar, 1985–1988. VI, 207 pages. 1990.

Vol. 1441: M. Coornaert, T. Delzant, A. Papadopoulos. Géométrie et théorie des groupes. X, 165 pages. 1990.

Vol. 1442: L. Accardi, M. von Waldenfels (Eds.), Quantum Probability and Applications V. Proceedings, 1988. VI, 413 pages. 1990.

Vol. 1443: K.H. Dovermann, R. Schultz, Equivariant Surgery Theories and Their Periodicity Properties. VI, 227 pages. 1990.

Vol. 1444: H. Korezlioglu, A.S. Ustunel (Eds.), Stochastic Analysis and Related Topics VI. Proceedings, 1988. V, 268 pages. 1990.

Vol. 1445: F. Schulz, Regularity Theory for Quasilinear Elliptic Systems and – Monge Ampère Equations in Two Dimensions. XV, 123 pages. 1990.

Vol. 1446: Methods of Nonconvex Analysis. Seminar, 1989. Editor: A. Cellina. V, 206 pages. 1990.

Vol. 1447: J.-G. Labesse, J. Schwermer (Eds), Cohomology of Arithmetic Groups and Automorphic Forms. Proceedings, 1989. V, 358 pages. 1990.

Vol. 1448: S.K. Jain, S.R. López-Permouth (Eds.), Non-Commutative Ring Theory. Proceedings, 1989. V, 166 pages. 1990.

Vol. 1449: W. Odyniec, G. Lewicki, Minimal Projections in Banach Spaces. VIII, 168 pages. 1990.

Vol. 1450: H. Fujita, T. Ikebe, S.T. Kuroda (Eds.), Functional-Analytic Methods for Partial Differential Equations. Proceedings, 1989. VII, 252 pages. 1990.

Vol. 1451: L. Alvarez-Gaumé, E. Arbarello, C. De Concini, N.J. Hitchin, Global Geometry and Mathematical Physics. Montecatini Terme 1988. Seminar. Editors: M. Francaviglia, F. Gherardelli. IX, 197 pages. 1990.

Vol. 1452: E. Hlawka, R.F. Tichy (Eds.), Number-Theoretic Analysis. Seminar, 1988–89. V, 220 pages. 1990.

Vol. 1453: Yu.G. Borisovich, Yu.E. Gliklikh (Eds.), Global Analysis – Studies and Applications IV. V, 320 pages. 1990.

Vol. 1454: F. Baldassari, S. Bosch, B. Dwork (Eds.), p-adic Analysis. Proceedings, 1989. V, 382 pages. 1990.

Vol. 1455: J.-P. Françoise, R. Roussarie (Eds.), Bifurcations of Planar Vector Fields. Proceedings, 1989. VI, 396 pages. 1990.

Vol. 1456: L.G. Kovács (Ed.), Groups – Canberra 1989. Proceedings. XII, 198 pages. 1990.

Vol. 1457: O. Axelsson, L.Yu. Kolotilina (Eds.), Preconditioned Conjugate Gradient Methods. Proceedings, 1989. V, 196 pages. 1990.

Vol. 1458: R. Schaaf, Global Solution Branches of Two Point Boundary Value Problems. XIX, 141 pages. 1990.

Vol. 1459: D. Tiba, Optimal Control of Nonsmooth Distributed Parameter Systems. VII, 159 pages. 1990.

Vol. 1460: G. Toscani, V. Boffi, S. Rionero (Eds.), Mathematical Aspects of Fluid Plasma Dynamics. Proceedings, 1988. V, 221 pages. 1991.

Vol. 1461: R. Gorenflo, S. Vessella, Abel Integral Equations. VII, 215 pages. 1991.

Vol. 1462: D. Mond, J. Montaldi (Eds.), Singularity Theory and its Applications. Warwick 1989, Part I. VIII, 405 pages. 1991.

Vol. 1463: R. Roberts, I. Stewart (Eds.), Singularity Theory and its Applications. Warwick 1989, Part II. VIII, 322 pages. 1991.

Vol. 1464: D. L. Burkholder, E. Pardoux, A. Sznitman, Ecole d'Eté de Probabilités de Saint- Flour XIX-1989. Editor: P. L. Hennequin. VI, 256 pages. 1991.

Vol. 1465: G. David, Wavelets and Singular Integrals on Curves and Surfaces. X, 107 pages. 1991.

Vol. 1466: W. Banaszczyk, Additive Subgroups of Topological Vector Spaces. VII, 178 pages. 1991.

Vol. 1467: W. M. Schmidt, Diophantine Approximations and Equations. VIII, 217 pages. 1991.

Vol. 1468: J. Noguchi, T. Ohsawa (Eds.), Prospects in Complex Geometry. Proceedings, 1989. VII, 421 pages. 1991.

Vol. 1469: J. Lindenstrauss, V. D. Milman (Eds.), Geometric Aspects of Functional Analysis. Seminar 1989-90. XI, 191 pages. 1991.

Vol. 1470: E. Odell, H. Rosenthal (Eds.), Functional Analysis. Proceedings, 1987-89. VII, 199 pages. 1991.

Vol. 1471: A. A. Panchishkin, Non-Archimedean L-Functions of Siegel and Hilbert Modular Forms. VII, 157 pages. 1991.

Vol. 1472: T. T. Nielsen, Bose Algebras: The Complex and Real Wave Representations. V, 132 pages. 1991.

Vol. 1473: Y. Hino, S. Murakami, T. Naito, Functional Differential Equations with Infinite Delay. X, 317 pages. 1991.

Vol. 1474: S. Jackowski, B. Oliver, K. Pawałowski (Eds.), Algebraic Topology, Poznań 1989. Proceedings. VIII, 397 pages. 1991.